Conceptual anomalies
in economics and statistics

Conceptual anomalies in economics and statistics

Lessons from the social experiment

LELAND GERSON NEUBERG
Department of Economics, Tufts University

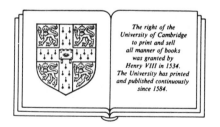

The right of the
University of Cambridge
to print and sell
all manner of books
was granted by
Henry VIII in 1534.
The University has printed
and published continuously
since 1584.

CAMBRIDGE UNIVERSITY PRESS

Cambridge
New York   New Rochelle   Melbourne   Sydney

Published by the Press Syndicate of the University of Cambridge
The Pitt Building, Trumpington Street, Cambridge CB2 1RP
32 East 57th Street, New York, NY 10022, USA
10 Stamford Road, Oakleigh, Melbourne 3166, Australia

First published 1989

Printed in the United States of America

Library of Congress Cataloging-in-Publication Data

Neuberg, Leland Gerson, 1943–
Conceptual anomalies in economics and statistics : lessons from
the social experiment / Leland Gerson Neuberg.

p.   cm.

Bibliography: p.

Includes index.

ISBN 0-521-30444-X

1. Economics, Mathematical.   2. Mathematical statistics.
I. Title.
HB135.N47 1989   88–19012
330′.01′51 – dc19   CIP

British Library Cataloguing in Publication Data

Neuberg, Leland Gerson
Conceptual anomalies in economics and statistics : lessons from
the social experiment.

1. Social policies. Concepts
I. Title
361.6′1′01

ISBN 0 521 30444 X

Contents

Preface

Conceptual Anomalies in Economics and Statistics synthesizes close to thirty years of study, research, and teaching in a variety of academic areas. Teachers, students, family members, an editor, schools, and funding institutions contributed indirectly to the writing of this book and deserve recognition.

Though I didn't realize it then, I started preparing to write this book at Chicago's Hyde Park High School during 1956–60, where Mrs. Eva Shull sparked the interest in mathematics evident in nearly every chapter.

During 1960–5, I studied humanities and science at the Massachusetts Institute of Technology (M.I.T.), concentrating in literature (with some philosophy) and mathematics (with some physics). Literature courses with Benjamin DeMott, Norman Holland, Louis Kampf, and William Harvey Youngren helped me develop the textual analytic writing style that permeates this book. A philosophy course with John Rawls introduced me to the work of David Hume, a discussion of which opens Chapter 4, and inspired me to study Mill and the modern philosophers on whose writings the book draws. Gian Carlo Rota's course in combinatorial analysis underpins Chapter 2. Courses in elementary classical, advanced classical, and quantum mechanics from Alan Lazarus, R. H. Lemmer, and Irwin Pless, respectively, suggested some of my examples in Chapter 5.

During 1965–7, I was a graduate student in mathematics at Northwestern University, and during 1967–8, I taught mathematics at Chicago City College, Southeast Branch. Students in a finite mathematics course that I conducted as a teaching assistant at Northwestern and in a logic course I taught at Chicago City College stimulated the interest in logical formalism which structures the book. A course in probability theory from Meyer Dwass at Northwestern lies behind parts of Chapter 4.

As a graduate student in mathematics (1968–70) and physics (1970–1) at the University of Illinois at Chicago Circle, courses in group theory from Norman Blackburn deepened my feeling for the role of deduction in formal mathematics. With these courses behind me, as a teaching assistant for a freshman physics course, my appreciation of the role of logic and deduction in classical mechanics grew and eventually became the basis for Chapter 5.

In 1971, I began a five-year stint as a graduate student in city and regional planning at the University of California, Berkeley. Outside field courses in mathematical statistics from Peter Bickel and Erich Lehmann would later inspire much of Part I. Disagreements with an instructor, Professor Qualls, of an outside field course in microeconomics regarding what precisely were the results of welfare economic theory would later crystallize into Chapter 8. From studying economics and statistics at Berkeley, I concluded that these fields, despite their mathematics, are far less capable of explaining society than physics is capable of explaining nature. This entire book seeks to elaborate this conclusion.

The income maintenance experiments first caught my attention as a graduate student at Berkeley searching for a dissertation topic. Many had heralded these studies as providing a methodological breakthrough, synthesizing modes of empirical inquiry in economics and statistics, and capable of putting social policy formation on a scientific footing. Skeptical of such claims, I nevertheless decided to postpone a careful study of the experiments until I had done some empirical econometric research of my own. My dissertation was an empirical microeconometric cost function analysis of electric power distribution. The struggle to keep this research deductively coherent eventually inspired Chapter 6. Discussions with two dissertation committee members – Steve Peck (economist) and Howard D'Abrera (statistician) – clarified many of the conceptual problems in economics and statistics which the book considers.

During a year as an NSF Postdoctoral Fellow at Berkeley, I taught a course in elementary statistics in the Statistics Department and revised and extended my dissertation research into two journal articles. For 1977–80, I was Assistant Professor of Urban and Policy Sciences at the State University of New York at Stony Brook's Harriman College. In my final year there I taught a course in microeconomic theory from Henderson and Quandt. These experiences further clarified my views on the foundations of statistics and economics, deepening my conviction that neither field offers much in the way of scientific help for social policy-making. I decided to write a book exploring these views and convictions.

In the spring of 1980, I received an individual grant from the Ford Foundation to write a book on social experimentation. The following academic year, I took leave of my position at the Harriman College and returned to M.I.T. as a Research Fellow at the invitation of Dick Larson, co-director of the Operations Research Center. Synthesizing more than twenty years of study in somewhat diverse fields and wrapping my argument around the conceptual aspects of the income maintenance experiments proved more difficult than I had anticipated. By the end of the first year, I had barely begun the task. So I resigned my position at the Harriman College to concentrate full time on writing the book.

My initially intended one-year visit at M.I.T. ended up lasting six years. Larry Susskind arranged for a second year, this time at the Laboratory of Architecture and Planning. There then followed four years at the Statistics Center arranged by Herman Chernoff and Roy Welsch, its co-directors. Use of M.I.T.'s library system allowed me to do the research which provided the basis of Chapters 3 and 7. At a talk during one of M.I.T.'s independent activities periods, I presented an early version of some of the material from Chapter 2 and received the invaluable criticism of Herman Chernoff.

Rewriting each section several times as I proceeded, I finished the first version of the manuscript in May 1983. From the comments of two Cambridge University Press readers, I concluded two things. First, the manuscript contained two books. I decided to put one of these books aside for later reconsideration. Second, I had to rewrite the remaining book almost from scratch. I began rewriting in the summer of 1984 and finished in August 1986, again rewriting each section several times as I proceeded. As Visiting Assistant Professor of Economics at Tufts University, from September 1986 through October 1987, I constructed a third version of the manuscript which improved the second but was not a complete rewriting. Small grants from the Tufts University Committee on Faculty Research Awards and the Department of Economics enabled me to complete the book's illustrations and index.

I would like to thank my wife, Donna, and daughter, Eva, for putting up with me during the seven years I was writing the book. Also, Donna's employment was our primary source of financial support for several of those years. For Eva, who loved reading books from an early age, the seven years must have been especially strange and frustrating. From age 8 to 15 she saw that her father's primary task in life seemed to be writing, but never finishing, one book. I hope that living through that ordeal doesn't ultimately discourage her from writing books herself.

Also, during the period of writing, I was frequently setting a delivery date, failing to make it, and living in fear that Cambridge University Press would finally say it would not accept further postponements. Yet each

Preface

time I asked for a resetting, my editor, Colin Day, graciously accepted the newly proposed date. Without his kind indulgence I would never have completed the project.

Finally, I would like to thank my parents, LeRoy and Sari. Seeing early that they had in me a son who liked school, they always encouraged and supported my studies. Without their financial aid at crucial points, I would never have gotten to the stage of writing the book, nor would I have completed it. But more important, they somehow taught me persistence, self-criticism, and yet strong belief in self. Without those three qualities I couldn't have finished the book. I'm not sure how one teaches such qualities, but I hope I have succeeded a little in teaching them to Eva by the example of writing the book.

Finally, I hope reading the book helps others to appreciate how weak is our current scientific understanding of society as much as writing it helped me to do so.

Leland Gerson Neuberg

Brookline, Massachusetts

Introduction

I.1 The idea of a controlled social experiment

I.1.1 *The historical origin of the income maintenance experiments*

Beginning in the early 1960s, the work of quantitative social scientists and applied statisticians began to play an increasing role in social policy deliberations in the United States. But their experience with the Coleman Report, which appeared in mid-1966, made many leading policy-oriented statisticians and social scientists doubt the value of large-scale observational studies (such as Coleman's) as aids in formulating social policy. Some began to argue for the statistician's classical Fisher-type controlled experiment as the needed precursor to public policy formulation.[1]

Meanwhile, rebellions in the inner cities replaced the civil rights movement of the early 1960s. In an economic expansion driven by the Vietnam War, these rebellions triggered one traditional response of the state: an expansion of the welfare system and a stepping-up of the "War on Poverty." But political liberals in the antipoverty programs and welfare rights groups began to argue for even more: a guaranteed income program to replace the demeaning welfare system. Economists of many political persuasions were amenable, for different reasons, to one form of guaranteed income: a negative income tax (NIT). But political conservatives in Congress balked, claiming that such a program would lead the poor to stop working.

In late 1966, Heather Ross, an M.I.T. economics doctoral candidate, working for a Washington antipoverty agency, made a proposal that eventually broke the political deadlock (Ross 1966). Aware of statisticians' reservations about observational studies, Ross suggested a controlled social experiment to test congressional conservatives' claim that

---

[1] Gilbert and Mosteller (1972) presented an influential argument for "controlled experiments."

an NIT would lead the poor to stop working. The proposal gained the assent of both liberals and conservatives and eventually grew into the four income maintenance experiments of 1968–80.

Researchers planned the four studies as random-assignment experiments like those routinely employed in agricultural and biomedical empirical inquiry. In agronomy, treatment-group fields receive specified quantities of fertilizer, and researchers compare treatment- and control-group crop yields. In biomedicine, treatment-group patients receive specified drug doses, and researchers compare treatment- and control-group responses. So in the income maintenance experiments, treatment-group families would receive an NIT with a (guarantee level, tax rate) amount, and researchers would compare the hours worked for the treatment and control groups. (Later they decided to compare divorce rates too.) Though the income maintenance researchers usually thought of themselves as methodological pioneers, there had been precedents for the use of random-assignment experiments in the study of social phenomena. For example, in the 1950s, psychologists studied "juvenile delinquents" in Cambridge-Somerville, Massachusetts. The treatment group received psychological counseling, and researchers compared treatment- and control-group recidivism rates.[2]

As data collection and analysis proceeded, much of the income maintenance research gradually moved away from its planners' original conception. For example, agricultural and biomedical experiments usually compare the mean responses of several randomization-created treatment groups. Instead, the income maintenance researchers began to employ linear (or even nonlinear) response models. Sometimes they modeled two responses simultaneously. And with the hours-worked response variable, the researchers often related the model's functional form to a utility function and budget constraint of the microeconomic theory of labor–leisure choice. Ultimately, the connection of much of the income maintenance data analysis to the randomization-created groups became obscure.[3]

---

[2] For a reevaluation of this study, see McCord and McCord (1959).
[3] For movement to a response model approach (i.e., "parameterized treatment") in the experiments' planning stage, see Orcutt and Orcutt (1968). Hausman and Wise (1979) provide an example of a simultaneous-equations or structural model. Keeley et al. (1978) provide an example of work based on the theory of labor–leisure choice. The work of Robins and West (1980) contains a simultaneous-equations or structural model and is also based on the theory of labor–leisure choice.

I.1.2 *Two senses of "experiment"*

Statisticians usually claim that a random-assignment study is a more reliable basis for inferring cause and effect than is an observational study. For example, Gilbert and Mosteller argue:

. . . suppose that we take a large group of people, give a randomly chosen half of them $1000 a year and give the other half nothing. If at the end of two years we find that the half with the extra money have increased the size of their dwellings considerably more than the others, few will argue that the increased income was not a cause. The random assignment of the experimental treatment triumphs over all the little excuses for not attributing the effect to the difference in treatment. (1972, p. 372)

Yet here and elsewhere statisticians usually leave vague what makes a random-assignment study a more reliable basis for inferring cause and effect than is an observational study.

In fact, the analysis that statisticians conduct with data from random-assignment studies often fails to involve any assumptions which depend on randomization. For example, in a simple (i.e., unstratified) random-assignment study, statisticians usually assume that two independent normal probability distributions with the same variances generated the treatment and control observations. Then they conduct analysis of variance – for example, test the hypothesis that the two normal-distribution means are equal. Yet such analysis nowhere depends on the fact that randomization occurred: One would draw the same inferential conclusions from the analysis had random assignment not occurred, that is, had the study been an observational one.

Part I of this book seeks to clarify the relationship between random assignment and cause-and-effect conclusions. First we construct an inferential logic based on the observations and random assignment. Then we uncover difficulties for such a logic posed by social phenomena in particular. Finally we construct a second inferential logic based on the observations and infinite populations, but not on randomization. Thus, we seek at once to construct a randomization-based logic, lay bare its conceptual difficulties, and classify the difficulties into those which inhere in the logic and those which appear only in its application to social phenomena. Hence, Part I is an immanent critique of social experiments in the random-assignment sense of "experiment."

However, "experiment" also can mean an empirical study which gets a set of observations to confront and possibly falsify a theory. Because some of the income maintenance research involved the microeconomic theory of labor–leisure choice, perhaps the four studies were "experiments" in this second, theory-testing sense. Further, if the theory-testing sense of "experiment" characterizes science more generally, then perhaps the income maintenance experiments were tests of a scientific theory of microeconomics.

In fact, however, many have in recent years questioned that the falsification of theories through tests even properly characterizes the natural sciences, let alone social sciences. If these skeptics are correct, then understanding the income maintenance studies as theory-testing experiments seems a hopeless task; that is, it is a way of misunderstanding them. Further, there is a body of microeconomic theory and empirical work – including the income maintenance studies – which *conceivably* conforms to a theory-testing logic. But there is another body of microeconomic work – including general equilibrium theory and stability analysis – which seems to follow an entirely different logic.

Part II seeks to construct two logics of economic inquiry. First we catalogue difficulties which arise in reconstructing both natural science and economics as generation and testing-falsification of theories. Presumably, none of these difficulties keeps economics from being science, because they don't keep natural science from being science. Then we search for difficulties which arise in reconstructing economics, but not natural science, as generation and testing-falsification of theories. *Conceivably* these latter difficulties include some which keep economics from being science. Finally, we study how the logic of general equilibrium theory and stability analysis differs from, and is related to, a theory-testing logic. Hence, Part II is an immanent critique of controlled social experiments as tests of theory.

I.1.3   *The role of the income maintenance experiments in this book*

The income maintenance experiments originated in the social conflict and urban unrest and upheavals of the 1960s and early 1970s. Dissipation of the unrest and transformation of the conflict followed the experiments' completion. So one could seek to understand the experiments in their historical context. Such a historical inquiry might reasonably ask: What social groups favored conducting the experiments and why? or, What distortions arose in transmission of the experimenters' findings to the political realm? or, Did the reported findings cause the eventual non-adoption of an NIT? And understanding the experiments' relation to subsequent events would require detailed knowledge of precise findings and results.

We do *not* seek such a historical understanding of the income maintenance experiments here.[4] Instead, our goal is a critical conceptual and methodological understanding of two modes of social inquiry. We focus on the income maintenance experiments as examples of those two modes of social inquiry. For our methodological inquiry, we ask, What sorts of hypotheses can randomization aid in testing, and what concept of cause and effect is implicit in such tests? or, How do controlled social experiments resemble and differ from agronomy and physics experiments? or, Are there unavoidable difficulties for controlled social experiments which make their results unreliable? We are less interested in the details of the income maintenance experiments' results and findings than in the details of their techniques as examples of the two modes' methods of social inquiry.

We begin the investigation of each mode of *social* inquiry by asking what are the most *general* conceptual difficulties of the mode's methods; that is, we ask, How is *any* randomization-based inference conceptually problematic? or, How does *any* science fail to be theory-testing experiments? Next we ask, Are there any *special* difficulties for a mode of inquiry applied in the *social* realm? Here we distinguish unavoidable difficulties in applying a mode's methods to social phenomena from experimenter errors. For example, the random-assignment mode's Hawthorne effect is unavoidable. But stratified assignment without a weighting of the strata in analyzing the data is an experimenter error. Or, the theory-testing mode's ad hoc models of differences between individuals are unavoidable given the present form of microeconomics. But the use of a labor-supply function which presupposes utility *minimization* is an experimenter error.

Part II also explores this question: Is microeconomics science in the same sense that natural sciences such as physics are? We note that microeconomic experiments such as those on income maintenance seem to share a set of conceptual difficulties with natural science experiments such as those of classical mechanics. Hence, those shared difficulties can't be a basis for concluding that microeconomics is not science. But microeconomics experiments such as those on income maintenance also exhibit a set of conceptual difficulties not evident in natural science experi-

---

[4] Neuberg (1986; in press) begins a historical critique of the income maintenance experiments.

ments such as those of classical mechanics. So the second set of conceptual difficulties *may* be a basis for concluding that microeconomics in its present form is an undeveloped science.

I.2   The book's structure

I.2.1   *The argument by chapters*

Part I develops a critical account of the statistical logics of controlled social experiments. Chapter 1 examines the philosophical basis of a first statistical logic of controlled experiments. We develop two distinct concepts of cause and effect and two of probability, and then we fuse one of the cause-and-effect concepts and one of the probability concepts into a notion of a gambling-device law (GDL). Chapter 2 demonstrates that the GDL notion is the philosophical basis of a first statistical logic of controlled experiments by constructing a theory of randomization-based causal inference based on it. The theory of causal inference includes an unbiased point estimate of the mean causal effect, a test of the hypothesis of no causal effects, and a test of the hypothesis of no mean causal effect. We also extend the theory of unbiased causal point estimation to a case of univariate regression. And throughout the chapter we focus on the theory's conceptual difficulties and stress the role of assumptions (in addition to the assumption that the observations are randomization-generated) in the inference.

Chapter 3 introduces the technique of double-blindness to the theory of unbiased causal point estimation and so extends the theory to situations involving a conscious object of inquiry. But Chapter 3 also stresses that certain features of a controlled *social* experiment introduce unavoidable biases into the causal point estimation of Chapter 2. For example, participants are blind neither to being in an experiment nor to their treatment, and they are volunteers and free to drop out of the experiment before its completion. And these features lead to Hawthorne effect, self-selection, and attrition biases. Chapter 4 develops the second statistical logic of controlled social experimentation. This second logic – a Neyman–Pearson theory logic – is based on the concept of an infinite population. We stress that randomization plays no role in the second logic (e.g., the logic's inferential statements based on data from a controlled social experiment would be the same had an observational study generated the data). And we note that weakening the infinite-population assumption of the second logic leads to weaker inferential conclusions.

Part II develops a critical account of two microeconomic logics, one of which is sometimes involved in controlled social experiments. Chapter 5 constructs a theory-testing logic of classical mechanics experiments. Then we note some difficulties (e.g., circular-like reasoning, nonfalsifiable laws, and approximation conventions) with an account of classical mechanics experiments as theory-testing. Chapter 6 constructs a theory-testing logic of microeconomic experiments, like those on income maintenance. Then we note precisely the same difficulties with an account of microeconomics as theory-testing, as with Chapter 5's account of classical mechanics as theory-testing. But we also note a second set of difficulties, which have no classical mechanical parallels, with an account of microeconomics as theory-testing. This second set of difficulties includes how to know that one observes a market equilibrium and how to deal with differences in individual consumer tastes and production technologies. If microeconomics fails to be science, then, it is because of the second set of difficulties, not because of those philosophical problems it shares with natural sciences like classical mechanics.

Chapter 7 considers the income maintenance studies as theory-testing microeconomic experiments and discovers the second set of difficulties catalogued in Chapter 6. But we also note a third set of difficulties – avoidable researcher errors. These include a failure to employ a labor-supply function that follows deductively from maximization of a utility function, and a failure to use an estimating procedure with statistical optimality properties. Chapter 8 develops the second microeconomic logic – a logic of general equilibrium theory and stability analysis. We construct this logic as a normative argument for a perfectly competitive private-ownership economy. Among the logic's difficulties, we note two of special importance. First, the logic fails to prove that a perfectly competitive private-ownership economy leads in any given amount of time to a Pareto-efficient allocation which is in the interest of society. Second, the logic fails to discover an assumption set on consumer preferences consistent with both introspection and stability analysis in its present form.

The Conclusion reconsiders in further depth those conceptual difficulties of controlled social experiments which are neither researcher errors nor general philosophical difficulties of science. We classify these middle-range or methodological difficulties into three families: the problems of complexity, behavior, and time. The problem of behavior includes the difficulties of atomization, individuality, consciousness, and will. The problem of time includes the difficulties of dynamics and equilibrium, time-dependent utility-production, and history. Chomsky's linguistics is an example of a social science which avoids the problems of behavior and time and is mastering the problem of complexity. We conclude that controlled social experiments in their present form are undeveloped science and are somewhat overrated tools for public policy decision-making.

I.2.2   *The reasons for formalization*

In the course of the book's argument we consider five distinct logics in some detail. Chapters 2 and 3 develop a randomization-based logic and Chapter 4 an infinite-population-based logic of statistical inference for data from controlled experiments. Chapters 5 and 6 construct logics of theory-testing experiments in classical mechanics and microeconomics, respectively. Chapter 8 develops two versions of the logic of a microeconomic argument for a perfectly competitive private-ownership economy. Although it is the custom to do so only for the two statistical logics and the last microeconomic logic, we present each of the five logics as a formalism. That is, we develop each logic as a quasi-axiomatic system of axioms, definitions, theorems, and so forth. Why do we stress formalization more than is usual?

Many economists claim that their work bears some resemblance to that of physicists. We therefore formalize the theory-testing logics of classical mechanics and microeconomics so as to be able to most carefully appraise the claim that economics and physics are similar. With both theories expressed in quasi-axiomatic form, similarities appear more sharply. For example, formalization reveals that the central law sketches in each theory follow deductively from optimization procedures. That is, Newton's Second Law in classical mechanics follows from God's or nature's minimization of the action (possibly subject to constraints). And the marginal utility and product conditions in microeconomics follow from consumer–producer maximization of utility-production functions subject to constraints.

But with both theories expressed in quasi-axiomatic form, differences also appear most sharply. For example, formalization reveals that in classical mechanics there is a single law sketch which when coupled with a particular force equation implies a prediction statement which can be confronted with observations. In contrast, formalization reveals that in microeconomics, many law sketches, when coupled with particular utility and production function equations, imply particular market supply and demand equations. These market supply and demand equations, in turn, imply prediction statements which can be confronted with observations.

The mathematician Kurt Gödel proved that any axiomatization of ordinary arithmetic could not be both complete and consistent.[5] Gödel's proof suggests that any attempt to put knowledge on the formal basis of an axiom system may run into conceptual difficulties. Each of our five logics is simultaneously a mathematical system and a set of words meant to say something about the world of our observations. So we formalize the logics in part to uncover and catalogue the conceptual difficulties which Gödel's proof suggests we will encounter. That is, formalizing a logic serves to clarify what aspects of the logic resist formalization. Hence, formalization enables us to compare logics not only on the basis of their formal structures but also on the basis of which of their aspects resist formalization.

For example, formalizing randomization-based tests of the hypothesis of no causal effects reveals that such tests depend on a formalization-resistant choice of test statistic; or formalizing theory-testing logics of classical mechanics and microeconomics identifies formalization-resistant points and suggests tentative conclusions about barriers to microeconomics becoming fully developed science; or, finally, formalizing frequently made normative arguments for a perfectly competitive private-ownership economy allows us to identify gaps and inconsistencies in those arguments.

I.2.3   *Some recurring themes*

The book's overall theme is a critique of the methods of controlled social experimentation. On this theme we argue by demonstration that if the methods are laid out in detail, their weaknesses will emerge from *within* that account of the techniques. For example, we make no argument against randomization in controlled experiments as some others have done.[6] Rather, we construct a formal logic for randomization-based inference in controlled experiments, and from within that logic itself the fact emerges that testing the hypothesis of no mean causal effect is problem-

---

[5] See Gödel (1970) for Gödel's proof and Findley (1942) for a conceptual explication of the proof.

[6] The earliest direct argument against randomization was made by the statistician Gosset ("Student" 1937). More recently, some philosophers have revived Gosset's original argument (Seidenfeld 1981; Levi 1983).

atic. Hence, our account of the methods of controlled social experimentation is an *immanent* critique.

In Part I we develop the theme that, contrary to the beliefs or hopes of many social statistical researchers and statisticians, empirical statistical inference always depends on assumptions which go beyond the observations. For example, randomization-based inference depends on the metaphysical assumption of deterministic causality; or our second logic of statistical inference depends upon the assumption of the existence of unobservable infinite populations. We call the need for assumptions, in statistical inference, that go beyond the observations the problem of empiricism.

In Part II we develop the theme that, contrary to the beliefs of some philosophers, a theory-testing logic of science fails to solve such "epistemological problems" as the problem of empiricism. Rather, in the theory-testing logic, the problem of empiricism becomes the problem of how words in axioms and theorems hook onto the world of our observations. For example, attaching the word "mass" in classical mechanics or the words "supply" and "demand" in microeconomics to the world of our observations involves circular-like reasoning.

Closely related to the problem of empiricism is another theme which runs contrary to the understanding of many statisticians and economists: Weakening the assumptions of statistical inference or economic theory weakens the inferential or deductive conclusions one may draw. For example, Chapter 4 shows that weakening the classical normal-probability-distributions assumption in a test of the hypothesis that two population means are equal makes significance probabilities asymptotic in the inferential conclusion. That is, one can speak, under the hypothesis, not of how likely the observations were, but only about how likely they would have been in the limit as sample sizes approached infinity. Chapter 8 shows that weakening the assumption of consumer-preference strong convexity to convexity weakens the argument for a perfectly competitive private-ownership economy. That is, the argument becomes this: Such an economy is consistent with, rather than eventually leading to, a Pareto-efficient allocation.

In Part II and in the Conclusion we pursue the theme of parallels in classical mechanics and microeconomics. We don't *directly* deny the claim of some economists that economics is a science of society on a par with physics' account of nature. Rather, in immanently critical fashion we develop certain similarities between classical mechanics and microeconomics. Where the similarities leave off then demonstrates that to the extent that microeconomics resembles physics, it is an undeveloped or logically incoherent physics. (By "logically incoherent physics" we mean a physics in crisis like that of the Bohr model of the hydrogen atom during the period between the collapse of classical and rise of quantum mechanics.)

Finally, statisticians and economists tend to see their fields of knowledge developing much as bricks are added to the foundation in the construction of a building. Not directly contradicting such a view, throughout the book we stress that with respect to their conceptual foundations, modern statistics and economics merely reformulate old ideas. For example, Fisher's randomization-based inference merely reformulates John Stuart Mill's Method of Difference along with its difficulties; or the Neyman–Pearson theory of inference, based on infinite populations, is a form of David Hume's induction, with its problems; or general equilibrium theory and stability analysis are versions of Adam Smith's argument, with its original anomalies preserved, for a perfectly competitive private-ownership economy. So contrary to what is suggested by the view of many statisticians and economists of their fields' development, "progress" in the *fundamental* aspects of their knowledge is painfully slow if it occurs at all.

I.2.4 *Notation and proofs*

The book's formal focus leads to strong reliance on symbolic notation. Sometimes a notational system spans more than one chapter (e.g., Chapters 2 and 3 involve essentially the same notation). But the book's focus on *several* logics precludes a single notational system. So despite notational overlap of chapters, we base each chapter on its own self-contained notational system. Also, the overall number of symbols is very large, and we stick as close as possible to standard notations when they exist. Consequently, a given symbol in different chapters sometimes has different meanings. For easy reference, a list of principal symbols arranged by chapter, in order of their first appearance, can be found at the back of the book.

Finally, we formalize a logic to clarify its structure, and so a theorem's proof interests us less than its statement form. Therefore, theorem proofs not given in the footnotes appear in the Appendix.

PART I

Statistical logics

On the insufficiency of the observations for inference in controlled experiments

CHAPTER 1

J. S. Mill and some philosophical underpinnings of controlled experimentation

1.1 Introduction: An ancient example

Writing in 1949, the psychologist Richard L. Solomon observed: "The history of the idea of a controlled experiment is a long one. Usually one goes back to J. S. Mill's canons for the concept of experimental controls" (p. 137). Recent historical research, however, reveals that one has to go a good deal further back than Mill for the origin of controlled experimentation. Consider the following passage from the Old Testament:

In the third year of the reign of Jehoiakim of Judah, Nebuchadnezzar king of Babylon came to Jerusalem and laid siege to it. The Lord delivered Jehoiakim king of Judah into this power, together with all that was left of the vessels of the house of God; and he carried them off to the land of Shinar, to the temple of his god, where he deposited the vessels in the treasury. Then the king ordered Ashpenaz, his chief eunuch, to take certain of the Israelite exiles, of the blood royal and of the nobility, who were to be young men of good looks and bodily without fault, at home in all branches of knowledge, well-informed, intelligent, and fit for service in the royal court; and he was to instruct them in the literature and language of the Chaldaeans. The king assigned them a daily allowance of food and wine from the royal table. Their training was to last for three years, and at the end of that time they would enter the royal service.

Among them there were certain young men from Judah called Daniel, Hananiah, Mishael and Azariah; but the master of the eunuchs gave them new names: Daniel he called Belteshazzar, Hananiah Shadrach, Mishael Meshach and Azariah Abed-nego. Now Daniel determined not to contaminate himself by touching the food and wine assigned to him by the king, and he begged the master of the eunuchs not to make him do so. God made the master show kindness and goodwill to Daniel, and he said to him, "I am afraid of my lord the king: he has assigned you your food and drink, and if he sees you looking dejected, unlike the other young men of your age, it will cost me my head." Then Daniel said to the guard whom the master of the eunuchs had put in charge of Hananiah, Mishael, Azariah and himself, "Submit us to this test for ten days. Give us only vegetables to eat and water to drink; then compare our looks with those of the young men who have lived on the food assigned by the king, and be guided in your treatment of us by what you see." The guard listened to what they said and tested them for ten days. At the end of ten days they looked healthier and were better nourished than all the young men who had lived on the food assigned them by the king. So the guard took away the assignment of food and the wine they were to drink, and gave them only the vegetables. (Daniel 1970, pp. 1251–2)

With a slight difference in language, the passage contains a report of a controlled experiment. For 10 days an experimental group was fed a new diet of "vegetables to eat and water to drink" while a control group consumed the standard diet of "food and wine from the royal table." At the end of 10 days the two groups were compared on the response variable "appearance of health and nourishment," and the controls were found wanting in relation to the experimentals.

True, the control-group size is a little vague (the experimental group appears to have contained four observations), and the scale of measurement of the response variable is unclear. Also, all the modern niceties are absent: blindness, randomization, and a statistical inferential framework into which to cast the results. Finally, in relation to the framework issue, given that both the controls and experimentals were "of the blood royal and of the nobility . . . young men of good looks and bodily without fault, at home in all branches of knowledge, well-informed, intelligent," one wouldn't want to make any inferential wagers on extensions of the experimental results to larger "populations." Still, considering that the report is from the Old Testament, one is inclined to make allowances for the somewhat primitive technique.

Clearly, then, the idea of a controlled experiment predates Mill's canons. Nevertheless, there is good reason to begin an inquiry into the conceptual basis of controlled social experimentation by considering some of Mill's canons, for philosophers regard Mill's Method of Difference as in some way underlying the controlled experiment. Thus, the philosopher J. L. Mackie explains:

The second well-known approximation to the difference observation is the standard controlled experiment, where what happens in the "experimental case" . . . is compared with what happens, or fails to happen, in a deliberately constructed "control case" which is made to match the experimental case in all ways thought likely to be relevant other than the feature . . . whose effects are under investigation. (1974, p. 72)

Unfortunately, Mackie fails to elaborate the sense in which the controlled experiment "approximates" Mill's Method of Difference. So an explanation of Mill's conceptions and their problems, together with an account of their relationship to the standard controlled experiment, appears appropriate to deepening our understanding of the conceptual basis of the controlled experiment's social variant.

1.2   The method of difference

1.2.1   *Mill's concept of causality*

Mill offers what he refers to as both "the canon which is the regulating principle of" and "the axioms implied in" his famous Method of Difference:

If an instance in which the phenomenon under investigation occurs, and an instance in which it does not occur, have every circumstance in common save one, that one occurring only in the former; the circumstance in which alone the two instances differ, is the effect, or the cause, or an indispensable part of the cause of the phenomenon. (1973, p. 391)

The method's canon then is a rule or criterion for the cause–effect relationship. Like an axiom, it defines cause and effect in terms of undefined entities (i.e., "instance," "phenomenon," "circumstance") through which the canon can be converted into a proposition about a situation in the world. Also like an axiom, the canon provides necessary (but not sufficient) conditions for the cause–effect relationship.[1]

Mill was not the first in what we have come to call the British empiricist tradition to offer an explicit concept of causality. Roughly 100 years earlier, David Hume had also suggested a concept of causality:

. . . we may define a cause to be *an object followed by another, and where all the objects, similar to the first, are followed by objects similar to the second.* Or, in other words, *if the first object had not been, the second never had existed.* (1955, p. 87)

Notice that Hume employs some subjunctive-tense language which Mill does not, while Mill allows for "an indispensable part of the cause" which Hume does not. Also, Hume is explicit on two other matters on which Mill's canon is not: A cause must precede its effect in time, and like causes must lead to like effects.[2]

---

[1] Put another way, the canon is a mere logical form: If a set of conditions holds for a phenomenon and a circumstance, then the circumstance is the cause of the phenomenon. What remains problematic in any situation is establishing that the set of conditions holds for the phenomenon and the circumstance.

[2] Elsewhere (1973, p. 344), Mill explicitly argues that a cause must precede its effect in time. Mackie (1974, Chapter 7) argues for a concept of "fixity" rather than time ordering as the essential asymmetry in the cause–effect relationship.

   Mill (1973) explicitly considers (and rejects) the notion that phenomena need not have causes (as we shall soon see). He also argues that the cause of a phenomenon in one situation need not be the cause of the same phenomenon in another situation (in Book III, Chapter X). Yet he never considers the possibility that like causes need not lead to like effects. Even though his Method of Difference canon is not explicit on determinism, all of Mill's discussions of causality, laws, and so forth (1973), *implicitly* follow the view that like causes *must* lead to like effects.

The feature of like effects following like causes – explicit in Hume and implicit in Mill – appears consistent with our usual sense of cause and effect. Yet a concept of causality need not have such a feature. With $P(X_t)$ meaning the probability of event X occurring at time t, the philosopher Patrick Suppes recently offered the following concept of causality: "The event $B_{t'}$ is a prima facie cause of the event A_t if and only if (i) $t' < t$, (ii) $P(B_{t'}) > 0$, (iii) $P(A_t|B_{t'}) > P(A_t)$" (1970, p. 12). Notice that unlike Hume and Mill, Suppes defines causality in terms of the undefined term "probability." "Probability" is a notion much disputed by philosophers, but *whatever* is meant by the term, Suppes's probabilistic conception of causality supposes that like causes need not lead to like effects.

Suppes's account of causality attempts to formalize a notion of causality which emerged historically in physics with the collapse of Newtonian mechanics and the rise of quantum mechanics in the late nineteenth and early twentieth centuries and was finally codified in the Schrödinger equation. Schrödinger himself contrasts the classical (Hume–Mill) and the quantum (Suppes) concept of causality:

According to classical physics, and especially mechanics, it would be necessary to undertake certain operations in order to take a mass point to a given place at the initial point of time and in order to impress upon it a given velocity. Thus we might take it between nippers, carry it to the place in question and push it in an appropriate direction. Quantum mechanics teaches us that if such an operation is undertaken with a mass point a great number of times, the same result does not invariably come about *even if the operation is always exactly the same.* (1935, p. 58, emphasis added)

In contrast to the statistical or probabilistic concept of causality of Schrödinger–Suppes, the Hume–Mill concept of causality is a *deterministic* one.

Suppes (1970, pp. 44–5) correctly argues that his concept generalizes Hume's first condition that a cause is *"an object followed by another, and where all the objects, similar to the first, are followed by objects similar to the second."* That is, Hume's first condition follows from Suppes's concept with $P(A_t|B_{t'}) = 1$. So as a generalization of Hume's first condition, Suppes's concept retains any features of Hume's first condition which are problematic.

Notice that in addition to being as problematic as Hume's first condition, Suppes's concept seems to contain no version of Hume's second condition: The first object is the cause of the second where *"if the first object had not been, the second never had existed."* Suppes seeks to remedy this absence by introducing a concept of prima facie negative cause:

"The event $B_{t'}$ is a prima facie negative cause of A_t if and only if (i) $t' < t$, (ii) $P(B_{t'}) > 0$, (iii) $P(A_t|B_{t'}) < P(A_t)$" (1970, p. 43). He then deduces that $c(B_{t'}) (= B_{t'}$'s not occurring) is a prima facie negative cause of A_t if and only if $B_{t'}$ is a prima facie cause of A_t (p. 44). So Suppes's account makes the probabilistic version of Hume's second condition *equivalent to* the probabilistic version of Hume's first condition. Such a result contradicts our intuitive sense of cause and effect: Causality is regular co-occurrence plus something *not* deducible from regular co-occurrence which is expressed in Hume's second condition.

Another consequence of Suppes's concept of prima facie cause is this: With $C_{t''}$ a prima facie cause of $B_{t'}$, and $B_{t'}$ a prima facie cause of A_t, $C_{t''}$ need not be a prima facie cause of A_t. Such an absence of transitivity also contradicts our intuitive sense of cause and effect: If C causes B, and B causes A, it should follow that C causes A.[3]

Because formalization of the quantum concept of causality leads to counterintuitive results, our explication of controlled experimentation in Chapter 2 will be built on the Hume-Mill rather than the Schrödinger-Suppes conception of causality. Given that controlled experiment is the heart of modern *statistics,* the choice of *deterministic* rather than probabilistic causality for our account's foundation may appear somewhat perverse. Yet we shall demonstrate that a coherent account of controlled experimentation doesn't require the counterintuitive quantum concept of causality.[4] Of course, we *shall* require a concept of *probability* for our explication. But that concept will be introduced *separate* from the notion of causality (following the classical tradition) rather than *within* the notion of causality (following the quantum tradition). And though the Schrödinger-Suppes concept of causality contradicts our unreflective sense of cause and effect, the Hume-Mill conception is not without its own difficulties.

1.2.2 *Inconsistency and circularity*

Like any other axiom, Mill's canon runs into difficulties when one seeks to apply it to a situation in the world. The problems arise because the canon provides only a necessary (but not sufficient) condition for causal-

---

[3] On the intransitivity of his prima facie cause concept, see Suppes (1970, pp. 58–9).

[4] Mackie (1974, Chapter 9) argues that deterministic causality can be used to explain many phenomena for which a probabilistic concept of causality at first appears required. Bohm (1957) gives a heretical account of quantum mechanics itself based on the Hume-Mill deterministic rather than the Schrödinger-Suppes probabilistic conception of causality.

ity. Suppose, for example, one has an instance where *A* with some other circumstances is followed by *B*, and another instance in which the same circumstances without *A* are followed by the absence of *B*. Mill then explains:

What, however does this prove? It proves that B, in the particular instance, cannot have had any other cause than A; but to conclude from this that A was the cause, or that A will on other occasions be followed by B, is only allowable on the assumption that B must have some cause; that among its antecedents in any single instance in which it occurs, there must be one which has the capacity of producing it at other times. This being admitted, it is seen that in the case in question that antecedent can be no other than A; but, that if it be no other than A it must be A, is not proved, by these instances at least, but taken for granted. (1973, p. 562)

Mill concludes: "The validity of all the Inductive Methods depends on the assumption that every event, or the beginning of every phenomenon, must have some cause; some antecedent, on the existence of which it is invariably and unconditionally consequent" (1973, p. 562). He labels that assumption the law of universal causation.

To the question On what does the law of universal causation rest? Mill ultimately offers a not completely satisfactory answer. He first identifies a form of induction – *inductio per enumerationem simplicem* (IPES or induction) which "consists in ascribing the character of general truths to all propositions which are true in every instance that we happen to know of" (1973, p. 312). Thus, for many centuries Europeans had concluded that all swans were white because they had seen white swans and they had never seen a swan which was not white. Of course, Europeans were wrong on the matter of swan color. Mill offers a sensible critique of IPES reasoning:

But though we have always a propensity to generalize from unvarying experience, we are not always warranted in doing so. Before we can be at liberty to conclude that something is universally true because we have never known an instance to the contrary, we must have reason to believe that if there were in nature any instances to the contrary, we should have known of them.[5]

---

[5] Mill (1973, p. 312). Notice that *inductio per enumerationem simplicem* amounts to inducing on the basis of something like Hume's first condition: *"an object followed by another, and where all objects, similar to the first, are followed by objects similar to the second."* Instead of repeated occurrence of one object following another, we have repeated occurrences of an object with a particular property (e.g., swans which are white). Mill's critique of IPES is that good induction requires something like Hume's second condition, too, where *"if the first object had not been, the second never had existed."*

He concludes that "for the accurate study of nature, we require a surer and a more potent instrument" (1973, p. 313) than IPES reasoning (i.e., we require his own form of Induction: the Method of Difference).

Yet though he offers an argument on the weakness of IPES, he also claims that "the law of causality rests on an induction by simple enumeration" (1973, p. 567). That is, we conclude that all events have causes from the fact that all events which we have seen have had causes. Mill senses he is being inconsistent and defends himself as follows:

Now the precariousness of the method of simple enumeration is in an inverse ratio to the largeness of the generalization. The process is delusive and insufficient, exactly in proportion as the subject-matter of the observation is special and limited in extent. As the sphere widens, this unscientific method becomes less and less liable to mislead; and the most universal class of truths, the law of causation for instance, and the principles of number and of geometry, are duly and satisfactorily proved by that method alone, nor are they susceptible of any other proof.[6]

Thus, his argument appears to be that IPES is no good for establishing a *particular* causal law (e.g., that C causes c), but is good for establishing the universal law of causality (i.e., that all events have causes).

As long as one accepts Mill's dubious distinction between special/limited and universal truths, he can be absolved of inconsistency: Induction is for particular truths, induction for universal ones. But then Mill's overall explanation of Induction (or the Method of Difference) remains circular. Thus, as he sees it, Induction assumes the law of universal causation. But if we know that all events have causes because all those we have seen have had causes, how do we know any particular event we've seen had a cause? Presumably, since the event is *particular,* by Induction. But then Induction assumes the law of universal causation which is itself based on Induction, and Mill's account of the Method of Difference is circular.

Mill then is trapped: Either his explication of the Method of Difference is inconsistent or it is circular. What drives him into the corner is his empiricism, that is, his desire to base Induction *solely* on what is observed. Suppose we observe two instances: ABC followed by abc and AB followed by ab. Then we can conclude from these observations, and the assump-

---

[6] Mill (1973, p. 569). He precedes the quoted passage with this rumination: "It would seem, therefore, that induction *per enumerationem simplicem* not only is not necessarily an illicit logical process, but is in reality the only kind of induction possible: since the more elaborate process depends for its validity on a law, itself obtained in that inartificial mode. Is there not then an inconsistency in contrasting the looseness of one method with the rigidity of another, when that other is indebted to the looser method for its own foundation? The inconsistency, however, is only apparent" (p. 567).

tion that c has a cause, that C is an indispensable part of the cause of c. As Mackie puts it in his reconstruction of Mill's methods: "Specific deterministic assumptions play a part that makes them in at least one way prior to the laws discovered with their help."[7] And our critique of Mill's attempt to base the prior deterministic assumption on a form of induction suggests that the required assumption cannot be based on observation. Particular causal knowledge then appears to rely on observation – but not solely on observation.

Recall that Mill's argument was that a weak form of induction – IPES – could be used to establish the universal law of causation – a law of laws – whereas any particular causal law required a strong form of induction – his Method of Difference – for its establishment. We showed that such an argument, as an attempt to ground Induction or the Method of Difference *solely* on observation, is problematic. As Mackie puts it: "By making explicit the need for assumptions we have abandoned any pretence that these methods in themselves solve or remove the problem of induction."[8] More than the Method of Difference is required, and what more is required seems *not* to be further observation.

1.2.3   A role for theory

We have seen now that insistence on basing causality solely on observations led Mill into a logically problematic argument. Insisting that causal-

---

[7] Mackie (1974, p. 316). Mackie's 1974 Appendix reconstructs Mill's method as *eliminative* methods. Such a view makes the need for a particular deterministic assumption with the Method of Difference plain enough. That is, the method merely eliminates, as possible causes of the phenomenon, those circumstances the canon's two instances have in common. To conclude that the remaining one circumstance is the cause of the phenomenon, one must assume that the phenomenon has a cause.

[8] Mackie (1974, p. 316). He continues: "If the requisite observations can be made, the ultimate justification for any conclusion reached by one of these methods will depend on the justification for the assumption used, and since this proposition is general in form, any reliance that we place on it or on its consequences will have to be backed by some other kind of inductive reasoning or confirmation or corroboration." Mackie certainly is right to argue that the methods merely reduce the problem of induction to justifying the assumption employed in using the method to draw a conclusion. His suggestions for how the assumption can be justified, though, seem questionable. By "other kind of inductive reasoning" he may mean IPES. But if our argument is correct, IPES *cannot* justify the assumption. The meaning of "confirmation or corroboration" is unclear. It thus appears safest to say that the conclusions we draw about cause and effect in a particular situation in the world are drawn on the basis of the observations *and* an assumption whose justification remains unclear since it seems not baseable on observation. In other words, we seem, at least tentatively, to have arrived at a position of skepticism with respect to causal knowledge.

ity is no more than regular, observable, co-occurrence led Mill's empiricist predecessor, Hume, to ignore a role for theory in scientific under-
tanding. Thus, Hume offers an example from physics:

We say, for instance, that the vibration of this string is the cause of this particular
sound. But what do we mean by that affirmation? We mean that this vibration is
followed by this sound, and that all similar vibrations have been followed by
similar sounds; or, that this vibration is followed by this sound, and that, upon the
appearance of one, the mind anticipates the senses and forms immediately an idea
of the other. We may consider the relation of cause and effect in either of these two
lights; but beyond these we have no idea of it. (1955, pp. 87–8)

Yet physicists *do* have a further idea of the cause and effect here. They
conceive of "sound waves" between the vibration and the sound. Perhaps
the "sound waves" are meant to be understood as "in the phenomena";
perhaps they are better seen as in the physicist's concepts of the phenomena. In either case, "sound waves" appear to be part of a theory which goes
beyond the observations and constitutes part of our concept of cause and
effect in the case of the vibrating string and sound.

Another feature of Mill's canon which is a difficulty in any particular
situation is the requirement that the two instances "have every circumstance in common save one." Mill qualifies the "every circumstance"
requirement as follows:

It is true that this similarity of circumstances needs not extend to such as are
already known to be immaterial to the result. And in the case of most phenomena
we learn at once, from the commonest experience, that most of the coexistent
phenomena of the universe may be either present or absent without affecting the
given phenomenon; or, if present, are present indifferently when the phenomenon does not happen and when it does. (1973, p. 392)

Yet replacing the "every circumstance" clause with "every material circumstance" merely begs the question: Which circumstances are material?
Mackie suggests the criterion of spatio-temporal nearness for materiality.
Yet though sometimes helpful, spatio-temporal nearness is not always
so – for example, in certain cases of radiation-caused cancers, the cause
may precede the (observed) effect by 20 years or more.[9]

---

[9] Mackie (1974, p. 74) emphasizes the tentative nature of spatio-temporal nearness as a
criterion for materiality. Mackie also stresses that Hume hesitantly included spatio-temporal nearness, or "contiguity," in our idea of causation and that "when he [Hume]
includes contiguity he does not mean that we require every cause to be contiguous with its
effect, but merely that where this is not the case, cause and effect are thought to be joined
by a chain of intermediate items, where each such item is the effect of its predecessor and
causes its successor, and is contiguous with both of these" (p. 19). (In fact, as we gain a

Prior misclassification of an immaterial circumstance as material and a material circumstance as immaterial can lead the Method of Difference to misidentify a phenomenon's cause. For suppose that C actually causes c while D is immaterial to c and that one is seeking the cause of c. Suppose further that one believes C is immaterial to c and D is material to c. If one has $ABCD$ followed by abc and AB followed by ab, one's prior conception of materiality will lead one to describe the situation as ABD followed by abc and AB followed by ab. Then the canon dictates the false conclusion that D causes c. Thus, we must assume both that a phenomenon has a cause *and* that the cause is on our list of material circumstances for the Method of Difference to correctly identify the cause. Perhaps then theory may have a role to play in identifying material circumstances.

The canon's phrase "have every circumstance in common save one" may be viewed as one of the axiom's undefined terms. In the more developed sciences, theory gives the phrase its meaning. For example, in classical statistical mechanical theory, one derives the entropy of an ideal gas assuming that the gas's N particles are distinguishable by their positions and momenta. Each particle is characterized by a six-vector of three position and three momentum components. Beyond the six-vector, the particles are regarded as "indistinguishable" or the "same." The quantum statistical mechanical theory of the ("same"?) ideal gas, however, views the gas's atoms as indistinguishable in a somewhat different sense:

A state of the gas is described by an N-particle wave function, which is either symmetric or anti-symmetric with respect to the interchange of any two particles. A permutation of the particles can at most change the wave function by a sign, and it does not produce a new state of the system. From this fact it seems reasonable that the Γ-space volume element $dpdq$ corresponds to not one but only $dpdq/N!$ states of the system. (Huang 1963, p. 154)

Thus, the quantum mechanical theory sees $N!$ "identical" versions of $dpdq$ distinguishable states where the classical mechanical theory sees $dpdq$ distinguishable states. In either the classical or quantum mechanical case, here it is theory which fills in the meaning for "have every circumstance in common save one" in a way which seems to go beyond the observations.

Footnote 9 *(cont.)*
better understanding of radiation-caused cancers, we may discover temporally intervening precancerous states or previously undetected cancerous states.) Mackie concurs with Hume's hesitancy to include contiguity even in this sense as a necessary feature of our idea of causation: "While we are happiest about contiguous cause–effect relations, and find 'action at a distance' over either a spatial or a temporal gap puzzling, we do not rule it out" (p. 19).

In the so-called social sciences, Mill's phrase "have every circumstance in common" becomes the famous ceteris paribus condition about which disputes usually arise. For social "science" theories have so far failed to conceptualize objects of inquiry which are the "same" in a way defined by the theory. They instead study objects such as consumers whose tastes differ and try to employ statistical techniques to make their objects somehow the same. For example, a set of more or less ad hoc criteria might be used to create "matched pairs" of consumers (i.e., pairs in which each pair has a set of tastes in common). But results of such studies are always open to reasonable challenge on the grounds that the matching is badly done. Consumer theory needs to conceptualize its objects of inquiry so that within the theory itself they are the same, before it can be regarded as science in the strong sense in which the classical and quantum mechanical theories of an ideal gas are science.

1.2.4 *Counterfactual conditionals*

The two instances of Mill's canon which "have every circumstance in common save one" certainly suggest the frequently employed controlled experimental techniques of matched pairing. But we shall make an aspect of causality *not* explicitly expressed in the canon central to our Chapter 2 explication of the controlled experiment. Remember that in Hume's discussion of causality he employed the phrase "if the first object had not been, the second never had existed." Mill, too, employs similar phraseology in an offhand example: "if a person eats of a particular dish, and dies in consequence, that is would not have died if he had not eaten of it, people would be apt to say that eating of that dish was the cause of his death" (1973, p. 327). Neither Hume nor Mill further develops the subjunctive-tense formulation at the center of their quoted remarks.

Notice that there is a pair involved in the Hume–Mill phrases. That is, there is both what was and what would have been if. But unlike in previous illustrations of the canon or in a matched-pairs experiment, the two elements of the pair here have different ontological statuses, for what would have been if, never actually occurs. So one and only one element of the pair is observable. Thus, by making a pair of instances, one of which doesn't happen, central to our conception of causality, we nudge the concept a little further away from one which is based solely on the observations.

Though neither Hume nor Mill develops the aspect of causality implicit in their subjunctive-tense phraseology, other philosophers have recently taken up and elaborated the idea. The Hume–Mill notion is that to say C causes c means (among possibly other things) that if C had not occurred under the circumstances, then c would not have occurred. Philosophers have labeled the phrase "if C had not occurred under the circumstances, then c would not have occurred" a counterfactual conditional – conditional because of the if–then form of the statement and counterfactual because of the reference to what did *not* occur. Though philosophers continue to debate the proper role of counterfactual conditionals in a concept of causality, there appears to be a growing consensus that a causal proposition *somehow* entails (at least implicitly) a counterfactual conditional.[10]

We shall make the counterfactual-conditional aspect of causality central to our explication of the controlled experiment. In such an experiment there are control and treatment groups. Each member of the treatment group has an observed response, as does each member of the control group. But each member of the treatment group also has a response that would have been observed had the member been in the control group. And each member of the control group has a response that would have been observed had the member been in the treatment group. Thus, for every observed response there is a counterfactual response which would have been observed had the responder been in the other group. We take the task of causal inference to be the making of statements about the union of the observed and counterfactual responses. To complete the foundation for such inference, we must expand the Hume–Mill deterministic conception of causality by adding to it a concept of probability, employing in the expansion notions which Mill himself provides.

1.3   A role for probability

1.3.1   *Mill's distinction between law and collocation*

Mill draws a distinction between "laws of causation and collocations of the original causes" (1973, p. 344), or between "laws" and "collocations," that will help us fuse the idea of probability with the Hume–Mill deterministic concept of causality. By "law," Mill means a causal relationship in his Method of Difference sense. So suppose that A causes B, C causes D, and E causes F. And suppose also that an instance consisting of ACE is

---

[10] Three recent discussions of counterfactual conditionals are Chapter 3 of Mackie (1973), Chapter 2 of Mackie (1974), and Chapter 1 of Goodman (1979).

arranged. Then *BDF* will follow. But we cannot say that *ACE causes BDF;* that is, we cannot say that a relationship of law exists between *ACE* and *BDF. BDF*'s following *ACE* results from three laws and the collocation of the three original causes *A, C,* and *E.*

Mill identifies "collocations consisting in the existence of certain agents or powers, in certain circumstances of place and time" (1973, p. 465). In our example, the collocation was the co-location of *several* (i.e., three) causes. But the collocation concept can also be employed in connection with a single original cause and a single relationship of law. Suppose one drops a mass from a height h. The Newtonian law of conservation of energy tells us that when the mass hits the ground, it will be traveling at a speed of $(2gh)^{1/2}$, where g is the gravitational constant. The law involved here may be expressed as follows: Original height h causes impact velocity $(2gh)^{1/2}$. But h in this formulation of the law is merely a placeholder. In any actual situation of a released mass, there will be a value for h, say h_0; h_0 is "the existence of certain agents or powers, in certain circumstances of time and place" (i.e., h_0 is a collocation). We explain the actual impact velocity $(2gh_0)^{1/2}$ in terms of the law, original height h causes impact velocity $(2gh)^{1/2}$, and the collocation h_0.

More generally, a Newtonian mechanical system is governed by a differential equation, or what philosophers call a deterministic law of working (DLW). But the system's state at any point in time is determined by the law of working *and* the system's original state. That is, the differential equation's solution requires a set of initial conditions. What in modern parlance we refer to as "the system's initial conditions" in Mill's discourse is called a "collocation of the original causes." To explain situations in the world, the concept of deterministic causal law needs an additional concept of initial conditions or collocation. And it is through the collocation concept that we can introduce a probabilistic element into the Hume–Mill deterministic notion of causality.

1.3.2 *Two concepts of probability*

We restrict attention in the first two chapters to a finite class of distinct elementary events or outcomes and suppose that each elementary event has a unique probability associated with it. We assume that these elementary events' probabilities possess the usual formal properties of a discrete probability measure: Each is ≥ 0, and their sum is 1. We define an event as any subclass of the class of elementary events, and we define the probability of an event as the sum of the probabilities of the elementary events comprising it. Our approach to probability so far, then, is axiomatic with the elementary events' probabilities as undefined terms. It remains to adopt a concept of probability which will give meaning to the undefined terms and thus provide a way of attaching the formalism to situations in the world.

Mill discusses what later philosophers have identified as two distinct concepts of probability. He first quotes Laplace:

Probability has reference partly to our ignorance, partly to our knowledge. We know that among three or more events, one, and only one, must happen; but there is nothing leading us to believe that any one of them will happen rather than the others. In this state of indecision, it is impossible for us to pronounce with certainty on their occurrence. It is, however, probable that any one of these events, selected at pleasure, will not take place; because we perceive several cases, all equally possible, which exclude its occurrence, and only one which favours it.

The theory of chances consists in reducing all events of the same kind to a certain number of cases equally possible, that is, such that we are *equally undecided* as to their existence; and in determining the number of these cases which are favourable to the event of which the probability is sought. The ratio of that number to the number of possible cases, is the measure of the probability; which is thus a fraction, having for its numerator the number of cases favourable to the event, and for its denominator the number of all the cases which are possible.[11]

Laplace argues that "the probability of an event is not a quality of the event itself, but a mere name for the degree of ground which we, or some one else, have for expecting it" (Mill 1973, p. 535). On Laplace's view,

---

[11] Laplace (1917, pp. 6–7) contains the passage Mill (1973) quotes on p. 534. Laplace, like Mill, believes in what Mill calls the law of universal causation: "Present events are connected with preceding ones by a tie based upon the evident principle that a thing cannot occur without a cause which produces it" (Laplace 1917, p. 3). Laplace however, unlike Mill, makes no argument that the law of universal causation is based on a form of induction. Instead, he merely identifies the principle as "this axiom, known by the name of *the principle of sufficient reason*. . . ." (Laplace 1917, p. 3).

Notice that Laplace follows in the wake of Descartes's classical French rationalism by understanding the universal determinism assumption as based on reason. Descartes (1955, p. 252) argues "that we exist, inasmuch as our nature is to think. . . ." And then he adds: "in addition to the notions we have of God and our thoughts, we shall likewise find within us a knowledge of many propositions which are eternally true, as for example, that nothing cannot be the cause of anything, etc." (pp. 252–3). Thus, for Descartes, the law of universal causation seems to have very nearly the staus of an innate idea.

Mill, in contrast to Laplace, follows in the wake of Berkeley's (1957) and Hume's (1955) classical British empiricism by viewing the universal determinism assumption as based on experience. In fact, neither reason nor experience seems an adequate basis for the particular deterministic assumption which the Method of Difference requires. In a sense, though, we base our choice of the Hume–Mill concept rather than the Schrö-dinger–Suppes concept on reason. That is, we reject the Schrödinger–Suppes concept because it has counterintuitive or unreasonable consequences.

then, a probability space consists of a set of possible elementary events together with our having no good reason to believe one of those events rather than another will occur. Mill adopts Laplace's concept, and Mackie identifies the concept as the classical equiprobable concept of probability (1973, pp. 160–3).

The main difficulty with Laplace's classical concept is that in locating probability in our belief about events rather than in the events themselves, the concept seems to lose an element of objectivity in at least one sense of the term "objective." That is, "the probability of an event to one person is a different thing from the probability of the same event to another, or to the same person after he has acquired additional evidence" (Mill 1973, p. 535). Thus, while we would like to be able to employ expressions like "the probability of event C," the classical good-reasons concept of probability appears to force us to expressions like "the probability of event C to person(s) X."

Mill informs us that in the first edition of his *Logic,* he argued that Laplace's concept was incomplete, though he subsequently abandoned the argument. His argument was as follows:

To be able (it has been said) to pronounce two events equally probable, it is not enough that we should know that one or the other must happen, and should have no grounds for conjecturing which. Experience must have shown that the two events are of equally frequent occurrence. Why, in tossing up a halfpenny, do we reckon it equally probable that we shall throw cross or pile? Because we know that in any great number of throws, cross and pile are thrown about equally often; and that the more throws we make, the more nearly the equality is perfect. We may know this if we please by actual experiment; or by the daily experience which life affords of the same general character. (Mill 1973, pp. 534–5)

It is possible to regard the view of probability Mill expresses here not as an addition to the classical view but as a probability concept distinct from the classical one. Thus, on the second probability view, a probability space consists of a set of elementary events together with the proportion of those events' occurrences in an "indefinitely long" series of trials. Mackie identifies this concept as the "limiting frequency concept [of probability] in an infinite or indefinite sequence" (1973, p. 173).

Mackie sees "one problem [with the limiting frequency concept] is whether the notion of convergence to a limit, which is at home in series determined by some mathematical formula, can be applied to an empirical sequence" (1973, pp. 173–4). Mackie argues as follows:

. . . if there were an empirical series which actually did continue indefinitely, then it could in fact exhibit a limiting frequency. Suppose, for simplicity, that a penny never wore away, and was tossed in the usual way, or the usual variety of ways, at intervals of ten seconds for ever. The results of such tosses would form an indefinitely extended sequence of Hs and Ts. . . . That is, a concrete empirical sequence, indefinitely extended, could in fact have a limiting frequency. Of course we with our finite lives and limited powers of observation could not observe or conclusively verify this, or conclusively falsify it either, but that is neither here nor there: the meaningfulness of a statement does not depend on the possibility of checking it. (1973, p. 174)

He concludes that the limiting frequency concept is in some sense meaningful. Yet his argument in a way begs the question of what is problematic about the limiting frequency concept. For the question really is this: What would it mean for there to be "an empirical series which actually did continue indefinitely"? We could not observe such a series and verify its existence, nor could we verify or observe (as Mackie himself suggests) that the series had a limit. Thus, as a formal concept, limiting frequency may be meaningful. But for the concept not to be vacuous, for it to apply to a situation in the world, requires an assumption not based on the observations. We must assume that "a concrete empirical sequence" extends indefinitely and *does* "in fact have a limit." And *that assumption* is what is problematic about the concept.

We shall require only Laplace's concept, not the limiting frequency concept, of probability for our account of the modern controlled experiment. The lack of objectivity inherent in the classical equiprobable concept will be partially circumvented by conceptualizing a situation where all reasonable observers would have no good reason to believe one of a set of possible events rather than another will occur. Thus, the qualifier "to person(s) X" will be dropable from "the probability of event C to person(s) X" not because we shall succeed in locating probability in event C itself but rather because "person(s) X" will be replaceable with "all reasonable observers." Before discarding the limiting frequency concept of probability, though, we must examine how Mackie weds it to Mill's concept of causality.

1.3.3   *Understanding a die roll*

A rolled die is a Newtonian mechanical system and so presumably is governed by a Newtonian deterministic law of working. Yet if anything evinces an intuitive notion of probability, it is a rolled die. What is puzzling and needs explaining is how a deterministic system comes to be a prototype for notions of probability. Mill explains that "in any great number of throws [of a coin], cross and pile are thrown about equally often" because of "the effect of mechanical laws on a symmetrical body acted upon by forces varying indefinitely in quantity and direction" (1973, p. 535). Mackie offers a somewhat more elaborate explanation of die rolling:

In the case of the die we can at least sketch an explanation [of the results of tossing it]. There is a considerable range of variation between possible ways of placing the die in the dice-box and throwing it onto the table. Any one such way leads deterministically, by strict laws of sufficient causation, to a certain result, a certain face's being uppermost when the die comes to rest. If the die is symmetrical, equal portions of this range of possible variation will thus lead to the results 1-up, 2-up, and so on. But if the die is asymmetrical, say if it is loaded to favour 6-up, a larger portion of this range of possible variation will lead, deterministically, to 6-up than to any other result. Add to all this the vital further fact that in a certain long run of throws of the die the actual ways of placing the die in the dice-box and throwing it were scattered fairly evenly over the whole range of possible ways, and we have an explanation why in this run the six faces came up about an equal number of times each (if the die was symmetrical) or unequally, with 6-up the most frequent (if it was loaded).
But this explanation depends on the conjunction of physical symmetry or asymmetry with deterministic laws of working and with a brute fact of collocation, the actual scatter of throwing methods used in the run. (1974, pp. 242–3)

Mackie's explanation then brings together Mill's concepts of collocation and causal law with the physical symmetry of the die and the limiting frequency concept of probability.

On Mackie's account, when a die is rolled, a given collocation [= initial (position, velocity)] in Mill's sense leads via a causal law (= a Newtonian deterministic law of working) in Mill's sense to a certain face up. Physical symmetry of the die suggests that the same number of possible collocations would lead to any given side up. In any long series of die rolls, the actual collocations will be scattered fairly evenly among the possible collocations; so a roughly equal number of collocations leading to each side up will occur. Thus, the equal limiting frequencies of occurring collocations leading to each face up, together with the die's physical symmetry and the causal law, explain the equal limiting frequencies of each face up. Mackie labels a gambling-device law (GDL) the fact that a long series of tosses of a symmetric die leads to a roughly equal number of each face up.

Notice that because he employs the limiting frequency concept of probability in his account, Mackie's explanation is of a long series of die rolls rather than of a single die roll. It should be possible to construct an explanation of a *single* die roll merely by substituting the classical good-reasons equiprobable concept of probability for the limiting frequency concept of probability in Mackie's explanation. In fact, though, some slight further elaboration seems called for.

Consider a single die roll. By Newton's laws, such a roll's outcome is determined as soon as the roll's initial (position, velocity) is known. An omniscient God can know the precise initial (position, velocity) of the roll, but we evidently cannot. The die's symmetry suggests that the same number of possible collocations would lead to any given side up. Yet the collocations leading to a given side up are not contiguous in initial (position, velocity) space. That is, suppose that $v_1 \neq v_0$. If initial (p_0, v_0) leads to 6 up, initial (p_0, v_1) might lead to 5 up, where v_1 and v_0 are indistinguishable with our available measuring devices. Thus, we have no good reason to believe that our roll's collocation falls in the class of collocations leading to a given side up. Hence, we have no good reason to believe one rather than another of the six possible outcomes will occur.

Our explanandum, like Mackie's, can be put in the form of a law: Whenever a symmetric die is rolled, we have no good reason to believe one rather than another of the possible outcomes will occur. We explain the explanandum in terms of the die's symmetry, our having no good reason to believe a collocation leading to one rather than another of the possible outcomes will occur, and the deterministic law of working governing the Newtonian mechanical system. We borrow the term gambling-device law (GDL) for our explanandum from Mackie.

We shall make the conceptual equivalent of a single die roll, rather than a long series of such rolls, central to our account of the controlled experiment. So our good-reasons probability GDL (rather than Mackie's limiting frequency probability GDL) will be the centerpiece of our explication of the controlled experiment. From here on, reference to a GDL will mean a good-reasons equiprobable GDL.

1.4   Summary and conclusion

We characterized Mill's Method of Difference as a way, based on the observations, of identifying a deterministic cause of a particular phenomenon, *if* the phenomenon has a deterministic cause *and* that cause can be gotten onto a list of cause candidates. We also argued that the assumption that the phenomenon has a deterministic cause *cannot* be based *solely* on observation and suggested that theory which in part extends beyond the observations might aid in getting the cause on a list of cause candidates.

Theory was also seen as sometimes providing the sense of "sameness" required in the two instances of Mill's canon, and the counterfactual conditional, which involves an instance which is in principle unobservable, was also understood as providing the sense of "sameness." Finally, we explained Mill's concept of (deterministic) causal law, collocation ($=$ initial conditions), and good-reasons probability (borrowed from Laplace) and then forged these concepts into a concept of GDL for understanding a die roll.

Our discussion began in the classical British empiricist tradition of Berkeley, Hume, and Mill. But by making a series of choices within that tradition or arguments against elements of that tradition, we have stressed that the tradition's concepts of cause, law, probability, and so forth, are not dependent *solely* on observations. Such a view of concepts, methods, and so forth, based solely on observation often seems to be what the tradition and many of its inheritors are striving for. We have begun an immanent critique of empiricism which stresses where the empiricist tradition itself seems to rely on something beyond the observations.

In choosing the Hume–Mill concept of cause over the Schrödinger–Suppes concept, we rejected a concept of cause which seeks to revise the classical concept to preserve reliance solely on observation. That is, in the late nineteenth and early twentieth centuries, situations were observed in which *apparently identical* phenomena C_1 and C_2 had *different* effects. The quantum concept of probabilistic causality was invented to make these situations conform to Berkeley's dictum that what is is what is perceived or observed to be.[12] In retaining the classical concept of causality, we implicitly assume that where C_1 and C_2 lead to different effects, the identity of C_1 and C_2 is only apparent, not actual.

Another place we have chosen against sole reliance on observations is by retaining the counterfactual element in causality, rather than standing

---

[12] The chief question Berkeley (1957) poses for himself is this: What is the relationship between what there is and what there is perceived to be? His answer (pp. 125–6): "The table I write on I say exists; that is, I see and feel it: and if I were out of my study I should say it existed; meaning thereby that if I was in my study I might perceive it. There was an odour, that is, it was smelt; there was a sound, that is, it was heard; a colour or figure, and it was perceived by sight or touch. This is all that I can understand by these and the like expressions. For as to what is said of the *absolute* existence of unthinking things, without any relation to their being perceived, that is to me perfectly unintelligible. Their *esse* is *percipi;* nor is it possible they should have any existence out of the minds of thinking things which perceive them." For Berkeley, then, what there is is what there is perceived (or observed) to be, period. So historically, when cause C was perceived or observed to be sometimes followed by effect c and sometimes by effect $d \neq c$, probabilistic causality was invented in the spirit of Berkeley's ontology.

Classical deterministic concept of causality	Mill's argument that particular determinism assumption is based on observation	Mill/Mackie view that previous observation and spatio-temporal nearness constitute list of cause candidates including actual cause	Causality as frequent co-occurrence plus counterfactual conditional	Good-reasons equiprobable concept of probability borrowed from Laplace by Mill	Limiting frequency concept of probability can be based solely on the observations
$\longrightarrow$			$\longrightarrow$	$\longrightarrow$	
	$\longrightarrow$	$\longrightarrow$			$\longrightarrow$
What there is is what there is observed to be: quantum concept of probabilistic causality	Rejection of Mill's argument that particular determinism assumption is based on observation	Theory beyond observation also helps constitute list of cause candidates including actual cause	Regularity theory of causality as frequent co-occurrence	Limiting frequency concept of probability discussed by Mill	Limiting frequency concept of probability cannot be based solely on the observations

$\longrightarrow$ Indicates choice that was made

Figure 1.1. Beginning an immanent critique of empiricism.

with regularity theorists, who follow Hume but ignore the clause "if the first object had not been, the second never had existed" in his definition. Also, we reject Mill's argument that the assumption of particular determinism that his Method of Difference requires can be based on observation. And we argued that the limiting frequency concept of probability cannot be attached to the world solely on the basis of the observations. Finally, instead of the limiting frequency probability concept usually favored by statistician inheritors of the empiricist tradition, we choose a good-reasons probability approach, which Mill borrowed from French rationalism, for the formulation of our account of the modern controlled experiment. Figure 1.1 depicts the beginning we have made of an immanent critique of empiricism.

Thus far, our immanent critique of empiricism has mainly considered concepts of causality and probability stemming from Berkeley, Hume, and Mill. We shall next extend the critique to the modern controlled experiment itself, a technique of empirical inquiry which inherits much of the classical empiricist tradition. Our account of the controlled experiment will center on the concepts of GDL and counterfactual conditional. Just as we did in the discussion of Mill's Method of Difference, we shall offer an explication of controlled experimental method which understands the method as based on the observations *and* on a set of assumptions which are not based solely on observation.

CHAPTER 2

R. A. Fisher, randomization, and controlled experimentation

2.1 Introduction: Fisher and the emergence of the modern approach

2.1.1 Assessing Fisher's historical role

In a 1976 paper on R. A. Fisher's work, the statistician L. Savage remarks: "Fisher is the undisputed creator of the modern field that statisticians call the design of experiments" (1976, p. 450). What does it mean, though, to say that Fisher "created" a field? Work on techniques of agricultural field experimentation began with Bacon in 1627.[1] By the end of the nineteenth century, that work included such modern techniques as matched-pairs designs, use of (at least particular) Latin-square layouts, and replication. And the use of probabilistic methods in astronomical observations began with the work of Legendre (in 1806) and Gauss (in 1809) on the theory of errors. By the end of the nineteenth century, these probabilistic methods included such staples of modern analysis of variance as fixed-effects and variance-components models. Finally, Darwin, in 1859, made the roughly modern notion of variation and variability central to subsequent work in evolutionary biology. By the dawn of the twentieth century there were three independent strands of research (in agriculture, astronomy, and biology), each containing elements of the statistician's modern experimental design.[2]

In the early part of the twentieth century the three independent strands of research began to come together. In 1908, "Student" proposed his t test, worked out some of its probability distributional properties (thus extending the theory of errors), and illustrated its use with a matched-pairs

---

[1] Of course, we saw in Section 1.1 that in the design of experiments, Bacon had a rather early predecessor. Stigler (1974, p. 431) argues for a somewhat later predecessor: "In the eleventh century, many modern principles of design were spelled out by the famous Arabic doctor, scientist, and philosopher Avicenna."

[2] For the history of agricultural field experimentation, see Fussell (1935) and Crowther (1936). On the early use of probabilistic methods in astronomy, see Scheffé (1956, pp. 256–7). Darwin (1958, especially Chapters I, II, and V) develops the concept of variation.

design. A 1910 paper by an agronomist and an astronomer illustrated the use of the probable error in comparing the means of two treatments.[3] R. A. Fisher was conversant with all three strands of research, and in roughly the period 1918–26 he achieved their decisive synthesis.

Fisher's "creation," then, of the statistician's field of experimental design was a synthesis of the earlier work of others in agriculture, astronomy, and biology. The synthesis entailed generalization (e.g., Fisher generalized the matched-pairs design to his technique of blocking, and Student's t test to his z test[4]). The synthesis also involved reinvention. For example, incorporating Darwin's notion of variation, Fisher reinvented variance-components analysis-of-variance models used by astronomers 50 years earlier.[5] Finally, the technique of randomization and its wedding to matched-pairs designs, blocking in general, and Latin-square layouts were inventions of which Fisher might best be called "undisputed creator."

Statistical Methods for Research Workers (in 1925) and *The Design of Experiments* (in 1935) codified Fisher's new synthesis.[6] These textbooks (and later texts modeled after them) became the basis of experimental practice, first in agriculture, later in biomedicine and psychology, and more recently in economics and sociology. We therefore begin our consideration of controlled social experimentation proper with a look at Fisher's synthesis. In particular, we focus on the technique of randomization which seems so central to Fisher's account of the modern controlled experiment, yet remains a problematic and sometimes controversial technique.

2.1.2 *Fisher's account of randomization and Neyman's immanent critique*

Fisher's published work, beginning in the early 1920s, contains a series of closely related arguments for randomization. He defends randomization's use with some of the traditional sorts of designs (e.g., Latin square and matched pairs). For example, he argues that "the term Latin Square should only be applied to a process of randomization by which one is selected at random out of the total number of Latin Squares possible"

---

[3] Cochran (1976, pp. 12–14) discusses Student's paper, and Wood and Stratton (1910) is the paper by an agronomist and an astronomer.

[4] Thus, R. A. Fisher (1966, p. 57) himself says: "The z-test may be regarded as an extension of the t-test, appropriate to cases where more than two variants are to be compared."

[5] Again, see Scheffé (1956, pp. 255–7).

[6] These two texts are R. A. Fisher (1970) and R. A. Fisher (1966), respectively.

(1972, p. 91). And he explains: "If a solution is chosen at random from such a set [of possible Latin squares] each plot has an equal probability of receiving any of the possible treatments" (1966, pp. 71–2). His critique of Darwin's 15-matched-pairs plant-growth experiment is that "the one flaw in Darwin's procedure was the absence of randomization" (1966, p. 44). If "their sites have been assigned to members of each pair independently at random, the 15 differences . . . would each have occurred with equal frequency with a positive or with a negative sign" (1966, p.45), so that "each of these 2^{15} combinations [i.e., possible sums] will occur by chance with equal frequency" (1966, p. 46).

Fisher also explains randomization in his famous lady-tasting-tea thought experiment, where a lady is given 8 cups of milk-tea, 4 each with milk and tea added first, and asked to identify the milk-first cups. "If . . . we assign, strictly at random, 4 out of the 8 cups to each of our experimental treatments, then every set of 4 . . . will certainly have a probability of exactly 1 in 70 of *being* the 4, for example, to which the milk is added first" (1966, p. 20). And he also adds that the physical act of randomization "in fact, is the only point in the experimental procedure in which the laws of chance, which are to be in exclusive control of our frequency distribution, have been explicitly introduced" (1966, p. 19). Finally, he defends randomization's use with his technique of blocking. "The purpose of randomization" is to provide "that any two plots, not in the same block, shall have the same probability of being treated alike, and the same probability of being treated differently in each of the ways in which this is possible" (1966, pp. 63–4).

On Fisher's account, randomization's role is always the same: to create an equiprobable distribution appropriate to the experiment's context – for example, equally likely particular Latin squares or 2^{15} equally likely sums (in Darwin's experiment). There is even a hint, in randomization's "being the only point in the experimental procedure in which the laws of chance . . . have been explicitly introduced," that a GDL governs (at least the thought experiment). Where Fisher's account appears sometimes to falter is in failing to link randomization's equiprobable distribution to subsequent statistical analysis. For instance, in the analysis of variance for his randomized-block example, Fisher suggests employing a z test (1966, pp. 52–8). Yet he fails to identify any role for the equiprobable distribution, which randomization creates, in the z test's calculations.

In 1935, Jerzy Neyman presented the earliest published critical discussion of the technique of randomization during a meeting of England's

Royal Statistical Society. Practical researchers had begun, largely at Fisher's urging, to employ randomization in their designs, followed by the *z* test in their statistical analyses. Unlike many later critiques, Neyman's does not challenge the wisdom of the physical act of randomization per se. Instead, he first reconstructs the logic of the randomized-block and Latin-square designs and then draws out of this logic the conclusion that these designs and the *z* test are not compatible. Neyman summarizes his argument:

> I noticed the inconsistency of assumptions underlying the *z*-test and the procedure of agricultural experiments. The *z* distribution is deduced from that of two sums of *independent* squares. The methods of Randomized Blocks and of Latin Squares are based on *restricted* sampling, and restricted sampling means the mutual dependence of results. From the point of view of statistical theory, the inconsistency in the application of the *z*-test to the data of Randomized Blocks and Latin Squares is therefore striking. (1935, p. 174)

Fisher's response to Neyman, during the paper's discussion, fails to address Neyman's critique.

We first construct a theory of randomization-based causal point estimation which relies both on certain assumptions not based on the observations and on the concept of counterfactual conditional. We next develop two immanent critiques of randomization-based causal hypothesis-testing. Finally, we extend the theory of randomization-based causal point estimation beyond its core.

2.2   Randomization-based causal inference and its difficulties

2.2.1   *The theory's assumption set and the first principle of equiprobability*

Viewed from the perspective of the treatment (control) group, the physical act of random assignment is conceptually equivalent to a die roll. Suppose there are N experimental subjects, and n $(N - n)$ of them are to be assigned to the treatment (control) group. The assignment can then result in $\binom{N}{n}$ $(\binom{N}{N-n} = \binom{N}{n})$ possible treatment (control) groups. From the perspective of the treatment (control) group *random* assignment amounts to rolling a die, each of whose $\binom{N}{n}$ $(\binom{N}{N-n})$ sides represents one of the possible treatment (control) groups, and then assigning subjects to treatment (control) in accordance with the die roll's outcome.

As we saw in the previous chapter, the die roll's outcome is governed by a GDL. That is, the die roll's outcome follows from the initial conditions and a Newtonian deterministic law of working. But since we have no good reason to believe one rather than another set of initial conditions will obtain, we therefore have no good reason to believe one rather than another of the possible treatment (control) groups will turn up. From the perspective of the treatment (control) group, random assignment creates the classical good-reasons equiprobable distribution on the space of possible outcomes for the treatment (control) group. With random assignment, *no observer* has good reason to believe one rather than another possible treatment (control) group will be chosen.

Once assignment has occurred, the experimenter applies the treatment, and the subjects respond. We assume that response is governed by a deterministic law of working and a known collocation in Mill's co-location sense. That is, suppose S is an N-vector, where S_i represents subject i, and A is an N-vector, where A_i represents the treatment given to subject i. Also, let T be an N-vector, where T_i represents the response of subject i. (For example, in an income maintenance experiment's economic component, T_i was the subject's change in hours worked from the period just before the experiment to a period after the experiment began.) Employing Mill's system of alphabetical characters, we represent the experimenter's co-locating subjects with their assigned treatments as SA; so what we assume is that SA causes T. The sense of "cause" here is the classical Hume–Mill deterministic sense (rather than the Schrödinger–Suppes quantum sense). That is, the same SA *always* results in the same T. No probabilistic element enters the assumptions here.

Thus far, we have spoken of the response vector T as if it were entirely unproblematic. In fact, though, what the experimenter has to work with is not the *actual* response vector T but rather the *reported* response vector, say $r(T)$. Especially (but not only) when an experimenter's reporting system entails subject reporting, it can go awry, so that $r(T) \neq T$. (Such may have been the case, for example, in the binary sociological response variable "marital dissolution" in some of the income maintenance experiments.) Nevertheless, we assume "perfect reporting," or $r(T) = T$, until Section 2.3.3. Notice that such an assumption has a Berkeleyan flavor: What is is what is reported (or perhaps perceived) to be. We take up the question of a possible epistemological basis of the perfect-reporting assumption in Chapter 3.

Our deterministic sense of cause here means that if SA causes T, $S'A'$ causes T', $S = S'$, and $A = A'$, then $T = T'$. But it does *not* mean (or imply) that if SA causes T, $S'A'$ causes T', $S_i = S'_i$, and $A_i = A'_i$ for some i, then $T_i = T'_i$. We do *add* such an assumption of "atomized individuals," however; that is, we assume that $\mathbf{T}(\mathbf{S}, \mathbf{A}) = (T_1(S_i, A_1),\ \ldots,\ T_N(S_N, A_N))$.[7] That is, we assume that subject i's response depends only on the treatment that subject i receives, not on the treatment (or response) of any subject $j \neq i$. The atomized-individuals assumption is simply the "no-externalities" assumption which has been a standard (but nevertheless problematic) one for economists, at least since Adam Smith. We postpone the question of the atomized-individuals assumption's plausibility and epistemological basis until Chapter 8 and the Conclusion.

Remember that $\mathbf{A}$ is the treatment vector; that is, A_i represents the treatment given to subject i. Until Section 2.3.3, we shall assume that all those in the treatment (control) group receive *exactly* the same treatment and that the treatment- and control-group treatments are distinct. That is, we assume that $A_i = A$ (B) for all i such that subject i is in the treatment (control) group and that $A \neq B$. Statisticians have labeled the assumption of no within-group variation of treatment the "no-technical-error" assumption. For the income maintenance experiments, "no technical error" means that each family in a given treatment group was offered an identical (guarantee level, negative tax rate) income plan. The no-technical-error assumption allows us to expand our representation of an individual's response through use of counterfactuals.

Until now, $\mathbf{T}$ has been the N-vector of factual or observed subject responses. We introduce the counterfactual element into our representation of an individual's response by viewing each subject as having a *pair* of responses. T_i is subject i's response to treatment, C_i is subject i's response to control, and $T_i - C_i$ is the ith causal effect for all $i = 1, \ldots, N$. Our assumptions then allow us to view random assignment as simultaneous (and, alas, closely linked) acts of random sampling (without replacement). That is, suppose the experimenter assigns n subjects at random to treatment [and so $(N - n)$ at random to control]. Such assignment is equivalent to drawing n of the N T_i's and $(N - n)$ of the N C_i's at random (without replacement) to be observed, *with the restriction* that T_i is observed if and only if C_i is not observed for all $i = 1, \ldots, N$. Thus, for every i, one and only one component of (T_i, C_i) is observed or factual – the other component is counterfactual. The task of causal inference becomes making statements about the union of factual and counterfactual

---

[7] Notice that we can write our deterministic causality assumption as $\mathbf{T} = \mathbf{T}(\mathbf{S}, \mathbf{A})$ and our atomized-individuals assumption as $\mathbf{T} = [T_1(S_1, A_1),\ \ldots,\ T_N(S_N, A_N)]$. Then "atomized individuals" implies deterministic causality. We keep the two assumptions separate to stress that the theory we shall construct presupposes a classical Hume–Mill deterministic (rather than a Schrödinger–Suppes probabilistic) causality.

responses on the basis of the factual responses (or observations) *and* our assumptions.[8]

Again suppose n subjects are assigned to the treatment group [so $(N - n)$ are assigned to the control group]. Let $t_1, \ldots, t_n$ and $c_1, \ldots, c_{N-n}$ be the actually observed treatment and control responses, respectively. Then our assumptions imply that $\{t_1, \ldots, t_n\} (\{c_1, \ldots, c_{N-n}\})$ is a subset of $\{T_1, \ldots, T_N\} (\{C_1, \ldots, C_N\})$. And if assignment is random, our assumptions are sufficient for the following.

The first principle of equiprobability (PEI)

$$p(t_1, \ldots, t_n; c_1, \ldots, c_{N-n})$$

$$= \begin{cases} 0 \text{ if } T_i \text{ and } C_i \text{ appear among the } n \text{ } t_j\text{'s and } (N-n) \text{ } c_k\text{'s,} \\ \quad \text{respectively, for any } i = 1, \ldots, N \\ 1/\binom{N}{N-n} \text{ otherwise.} \end{cases}$$

Here p means probability in the classical good-reasons sense. Thus, PEI means that any observer has no good reason to believe one rather than another of the $\binom{N}{n}$ possible sets of observations will occur.

Remember that the assumptions behind PEI include the following: Assignment and response are governed by deterministic laws of working. And if our argument in Chapter 1 is valid, these two assumptions cannot be based on observation (i.e., they are ontological, not empirical, assumptions). PEI also depends on the perfect-reporting and atomized-individuals assumptions. The epistemological basis of these two assumptions is not obviously empirical either, but we postpone that issue until later. And PEI depends on the no-technical-error assumption. Finally, PEI depends on two collocations (in two different senses). PEI's random assignment is conceptually equivalent to a die roll in which we have no good reason to believe one rather than another precise set of initial conditions occurs. And PEI assumes that the experimenter actually assigns subjects to treatment and control in strict accordance with the outcome of the die roll. We turn next to construction of a theory of causal inference based on PEI.

---

[8] A counterfactual element has often lurked beneath the surface of statistical studies which aim at discovering causal relationships. See, for example, Neyman (1935). But Rubin (1978, especially p.38) *explicitly* formulates such a counterfactual causal perspective. Our use of counterfactuals follows Rubin's precedent.

2.2.2 *Core of a theory of causal point estimation*

One historical source of confusion over randomization's role in inference is the frequent failure of Fisher's 1925 and 1935 textbooks to represent situations with the formal symbols of mathematics. To avoid confusion, we employ a formal, quasi-axiomatic approach wherever possible.[9] To avoid distraction, though, from the conceptual role of randomization in inference, we relegate theorem proofs to the Appendix. We begin with four definitions.

Definition: Any function $h(t_1, \ldots, t_n; c_1, \ldots, c_{N-n})$ is an *estimate*.

Definition: If h is an estimate, then $E[h(t_1, \ldots, t_n; c_1, \ldots, c_{N-n})] = \Sigma h(t_1, \ldots, t_n; c_1, \ldots, c_{N-n})p(t_1, \ldots, t_n; c_1, \ldots, c_{N-n})$ is the *expectation* of h.

Here Σ is over the finite class of possible sets of observations $t_1, \ldots, t_n$; $c_1, \ldots, c_{N-n}$. Recall that under PEI, for example, there are $\binom{N}{n}$ such possible sets, each having probability $1/\binom{N}{n}$ of occurring.

Definition: If h is an estimate, H^* is a function of the T_i's and C_i's and $E(h) = H^*$, then h is an *unbiased estimate* of H^*.

Definition

$$\overline{T} = \sum_{i=1}^{N} T_i/N \text{ is the mean of the population of } T_i\text{'s},$$

$$\overline{C} = \sum_{i=1}^{N} C_i/N \text{ is the mean of the population of } C_i\text{'s},$$

$$\overline{t} = \sum_{i=1}^{n} t_i/n \text{ is the mean of the sample of } t_i\text{'s, and}$$

$$\overline{c} = \sum_{i=1}^{N-n} c_i/(N-n) \text{ is the mean of the sample of } c_i\text{'s}.$$

---

[9] The main explicit axiom of our theory is the previous section's PEI. (Two other explicit axioms also appear in Section 2.3's extension of the theory.) But the assumptions of the previous section are also behind the theorem in the sense that they are implicit in the axioms and/or are required (in addition to the axioms) for the theorems' proofs.

The principal theorem of the randomization-based theory of causal point estimation then follows.

Theorem 2.1. *PEI implies* $\bar{t} - \bar{c}$ *is an unbiased estimate of* $\bar{T} - \bar{C}$.

Our theory thus centers on an unbiased estimate of the difference between treatment and control response means. The physical act of randomization induces the classical good-reasons equiprobable distribution on the class of $\binom{N}{n}$ possible sets of observations. What occurs, or what we observe, is a single $t_1, \ldots, t_n; c_1, \ldots, c_{N-n}$, from which we calculate $\bar{t} - \bar{c}$, which, by deduction, is an unbiased estimate of $\bar{T} - \bar{C}$.

The next logical step in the development of a randomization-based theory of causal point estimation is conceptualizing the variability of the estimate $\bar{t} - \bar{c}$ over possible sets of observations. Variance and covariance concepts are helpful for the task.

Definition

$$S_T^2 = \sum_{i=1}^{N} (T_i - \bar{T})^2/(N-1) \text{ is the variance of the population of}$$
$$T_i's,$$

$$S_C^2 = \sum_{i=1}^{N} (C_i - \bar{C})^2/(N-1) \text{ is the variance of the population of}$$
$$C_i's, \text{ and}$$

$$S_{TC} = \sum_{i=1}^{N} (T_i - \bar{T})(C_i - \bar{C})/(N-1) \text{ is the covariance of the}$$
population of $(T_i, C_i)'s$.

Definition: If h_j for $j = 1, 2$ are estimates, then $\mathrm{Var}(h_j) = E[h_j - E(h_j)]^2$ for $j = 1, 2$, and $\mathrm{Cov}(h_1, h_2) = E[(h_1 - E(h_1))(h_2 - E(h_2))]$.

The theorem which expresses the variability (under PEI) of the estimate $(\bar{t} - \bar{c})$ over possible sets of observations can then be deduced.

Theorem 2.2. *PEI implies*

(i) $\mathrm{Cov}(\bar{t}, \bar{c}) = -S_{TC}/N$,

(ii) $\mathrm{Var}(\bar{t} - \bar{c}) = [(N-n)/nN]S_T^2 + [n/(N-n)N]S_C^2 + (2/N)S_{TC}$.

Notice that (ii) of the theorem gives an expression for the variability of $(\bar{t} - \bar{c})$ in terms of population and sample sizes and population variances and covariance.

2.2.3 Rediscovery of Neyman's dependency difficulty

Proceeding to develop the randomization-based theory of causal point estimation, we seek an unbiased point estimate of $\mathrm{Var}(\bar{t} - \bar{c})$. Three new estimates seem promising aids.

Definition

$$s_t^2 = \sum_{i=1}^{n} (t_i - \bar{t})^2/(n-1),$$

$$s_c^2 = \sum_{i=1}^{N-n} (c_i - \bar{c})^2/(N-n-1), \text{ and}$$

$$s_p^2 = [(N-n)/nN]s_t^2 + [n/(N-n)N]s_c^2.$$

The new definition allows the deduction of another theorem.

Theorem 2.3. *PEI implies* $E[s_p^2] = \mathrm{Var}(\bar{t} - \bar{c}) - (2/N)S_{TC}.$

So to exhibit an estimate of $\mathrm{Var}(\bar{t} - \bar{c})$ unbiased under PEI, we need only exhibit an estimate of S_{TC} unbiased under PEI. But an estimate of S_{TC} unbiased under PEI is not exhibitable because $S_{TC} = \Sigma_{i-1}^{N}$ $(T_i - \bar{T})(C_i - \bar{C})/(N-1)$, and because T_i is observed iff C_i is not observed, we observe not a single (T_i, C_i) pair on which to base an estimate.

In fact, the problem that s_p^2 is not an unbiased estimate of $\mathrm{Var}(\bar{t} - \bar{c})$ under PEI stems from Neyman's dependency difficulty. If $t_1, \ldots, t_n$ were drawn at random (without replacement) from the T_i's and $c_1, \ldots, c_{N-n}$ were drawn at random (without replacement) from the C_i's, *subject to no restriction*, the probability of any possible set of observations would be $1/[\binom{N}{n}]^2$; that is, $p(t_1, \ldots, t_n; c_1, \ldots, c_{N-n})$ would equal $p(t_1, \ldots, t_n) \times p(c_1, \ldots, c_{N-n})$. So we would have a case of two independent random samples; $\mathrm{Cov}(\bar{t}, \bar{c})$ would equal 0, and $E[s_p^2]$ would equal $\mathrm{Var}(\bar{t} - \bar{c})$. But with PEI (or randomization), we *don't* have two unrestricted samples. The sampling is restricted by the following condition: T_i may appear in the t_j's iff C_i does not appear in the c_k's. Neyman reminds us that "restricted sampling means the mutual dependence of results." That is, under PEI, the probability of any possible set of observations is $1/\binom{N}{n}$, so that

$$p(t_1, \ldots, t_n; c_1, \ldots, c_{N-n}) = p(t_1, \ldots, t_n)$$
$$= p(c_1, \ldots, c_{N-n}) \neq p(t_1, \ldots, t_n) \times p(c_1, \ldots, c_{N-n}),$$

and we don't have independence of $t_1, \ldots, t_n$ and $c_1, \ldots, c_{N-n}$. In fact, under PEI, $\mathrm{Cov}(\bar{t}, \bar{c}) = -S_{TC}/N$, and $E[s_p^2] = \mathrm{Var}(\bar{t} - \bar{c}) - (2/N)S_{TC}$.

Let us next study the bias of s_p^2 as an estimate of $\mathrm{Var}(\bar{t} - \bar{c})$ under PEI. To do so we require another definition.

Definition

$$B = E(s_p^2)/\mathrm{Var}(\bar{t} - \bar{c}),$$

$$R_{TC} = S_{TC}/(S_T S_C),$$

$$M = [(N - n)/n] \text{ (the ratio of control- to treatment-group size),}$$
and

$$Q = S_T/S_C.$$

The new definition allows us to deduce another theorem.

Theorem 2.4. *Under PEI, $B = 1 - 2R_{TC}/[MQ + (1/MQ) + 2R_{TC}]$.*

Notice from the theorem that B has its largest value (∞) where $R_{TC} = -1$ and $M = 1/Q$, its smallest value $(1/2)$ where $R_{TC} = 1$ and $M = 1/Q$, and is 1 [i.e., s_p^2 is an unbiased estimate of $\mathrm{Var}(\bar{t} - \bar{c})$] only if $R_{TC} = 0$. Notice also that for given $R_{TC} \neq 0$ and $Q \neq 0$, B is made as close as possible to 1 by putting all but one subject in the treatment (or control) group. In practical situations, B may be substantially different from 1. Thus, in one of the income maintenance experimental studies (Hausman and Wise 1979), N was 585 and n was 334, so that M was about 3/4. Viewing the study from the perspective of our randomization-based theory of causal point estimation, then, and supposing $S_T^2 = S_C^2$, the study's B could have been anywhere from about 25/49 to 25.

One "solution" to the difficulty of s_p^2's bias as an estimate of $\mathrm{Var}(\bar{t} - \bar{c})$ under PEI would be simply to assume $S_{TC} = 0$. Such an assumption, though in principle not based on the observations, because no (T_i, C_i) is observable, would make s_p^2 an unbiased estimate of $\mathrm{Var}(\bar{t} - \bar{c})$ by making $\mathrm{Cov}(\bar{t}, \bar{c}) = 0$. But such an assumption would not help much in advancing our theory toward a test of something like the hypothesis H: $\overline{T} = \overline{C}$ against an alternative, say, A: $\sim H$, for $\text{Cov}(\overline{t}, \overline{c}) = 0$ doesn't imply that $t_1, \ldots, t_n$ and $c_1, \ldots, c_{N-n}$ are independent samples from the T_i's and C_i's, respectively, which is what the usual test of something like $\overline{T} = \overline{C}$ supposes.[10] In fact, under PEI, $t_1, \ldots, t_n$ and $c_1, \ldots, c_{N-n}$ are *not independent* samples from the T_i's and C_i's, respectively. Neyman's dependency difficulty then appears to impede a randomization-based test of a hypothesis like H: $\overline{T} = \overline{C}$. But Neyman's dependency difficulty doesn't hinder the testing of another kind of hypothesis.

2.2.4 *Randomization-based testing of a sharp causal hypothesis*

Consider now the hypothesis H_0: $T_i = C_i$ for all $i = 1, \ldots, N$.[11] H_0 means that each subject's treatment response is identical with that subject's control response. Remember that we can observe not a single (T_i, C_i) pair. Still, we would ideally like to have a procedure which is a priori capable of concluding, on the basis of randomization and what we do observe, that H_0 is false. In fact, though, the closest we can get to such an ideal procedure is a procedure which is a priori capable of concluding, on the basis of randomization and what we do observe, that either H_0 is false or an unusual event has occurred.

Formally, on the basis of convention we choose a number α between 0 and 1 called the *significance level*. Let h be an estimate. Recall that there are $\binom{N}{n}$ possible sets of observations to each of which there corresponds a value of h. Suppose h_0 is the actually observed value of h.

Definition: $N(h_0) =$ the number of possible sets of observations for which $|h - E[h]| \geq |h_0 - E[h]|$.

Then $N(h_0)$ is the number of possible sets of observations with values of h at least as extreme as that observed (i.e., h_0).

---

[10] The usual test of a hypothesis like H requires $t_1, \ldots, t_n$ ($c_1, \ldots, c_{N-n}$) to be independent too. But from PEI (and Lemma 2.1), $t_1, \ldots, t_n$ ($c_1, \ldots, c_{N-n}$) are not independent, i.e., for any possible $t_1, \ldots, t_n$ ($c_1, \ldots, c_{N-n}$), $p(t_1, \ldots, t_n)$ [$p(c_1, \ldots, c_{N-n})$] $= 1/\binom{N}{n}$ [$1/\binom{N}{N-n}$] $\neq (1/N)^n$ [$(1/N)^{N-n}$] $= \Pi_{i=1}^n p(t_i)$ [$\Pi_{i=1}^{N-n} p(c_i)$]. This *intra*-sample (in contrast to Neyman's *inter*-sample) dependency difficulty stems from the fact that the sampling implicit in randomization is without, rather than with, replacement.

[11] Our account of testing H_0 is similar to that of Rubin (1980), except that we do not include any pairing.

Definition: h_0 is an *unusual* occurrence of the estimate h iff $p(|h - E[h]| \geq |h_0 - E[h]|) \leq \alpha$.

We seek then the following theorem:

Theorem 2.5. *Suppose PEI holds,*

(i) *If $[N(h_0)/\binom{N}{n}] \leq \alpha$ for H_0, then H_0 is false or h_0 is an unusual occurrence of h,*

(ii) *If $[N(h_0)/\binom{N}{n}] > \alpha$ for H_0, then H_0 is false or h_0 is not an unusual occurrence of h.*

Albeit somewhat opaquely, Fisher (1966, pp. 44–8) was the first to suggest tests such as those characterized by the theorem, and Fisher's choice of h was $\bar{t} - \bar{c}$. Basu (1980, especially pp. 577–8) has recently offered an immanent critique of such tests: They depend rather strongly on one's choice of h, which is essentially an arbitrary matter.

Example 2.1. Let $n = 4$ and $N = 8$, so that $\binom{N}{n} = 70$. Define $\tilde{t} =$ the median of the t_i's, and $\tilde{c} =$ the median of the c_j's. Then PEI and H_0 imply $E(\bar{t} - \bar{c}) = E(\tilde{t} - \tilde{c}) = 0$. Suppose we observe t_i's 30, 18, 9, and 15 and c_j's 3, 5, 7, and 1, so that $(\bar{t} - \bar{c})_0 = 14$ and $(\tilde{t} - \tilde{c})_0 = 12.5$. Of the 70 possible values of $\bar{t} - \bar{c}$ under H_0, only 2 are as extreme as 14, while of the 70 possible values of $\tilde{t} - \tilde{c}$ under H_0, 4 are as extreme as 12.5. So choosing the conventional $\alpha = .05$, we find under H_0 that $N((\bar{t} - \bar{c})_0)/\binom{N}{n} = 2/70 \leq .05$, while $N((\tilde{t} - \tilde{c})_0)/\binom{N}{n} = 4/70 > .05$.

Thus, the dependency of the test on an arbitrary choice of h *appears* to lead to an anomaly: opposite test conclusions for choice of $h = \bar{t} - \bar{c}$ and $h = \tilde{t} - \tilde{c}$.

In fact, the conclusions for the two choices of h in the example are *not* contradictory. Let $U(\bar{t} - \bar{c})$ $(U(\tilde{t} - \tilde{c}))$ mean $(\bar{t} - \bar{c})_0$ $((\tilde{t} - \tilde{c})_0)$ is an unusual occurrence of $(\bar{t} - \bar{c})$ $((\tilde{t} - \tilde{c}))$. Then with $h = (\bar{t} - \bar{c})$ $((\tilde{t} - \tilde{c}))$, we conclude $\sim H_0 \vee U(\bar{t} - \bar{c})$ $(\sim H_0 \vee \sim U(\tilde{t} - \tilde{c}))$, and $[\sim H_0 \vee U(\bar{t} - \bar{c})] \wedge [\sim H_0 \vee \sim U(\tilde{t} - \tilde{c})]$ is *not* a contradiction. Still, the two sets of conclusions are *different* if not contradictory. Thus, an element of nonobjectivity reenters the theory in the sense that different observers choosing different h's can draw different (though not contradictory) conclusions from the same observations. In the standard Neyman–Pearson theory, concepts of alternative, power, and desirable properties of tests sometimes lead the theory itself to select an h to employ in testing a given hypothesis. Whether the arbitrariness of h's choice in our randomization-

The asymptotic test result then follows.

Theorem 2.6. *Suppose that as N approaches* ∞, $\overline{T}_N$, S^2_{NT}, F^2_{NT}, $\overline{C}_N$, S^2_{NC}, F^2_{NC} *approach T*, $V^2_T (>0)$, Y^2_T, C, $V^2_C (= V^2_T)$, Y^2_C *(all* $<\infty$), *respectively; that R with* $|R| < 1$ *is the limit of* R_{NTC}; *and that* $\max(T_{Ni} - \overline{T}_N)^2$ *and* $\max(C_{Ni} - \overline{C}_N)^2$ *are bounded. Also suppose that n/N approaches* γ *as n and* $(N - n)$ *approach* ∞. *Then if* $T = C$, *the limiting probability distribution, under PEI for every N, of* z_N *is* $N[0, 1 - 2\gamma(1 - \gamma)(1 - R)]$ *as n and* $(N - n)$ *approach* ∞.[12]

Notice first that the theorem's assumptions are about limits of hypothetical infinite sequences, and that a physical act of randomization creates PEI for only one member of the hypothetical infinite sequence of population responses. Also, the assumption that $|R| < 1$ is about the limit of a hypothetical infinite sequence none of whose members is observable. Hence, the theorem's assumptions certainly go beyond what is observable.

Note second that the theorem's null hypothesis is that the *limit* of the mean causal effect is 0. And the theorem concludes that z_N's limiting distribution is $N(0, \sigma^2)$, where $0 \le \sigma \le 1$, with σ equal to 1 only if the ratio of treatment- to control-group sizes becomes 0 or ∞ in the limit. In contrast, when the observations are independent, σ always equals 1. So if p_0 is the test result's p-value assuming independence, then one can say only that the test result's p-value under randomization is $\le p_0$. We turn next to some extensions of the core of our theory of randomization-based causal point estimation.

2.3   More randomization-based causal point estimation

2.3.1   *The principle of blocked equiprobability and stratification's role in causal inference*

Assume that the assumptions of Section 2.2.1 still hold, but that now assignment is no longer *simple* random. Suppose the N (T_i, C_i)'s are labeled (T_{11}, C_{11}), . . . , (T_{1N_1}, C_{1N_1}) and (T_{21}, C_{21}), . . . , (T_{2N_2}, C_{2N_2}), where the N_1 (T_{1i}, C_{1i})'s are a first block and the $N_2 (= N - N_1)$ (T_{2i}, C_{2i})'s are a second block. Suppose n_1 subjects from block one are assigned to the treatment group [so $(N_1 - n_1)$ from block one are assigned to the control group]. Also suppose $n_2 (= n - n_1)$ subjects from block two are assigned to the treatment group [so $(N_2 - n_2)$ from block two are assigned to the

Theorem 2.6 is essentially a conjecture of Copas (1973) proved in Neuberg (1987).

based test then is due to an underdeveloped theory or an unre
weakness in our theory remains unclear. In either case, the
theorem makes clear randomization's role in testing H_0.

2.2.5 Randomization-based testing of the hypothesis of no causal effect

Section 2.2.3 concluded that Neyman's dependency difficult
randomization-based test of a hypothesis like H: $\overline{T} = \overline{C}$. An
test of such a hypothesis is, however, possible. So far we hav
only a single population of N experimental subjects with re
represented as (T_i, C_i). Under PEI, n of the T_i's and $(N -$
were chosen as treatment- and control-group responses,
Now, however, consider a hypothetical infinite sequence c
of increasing size. That is, let the populations be $\Pi_1 = \{(T$
$\{(T_{21}, C_{21}), (T_{22}, C_{22})\}, \ldots, \Pi_N = \{(T_{N1}, C_{N1}), (T_{N2}, C_{N2}$
$C_{NN})\}$. For each N, under PEI, $n(N)$ of the T_{Ni}'s and $(N - n)$
are chosen as treatment- and control-group responses, res
the remainder of this section only, a subscript N indicates
of the sequence.

Note that only one population of the infinite sequen
response pairs of the actual experimental subjects. For th
actual physical act of random assignment selects the tre
trol-group responses. The other populations are hypotl
each of those populations the physical act of random
chooses the treatment- and control-group responses is

We next require a definition.

Definition

$$z_N(t_{N1}, \ldots, t_{Nn}; c_{N1}, \ldots, c_{N(N-n)})$$

$$= (\bar{t}_N - \bar{c}_N) \Big/ \Big[\Big(\sum_{i=1}^{n} (t_{Ni} - \bar{t}_N)^2 / n(n-1) \Big)$$

$$+ \Big(\sum_{i=1}^{N-n} (c_{Ni} - \bar{c}_N)^2 / (N-n)(N-n- \Big)$$

$$U_{NT} = \sum_{i=1}^{N} T_{Ni}^2 / N, \quad U_{NC} = \sum_{i=1}^{N} C_{Ni}^2 / N,$$

$$F_{NT}^2 = \sum_{i=1}^{N} (T_{Ni}^2 - U_{NT})^2 / N - 1), \quad \text{and}$$

$$F_{NC}^2 = \sum_{i=1}^{N} (C_{Ni}^2 - U_{NC})^2 / (N-1).$$

control group]. Let $t_{11}, \ldots, t_{1n_1}$ $(c_{11}, \ldots, c_{1(N_1 - n_1)})$ and $t_{21}, \ldots, t_{2n_2}$ $(c_{21}, \ldots, c_{2(N_2 - n_2)})$ be the actually observed responses for the treatment (control) group from blocks one and two, respectively. Then our assumptions imply that $\{t_{j1}, \ldots, t_{jn_j}\}$ $(\{c_{j1}, \ldots, c_{j(N_j - n_j)}\})$ is a subset of $\{T_{j1}, \ldots, T_{jN_j}\}$ $(\{C_{j1}, \ldots, C_{jN_j}\})$ for $j = 1, 2$.

If assignment is *stratified* random (i.e., random within each block), then our assumptions are sufficient for the following.

Principle of blocked equiprobability (PBE)

$$p(t_{11}, \ldots, t_{1n_1}; t_{21}, \ldots, t_{2n_2}; c_{11}, \ldots, c_{1(N_1 - n_1)}; c_{21}, \ldots, c_{2(N_2 - n_2)})$$

$$= \begin{cases} 0 \text{ if } T_{ji} \text{ and } C_{ji} \text{ appear among the } n_j \ t_{kl}\text{'s } and \ (N_j - n_j) \ c_{rs}\text{'s,} \\ \quad \text{respectively, for any } i = 1, \ldots, N_j \text{ for } j = 1, 2 \\ 1/\binom{N_1}{n_1}\binom{N_2}{n_2} \text{ otherwise.} \end{cases}$$

Here p means probability in the classical good-reasons sense. Thus, PBE means that any observer has no good reason to believe one rather than another of the $\binom{N_1}{n_1}\binom{N_2}{n_2}$ possible sets of observations will occur. Since PBE is conceptually equivalent to PEI independently on each block, PBE relies on the same assumptions (some nonempirical) as PEI.

Since each block is a (sub)population of pairs, we require first a new definition.

Definition

$$\bar{T}_j = \sum_{i=1}^{N_j} T_{ji}/N_j, \qquad \bar{C}_j = \sum_{i=1}^{N_j} C_{ji}/N_j,$$

$$\bar{t}_j = \sum_{i=1}^{n_j} t_{ji}/n_j, \quad \text{and} \quad \bar{c}_j = \sum_{i=1}^{N_j - n_j} c_{ji}/n_j \quad \text{for } j = 1, 2.$$

We then have an analogue of Theorem 2.1.

Theorem 2.7. *PBE implies that $\bar{t}_j - \bar{c}_j$ is an unbiased estimate of $\bar{T}_j - \bar{C}_j$ for $j = 1, 2$.*

One reason to block (or stratify) is to be able to make an unbiased comparison of the (sub)population response means for treatment and control for a given stratum. For example, in two of the income maintenance experiments, subjects were blocked into male- and female-headed families so that unbiased comparisons could be made within each of these strata of interest.[13]

We can also deduce another theorem.

Theorem 2.8. *PBE implies that if $n_j/N_j = n/N$ for $j = 1, 2$, then $\bar{t} - \bar{c}$ is an unbiased estimate of $\bar{T} - \bar{C}$.*

So the difference in observed treatment and control response means is an unbiased estimate of the difference in overall population treatment and control response means under either simple random or *proportional* stratified random assignment. Might there not be a reason, though, besides interest in the separate strata, to prefer proportional stratified random to simple random assignment?

Let $\mathrm{Var}_{\mathrm{PBE}}(\bar{t} - \bar{c})$ $(\mathrm{Var}_{\mathrm{PEI}}(\bar{t} - \bar{c}))$ be $E[(\bar{t} - \bar{c}) - E(\bar{t} - \bar{c})]^2$ under PBE (PEI). We can then prove another theorem.

Theorem 2.9. *If $n_j/N_j = n/N$ and $[(N_j - 1)/(N - 1)] \approx (N_j/N)$ for $j = 1, 2$, then*

$$\mathrm{Var}_{\mathrm{PEI}}(\bar{t} - \bar{c}) \approx \mathrm{Var}_{\mathrm{PBE}}(\bar{t} - \bar{c}) + [1/(N-1)n]$$
$$\sum_{j=1}^{2} N_j[((N-n)/n)^{1/2}(\bar{T}_j - \bar{T}) + (n/(N-n))^{1/2}(\bar{C}_j - \bar{C})]^2.$$

Theorem 2.9 gives a second reason for blocking (or stratification). Suppose that the theorem's hypothesis holds. Then as long as $\bar{T}_j \neq \bar{T}$ or $\bar{C}_j \neq \bar{C}$ for some $j = 1, 2$, $\mathrm{Var}_{\mathrm{PEI}}(\bar{t} - \bar{c}) > \mathrm{Var}_{\mathrm{PBE}}(\bar{t} - \bar{c})$. Thus, we block proportionally to lower Var of our (unbiased) estimate of the difference in overall population treatment and control response means. Theorem 2.8 assures us that by so blocking (with random assignment occurring independently in each block), our estimate has the same unbiasedness it would have under simple random assignment.

2.3.2 The second principle of equiprobability and another extension of causal point estimation

So far our theory of causal point estimation encompasses only the experiment's subjects. But suppose the experiment's N subjects have been drawn at random from a population of $\mathcal{N}$ $(\geq N)$ potential subjects. The

---

[13] These were the Gary and Seattle-Denver experiments.

physical act of random selection is conceptually equivalent to rolling a die each of whose $\binom{N}{N}$ sides represents one of the possible sets of experimental subjects, and then choosing the set of subjects in accordance with the die roll's outcome. The die roll's outcome is governed by a GDL, and so random selection creates the classical good-reasons equiprobable distribution on the space of possible subject-set outcomes. With random selection, *no observer* has good reasons to believe one rather than another possible subject set will be chosen.

Suppose that the assumptions of Section 2.2.1 still hold *and* that a subject's response does not depend on who else is selected for the experiment (another atomized-individuals assumption). Then we may represent the population of potential subject responses as $(\mathcal{T}_1, \mathcal{C}_1), \ldots, (\mathcal{T}_N, \mathcal{C}_N)$, where $\mathcal{T}_i(\mathcal{C}_i)$ is the response potential subject i would have if placed in the experiment's treatment (control) group. Our assumptions imply that $\{(T_1, C_1), \ldots, (T_N, C_N)\}$ is a subset of $\{(\mathcal{T}_1, \mathcal{C}_1), \ldots, (\mathcal{T}_N, \mathcal{C}_N)\}$, and if selection is random, we have the following.

The second principle of equiprobability (PEII)

$$p((T_1, C_1), \ldots, (T_N, C_N)) = 1/\binom{N}{N}.$$

Again, p means probability in the classical good-reasons sense. Thus, PEII means that any observer has no good reason to believe one rather than another of the $\binom{N}{N}$ possible subject response sets will occur.

Next we extend two earlier definitions to the selection-assignment context.

Definition

$$\overline{\mathcal{T}} = \sum_{i=1}^{N} \mathcal{T}_i/N \text{ is the mean of the population of } \mathcal{T}_i\text{'s,}$$

$$\overline{\mathcal{C}} = \sum_{i=1}^{N} \mathcal{C}_i/N \text{ is the mean of the population of } \mathcal{C}_i\text{'s.}$$

Definition: If h is an estimate, $\mathcal{H}^*$ is a function of the $\mathcal{T}i$'s and $\mathcal{C}_i$'s, and $E(h) = \mathcal{H}^*$, then h is an *unbiased estimate of $\mathcal{H}^*$*.

The extension of our theory of randomization-based causal point estimation then follows.

Theorem 2.10. *PEI and PEII imply that* $\bar{t} - \bar{c}$ *is an unbiased estimate of* $\bar{\mathcal{T}} - \bar{\mathcal{C}}$.

The physical acts of random selection and assignment induce the classical good-reasons equiprobable distribution on the class of $\binom{X}{N}\binom{N}{n}$ possible sets of observations. What occurs, or what we observe, is a single set of observations from which we calculate $\bar{t} - \bar{c}$, which, by deduction, is an unbiased estimate of $\bar{\mathcal{T}} - \bar{\mathcal{C}}$.

The addition of random selection to our theory suggests some further possible extensions. We might consider *stratified* random selection (a technique employed in the income maintenance experiments). Then in addition to the (simple, simple) random (selection, assignment) combination of the present section there would be three others to consider: (simple, stratified), (stratified, simple), and (stratified, stratified). In each of these cases, though, the story would be similar: Stratification or blocking (with certain assumptions) lowers the variance of an unbiased estimate of the difference in two population means. Also, with random selection, the two previous sections' tests are easily extended to the $(\mathcal{T}_i, \mathcal{C}_i)$ pairs. But rather than focus on the distracting symbolic details of the cited possible extensions, which would break no new conceptual ground, we turn instead to relaxation of two assumptions.

2.3.3   *Relaxing the perfect-reporting and no-technical-error assumptions*

Thus far we have assumed a perfect reporting system. That is, if $r(t_i)$ and $r(c_j)$ are the reported values of the responses t_i and c_j, respectively, then we have assumed $r(t_i) = t_i$ and $r(c_j) = c_j$ for all $i = 1, \ldots, n$ and $j = 1, \ldots, N - n$. Suppose, however, that the other assumptions of Section 2.2.1 hold, but that there is some error in the reporting, so that $r(t_i) = t_i + \epsilon_i$ and $r(c_j) = c_j + \eta_j$ for all $i = 1, \ldots, n$ and $j = 1, \ldots, N - n$, where the ϵ_i's and η_j's are reporting-error terms. Let $p(\epsilon_1, \ldots, \epsilon_n; \eta_1, \ldots, \eta_{N-n})$ be the probability of the error terms $\epsilon_1, \ldots, \epsilon_n;$ $\eta_1, \ldots, \eta_{N-n}$ occurring, where p has the usual formal properties, but the *meaning* of probability and the space of possible outcomes is left unspecified until Chapter 3 $[p(0, \ldots, 0; 0, \ldots, 0) = 1$ corresponds to the case of perfect reporting].

Consider next a definition.

Definition

$$\bar{\epsilon} = \sum_{i=1}^{n} \epsilon_i/n, \quad \bar{\eta} = \sum_{i=1}^{N-n} \eta_i/(N-n),$$

$$\bar{r}(t) = \sum_{i=1}^{n} r(t_i)/n, \quad \text{and} \quad \bar{r}(c) = \sum_{i=1}^{N-n} r(c_i)/(N-n).$$

Assume that $E(\bar{\epsilon} - \bar{\eta}) = 0$. We note that the plausibility of this assumption is based neither on the observations nor on the physical act of randomization, but postpone until Chapter 3 further consideration of a possible epistemological basis for the assumption. We can now extend our theory of randomization-based causal point estimation.

Theorem 2.11. *PEI and $E(\bar{\epsilon} - \bar{\eta}) = 0$ imply that $\bar{r}(t) - \bar{r}(c)$ is an unbiased estimate of $\bar{T} - \bar{C}$.*

Thus, Theorem 2.1 extends to a situation with reporting error.

Thus far we have also assumed no technical error [i.e., that treatment (control) was *identical* for every individual in the treatment (control) group]. Suppose, however, that the other assumptions of Section 2.2.1 hold, but that individual treatments in a group vary somewhat from the ideal treatment. Then we can write the treatment (control) responses as

$$t_i^* = t_i + \zeta_i \, (c_j^* = c_j + \xi_j) \quad \text{for all } i = 1, \ldots, n \, (j = 1, \ldots, N-n),$$

where the ζ_i's (ξ_j's) are technical-error terms. Let $p(\zeta_1, \ldots, \zeta_n; \xi_1, \ldots, \xi_{N-n})$ be the probability of the error terms $\zeta_1, \ldots, \zeta_n; \xi_1, \ldots, \xi_{N-n}$ occurring, where p has the usual formal properties, but the *meaning* of probability and the space of possible outcomes is left unspecified for now. Chapter 4 considers the usual meaning for the probability of technical-error terms [$p(0, \ldots, 0; 0, \ldots, 0) = 1$ corresponds to the case of no technical error].

Next we require a definition.

Definition

$$\bar{\zeta} = \sum_{i=1}^{n} \zeta_i/n, \quad \bar{\xi} = \sum_{i=1}^{N-n} \xi_i/(N-n),$$

$$\bar{t}^* = \sum_{i=1}^{n} t_i^*/n, \quad \text{and} \quad \bar{c}^* = \sum_{i=1}^{N-n} c_i^*/(N-n).$$

Assume that $E(\bar{\zeta} - \bar{\xi}) = 0$. Again, the plausibility of this assumption is based neither on the observations nor on the physical act of randomization. We can now again extend our theory of randomization-based causal point estimation.

Theorem 2.12. *PEI and* $E(\bar{\zeta} - \bar{\xi}) = 0$ *imply that* $\bar{t}^* - \bar{c}^*$ *is an unbiased estimate of* $\bar{T} - \bar{C}$.

Thus, Theorem 2.1 extends to a situation with technical error.

Notice that the extensions of our theory of randomization-based causal point estimation to include reporting and technical errors are quite similar. The two extensions suggest other straightforward extensions. Reporting and technical errors could both be allowed in the same situation, one or the other or both types of errors could be allowed in situations involving random selection and stratification, and so forth. Each such extension would mire us in details of symbols in order to reiterate the same two conceptual points. Randomization induces a classical good-reasons equiprobable distribution on the space of possible observations. Then from the equiprobable distribution and an assumption about (reporting and/or technical) error-term expectations, one can deduce that the difference in reported observed means is an unbiased estimate of the difference in population means. We therefore forgo such conceptually routine extensions and conclude our discussion of randomization-based causal point estimation with a situation requiring some conceptual novelties.

2.3.4   *Causal univariate regression*

Now suppose the assumptions of Section 2.2.1 hold for any response variables. So far we have represented our experiment's subject as a response pair (T_i, C_i). But now we expand that representation slightly and see each subject as a quadruple $(T_i, X_{Ti}, C_i, X_{Ci})$, where X_{Ti} (X_{Ci}) is a second treatment (control) response variable. An actually observed treatment (control) response now becomes (t_i, x_{ti}) $((c_i, x_{Ci}))$. We define means as earlier.

Definition

$$\bar{X}_T = \sum_{i=1}^{N} X_{Ti}/N, \qquad \bar{X}_C = \sum_{i=1}^{N} X_{Ci}/N,$$

$$\bar{x}_t = \sum_{i=1}^{n} x_{ti}/n, \quad \text{and} \quad \bar{x}_c = \sum_{i=1}^{N-n} x_{ci}/(N-n).$$

We are interested primarily in a special case.

Definition: If $X_{Ti} = X_{Ci}$ for all $i = 1, \ldots, N$, we say the X_{Ti}'s (or X_{Ci}'s) are a *trait* (with respect to treatment/control).

For example, in the income maintenance experiments, those in the treatment group received a negative income tax, and those in the control group did not. So such factors as age, race, and previous year's income of household head were traits in the income maintenance experimental context. A trait, then, is a "response" variable that can't respond.

With the expanded representation, we can define some population parameters.

Definition

$$\alpha_T = \sum_{i=1}^{N} T_i/N, \qquad \alpha_C = \sum_{i=1}^{N} C_i/N,$$

$$\beta_T = \sum_{i=1}^{N} (X_{Ti} - \overline{X}_T)T_i / \sum_{i=1}^{N} (X_{Ti} - \overline{X}_T)^2, \text{ and}$$

$$\beta_C = \sum_{i=1}^{N} (X_{Ci} - \overline{X}_C)C_i / \sum_{i=1}^{N} (X_{Ci} - \overline{X}_C)^2.$$

Notice that $\alpha_T = \overline{T}$ and $\alpha_C = \overline{C}$, but also that α_T and β_T (α_C and β_C) are ordinary-least-squares (OLSQ) regression coefficients for regressing T_i (C_i) on X_{Ti} (X_{Ci}). That is, if $\tau_i = T_i - A_T - B_T X_{Ti}$ ($C_i - A_C - B_C X_{Ci}$) for $i = 1, \ldots, N$ for some A_T and B_T (A_C and B_C), then $A_T = \alpha_T$ and $B_T = \beta_T$ ($A_C = \alpha_C$ and $B_C = \beta_C$) minimizes $\sum_{i=1}^{N} \tau_i^2$, or α_T and β_T (α_C and β_C) are the intercept and slope, respectively, of the "best" (in the sense of minimizing $\sum_{i=1}^{N} \tau_i^2$) straight line which can be drawn through the pairs (T_i, X_{Ti}) ((C_i, X_{Ci})) for $i = 1, \ldots, N$.

We can also define some new estimates.

Definition

$$\hat{\alpha}_T = \sum_{i=1}^{n} t_i/n, \qquad \hat{\alpha}_C = \sum_{i=1}^{N-n} c_i/(N - n),$$

$$\hat{\beta}_T = \sum_{i=1}^{n} (x_{ti} - \overline{x}_t)t_i / \sum_{i=1}^{N} (X_{Ti} - \overline{X}_T)^2 n, \text{ and}$$

$$\hat{\beta}_C = \sum_{i=1}^{N-n} (x_{ci} - \overline{x}_c)c_i / \sum_{i=1}^{N} (X_{Ci} - \overline{X}_C)^2 (N - n).$$

Notice that $\hat{\alpha}_T = \bar{t}$ and $\hat{\alpha}_C = \bar{c}$. Note also that because $\Sigma_{i=1}^{N}(X_{Ti} - \bar{X}_T)^2 n$ $(\Sigma_{i=1}^{N}(X_{Ci} - \bar{X}_C)^2(N - n))$ rather than $\Sigma_{i=1}^{n}(x_{ti} - \bar{x}_t)^2$ $(\Sigma_{i=1}^{N-n}(x_{ci} - \bar{x}_c)^2)$ appears in $\hat{\beta}_T (\hat{\beta}_C)$, $\hat{\beta}_T (\hat{\beta}_C)$ is *not* the OLSQ regression coefficient for regressing t_i (c_i) on x_{ti} (x_{ci}). The extension of Theorem 2.1 to causal univariate regression then follows.

Theorem 2.13. *PEI implies*

    (i) $\hat{\alpha}_T - \hat{\alpha}_C$ *is an unbiased estimate of* $\alpha_T - \alpha_C$,

    (ii) *if the* X_{Ti}*'s (or* X_{Ci}*'s) are a trait, then* $\hat{\beta}_T - \hat{\beta}_C$ *is an unbiased estimate of* $\beta_T - \beta_C$.

The income maintenance experiments actually involved multiple traits and more than two treatment groups [each characterized by a different (guarantee level, negative tax rate)]. But extending all our randomization-based causal point estimation results to a situation with multiple traits (via multivariate regression) and/or more than two groups would again be perfectly straightforward.

2.4   Summary and conclusion

From Chapter 1 we took the assumption that rolling an $\binom{N}{n}$-sided die is governed by a GDL. We then assumed that subjects were placed in groups strictly in accordance with the outcome of a physical act of randomization. We supposed that subject response is governed by a deterministic law of working, with "atomized individuals" and "no technical error" properties, and that there is "perfect reporting." Finally, we viewed each individual as consisting of *two* responses (treatment and control) – one and only one of which is observed. Our assumptions implied that there are $\binom{N}{n}$ equiprobable (in the classical good-reasons sense) possible sets of observations.

Upon our assumptions and their implied randomization-induced equiprobable distribution, we constructed a randomization-based theory of causal point estimation. The theory's principal result is that the difference in group observed response means is an unbiased estimate of the difference in population response means, where a population consists of the union of observed and counterfactual responses. Blocking or stratification provides a way of lowering the variability of the difference in group observed response means over allowed sets of observations. The theory's principal result extends to a situation in which random selection precedes assignment and to univariate regression. Finally, relaxing the perfect-reporting or no-technical-error assumption requires adding, to the theory's lexicon, a meaning for probability not created by randomization.

We also argued that hypothesis-testing is the Achilles' heel of randomization-based causal inference. Results of testing the hypothesis that each individual's treatment response equals that individual's control response depend on a rather arbitrary choice of estimate. And the dependence of treatment and control responses, noted by Neyman in 1935, allows statement of only an upper bound for the p-value of the results of a randomization-based test of no limiting mean causal effect. Table 2.1 outlines the theory of randomization-based causal inference and its difficulties.

Table 2.1. *A randomization-based theory of causal inference*

Assumptions

Physical acts
SRA (simple random assignment)
    1. GDL-governed
    2. Initial-conditions equivalency classes equiprobably distributed (in classical sense)
PSRA (proportional stratified random assignment)
    1. GDL-governed
    2. Both initial-conditions equivalency classes equiprobably distributed (in classical sense)
SRS (simple random selection)
    1. GDL-governed
    2. Initial-conditions equivalency classes equiprobably distributed (in classical sense)

Experiment (E)
    1. DLW-governed
    2. Assignment strictly by SRA outcome
    3. Assignment strictly by PSRA outcome
    4. Selection strictly by SRS outcome
    5. Responses assignment-independent (atomized individuals I)
    6. Responses selection-independent (atomized individuals II)
    7. No technical error
    8. Expectations technical-error means equal
    9. Perfect reporting
    10. Expectations reporting-error means equal
    11. Limits of hypothetical sequences exist
    12. Treatment–control responses not perfectly correlated

Table 2.1 *(cont.)*

Theory

Causal point estimation results
Under any one of the five alternative assumption sets, the difference in reported observed response means is an unbiased estimate of the difference in population response means.
Alternative assumption sets
1. SRA1,2; E1,2,5,7,9 (2.2.2, 2.3.4)[a]
2. SRA1,2; E1,2,5,7,10 (2.3.3)
3. SRA1,2; E1,2,5,8,9 (2.3.3)
4. SRA1,2; SRS1,2; E1,2,4,5,6,7,9 (2.3.2)
5. PSRA1,2; E1,3,5,7,9 (2.3.1)

Under the assumptions, the difference in reported observed slope coefficient estimates is an unbiased estimate of the difference in population OLSQ slopes.
Assumptions: SRA1,2; E1,2,5,7,9 (2.3.4)

Var_{PSRA}(difference in reported observed response means) $<$ Var_{SRA} (difference in reported observed response means).
Assumptions
PSRA1,2 (or SRA1,2); E1,3,5,7,9 (or E1,2,5,7,9) (2.3.1)

Causal hypothesis-testing results and difficulties
Population means are equal
Assumptions: SRA1,2; E1,2,5,7,9,11,12 (2.2.3, 2.2.5)[a]
Difficulty: Results are only an upper bound for p-value

For any individual, the treatment response is the same as the control response
Assumptions: SRA1,2; E1,2,5,7,9 (2.2.4)
Difficulty: Results depend on arbitrary choice of estimate, which blocks their objectivity (in an interexperimenter invariant sense)

[a] Numbers in parentheses refer to sections of this book.

Recall finally from Section 2.1.1 that Fisher originally developed his technique of randomization in *agricultural* experiments, where the objects of inquiry were fields, fertilizers, crops, and so forth. Even with such objects of inquiry, Chapter 2's theory of randomization-based causal *point estimation* faces the conceptual difficulty of reliance on assumptions in principle not baseable on the observations. As soon, though, as the object of inquiry becomes a conscious human being – as in the case of a social experiment – a whole host of further conceptual difficulties, to which we turn in the next chapter, arise for Chapter 2's theory of randomization-based causal point estimation.

CHAPTER 3

Some special difficulties of controlled social experiments

3.1 Introduction: The nature of the object of inquiry and difficulties for causal inference

Fisher first employed his technique of randomization, together with the theory of causal inference formalized in the previous chapter, in agricultural experiments, where the group members were fields, or sections of fields, of crops, and the treatments were various sorts or amounts of soil enrichment or fertilizer. Only with concepts of "field" and "crop" radically at odds with common sense, then, could we say that Fisher's objects of inquiry were conscious or willful. Within a few decades of its introduction, though, randomization became a standard feature of biomedical drug-testing experiments, where the group members are human subjects and the treatments are various sorts or amounts of drugs. Thus, the human subjects of such drug-testing experiments are certainly conscious, in any ordinary sense of the term, and also creatures of will.

Physicians had early recognized that subject awareness in a medical experiment in which subjects had to be relied on to report results might lead to difficulties. For example, knowledge that they were taking a new aspirin tablet might lead subjects to report greater relief from their headache conditions than had actually occurred. To counteract such misreporting, physicians employ a so-called double-blind experiment, in which a treatment group receives the aspirin, and the control group gets a "placebo," and neither subjects nor physicians know which group subjects are in. Recent historical research illuminates the origin of such techniques:

A few years later, both the medical Society in Vienna in 1844 and the Private Association of Austrian Homeopathic Physicians in 1856/57 were carrying out experiments on healthy humans in which both the testers and the tested were unaware of the medicine being used. The first beginning of blind testing methods are, therefore, to be found in the middle of the last century. (Haas, Fink, and Härtfelder 1963, p. 4)

In the period after the 1920s, drug testers combined Fisher's technique of randomization for control/treatment-group assignment with double-blind techniques which had been undergoing refinement for at least 75 years.

Notice that double-blindness seeks to frustrate misreporting by a conscious object of inquiry by denying that object of inquiry knowledge of its treatment/control status. Another difficulty with subject consciousness was first conceptualized in a series of classic studies in social psychology conducted jointly by researchers from the Harvard Graduate School of Business Administration and the management of the Western Electric Company at the latter's Hawthorne works in Chicago during 1926–32. The broad aim of these studies was to see how improved working conditions might affect worker productivity. In the chief study, the productivity of a small group of workers was first noted without their knowledge that they were being studied. Then, with the cooperation of the workers, management and experimenters introduced a series of working-condition improvements over a 2-year period and studied changes in worker productivity.[1]

The workers were apparently unaware throughout the study that their productivity was being monitored. For the first year or so, changes in working conditions were introduced at various points. Though each change could not be regarded as an improvement over previous conditions, productivity rose throughout the period. After about a year, with worker agreement, working conditions were restored for a 12-week period to what they had been at the study's outset, and "the history of the twelve-week return to the so-called original conditions of work is soon told. The daily and weekly output rose to a point higher than at any other time and in the whole period 'there was no downward trend'" (Mayo 1933, p. 65). Then there was a 31-week period of improved working conditions during which "their output rose again to even greater heights" (Mayo 1933, p. 65). Mayo explains:

It had become clear that the itemized changes experimentally imposed . . . could not be used to explain the major change – the continually increasing production. This steady increase as represented by all the contemporary records seemed to ignore the experimental changes in its upward development. (1933, p. 65)

---

[1] Roethlisberger, Dickson, and Wright (1939) provide a report of the Hawthorne Works studies by the original researchers; Whitehead (1938) reanalyzes some of the data; Mayo (1933, especially Chapters III and IV) situates the Hawthorne experiment results within the early industrial-relations literature.

A later explanation of the study's result goes as follows:

The girls observed that management was concerned with their welfare; they had been singled out for a great deal of attention by both management and researchers, and they were appreciative of their new found importance. This reaction has come to be widely known as the Hawthorne effect – a confounding variable in field experiments of this type. (Sills 1968, p. 241)

The explanation's view of management seems a trifle benign – after all, they were apparently seeking ways to increase worker productivity. Still, in contrast to fields, human subjects are of necessity conscious of being experimental objects and may behave differently – for a wide array of reasons – when under scrutiny than when not. If we call such scrutiny-induced differences in behavior the Hawthorne effect, then that effect may frustrate the ability of Chapter 2's theory of causal inference to discover cause and effect with a randomized, controlled experiment.

The technique of double-blindness and the Hawthorne effect are both related to the fact that in contrast to Fisher's original objects of inquiry, human subjects are conscious. But human subjects are also willful and social objects of inquiry – facts which also seem to frustrate the ability of Chapter 2's theory of causal inference to discover cause and effect with a randomized, controlled experiment. For example, explicit or implicit social rules for the ethical treatment of human subjects bar controlled experiments based on involuntary random selection from a larger population and so preclude the previous chapter's inference to such a larger population. Also, volunteer subjects may change their minds about participating during an experiment, and again social rules for the ethical treatment of human subjects insist that such volunteers be allowed to exercise their will and drop out. But then such attrition mars even the previous chapter's causal inference based only on volunteer subjects.

So far we have cited causal inferential difficulties in controlled experiments with conscious, willful, social subjects that arise because such subjects differ from the fields of crops for which Fisher developed the theory of causal inference. Other difficulties with causal inference in social experiments arise, however, when researchers or inquirers fail to match statistical analytic approaches to their employed techniques of random assignment. For example, the previous chapter's account suggested a statistical analytic approach which will yield an unbiased estimate of the average causal effect *if* proportional stratified random assignment has been employed. Sometimes, however, social experimenters will use non-proportional stratified random assignment and fail to properly modify their statistical analytic technique, and yet incorrectly write as if their estimate was an unbiased estimate of the average causal effect.

Sections 3.2 and 3.3 look at causal inferential difficulties in controlled social experiments at general and particular levels, respectively. Section 3.2.1 studies problems stemming from social subject consciousness, and Section 3.2.2 looks at a conflict between truth and ethics in social control experimentation. Section 3.2.3 formalizes a method of statistical analysis that when employed with unbalanced or nonproportional stratified-random-dom-assignment designs yields an unbiased estimate of the average causal effect. Sections 3.3.1–3.3.3 look at difficulties when the response is marital status, with sample selection, and with disproportional stratified assignment in the income maintenance experiments. Section 3.4 presents a summary and conclusions. Section 3.4.1 focuses on the difficulties unearthed, and Section 3.4.2 suggests a shift in perspective as a way around some of these difficulties. In this chapter, as elsewhere in the book, all proofs not cited in the footnotes are in the Appendix.

3.2   Impediments to causal point estimation in controlled social experiments

3.2.1   *Social subject consciousness*

Consider again the situation of Section 2.3.3. That is, let (T_i, C_i) for $i = 1, \ldots, N$ be the (treatment, control) response of the experiment's N participants, and let $\overline{T}$ and $\overline{C}$ be the means of the T_i's and C_i's, respectively. The item of inferential interest is $\overline{T} - \overline{C}$. Let $t_1, \ldots, t_n$ and $c_1, \ldots, c_{N-n}$ be the treatment and control responses that would be observed if reporting were perfect, and let $r(t_i)$ and $r(c_j)$ be the reported values of t_i and c_j, respectively. Suppose also that $r(t_i) = t_i + \epsilon_i$ and $r(c_j) = c_j + \eta_j$ for $i = 1, \ldots, n$ and $j = 1, \ldots, N - n$, where the ϵ_i's and η_j's are reporting-error terms, and let $\overline{\epsilon}$ $(\overline{\eta})$ be the mean of the n $(N - n)$ ϵ_i's $(\eta_j$'s) and $\overline{r}(t)$ $(\overline{r}(c))$ be the mean of the n $(N - n)$ $r(t_i)$'s $(r(c_j)$'s). And suppose PEI holds; that is,

$$p(t_1, \ldots, t_n; c_1, \ldots, c_{N-n})$$
$$= \begin{cases} 0 \text{ if } T_i \text{ and } C_i \text{ appear among the } n \ t_j\text{'s and } (N - n) \ c_k\text{'s,} \\ \quad \text{respectively, for any } i = 1, \ldots, N \\ 1/\binom{N}{n} \text{ otherwise.} \end{cases}$$

is the probability of $t_1, \ldots, t_n; c_1, \ldots, c_{N-n}$. And suppose that $F(\epsilon_1, \ldots, \epsilon_n; \eta_1, \ldots, \eta_{N-n})$ is the joint probability distribution of the reporting-error terms.

In the Appendix, we prove the following.

Theorem 2.11. *PEI and $E(\overline{\epsilon} - \overline{\eta}) = 0$ imply that $\overline{r}(t) - \overline{r}(c)$ is an unbiased estimate of $\overline{T} - \overline{C}$.*

We also argued in Section 2.3.3 that PEI is made plausible by the physical act of randomization and that PEI's meaning is that any observer has no good reason to believe one rather than another of the $\binom{N}{n}$ possible sets of observations will occur. And we stressed that no meaning has yet been given to the concept of probability behind $F(\epsilon_1, \ldots, \epsilon_n; \eta_1, \ldots, \eta_{N-n})$; that is, so far we can say only that F satisfies certain probability axioms, but not how the axioms' undefined terms attach to the world of our observations. Hence, we haven't offered an epistemological basis for $E(\overline{\epsilon} - \overline{\eta}) = 0$.

Now we seek to replace the condition $E(\overline{\epsilon} - \overline{\eta}) = 0$ with a condition that is sufficient for $E(\overline{\epsilon} - \overline{\eta}) = 0$ and yet can be given an epistemological basis that we can articulate. Consider the following.

Principle of reporting-error symmetry (PRES)

$$F(-\epsilon_1, \ldots, -\epsilon_n; -\eta_1, \ldots, -\eta_{N-n})$$
$$= 1 - F(\epsilon_1, \ldots, \epsilon_n; \eta_1, \ldots, \eta_{N-n}).$$

PRES is certainly sufficient for $E(\overline{\epsilon} - \overline{\eta}) = 0$.

Lemma 3.1. *PRES implies $E(\overline{\epsilon} - \overline{\eta}) = 0$.*

So PRES may be substituted for $E(\overline{\epsilon} - \overline{\eta}) = 0$ in Theorem 2.10.

Theorem 3.1. *PRES and PEI imply that $\overline{r}(t) - \overline{r}(c)$ is an unbiased estimate of $\overline{T} - \overline{C}$.*

Next, what of the epistemological basis of PRES? Notice that PRES means that the reporting-error distribution is symmetric about **0**. Double-blindness makes PRES plausible because double-blindness means that neither a subject nor the experimenters know the treatment/control status of the subject. Without such experimenter/subject knowledge, any observer has no good reason to believe that subject i/experimenter report-ing error will be to one or the other side of 0. So double-blindness is the epistemological basis of PRES. Also, double-blindness attaches PRES to the world of observations by assuring no good reason for an observer to prefer one belief to another, just as randomization did for PEI. But whereas randomization fully specified $p(t_1, \ldots, t_n; c_1, \ldots, c_{N-n})$, double-blindness merely places a condition on F.

Double-blindness provides the epistemological basis for extending our theory of causal point estimation to a situation in which the rather im-plausible assumption of perfect reporting is dropped. Recently, however, the statistician Donald Rubin offered the following critique of double-blindness:

> . . . we cannot attribute cause to one particular action in the series of actions that define a treatment. Thus treatments that appear similar because of a common salient action are not the same treatment and may not have similar causal ef-fects . . . treatments given under double-blind conditions are different treat-ments than those given under blind or simple conditions and may have different causal effects . . . in an experiment to compare aspirin and a prescription drug, simple versions of the treatments may be important because in practice a patient will usually know if his doctor is recommending aspirin rather than a prescription drug. (Rubin 1978, p. 40)

Rubin's argument is that the subjects' knowledge of their treatment/con-trol status should be regarded as one component of the treatment, so that double-blind and simple studies differ on this component and hence may have different causal effects. In our formalism, though PEI and PRES imply that $\bar{r}(t) - \bar{r}(c)$ is an unbiased estimate of $\bar{T} - \bar{C}$, $\bar{T} - \bar{C}$ may differ in the double-blind and simple studies. Put another way, what we might call a double-blind effect may result from the subjects' nonknowl-edge of their treatment/control status, just as a Hawthorne effect may result from their knowledge that they are under experimental scrutiny.

So far we have formalized double-blindness's role in causal inference based on controlled experiments and likened the double-blind effect of Rubin's critique to the Hawthorne effect. But what bearing do these considerations have on controlled *social* experiments like the income maintenance experiments? In such experiments, in contrast to drug stud-ies, subjects' treatment/control status (in the case of an income mainte-nance experiment, whether or not they are on an NIT plan) *cannot* be concealed from them. Thus, there can be no double-blind effect in a controlled social experiment, but also the assumption of a symmetric reporting-error distribution in such an experiment of *necessity* lacks the epistemological basis of double-blindness. Because social subjects of necessity know that they are objects of experimental inquiry, a controlled social experiment is of necessity prey to the Hawthorne effect. Finally, social subjects know what they report and can monitor the news media and so are likely to have some understanding of the experiment's purpose of aiding a decision on whether or not to introduce a particular social program.

In the case of the income maintenance experiments, subjects were aware that their reported hours worked and marital status were experimenter-monitored, and if they read the papers or watched TV they probably understood that whether or not a social program in their interest would be adopted might be influenced by their experiment's outcome. The members of the treatment (control) group thus had an incentive to work more (less) and divorce less (more) than they would have had in the absence of such monitoring – the Hawthorne effect. And because of the absence of double-blindness, members of the treatment (control) group also could (and had an incentive to) *report* more (less) work and less (more) divorce than actually occurred. Thus, neither PRES nor the unbiasedness of $\bar{r}(t) - \bar{r}(c)$ as an estimate of $\bar{T} - \bar{C}$ is very plausible. Thus, social subject consciousness in a controlled social experiment seems to frustrate the inference of cause and effect with Chapter 2's theory of causal point estimation. We turn next to a conflict between truth and ethics in controlled social experimentation that further frustrates such inference.

3.2.2 *The good and the true in conflict*

At least since the time of Plato, philosophers have frequently conceived the good and the true as in harmony. But in a controlled experiment with human subjects, the two notions seem to be at odds with each other. In Section 2.3.2 we saw that true causal inference to a population larger than the experiment's participants requires that the participants be selected at random from that population. Yet the ethics of individual liberty require that only volunteers participate in a human-subject controlled experiment. If these ethics are followed, a self-selection bias with respect to any larger population enters the point estimate of the mean of all causal effects. But if random selection from the larger population is followed, the ethics of individual liberty are obviously contravened.

The good/truth dilemma is real enough for physicians and biostatisticians involved in human-subject controlled experiments. "Many physicians who participate in clinical trials say privately that if they were to

A. Conventional design

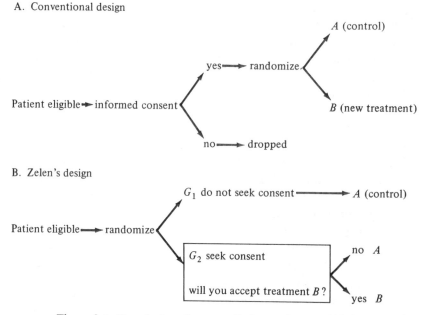

Figure 3.1. Two designs for controlled experiments with human subjects.

comply with the letter and spirit of the federal regulations [on "informed consent"], very few patients would enter randomized studies" (Zelen 1979, p. 1243). Researchers, for example, testing new drug treatments for various forms of cancer are forced to get people into their studies by not fully informing them or to employ subjects who are *very* near death and whose responses to treatments may poorly represent the responses of patients at earlier stages of the disease.

In the conventional approach to a test of a "best standard" treatment against a new experimental one, all patients deemed medically eligible are fully informed of the nature of the experiment. They are then asked to volunteer to be randomly assigned to a (new experimental) treatment group or (best standard treatment) control group and to have their responses to treatment studied. Those who refuse are dropped from consideration, and those who agree are assigned to treatment and control groups whose responses are subsequently compared. Figure 3.1A diagrams the conventional approach.

The conventional approach respects the ethics of individual liberty in the sense that those who refuse to participate in the experiment are neither randomized into a treatment nor studied further. The conventional approach's inferential problem is that the difference in treatment/control response means may poorly represent the mean of medically eligible population causal effects because of self-selection bias in the volunteering. At least, however, the difference in treatment/control response means is an unbiased estimate of the mean of all causal effects in the volunteers (see Theorem 2.1).

The biostatistician Marvin Zelen recently proposed a design which aims at overcoming the conventional approach's inferential difficulty. *All* patients deemed medically eligible are randomized into two groups – G_1 and G_2 – at least at the outset. G_1 is given treatment A, the best standard treatment, whereas those in G_2 are given a choice between treatment A and treatment B, the new experimental one, and the responses to treatment of *all* patients deemed medically eligible are subsequently studied. Figure 3.1B diagrams Zelen's design. By involving *all* patients deemed medically eligible, Zelen's proposed design *seems* to overcome the self-selection inferential difficulty of the conventional approach.

Zelen's design, however, appears to violate the ethics of individual liberty in two ways. First, those assigned to G_1 are offered no choice of treatment; they *must* take treatment A. Zelen defends this feature with the argument that those in G_1 receive the treatment that they would have gotten had there been no experiment (i.e., the *best* standard treatment). Still, since there *is* an experiment in which some participants receive treatment B, one could argue that *all* patients deemed medically eligible should be given *some* sort of treatment choice. Second, those assigned to G_1 participate in an experiment in which their responses are studied without their permission. Zelen defends this feature by arguing that it is no different from researchers doing a retrospective study of anonymous patients, for which patient permission is not always sought. Still, one could argue that patient permission for such a study *should* always be sought. Thus, from the perspective of the ethics of individual liberty, Zelen's proposed design is at best questionable.

Moreover, Zelen's design actually fails to overcome the self-selection inferential difficulty of the conventional approach. He argues:

The analysis of this new design requires that Group G_1 (receiving only treatment A) is compared with Group G_2 (receiving treatment A or B). In other words, the comparison must be made with all patients in Group G_2, regardless of which treatment each received. It is clear that including all patients dilutes the measurable effect of treatment B. Nevertheless, all patients must be included if the analysis is to provide a valid comparison with treatment A. (1979, p. 1244)

But his reasoning here is confusing. It is true that since his design randomizes all patients deemed medically eligible into G_1 and G_2, randomization-based causal point estimation (as in Section 2.2.2) can occur only by comparing G_1 and G_2 responses. But such a comparison *doesn't* "provide a valid comparison [of anything] with treatment A," contrary to what Zelen seems to suggest.

Because Zelen's design randomizes and studies *all* patients deemed medically eligible, it avoids the self-selection bias of the conventional design. But the goal of causal inference is an unbiased estimate of the mean of A/B causal effects. The difference in G_1 and G_2 means, though an unbiased estimate of the mean of G_1/G_2 causal effects, is not an unbiased estimate of the mean of A/B causal effects, because the presence of some treatment-A subjects in G_2 "dilutes the measurable effect of treatment B." Yet neither is the difference in A and B means an unbiased estimate of the mean of A/B causal effects, because all possible (A, B) samples have not been made equally probable by random assignment of subjects to treatments A and B.

Zelen's design merely replaces the self-selection bias of the conventional design with what one might call a dilution bias. With the conventional design, the difference in participants' A and B means is at least an unbiased estimate of the mean of participants' A/B causal effects, though not an unbiased estimate of the mean of A/B causal effects of all patients deemed medically eligible. With Zelen's design we get an unbiased estimate for the mean of *no* population's A/B causal effects, but only an unbiased estimate of the mean of the G_1/G_2 causal effects of all patients deemed medically eligible – an item of no real inferential interest. Thus, Zelen's design is actually inferentially *inferior* to the conventional design.

We have argued that human-subject controlled experiments suffer from a conflict between the good and the true and that the conventional design of such experiments respects the ethics of individual liberty at the cost of introducing a self-selection bias into causal inference to any population larger than the participants. One proposed alternative design succeeds in replacing the conventional design's self-selection bias with a more troublesome dilution bias and also may violate the ethics of individual liberty. So it isn't clear that a resolution of the human-subject controlled experiment's good–truth conflict exists. Next we formalize an approach to causal inference with unbalanced stratified random assignment.

3.2.3 *Unbiased causal point estimation with disproportional stratified random assignment*

Recall that in Section 2.3.1 we extended the theory of causal point estimation to a situation where proportional stratified random assignment has occurred. Suppose the $N(T_i, C_i)$ subject responses are relabeled $(T_{11}, C_{11}), \ldots, (T_{1N_1}, C_{1N_1})$ and $(T_{21}, C_{21}), \ldots, (T_{2N_2}, C_{2N_2})$, where the N_1 (T_{1i}, C_{1i})'s are a first block and the N_2 $(=N - N_1)(T_{2i}, C_{2i})$'s are a second block. Suppose n_1 subjects from block 1 are assigned to the treatment group [so $(N_1 - n_1)$ from block 1 are assigned to the control group]. Also suppose $n_2 = (n - n_1)$ subjects from block 2 are assigned to the treatment group [so $(N_2 - n_2)$ from block 2 are assigned to the control group]. Then if $t_{11}, \ldots, t_{1n_1}$ $(c_{11}, \ldots, c_{1(N_1-n_1)})$ and $t_{21}, \ldots, t_{2n_2}$ $(c_{21}, \ldots, c_{2(N_2-n_2)})$ are the actually observed treatment-group (control-group) responses from blocks 1 and 2, respectively, $\{t_{j1}, \ldots, t_{jn_j}\}$ $(\{c_{j1}, \ldots, c_{j(N_j-n_j)}\})$ is a subset of $\{T_{j1}, \ldots, T_{jN_j}\}$ $(\{C_{j1}, \ldots, C_{jN_j}\})$ for $j = 1, 2$.

We also had that the physical act of *stratified* random assignment means that PBE holds; that is,

$$p(t_{11}, \ldots, t_{1n_1}; t_{21}, \ldots, t_{2n_2}; c_{11}, \ldots, c_{1(N_1-n_1)}; c_{21}, \ldots, c_{2(N_2-n_2)})$$

$$= \begin{cases} 0 \text{ if } T_{ji} \text{ and } C_{ji} \text{ appear among the } n_j \ t_{kl}\text{'s } and \\ \quad (N_j - n_j) \ c_{rs}\text{'s, respectively, for any } i = 1, \ldots, N_j \\ \quad \text{for } j = 1, 2 \\ 1/\binom{N_1}{n_1}\binom{N_2}{n_2} \text{ otherwise.} \end{cases}$$

And we had that $\bar{T}$, $\bar{T}_j$, and $\bar{t}_j$ are the means of the N T_i's, N_j T_{ji}'s, and n_j t_{kl}'s, respectively, and that $\bar{C}$, $\bar{C}_j$, and $\bar{c}_j$ are the means of the N C_i's, N_j C_{ji}'s, and $(N_j - n_j)$ c_{rs}'s, respectively. Section 2.3.1's conclusion was that if the stratified random assignment is *proportional* [i.e., $(n_j/N_j) = (n/N)$ for $j = 1, 2$], then $\bar{t} - \bar{c}$ is an unbiased estimate of $\bar{T} - \bar{C}$, just as it was in the case of *simple* random assignment (see Theorem 2.8).

What we now wish to formalize are some results when the proportionality condition is dropped. Consider a definition first.

Definition

$$\bar{t}^+ = (N_1\bar{t}_1 + N_2\bar{t}_2)/N \quad \text{and} \quad \bar{c}^+ = (N_1\bar{c}_1 + N_2\bar{c}_2)/N.$$

We can next state a theorem which drops the proportionality condition.

Theorem 3.1. *Let PBE hold, and suppose that* $(n_j/N_j) \neq (n/N)$ *for* $j = 1, 2$. *Then*

(i) $\bar{t} - \bar{c}$ *will not generally be an unbiased estimate of* $\bar{T} - \bar{C}$,

(ii) $E(\bar{t} - \bar{c}) - (\bar{T} - \bar{C}) = ([(n_1 N - nN_1)\bar{T}_1 + (n_2 N - nN_2)\bar{T}_2]/nN) - ([(nN_1 - n_1 N)\bar{C}_1 + (nN_2 - Nn_2)\bar{C}_2]/N(N - n))$,

(iii) $\bar{t}^+ - \bar{c}^+$ *is an unbiased estimate of* $\bar{T} - \bar{C}$.

Notice that (i) tells us that when stratified random assignment has been disproportional, the simple difference in observed treatment and control means will not generally be an unbiased estimate of the mean of all causal effects. Part (ii) tells us the bias of $\bar{t} - \bar{c}$ as an estimate of $\bar{T} - \bar{C}$, but in terms of $\bar{T}_1, \bar{T}_2, \bar{C}_1$, and $\bar{C}_2$, which are not observable.

Theorem 3.1(iii) provides an unbiased estimate of the mean of all causal effects in the case of disproportional stratified random assignment. Notice that since

$$\bar{t}^+ - \bar{c}^+ = (N_1/N)(\bar{t}_1 - \bar{c}_1) + (N_2/N)(\bar{t}_2 - \bar{c}_2),$$

the unbiased estimate is a weighted sum of the difference in observed treatment and control means *on each block,* where a block's weight is its proportion of all subjects. Notice also that we may write $\bar{t} - \bar{c}$ as $[(n_1/n)\bar{t}_1 - (N_1 - n_1)\bar{c}_1/(N - n)] + [(n_2/n)\bar{t}_2 - (N_2 - n_2)\bar{c}_2/(N - n)]$. Thus, if a study employs unbalanced random assignment and reports $\bar{t} - \bar{c}, n_1, n_2, N_1$, and N_2, not only is $\bar{t} - \bar{c}$ not an unbiased estimate of the mean of all causal effects, but also one cannot calculate an unbiased estimate from the reported information. When unbalanced random assignment has been employed and only the difference in observed treatment and control means is reported, we label the bias of the reported estimate of the mean of all causal effects a failure-to-weight bias.

So far we have considered four sorts of biases which may prevent our Chapter 2 causal inferential formalism from applying to data from a controlled social experiment. The subjects of such an experiment *know* that they are experimental subjects and so may respond with a Hawthorne-effect bias that would not be present in the absence of an experiment. Also, respecting their subjects' privacy, and/or because of difficulties of independent verification, researchers may employ subject-reported response variables. But then social subjects, of necessity knowing their treatment/control status, may respond with a self-reporting bias. And researchers respecting the ethics of individual liberty usually will employ only volunteers as subjects, introducing a self-selection bias into their results. Finally, researchers often will employ disproportional random assignment, but report only the difference of simple observed response means, introducing a failure-to-weight bias into their reported results. Section 3.3 considers the income maintenance experiments, whose causal inferential possibilities are marred by these four and other difficulties.

3.3 Impediments to causal point estimation in the income maintenance experiments

3.3.1 *Difficulties when the response variable is marital status*

In Section 3.2.1 we cited the presence of Hawthorne-effect and self-reporting biases in the income maintenance experiments and speculated on the direction these biases would take with two response variables, one of which was marital status. We argued that knowledge that they were in an experiment on whose outcome the fate of a social program in their economic interest might depend may have led treatment-group subjects to divorce less than they would have had the situation not been an experiment (Hawthorne effect). And the same knowledge could have led treatment-group subjects to underreport divorce (self-reporting$_1$ bias). With the marital-status response variable, though, five additional difficulties arise for causal point estimation.

The response variable of initial and prime interest in all four experiments was work effort. Also, the researchers probably were loath to breach the privacy of subjects on their marital situations and were likely aware that in the low-income social classes they were studying, cohabitation rather than legal marriage is a frequent arrangement. Apparently for those three reasons the researchers in all four experiments employed less than legal definitions of marriage and divorce. For example, the Seattle-Denver researchers defined marriage as coresidence with pooled resources (in more than 95% of the cases, self-reported as 1 year or more), and divorce as separation reported by either party to be permanent. They pointed out that "our design suffers a weakness with respect to possible differences in rates of formation and dissolution between legal marriages and consensual unions. We cannot distinguish between the two types of unions" (Hannan, Tuma, and Groenveld 1977, p. 1195). The other three experiments' designs suffered similar weaknesses.

The New Jersey experiment's published findings on marital status suffered from an additional difficulty. The researchers seem to have implicitly supposed that the family stability goal of income maintenance was encouraging the presence of some (not necessarily the same) coresident male in the family rather than preserving the initial consensual or legal union. For they "have taken all families classified at the outset of the experiment as nuclear . . . and subdivided them into two groups: those which remained nuclear . . . and those which changed to female-headed family units" (Knudson, Scott, and Shore 1977, p. 258). The New Jersey experiment's published findings give a poor estimate of an NIT's effect on a given family's stability.

The Seattle-Denver researchers' published findings on marital status suffer from several other difficulties. First, the researchers explain,

While most of the dissolutions we recorded were relatively permanent, over the first year of the experiment 8.5% of the dissolutions recorded for blacks, 7.3% for whites, and 12.3% for Chicanas resulted in reconciliation within six months. According to our procedures, each reconciliation was recorded as a remarriage and the initial separation as a dissolution. Since the initial separation and reconciliation was sometimes brief, the reader may question whether we are making overly fine distinctions. (Hannan, Tuma, and Groenveld 1977, pp. 1195–6)

So the researchers recorded a separation when either party reported a permanent separation, even when they had a subsequent report of a reconciliation. By ignoring reconciliation, the Seattle-Denver experiment's published findings become a poor estimate of an NIT's effect on permanent separations.

During the period of the four experiments, welfare was available in many states only to single-parent families. Many argued that such arrangements encouraged marital dissolutions; that is, husbands deserted wives so that the latter could go on welfare. Based on this argument, some claimed that an NIT should strengthen family stability as long as its benefit structure provided no economic incentive to dissolve. Only the benefit structures of three of the experiments were dissolution-neutral, however. In case of a dissolution in Gary, the wife and child kept the guarantee, and the husband got nothing. In the New Jersey and Rural experiments the guarantee was rescheduled, with the sum of the wife-children guarantee and husband guarantee equal to the family guarantee before the dissolution. Seattle-Denver, however, rescheduled the guarantee with the sum of the wife-children and husband guarantees greater than the family guarantee before the dissolution. So Seattle-Denver's benefit structure was not dissolution-neutral. On the contrary, the experiment's benefit structure encouraged both dissolution and subject-reporting of dissolutions when none had occurred.[2]

All told, we have so far identified seven biases in the income maintenance experiments' published estimates of changes in marital status with an NIT. Two of the biases stem from subject consciousness and inhere in social experiments more generally. The Hawthorne effect stems from subjects' knowledge that they are part of an experiment, and self-reporting$_1$ bias stems from that same knowledge plus experimenters' reliance on subject-reporting. All four experiments also suffered from cohabitation bias (i.e., definitions of marriage and divorce which employed coresidence and its cessation). In addition, New Jersey failed to count a separation when it was followed by further unions (i.e., suffered from multiple-union bias), and Seattle-Denver counted a separation when it was followed by a reconciliation (i.e., suffered from reconciliation bias). Finally, Seattle-Denver suffered from both nonneutral-benefit bias and self-reporting$_2$ bias stemming from the benefit structure's economic incentives to dissolve and report dissolutions when none had occurred.

We argued earlier that the Hawthorne effect and self-reporting$_1$ biases would tend to underestimate divorce rate increases with an NIT. The directions of cohabitation and multiple-union biases are unclear. The Seattle-Denver researchers gauged reconciliation bias: "If one were to discount the impacts of income maintenance on dissolution in order to take the effect on reconciliation into account, the impact should be reduced approximately 20%–30%."[3] Finally, nonneutral-benefit and self-reporting$_2$ biases clearly overestimate the separation that would occur with a neutral-benefit-structure NIT. So even without considering any further inferential difficulties from which the experiments suffered, the

---

[2] Kershaw and Fair (1976, pp. 83–6) provide a comparison of how the four experiments rescheduled benefits with a dissolution. Sawhill et al. (1975, pp. 48–9) argue that because of taxes, even the New Jersey (and, because the rules were the same, Rural) experiment's benefit structure was *not* dissolution-neutral, but rather encouraged dissolutions and their reporting when none had occurred. Hannan, Tuma, and Groenveld (1977, p. 1194) provide a vague discussion of Seattle-Denver's benefit rescheduling with a dissolution that fails to note the rescheduling's dissolution-nonneutrality. Kurz, Spiegelman, and Brewster (1973, pp. 41–3), however, make abundantly clear the dissolution-nonneutrality of the experiment's benefit structure. Bishop (1980, pp. 320–1) gives a discussion of Gary's rescheduling of benefits with a dissolution.

[3] From Hannan, Tuma, and Groenveld (1977, p. 1196, fn. 9). For the details of how the Seattle-Denver researchers estimated reconciliation bias, see Hannan, Tuma, and Groenveld (1976, pp. 127–9).

relationship between the unbiased causal point estimates of Chapter 2's formalism and the actual marital-status change estimates the experimenters reported is anyone's guess. We turn next to some sample selection biases from which the income maintenance experiments suffered.

3.3.2   *Sample selection difficulties*

In Section 2.3.2 we developed the formalism for an unbiased estimate of the mean of a population's causal effects based on the observations of a randomly selected sample. We also suggested the formalism for the same inference when the sampling procedure was *stratified* random. Formalism for the inference is possible with other sampling schemes as well – such as random sampling of location sites first from a population of possible sites and simple or stratified random sampling of subjects at the selected sites (i.e., cluster sampling design). The income maintenance experiments, however, followed no formalizable random sampling scheme.

New Jersey was chosen for the first experiment's site because the state then had no welfare program for male-headed families, so that the work-effort response for those receiving an NIT could be compared with that for a control group having no income transfers available. Later, some Pennsylvania sites were added to the New Jersey experiment to increase the number of white families under study. The Rural experiment added rural poor to the inquiry – whites in Iowa and blacks and whites in North Carolina. Gary, Seattle, and Denver were chosen for the last two experiments in part because these cities had had successful Model Cities programs which could supply interviewers. Gary added a midwestern site and Seattle-Denver a western site. When Seattle developed high unemployment because of an aerospace industry downturn, "the Seattle Experiment was extended to Denver, a more representative western city" (Spiegelman and Yaeger 1980, p. 467). So the experiments' sites were not randomly chosen from a population of possible sites.

Within New Jersey, cities were chosen on the basis of such criteria as proximity to the Princeton researchers, nondependence of the city's residents on other cities' labor markets, anticipated absence of riots, and so forth. Within cities, all volunteer families in blocks selected at random from the 1960 census "poverty tracts" who, by means of screening and preenrollment interviews, were found to satisfy the eligibility criteria (e.g., previous year's income below a certain level) were enrolled in the study. During the three-year experiment, a number of the initial volunteer families dropped out of the study.[4] For the Rural experiment, the researchers said that "families were selected randomly from within the experimental sites and if eligible [e.g., previous year's income below a certain level], were assigned randomly" (U.S. Department of Health, Education, and Welfare 1976, p. 4). No mention was made of families who refused to participate, though some must have. Again, some starting families dropped out of the study over the three-year experiment.

In Gary, researchers enrolled families from "the Model Cities project area and other low-income neighborhoods" (Kehrer et al. 1975, p. 12). All volunteers with the previous year's income below, and a portion of volunteers with the previous year's income above, a certain level were enrolled. Eventually some families dropped out of the study during its three-year course. In Seattle, researchers set eligibility on the basis of previous year's income below a certain level and identified census tracts with median incomes below that level. All of the public housing project blocks, 50% of the Model Cities area, and $\frac{1}{3}$ of the remaining blocks were chosen at random, and about 50% of eligible volunteers from chosen blocks were assigned to treatment groups on a stratified random basis.[5] In Denver, "on-site surveying, with the assistance of Census tract map and demographic data, led to the establishment of the sample area boundaries . . . centering on the Model Cities area" (Muraka and Spiegelman 1978, p. 19). Eighty percent of all housing units within the boundary were chosen at random; around 60% of volunteer families in these units who were not ruled ineligible by an initial screening interview were chosen at random for a preexperimental interview, and about 70% of volunteers not ruled ineligible by the preexperimental interview were assigned to treatment groups on a stratified random basis. As in the other three experiments, some Seattle-Denver starting families dropped out during the experiment.

At least three sorts of sample selection inferential difficulties are evident from the selection procedure of the income maintenance experiments. First is the problem of self-selection from the use of volunteers. It is difficult to be sure of the direction and size of any self-selection bias with a given response variable. A New Jersey interviewer reported one man's reason for refusing to participate:

He's a proud young man who finally insisted that he did not believe in taking money for nothing. He lived on what he, himself alone, earned. He had a family,

---

[4] Kershaw and Fair (1976, Chapter 2) give a detailed account of the New Jersey experiment selection process.

[5] See Muraka and Spiegelman (1978, pp. 7–18) for more details on Seattle sample selection.

Table 3.1. *Refusal in New Jersey (number of families)*

Screening interview		*Eligibility from preenrollment*	
Refused interview	7,958	Ineligible for enrollment	828
Completed interview	27,350	Eligible for enrollment	1,511
Eligibility from screening		*Enrollment*	
Ineligible for preenrollment	24,226	Refused enrollment	62
Eligible for preenrollment	3,124	Enrolled	1,216
Preenrollment interview			
Refused interview	211		
Completed interview	2,341		

and it was his responsibility, not anyone else's to take care of them. No form of welfare would be acceptable to this man despite the guise of "basic income" and "experiment." This man is sincerely obsessed with the Protestant ethic – he makes speeches about the value of work in his neighborhood. Says it's a good program but not for him. Wishes us luck. (Kershaw and Fair 1976, pp. 41–3)

If this man is at all typical of refusers, then the experiments' reported work-effort response to an NIT should *overestimate* such response be-cause of self-selection bias.

Whatever its direction, any self-selection bias was not likely to have been extensive unless large numbers of individuals refused to participate in the experiments. Figures on refusals are not available for the Gary and Rural experiments and are incomplete for Seattle-Denver. Table 3.1 con-tains the refusal figures for the New Jersey experiment.[6] Notice that the difficulty with these figures is that we don't know what proportion of those refusing the screening and preenrollment interviews would have proved eligible for the experiment. If we suppose that at each stage that propor-tion was the same as for those *completing* the interview, then 39% of those who were eligible refused to volunteer for the experiment.[7] So if refusers were systematically more work-oriented than volunteers, the experiment could have substantially overestimated work reduction with an NIT.

In the New Jersey, Rural, and Seattle-Denver experiments, the sample distribution of previous year's family income was truncated at a certain level (different in each experiment). In Gary, those with previous year's family income higher (lower) than a certain level were sampled dispropor-

[6] Table 3.1 is based on Table 2.1 of Kershaw and Fair (1976, p. 31).
[7] That is,

$$1/1 + [1{,}216/(1{,}513/(1{,}513 + 828))[7{,}958(3{,}124/(3{,}124 + 24{,}226)) + 211] + 62] \approx 39\%.$$

Table 3.2. *Attrition in the income maintenance experiments*

Study	Treatment groups[a] (%)	Control group[a] (%)	Total (%)
New Jersey	16	25	20
Rural	NA	NA	10
Gary	12	23	17
Seattle-Denver	6*	8*	15

[a] NA means that the figure is not available; an asterisk indicates that the figure is for the first half of the three-year experiment.

tionately lower (higher).[8] So from the perspective of Chapter 2's formalism, sampling in the experiments was disproportionately stratified from two strata, where in three of the experiments *no* individuals were drawn from one stratum. Hence, in those three experiments, *no* weights could have assured unbiased causal point estimation, and in Gary no weights were employed to correct for the lower sampling ratio of the group with higher previous year's income. The magnitude and direction of any truncation bias for a given response variable depend on the variable and previous year's income. For example, if higher-income groups tended to have more stable marriages, then the experiments would overestimate dissolution increases with an NIT.

Table 3.2 indicates the amount of attrition the experiments experienced.[9] If the probability of attrition depends on the response variable, then the estimated treatment/control response difference will be biased. For example, the Seattle-Denver researchers argued that "attrition will lead to biased estimates of impacts whenever the relationship between experimental treatment and the probability of attrition depends on whether a marital-status change has occurred" (Hannan, Tuma, and

---

[8] To be eligible for the New Jersey and Rural experiments, "total family income could not exceed 150 percent of the poverty line [in the year prior to the experiment]" (Kershaw and Fair 1976, p. 8). Seattle-Denver truncated the two-parent families at the previous year's "income (family of four equivalent) in 1970–1971 dollars [of $11,000]" (Keeley, Spiegelman, and West 1980, p. 15). Gary sampled only a portion of those otherwise eligible families whose previous year's income was greater than 240% of the poverty line (Kehrer et al. 1975, pp. 12, 15).

[9] Table 3.2 is based on attrition figures from Kershaw and Fair (1976, p. 105), U.S. Department of Health, Education, and Welfare (1976, p. viii), Kehrer et al. 1975, pp. 15, 19–20), Hannan, Tuma, and Groenveld (1976, Tables 8-1 and 8-2, pp. 104–5), and Spiegelman and Yaeger (1980, p. 474).

Groenveld 1977, p. 1200). In fact, the Seattle-Denver researchers gauged that the dissolution rate among control families was greater for those who dropped out, so that their estimate of the control dissolution rate was low and hence "interactive associations between treatment, marital status change, and attrition may have caused us to overestimate the experimental response" (Hannan, Tuma, and Groenveld 1976, p. 109). The argument that attrition bias may have led to overestimation of the dissolution response to an NIT applies to the other three experiments too.[10]

We have identified three sample selection difficulties for Chapter 2's formalism for causal inference to a population larger than the sample of observations in the income maintenance experiments. Self-selection rather than random selection makes the formalism's PEII implausible. Disproportionate stratified selection, with failure to weight, blocks the unbiased estimation of Theorem 3.2(ii)'s extension to a situation of disproportionate stratified selection. Attrition makes the formalism's PEI implausible. The magnitude and direction of any biases resulting from the sample selection difficulties depend on the nature of the response in those logically possible samples ruled out by self-selection or attrition or implicitly weighted against by failure to weight. We turn next to the problem of failure to weight when assignment is stratified random.

3.3.3   *Disproportional stratified random assignment*

In Section 3.2.3 we argued that if stratified random assignment is disproportional, then the difference in observed treatment and control means is *not* an unbiased estimate of the mean of all causal effects. Further, we showed that an unbiased estimate of the mean of all causal effects can be constructed by properly weighting the observed strata means in each group. So if researchers report only the difference in observed treatment and control means, not the observable strata means, in each group, then one cannot retrieve an unbiased estimate of the mean of all causal effects from what is reported. Three of the four income maintenance experiments suffered from such failure-to-weight bias.

The New Jersey researchers assigned families to eight treatment groups [each characterized by a (guarantee, tax rate) pair] and a control group on a stratified random basis, where the strata were three income classes for

---

[10] Two attrition researchers with the Gary experiment data emphasize that "if attrition is related to endogenous variables, biased estimates result" (Hausman and Wise 1979, p. 462).

Table 3.3. *New Jersey assignment design (number of families)*

Experimental plan (guarantee, tax rate)	Income stratum			Total
	Low	Medium	High	
A (50, 30)	5	31	12	48
B (50, 50)	29	37	5	71
C (75, 30)	30	14	50	94
D (75, 50)	5	57	36	98
E (75, 70)	13	51	0	64
F (100, 50)	22	34	20	76
G (100,70)	11	26	33	70
H (125, 50)	50	8	80	138
Control	238	165	247	650
Total	403	423	483	1,309

Note: The guarantee is given as a percentage of the poverty level.

the previous year. Table 3.3 depicts the New Jersey assignment design.[11] Notice that the stratified assignment between the control and *any* treatment group is disproportional. For example, of the subjects in treatment E and control, 9% are in E; yet 7%, 16%, and 0% of those in the low-, medium-, and high-income strata, respectively, are assigned to E. Hence, the difference in observed treatment E and control change in work-effort response means is *not* an unbiased estimate of the mean of all change in work-effort causal effects. And since the researchers failed to report the observed strata means in each group, an unbiased estimate is irretrievable.

The Gary researchers assigned two-parent families to four treatment groups [each characterized by a (guarantee, tax rate) pair] and a control group on a stratified basis, where the 10 strata were 5 previous year's income classes by the 2 location classes – inside and outside the Model Cities project area. Table 3.4 depicts the Gary assignment design.[12] Again notice that the stratified assignment between the control and *any* treatment is disproportional. For example, of the families in treatment C and control, 28% are in C; yet 20%, 41%, 13%, 33%, 29%, 20%, 44%, 15%, 46%, and 16% of those in each of the 10 strata are assigned to C. So again the difference in observed treatment C and control change in work-effort response means is *not* an unbiased estimate of the mean of all change in

[11] Table 3.3 is based on Table 6.1 of Kershaw and Fair (1976, p. 104).
[12] Table 3.4 is based on Table 3 of Kehrer et al. (1975, p. 16).

Table 3.4. *Gary assignment design for two-parent families*

Experimental plan (guarantee, tax rate)	Model Cities: poverty-level income stratum					Outside Model Cities: poverty-level income stratum					Total
	1	2	3	4	5	1	2	3	4	5	
A (4,300, 40)	5	4	7	12	10	5	7	9	23	4	86
B (4,300, 60)	2	7	6	12	5	4	5	9	19	9	78
C (3,300, 40)	2	5	8	38	10	2	7	9	45	5	131
D (3,300, 60)	4	5	19	14	6	1	6	23	22	8	108
Control	8	7	60	78	24	8	9	53	52	27	326
Total	21	28	100	154	55	20	34	103	161	53	729

Note: The guarantee is a dollar value for a family of four.

work-effort causal effects. And since the researchers failed to report the observed strata means in each group, an unbiased estimate is irretrievable.

The Seattle-Denver researchers employed the most complicated assignment design of all. Two-parent families were assigned to 88 treatment groups [each characterized by a (guarantee, initial tax rate, rate of decline of tax rate with additional $1,000 of earned income, percentage of manpower training subsidized, treatment duration) vector] on a stratified basis, where the 36 strata were 6 previous year's income classes by 2 cities by 3 racial groups.[13] Again, stratified assignment between *any* two treatment groups was disproportional. For example, of the families in treatment A (4,800, .7, .000025, 50%, 3 years) and B (0, 0, 0, 0%, 3 years) (i.e., the control), 9% were in A; yet none of those in the Denver–black–lowest-previous-year's-income stratum was assigned to A.[14] So again the difference in observed treatment A and B change in work-effort response means is *not* an unbiased estimate of the mean of all change in work-effort causal effects. And since the Seattle-Denver researchers failed to report the observed strata means in each group, just as in New Jersey and Gary, an unbiased estimate is irretrievable.

[13] For more details of the assignment in Seattle-Denver, see Keeley, Spiegelman, and West (1980, pp. 12–17) and Conlisk and Kurz (1972).

[14] The example is based on figures from Table 6, pp. 22–4, and pp. 28–9 of Conlisk and Kurz (1972).

3.4 Summary and conclusion

3.4.1 *The inferential difficulties*

Chapter 2 developed a formalism for causal point estimation based on probability distributions created by physical acts of (possibly stratified) random selection and assignment. We began the present chapter by arguing that certain features of controlled *social* experiments (e.g., subject consciousness and the ethics of individual liberty) interfere with the application of Chapter 2's inferential formalism. First, knowing they are experimental objects of inquiry may lead subjects to respond differently than they would in a nonexperimental situation (the Hawthorne effect). For example, treatment-group families in the income maintenance experiments may have worked more than they would have in a nonexperimental situation. Second, controlled *social* experiments are of necessity not double-blind, because subjects know which treatment they receive; so when there is subject-reporting of responses, there may be systematic misreporting (self-reporting bias). For example, treatment-group families in the income maintenance experiments may have reported more work than actually occurred. Third, to respect the ethics of individual liberty, controlled social experiments must rely on volunteers rather than random selection, which leads to possible self-selection bias. Finally, controlled social experiments often employ stratified assignment, but fail to report the observed strata means in each group, which leads to failure-to-weight bias.

Next we turned to the income maintenance experiments, beginning with change-in-marital-status response, where we identified seven inferential difficulties. First, knowledge that they were objects of inquiry in an experiment on whose outcome the fate of a social program in their economic interest might depend may have led treatment-group families to divorce less than had the situation not been an experiment (the Hawthorne effect). Second, the same knowledge may have led subject families to report less divorce than there was (self-reporting$_1$ bias). Third, all four experiments failed to employ legal definitions of marriage and divorce, leading to possible cohabitation bias. Fourth, the New Jersey experiment measured only family changes from nuclear to female-headed, not changes of partners, leading to multiple-union bias in the measure of an NIT's effect on dissolutions. Fifth, Seattle-Denver counted quickly reconciled separations as dissolutions, leading to an overestimate of an NIT's effect on dissolution increases. Finally, Seattle-Denver's benefit structure

Table 3.5. *Some inferential difficulties*

Social experiments generally	Income maintenance experiments' response with an NIT	
	Marital status	Work effort
1. Hawthorne effect	1. Underestimate dissolutions	1. Underestimate work reduction
2. No double-blindness + self-reporting$_1$	2. Underestimate dissolutions	2. Underestimate work reduction
3. Self-selection	3. —	3. Overestimate work reduction
4. Failure to weight strata, with disproportional stratified assignment	4. No direction predicted	4. No direction predicted

Additional difficulties for marital status
1. Legal definitions of marriage and divorce not employed: no direction predicted
2. Multiple unions counted as no dissolution (New Jersey): no direction predicted
3. Quickly reconciled separations counted as dissolutions (Seattle-Denver): overestimate dissolutions
4. Nonneutral-benefit structure (Seattle-Denver): overestimate dissolutions
5. Self-reporting$_2$: overestimate dissolutions
6. Failure to weight strata, with disproportional stratified selection: overestimate dissolutions
7. Attrition: overestimate dissolutions

contained economic incentives to dissolve and report dissolutions where none had occurred; so the experiment suffered from nonneutral-benefit and self-reporting$_2$ biases that overestimated dissolution increases with an NIT.

Next we considered three types of sample selection inferential difficulties in the income maintenance experiments with different response variables. In all four experiments, self-selection or reliance on volunteers led to overestimation of work-effort response with an NIT if refusers tended to be "sincerely obsessed with the Protestant ethic." And stratified sample selection, with higher-income strata underweighted, led to overestimation of dissolution increases with an NIT if higher-income groups tended to have more stable marriages. Also, if the relationship between experimental treatment and the probability of attrition depends on the response variable, then the estimated treatment/control response difference will be biased. For example, the Seattle-Denver researchers gauged that attrition bias led to an overestimate of increase in marital dissolutions with an NIT, and their argument applies to the other three experiments too. Finally, three of the four experiments employed disproportional stratified random assignment and failed to report observed strata means in each group. Hence, these three experiments experienced failure-to-weight bias in their estimates of treatment/control response (change in either work effort or marital status with an NIT).

Table 3.5 lists the inferential difficulties we have identified. Notice that in the income maintenance experiments, with either marital status or work-effort response, there is at least one overestimation and one under-estimation difficulty. Hence, with the assumptions we have made so far, we cannot gauge whether the published estimates of these responses underestimate or overestimate the mean of all causal effects. We can merely note that formal unbiased causal point estimation of the mean of all causal effects is built on three no-good-reasons equiprobable assumptions, each made plausible by a physical act – random assignment and selection (possibly stratified) and double-blindness. But the inferential difficulties we have catalogued destroy (in the income maintenance experiments in particular and in social experiments generally) the physical acts' epistemological basis of the no-good-reasons equiprobable assumptions which are sufficient for unbiased estimation of the mean of all causal effects.

3.4.2 *Remedying the inferential difficulties*

Those who analyzed the data generated by the income maintenance experiments occasionally sought a remedy for one of the inferential difficulties we have identified. For example, two of the Seattle-Denver researchers asked us to

Consider the following version of the dummy-variable model in which the permanent differences between experimentals and controls are assumed to be random: (12) $y_p = F\gamma + \mu + \epsilon_p$ (13) $y_t = \tau + F\gamma + F\beta + \mu + \epsilon_t$ where y_p = value of the dependent variable in the preexperimental period, y_t = value of the dependent variable during the experimental time period, F = a dummy variable for treatment (1 for experimentals, 0 for controls), μ = permanent component of the dependent variable that persists over time, γ = permanent difference in the dependent variable between experimentals and controls, τ = a time effect, β = the experimental effect, and ϵ_p, ϵ_t = random error terms, assumed to be uncorrelated. (Keeley and Robins 1980, p. 492)

Notice that the researchers fail to make any connection between random and uncorrelated error terms and any physical act of randomization which occurred. The meaning of probability behind their model is thus unclear. The two researchers next argue that

> If the assumptions of this model are valid, an unbiased estimate of the experimental effect can be obtained by regressing $y_t - y_p$ on a treatment dummy and a control term. To see this, rewrite the model in regression format by subtracting equation (12) from equation (13): (14) $y_t - y_p = \tau + F\beta + \epsilon_t - \epsilon_p$. The least-squares estimate of the experimental effect, β, is unbiased because F is uncorrelated with $\epsilon_t - \epsilon_p$ (i.e., the permanent differences in $y_t - y_p$ between experimentals and controls are assumed to be random). (Keeley and Robins 1980, p. 493)

Notice, however, that the researchers' model fails to model the random assignment which occurred, for participants were assigned to *many* treatment groups, while F of the researchers' model collapses all of those groups into *one*. Hence, the researchers' β corresponds to no populations of interest in randomization-based causal point estimation.

The two Seattle–Denver researchers recognized the failure-to-weight bias that stems from unaccounted-for *stratified* assignment, arguing that

> If the dependent variable of interest, y, is labor supply, then the following equations are an approximation of the assignment process: (15) $F = E\pi + v$ (16) $E = y_p\alpha + u$ where $E = $ a variable representing normal income [i.e., previous year's income], and v, $u = $ random error terms assumed to be uncorrelated with each other and with ϵ_t and ϵ_p. . . . Because of the assignment process, the least-squares estimate of β in equation (14) is biased. The bias arises because the ϵ_p component of the error term is correlated with the treatment dummy. The bias is given by: (17) $E(b - \beta) = -\sigma_{\epsilon_p}^2 \alpha\pi/\sigma_F^2$ where σ^2 represents the variance of the relevant subscripted variable, and b is the least-squares estimate. (Keeley and Robins 1980, p. 493)

So the researchers' proposed remedy for failure-to-weight bias is problematic for three reasons. First, their approximation of the assignment process is dubious (see Section 3.3.3). Second, the bias of b as an estimate of β is a function of four unobservable quantities: $\sigma_{\epsilon_p}^2$, α, π, and σ_F^2. Finally, β itself is not a parameter which characterizes the populations of interest in the causal point estimation of Chapter 2's formalism.

Three studies by two independent researchers with New Jersey and Gary experiment data sought to remedy other inferential difficulties. Two studies of truncation bias with an earnings response variable using New Jersey data asked us to

> Assume that in the population the relationship between earnings and exogenous variables is of the form (1.1) $Y_i = X_i\beta + \epsilon_i$, where Y is earnings; X is a vector of exogenous variables including education, intelligence, etc. . . . ; i indexes indi-

viduals; $\boldsymbol{\beta}$ is a vector of parameters; and ϵ is a disturbance term with expected value zero and variance σ^2 for each individual. Thus Y_i is distributed normally with mean $\mathbf{X}_i\boldsymbol{\beta}$ and variance σ^2, $N(0, \mathbf{X}_i\boldsymbol{\beta})$.[15]

In a study of attrition bias using Gary data, the same researchers begin by supposing that

$y_{it} = \mathbf{X}_{it}\boldsymbol{\beta} + \epsilon_{it}$ $(i = 1, \ldots, N; t = 1, 2)$, where i indexes individuals and t indexes time periods. . . . The residual in the specification is then decomposed into two orthogonal components, an individual effect μ_i, which is assumed to be drawn from an iid [independent, identically distributed] distribution and to be independent of the $\mathbf{X}_{it}$'s, and a time effect, η_{it}, which is assumed to be a serially uncorrelated random variable drawn from an iid distribution. Thus, the assumptions on ϵ_{it} are: (1.2) $\epsilon_{it} = \mu_i + \eta_{it}$, $E(\epsilon_{it}) = 0$, $V(\epsilon_{it}) = \sigma_\mu^2 + \sigma_\eta^2 = \sigma^2$, $\epsilon_{it} \sim N(0, \sigma^2)$.[16]

Notice that independence assumptions assure that the populations involved in the three studies are not the finite ones of Chapter 2's causal inferential formalism. (See Sections 2.1.2, 2.2.3, and 2.2.5 on the dependency problem in causal inference.) So the $\boldsymbol{\beta}$ parameters of the three studies are not those of interest in causal inference. Further, normal-probability-distribution assumptions assure that the probability meaning behind the studies cannot be a no-good-reasons equiprobable one whose epistemological basis is physical acts of random assignment and selection.[17]

None of the studies based on the income maintenance experiments sought remedies for more than two of the inferential difficulties we have identified with controlled social experiments.[18] So even if those remedies

---

[15] From Hausman and Wise (1976, p. 423; 1977, p. 920).

[16] Hausman and Wise (1979, p. 457). Hausman and Wise (1981) offer a remedy for truncation bias with an income response variable in the Gary experiment that also deserts Chapter 2's finite-population, counterfactual, random-assignment-selection-induced, no-good-reasons equiprobable distribution: "Assume that in the population, income $Y = \mathbf{X}\boldsymbol{\beta} + \epsilon$, with Y given $\mathbf{X}$ distributed with mean $\mathbf{X}\boldsymbol{\beta}$, variance σ^2, and density function denoted by $f(Y/\mathbf{X})$ . . . where f is the normal density function $N(\mathbf{X}\boldsymbol{\beta}, \sigma^2)$" (p. 368).

[17] That is, a normal distribution presupposes an infinite population, whereas a no-good-reasons equiprobable distribution presupposes finite populations. For elaboration of this point, see Section 4.2.2.

[18] For example, with a labor-supply-type response (e.g., hours worked or earnings) we have identified at least five sorts of bias: attrition, truncation, self-selection, self-reporting, and failure to weight. *No* studies sought to remedy self-selection bias. Studies that sought to remedy only one of the other biases were the following: Hausman and Wise (1979) (attrition, Gary); Hausman and Wise (1976) (truncation, New Jersey); Hausman and Wise (1977) (truncation, New Jersey); Hausman and Wise (1981) (truncation, Gary); Keeley and Robins (1980) (failure to weight, Seattle-Denver); Nicholson (1977) (self-reporting, New Jersey); Halsey (1980) (self-reporting, Seattle-Denver). Apparently the only study that sought to remedy two of the biases simultaneously was Robins and West (1980) (attrition and failure to weight, Seattle-Denver).

were in some sense successful, any such study still suffered from a host of inferential problems. More interesting, however, is the fact that in pursuing a remedy, any such study abandoned the Chapter 2 causal inferential framework. Finite populations, counterfactuals, and randomization-based no-good-reasons equiprobability have no role to play in what inference did occur in the studies. The basis of the studies' inference was an entirely different theory, involving the notion of an infinite population, on which we focus critical attention in the next chapter.

Hume's problem of induction in modern statistical inference and controlled experimentation

4.1 Introduction: Hume's problem in the modern approach

In the first three chapters we have given considerable attention to the notion of causality. But Hume originally came to the notion of causality by posing another question. He first asks himself, *"What is the nature of all our reasonings concerning matters of fact?"* and answers, "they are founded on the relation of cause and effect." He then asks, *"What is the foundation of all our reasonings and conclusion concerning that relation* [of cause and effect]?" and answers, "in one word, *experience*" (1955, p. 46). Thus, in arguing that our reasoning about matters of fact is based on "in one word, experience," Hume appears to adopt an empiricist perspective.

But Hume's empiricism is not as thoroughgoing as that of his successor Mill. Hume continues his argument:

We have said that all arguments concerning existence [or matters of fact] are founded on the relation of cause and effect, that our knowledge of that relation is derived entirely from experience, and that *all our experimental conclusions proceed upon the supposition that the future will be conformable to the past.* To endeavor, therefore, the proof of this last supposition by . . . arguments regarding existence [or matters of fact], must be evidently going in a circle and taking that for granted which is the very point in question. (1955, pp. 49–50; emphasis added)

Hume then anticipates and rejects as circular Mill's later argument that Induction can be based on induction (see Section 1.2.2). *The* question, then, for Hume, is, On what do we base "the supposition that the future will be conformable to the past"? He answers only that having experienced or "observed similar objects or events to be constantly conjoined together," we are in the "*custom* or *habit*" (1955, p. 56) of concluding that they will continue to be so conjoined. We take Hume's problem of induction, then, to be, On what (if anything), in addition to experience, do we base reasoning about matters of fact?

In modern statistical inference, one begins with a sample of observed matters of fact. One assumes that the sample comes from a population about which one has partial knowledge. On the basis of the assumption and the observations, one infers the complete nature of the population. One could next deduce knowledge of future matters of fact, though this step in the reasoning usually is not spelled out. In the context of modern statistical inference, then, Hume's question seems to be, On what do we base both the supposition that the observations come from a population and the assumed partial knowledge of that population?

To fix ideas, we begin with the example, in the statistician's lexicon, of a classical two-sample situation. First we clarify the meaning of the statistician's terms. Next we examine some ways in which the statistician's undefined terms might be attached to the world. Then we examine several essential concepts of the Neyman–Pearson theory – *the* modern theory of statistical hypothesis-testing. Afterward, we consider some traditional problems associated with the classical two-sample situation along with their Neyman–Pearson theory solutions. Finally we study various historical approaches – rank-based, asymptotic, approximation theorem, and consistent nonparametric regression – to weakening the classical normality assumption. We concentrate on what changes occur in the theory results when the classical normality assumption is weakened.

4.2   Meaning, reference, and some theory results in modern statistical inference

4.2.1   *A classical situation and an objection*

Let $t_1, \ldots, t_n$ be a sample from a normal population with mean μ_t and variance σ_t^2, and let $c_1, \ldots, c_m$ be an independent sample from a normal population with mean μ_c and variance σ_c^2. Such is roughly the language statisticians often employ to describe certain situations in the world. We ended Chapter 3 by concluding that whatever the statistician's language means here, it does *not* mean the same as similar language in Chapter 2's theory of causal inference, for the populations there were finite, whereas here they seem not to be, and the samples there were not independent, whereas here they are.

The philosopher Ian Hacking has questioned the meaning of our statistician's language here:

Since a population is a class of distinct things, infinite populations, such as the population of natural numbers, do make sense. But they will not be considered in the sequel. I am not sure that it makes sense to speak of sampling a population of imaginary things, unless what is meant is something like writing down the names of Cerberus, Bucephalus and Pegasus or else drawing the names from a hat. At any rate we shall not consider sampling infinite populations most of whose members are imaginary; in the same spirit we shall pass by the hypothetical infinite populations which Fisher believed to be the foundation of all statistics. (1965, p. 122)

Notice that Hacking seems to believe that infinite populations (at least *countably* infinite populations) "do make sense." He is troubled, though, about the notion of *sampling* from such a population. In fact, however, what is troublesome about the notion of sampling from an infinite population is closely related to what *sort* of sense such a population makes.

It surely makes sense to write and speak about the class of natural (or even real) numbers and to construct a mathematical theory (i.e., real analysis) of that class. What is less clear is how such a class makes sense in connection with perception or observation. The difficulty is how to attach the infinite class of our speech and writing to the object world of our perception and observation. For example, the problem with *sampling* from a (countably or uncountably) infinite population begins with the fact that we *cannot* place a sign for each member of the population on a slip of paper in a hat.

Naively, we might think that the problem of attaching what we write about to what we observe is somehow avoidable through the semantic move of scrapping the words "sample" and "population" with their connotation of drawing things out of a hat. In fact, statisticians do have an alternate way of expressing the standard two-sample situation. Let $t_1 \ldots, t_n$ be observations from n independent, identical normal probability distributions, with means μ_t and variances σ_t^2, and let $c_1, \ldots, c_m$ be independent observations from m independent, identical normal probability distributions, with means μ_c and variances σ_c^2. With the new language, the problem of concept attachment translates into giving a meaning to "probability distribution" in a somewhat different situation than that studied in Section 1.3.2.

On the formal or axiomatic level, we may proceed as in Section 1.3.2. Let T_i (C_i) range over the possible values of t_i (c_i) for every i. What, then, do we mean by saying that t_i (c_i) is an observation from a normal probability distribution with mean μ_t (μ_c) and variance σ_t^2 (σ_c^2)? We mean, first of all, that the probability that $T_i \leq t_i$ $(C_i \leq c_i)$, where t_i (c_i) is any real number, satisfies the following.

Axiom 4.1

$$p(T_i \le t_i) \quad (p(C_i \le c_i)) = \left[1 \Big/ \left(2\prod\right)^{1/2} \sigma_t\right] \int_{-\infty}^{t_i} e^{-(1/2)[(T_i-\mu_t)/\sigma_t]^2} \, dT_i$$
$$\left(\left[1 \Big/ \left(2\prod\right)^{1/2} \sigma_c\right] \int_{-\infty}^{c_i} e^{-(1/2)[(C_i-\mu_c)/\sigma_c]^2} \, dC_i\right).$$

And by independence, then, we mean merely that probabilities multiply.

Axiom 4.2

$$p(T_1 \le t_1, \ldots, T_n \le t_n; C_1 \le c_1, \ldots, C_m \le c_m)$$
$$= \prod_{i=1}^{n} p(T_i \le t_i) \prod_{j=1}^{m} p(C_j \le c_j).$$

But as we saw in Section 1.3.2, though such formal axioms solve what might be called the problem of probability's meaning and independence's meaning, they fail to solve what we might label the problem of probability's reference. That is, the question remains open as to how the probability concept which satisfies our written axioms attaches to the object world of our perception and observation.

4.2.2   *A limiting relative-frequency approach to the problem of probability's referent*

To attach Axioms 4.1 and 4.2 to the object world of our perception and observation, we first try a limiting relative-frequency approach.[1] Imagine r repetitions of the process which generates the $n + m$ observations $t_1, \ldots, t_n; c_1, \ldots, c_m$ with the *same* set (a subset of R^{m+n}) of possible outcomes for each repetition. Allow either the classical Hume–Hill or the quantum Schrödinger–Suppes concept of causality to underlie the repetitions (see Section 1.2.1). On the classical view, the repetitions are not identical; their initial conditions range over a set of elements, and a DLW allows a set of two or more possible outcomes. On the quantum view, the

---

[1] Other accounts of "normal probability distribution" are certainly possible. We could, for example, adopt an account on what Mackie (1973, pp. 179–87) has labeled an "objective chance" or "propensity" view of probability. Such a view, though, holds that causes in some sense "necessitate" their effects and thus parts with Hume *from the start*. Limiting relative frequency at least begins with a proportion of observations falling in a certain range and so seems more compatible with our goal of *immanent* critique of empiricism.

repetitions may be identical; they still each have the same set of possible outcomes.

The result of any repetition is an $n + m$ vector $(t_1, \ldots, t_n; c_1, \ldots, c_m)$, and $(T_1, \ldots, T_n; C_1, \ldots, C_m)$ ranges over the possible $(t_1, \ldots, t_n; c_1, \ldots, c_m)$'s. Then let $\pi_r(T_1 \leq t_1, \ldots, T_n \leq t_n; C_1 \leq c_1, \ldots, C_m \leq c_m)$ be the proportion of their imagined repetitions which result in $(t_1, \ldots, t_n; c_1, \ldots, c_m)$'s for which $T_i \leq t_i$ and $C_j \leq c_j$ for all $i = 1, \ldots, n; j = 1, \ldots, m$. Also let $\pi_r(T_i \leq t_i) (\pi_r(C_j \leq c_j))$ be the proportion of the r imagined repetitions for which $T_i \leq t_i (C_j \leq c_j)$ for any $i = 1, \ldots, n$ $(j = 1, \ldots, m)$. Notice that though the repetitions are imagined, they could, in principle at least, be carried out so that $\pi_r(T_1 \leq t_1, \ldots, T_n \leq t_n; C_1 \leq c_1, \ldots, C_m \leq c_m)$, $\pi_r(T_i \leq t_i)$, and $\pi_r(C_j \leq c_j)$ could be based strictly on what is observable.

Next, however, suppose that $\pi_r(T_i \leq t_i) (\pi_r(C_j \leq c_j))$ approaches $p(T_i \leq t_i) (p(C_j \leq c_j))$ (of Axiom 4.1) as r approaches ∞. Just as in die rolling we *cannot* observe an *infinite* number of repetitions of the process that generates our observations, so here, as in Section 1.3.2, our supposition that the limit of $\pi_r(T_i \leq t_i) (\pi_r(C_j \leq c_j))$ exists and has the value of Axiom 4.1 goes beyond what is in principle observable. Suppose also that the limit of $\pi_r(T_1 \leq t_1, \ldots, T_n \leq t_n; C_1 \leq c_1, \ldots, C_m \leq c_m)$ as r approaches ∞ also exists. Then Axiom 4.2 attaches to the world of our observation and perception as

$$\lim_{r \to \infty} \pi_r(T_1 \leq t_1, \ldots, T_n \leq t_n; C_1 \leq c_1, \ldots, C_m \leq c_m)$$

$$= \prod_{i=1}^{n} \lim_{r \to \infty} \pi_r(T_i \leq t_i) \times \prod_{j=1}^{m} \lim_{r \to \infty} \pi_r(C_j \leq c_j).$$

So Axiom 4.2's attachment to the world of our observation and perception also goes beyond what is in principle observable.

Notice that our limiting relative-frequency account (LRFA) of "normal probability distribution" shares a difficulty with our LRFA of the equiprobability of a die roll's outcomes (see Sections 1.3.2 and 1.3.3): the nonobservability of the "indefinitely extended sequence." But our LRFA of "normal probability distribution" has an added problem that its die-roll counterpart lacks. The process of rolling a die has only a finite number of possible outcomes; so, for example, we can place or observe an ordinary die in each of its six possible outcomes before rolling it. But on our LRFA, T_i for any i or C_j for any j must have more than a finite number of possible outcomes on any repetition. Suppose T_i had only a finite number of possible outcomes ranked by their real values. Then, if $u < v$ are two such adjacent outcomes,

$$\lim_{r \to \infty} \pi_r(T_i \leq u) = \lim_{r \to \infty} \pi_r[T_i \leq (u + v)/2],$$

whereas $p(T_i \leq u) \neq p[T_i \leq (u + v)/2]$, which is a contradiction. Thus, our LRFA of "normal probability distribution" seems to require an infinite set of possible outcomes.

Historically, the normal probability distribution first gained importance in astronomers' measurements of interplanetary and interstellar distances. Astronomers assumed that their repeated measurements of an interplanetary or interstellar distance followed a normal probability distribution whose mean was the true distance. The normal probability distribution of their writings makes sense with respect to the other concepts of the Gaussian theory of errors that they developed. But in attaching their writings to the world of their observations and perceptions, on our LRFA, the astronomers postulated the existence of an entity generating their observations that is itself unobservable in principle. Thus, on our LRFA, our semantic maneuver from "population" and "sample" to "probability distribution" ends where it started: with a need for an infinite (and therefore in principle unobservable) set to attach what we write and speak to what we observe.

4.2.3   *A hypothetical approach to the problem of the normal probability distribution's referent*

We have argued that to attach the written notion of observations from independent, identical normal probability distributions to the world of our observation and perception requires going beyond what is in principle observable. But is it possible to construct an account of "normal probability distribution" that does *not* require an infinite and therefore in principle unobservable set of outcomes to attach what we write and speak to what we observe and perceive? The previous section's LRFA of "normal probability distribution" relied on imagined repetitions (of the process generating an observation), each of which had the *same* set of possible outcomes. If we can reconceptualize the repetitions so that the set of possible outcomes grows with the number of repetitions, perhaps we can circumvent the need for an infinite set of outcomes on any one of the repetitions.

Let's begin by imagining a process whose set of possible outcomes is $\{0, 1\}$ – say flipping a coin, where 0 is heads and 1 is tails. Suppose that the process is repeated r times, that the underlying concept of causality is either classical or quantum, that the probability of a 1 on any outcome is p, and that the repetitions are independent (see Axiom 4.2). Then p is the limit of the proportion of 1's occurring in an imagined sequence, "indefinitely extended," of coin flips. Let S_r be the sum of the r results. Then, by the de Moivre–Laplace Central Limit Theorem, $p([(S_r - rp)/(rp(1 - p))^{1/2}] \leq z)$ approaches $\int_{-\infty}^{z}(1/2\pi)^{1/2}e^{-(1/2)X^2}\,dX$ as r approaches ∞, where the density involved in the integral is that of a normal distribution with $\mu_t = 0$ and $\sigma_t^2 = 1$ (see Axiom 4.1). Notice that by concentrating not on the result of one flip (which has two possible outcomes) but on $(S_r - rp)/(rp(1 - p))^{1/2}$ (which has $r + 1$ possible values), we have reconceptualized the repetitions so that the set of possible outcomes grows with the number of repetitions. Thus, we get a "normal probability distribution" without an infinite set of outcomes on any repetition.

But how can we attach the process about which we've just written to the physical act of measuring the distance between two planets or the change in hours worked from one year to the next? As Hacking suggests, "if one conceives of an error on a measurement as produced by a large number of independent events binomially distributed, one is led to suppose the distribution must be more or less Normal, in virtue of de Moivre's theorem" (1965, p. 72). Thus, if t is the result of the measurement process, we must simply make a brute-force assumption that $t = \sigma_t[(S_r - rp)/(rp(1 - p))^{1/2}] + \mu_t$, where $(S_r - rp)/(rp(1 - p))^{1/2}$ results from a process like the one we wrote about.

The key word in Hacking's suggestion is "conceive." Our brute-force assumption certainly goes beyond the observations (e.g., we *see* no Bernoulli trial-results in making a single measurement of interplanetary distance). Thus, the coin flips presumed behind the observation are strictly hypothetical, conceptual, or "as if." So our account here of "normal probability distribution with mean μ_t and variance σ_t^2" avoids the use of an infinite (and hence in principle unobservable) set of outcomes for each repetition. But our account hardly resolves the problem of the normal probability distribution's referent on the basis of what is observable.

4.2.4   *The Neyman–Pearson problematique*[2]

Before proceeding further, we must review a few concepts of the Neyman–Pearson theory – the theory behind controlled experimental and

---

[2] Hacking (1965, Chapters VI and VII) and Lehmann (1950) give more detailed but still succinct accounts of the Neyman–Pearson problematique that stress the problematique's relationship to the likelihood-ratio principle.

other modern statistical tests of hypotheses. We begin with a set of observations which we assume were generated by a member of a certain class of probability distributions $\mathscr{P} = \{P_\theta : \theta \in \Theta\}$. That the generator of our observations is an element of $\mathscr{P}$ is an assumption which goes beyond the observable. Our aim is to test a hypothesis that partitions $\mathscr{P}$ (Θ) into two subclasses: one H (Θ_H) for which the hypothesis is true, and one K (Θ_K) for which it is false. We call H the hypothesis and K the alternative.

The assumption that the generator of our observations is a member of $\mathscr{P}$ defines the space of *possible* observations. A test procedure ϕ for H is a partition of the space of possible observations into a rejection region R_ϕ and an acceptance region A_ϕ. For a test procedure ϕ for H if our actual observations fall in R_ϕ (A_ϕ), we reject (accept) the hypothesis.[3] But what test procedure for H should we use? Since the hypothesis is either true or false, a test can make two kinds of errors: rejecting the hypothesis when it is true (an error of the first kind) or accepting the hypothesis when it is false (an error of the second kind). It seems reasonable to choose a test procedure which keeps the probabilities of the two kinds of errors low. The problem is that for a given number of observations, one cannot simultaneously control the two probabilities.[4]

The Neyman–Pearson theory approach to controlling the two probabilities begins by restricting attention to a subclass of the class of test procedures for H. For any test procedure ϕ for H, let p_θ(observations $\in R_\phi$) be the probability of accepting the hypothesis given θ. We then choose an α such that $0 < \alpha < 1$, called the significance level, and consider only tests ϕ for H for which p_θ(observations $\in R_\phi$) $\leq \alpha$ for all

---

[3] Recall that in Section 2.2.4 we argued that rejecting a hypothesis H really meant concluding that either H was false or an unusual event occurred. In fact, strictly speaking, in that section, rejecting H meant concluding that either H was false or an unusual event occurred *or* the equiprobable distribution didn't hold. But we regarded the physical act of randomization as sufficient reason for believing that the equiprobability distribution *did* hold and so could eliminate its not holding from the propositions of the multiple disjunction "rejecting H."

In this chapter, rejecting H really means concluding that H is false or that an unusual event occurred or *that it is false that the generator of the observations is a member of $\mathscr{P}$.* Call the italicized proposition $\sim P$. Here we have no physical-acts basis for believing P. In the remainder of the chapter, we shall argue that in all situations of interest to us, the truth of P goes beyond what is observable. If our argument is correct, and physical acts and the observable are the only fully acceptable bases for knowledge, then the Neyman–Pearson "rejecting H" should retain $\sim P$. We return to this point in Part II.

[4] The problem of controlling the two probabilities is in essence a slightly more complex version of the mathematical problem of controlling two distinct functions of the same argument. The difficulty is that two such functions generally don't have minima at the same value of the argument.

$\theta \in \Theta_H$. Such tests are called level-α tests for H. Customarily, α is a small number (e.g., .05, .025, .01, or .005). Notice that the first restriction introduces, by convention, a ceiling of α on the probability of an error of the first kind.[5]

Next, for any test procedure ϕ for H, let $\beta_\phi(\theta) = p_\theta(\text{observations} \in A_\phi)$ be the probability of accepting the hypothesis for $\theta \in \Theta_K$, and call $1 - \beta_\phi(\theta)$ the power of the test for H against the alternative θ. Let Φ_α^H be the class of all level-α tests for H. If H and K each contain only a single distribution (say p_{θ_H} and p_{θ_K}), then the fundamental Neyman–Pearson theory result is the existence and exhibition of a $\phi^* \in \Phi_\alpha^H$ such that $1 - \beta_{\phi^*}(\theta_K) \geq 1 - \beta_\phi(\theta_K)$ or $\beta_{\phi^*}(\theta_K) \leq \beta_\phi(\theta_K)$ for all $\phi \in \Phi_\alpha^H$. Such a ϕ^* for H is called a most powerful level-α test against the alternative θ_K. For H and K that are not singletons, the most powerful level-α test for H generally differs with the alternative $\theta \in \Theta_K$. For certain classes $\mathscr{P}$, one of whose members is assumed to have generated the observations and certain H's, though, one can exhibit a $\phi^* \in \Phi_\alpha^H$ that is most powerful among all $\phi \in \Phi_\alpha^H$ for any $\theta \in \Theta_K$. Such a ϕ^*, when it exists, minimizes among all Φ_α^H the probability of an error of the second kind for any $\theta \in \Theta_K$ and so is an optimal level-α test for H. We call such a ϕ^* a uniformly most powerful (UMP) level-α test for H.

For ($\mathscr{P}$, H)'s for which no UMP level-α test for H exists, we may still be able to exhibit an optimal level-α test for H by restricting attention to a subclass of Φ_α^H. Suppose $\phi \in \Phi_\alpha^H$ and $1 - \beta_\phi(\theta) < \alpha$, so that $\beta_\phi(\theta) > 1 - \alpha$ for some $\theta \in \Theta_K$. For such ϕ and θ, the probability of accepting the hypothesis is greater when it is false than when it is true. Because such a situation clearly seems undesirable, let's restrict attention to the subclass of Φ_α^H for which $1 - \beta_\phi(\theta) \geq \alpha$ for all $\theta \in \Theta_K$. Call this subclass Φ_α^{HU}, the class of all *unbiased* level-α tests for H. A test $\phi^* \in \Phi_\alpha^{HU}$ that is most powerful among all $\phi \in \Phi_\alpha^{HU}$ for any $\theta \in \Theta_K$ is called a uniformly most powerful unbiased (UMPU) level-α test for H. Since such a ϕ^*, when it exists, minimizes among all $\phi \in \Phi_\alpha^{HU}$ the probability of an error of the second kind for any $\theta \in \Theta_K$, if no UMP level-α test for H exists, ϕ^* is an optimal level-α test for H.

When an optimal (UMP or UMPU) level-α test exists for a given ($\mathscr{P}$, H), its rejection–acceptance regions are defined in terms of the possi-

---

[5] Restricting attention to level-α tests, then, is *roughly* like reformulating the problem of controlling two distinct functions of the same argument to controlling one function subject to a constraint on the other. Notice that the reformulation favors the probability of an error of the first kind – i.e., assures by the constraint that it will be below α.

ble values of its associated test statistic (= function of the observations). For the particular two-sample situation and H in which we are interested, for the strongest $\mathscr{P}$ we shall consider, though no UMP level-α test for H exists, there is a UMPU (and therefore optimal) level-α test.

Notice that unlike the sorts of tests of Section 2.2.4, the test statistic for a Neyman–Pearson theory optimal level-α test of H need not be arbitrary, but may be determined by the Neyman–Pearson problematique of controlling the probabilities of the two kinds of errors. Of course, capping the probability of type I error at a conventional level and minimizing the probability of type II error is only one of several possible approaches to the problem of controlling the probabilities of the two kinds of errors. Hence, though choice of the Neyman–Pearson theory test statistic may not be arbitrary, the Neyman–Pearson approach to controlling the probabilities of the two kinds of errors is itself in a sense arbitrary. We now turn to the particular two-sample situation and H in which we are interested.

4.2.5 *Some Neyman–Pearson theory results for the classical two-sample situation*

We return now to our observations $t_1, \ldots, t_n; c_1, \ldots, c_m$ and assume once again that they have all been independently generated – the t's by identical normal distributions with means μ_t and variances σ_t^2 and the c's by identical normal distributions with means μ_c and variances σ_c^2. (Remember that in Sections 4.2.2 and 4.2.3 we argued that these assumptions go beyond what is observable in connecting what we write about to what we can see.) We now add a further assumption: $\sigma_t^2/\sigma_c^2 = c_0$ for some known $c_0 \in R^1$; in fact, without loss of generality we assume $c_0 = 1$ or $\sigma_t^2 = \sigma_c^2 = \sigma^2$ (unknown), with $0 < \sigma < \infty$. (Either because we can't see the normal distribution or because the optimal test of $\sigma_c^2/\sigma_t^2 = c_0$ employs the other assumptions, our additional assumption also appears to go beyond what is observable.)[6] Call the assumptions of this paragraph *Axiom 4.3.*

The hypothesis we wish to test is H: $\mu_c - \mu_t = 0$, and the alternative is K: $\mu_c - \mu_t > 0$.[7] Let $\bar{t}$, $\bar{c}$, s_t^2, and s_c^2 be as in Chapter 2:

---

[6] That is, the optimal test of $\sigma_c^2/\sigma_t^2 = c_0$ presumes independent observations from normal distributions (Lehmann 1959, pp. 169–70). So even if the assumption $\sigma_c^2/\sigma_t^2 = c_0$ is based on such a test, it is still an assumption that goes beyond the observable.

[7] In fact, the results we present all still hold if the null hypothesis is changed to $\mu_c - \mu_t \leq 0$.

Definition

$$\bar{t} = \sum_{i=1}^{n} t_i/n, \qquad \bar{c} = \sum_{i=1}^{n} c_i/m,$$

$$s_t^2 = \sum_{i=1}^{n} (t_i - \bar{t})^2/(n-1), \text{ and}$$

$$s_c^2 = \sum_{i=1}^{n} (c_i - \bar{c})^2/(m-1).$$

Now we shall consider a new statistic and a test based on it.

Definition: $t^* = (mn/(m+n))^{1/2}[(\bar{t} - \bar{c})/s_p]$, where $s_p^2 = [1/(m+n-2)][(n-1)s_t^2 + (m-1)s_c^2]$.

Definition: Let ϕ^+ be a test procedure for H that rejects the hypothesis iff $t^* \geq C_\alpha$, where C_α is determined by $\int_{C_\alpha}^{\infty} t_{(m+n-2)}(y)\,dy = \alpha$, where $t_{(m+n-2)}$ is the probability density function for Student's t distribution with $(m+n-2)$ degrees of freedom.

The Neyman–Pearson theory results for our classical two-sample situation are given in the following two theorems.

Theorem 4.1. *Axiom 4.3 implies that ϕ^+ is a UMPU level-α test for H.*

Theorem 4.2. *Let $\delta = (\mu_c - \mu_t)/\sigma[(1/n) + (1/m)]^{1/2}$. Then Axiom 4.3 implies that the power of the test procedure ϕ^+ for H against any $(\mu_c - \mu_t) \in K$ is given by $1 - \beta_{\phi^+}(\delta) = \int_{-\infty}^{C_\alpha} t_{(\delta,(m+n-2))}(y)\,dy$, where $t_{(\delta,(m+n-2))}$ is the probability density function for the noncentral t distribution with noncentrality parameter δ and $(m+n-2)$ degrees of freedom.*[8]

Notice that the logic, if not the mathematics, of the Neyman–Pearson theory results is quite simple. We merely choose by convention an α to

---

[8] Proofs of Theorems 4.1 and 4.2 may be culled from Lehmann (1959, pp. 170–3). In this chapter, we present results as formal theorems so that the careful reader, by comparing the theorems in one section with those in another, can see how shifting the assumptions changes the theoretical results. Finally, notice that $1 - \beta_{\phi^+}(\delta)$ is a monotonically increasing function of δ. This result is intuitively pleasing: The further the real state of affairs from the hypothesis, the lower the probability of accepting the hypothesis.

cap our probability of an error of the first kind. From a table of Student's t distribution we read the value C_α corresponding to α. Our test is then: Reject the hypothesis iff the observed value of $t^* \geq C_\alpha$. Theorem 4.2 then assures us that *if the assumptions of Axiom 4.3 hold,* our test is a UMPU level-α (and therefore optimal) test for H. And Theorem 4.2 assures us that the probability of an error of the second kind can be read from a table of the noncentral t distribution *if the assumptions of Axiom 4.3 hold.*

Now consider the normality assumption of Axiom 4.3. The statistician P. J. Huber has argued that, historically, normality became a dominant assumption in part because normality gives the frequently used sample mean certain optimality properties in estimation problems. In testing problems such as the classical two-sample problem of interest to us, normality appears to have a similar rationale: It assures the existence of an optimal (in a Neyman–Pearson theory sense) test. Huber sensibly argues that such a justification of normality "borders on dogmatism" (1972, p. 1043): One assumes normality because one has a theory that gives results for normally distributed observations. He points out: "By 1960 one had recognized that one never has a very accurate knowledge of the true underlying distribution."[9] We turn, then, to the question of what happens to the Neyman–Pearson theory results for our classical two-sample situation when we weaken the normality assumption.

4.3   Weakening the classical normality assumption

4.3.1   *From normal to continuous distributions*[10]

Return to our observations $t_1, \ldots, t_n; c_1, \ldots, c_m$ and again assume that they have all been independently generated. But now suppose that the t's have been generated by identical continuous distributions $F(t)$ and the c's by identical continuous distributions $G(c)$. Suppose further that $G(x) = F(x - \Delta)$ for any $x \in R^1$ for some fixed $\Delta \in R^1$. The hypothesis we wish to test is $H: \Delta = 0$, and the alternative is $K: \Delta > 0$. Notice that the assumptions, hypothesis, and alternative are precisely those of Section 4.2.5, except that the assumption of normal distributions has been re-

---

[9] Huber (1972, p. 1045). Bickel and Doksum (1977, p. 344) appear to agree with Huber: "little justification for the exact normality of measurements of continuous type responses . . . can be given."

[10] Lehmann (1975, Chapter 2) gives a more detailed discussion of the rank-based methods considered in this section.

placed by the weaker assumption of continuous distributions, and the $0 < \sigma^2 < \infty$ restriction has been dropped.

When one assumes that a distribution is continuous, one certainly assumes less than when one assumes that it is normal. Yet the continuous assumption, like the normal assumption, is one that goes beyond the observable in attaching what we write about to what we can observe, for, formally, a continuous probability distribution $F(t)$ satisfies the following.

Axiom 4.4. *$F(t)$ is the limit of $F(t')$ as t' approaches t for any $t \in R^1$.*

Consider an LRFA of "continuous probability distribution," with each repetition having the same set of possible outcomes, and $\pi_r(T \le t)$ being the proportion of r imagined repetitions for which $T \le t$. Then the supposition that $\lim \pi_r(T \le r)$ as r approaches ∞ exists and satisfies Axiom 4.4 goes beyond what is in principle observable. And by the argument of Section 4.2.2 for the normal case, the set of possible outcomes on each repetition is infinite (and therefore not observable).

How can we choose a test procedure for H? Following the Neyman–Pearson approach, we first restrict attention to all level-α tests for H, for some α chosen by convention. Generally, for our H and K, a test ϕ's probability of an error of the first kind depends on what F actually is. It is reasonable, then, to seek a subclass of level-α tests for H whose probability of an error of the first kind does *not* depend on what F actually is. Suppose $R_1, \ldots, R_n; S_1, \ldots, S_m$ are the ranks of the observations $t_1, \ldots, t_n; c_1, \ldots, c_m$ (smallest rank is 1, second smallest is 2, etc.), respectively, among the $m + n$ observations. If $\phi = \phi(S_1, \ldots, S_m)$, then we call ϕ a rank test and let Φ^{HR} be the class of all rank tests for H.

We claim that if $\phi \in \Phi^{HR}$, then ϕ's probability of an error of the first kind doesn't depend on what F is, for by F's continuity, the vector $(S_1, \ldots, S_m)$ has $\binom{m+n}{m}$ possible values whose possibilities are not zero. And under $H: \Delta = 0$, the t's and c's are identically distributed, so that each of the $\binom{m+n}{m}$ $(S_1, \ldots, S_m)$'s has probability $1/\binom{m+n}{m}$ of occurring *no matter what F is*. [Notice that $(S_1, \ldots, S_m)$ then has a classical good-reasons equiprobable distribution, where H, rather than any physical act as in Chapter 2, makes the distribution plausible.] Since the rejection region of any $\phi \in \Phi_\alpha^{HR}$ (= the class of all level-α rank tests for H) consists merely of some $[\alpha\binom{m+n}{m}]$ (where [ ] means greatest integer $\le$) of the $\binom{m+n}{m}$ values of $(S_1, \ldots, S_m)$ whose probabilities are not zero, Φ_α^{HR} is the same for any continuous F.

Though we can cap the probability of an error of the first kind for any continuous F by first restricting attention to Φ^{HR}, the probability of an error of the second kind for any $\phi \in \Phi_\alpha^{HR}$ is $\beta_\phi(F, \Delta)$ (i.e., depends on the distribution F as well as on Δ). And there is no $\phi \in \Phi_\alpha^{HR}$ that is uniformly most powerful against all choices of (F, Δ). Things are not improved if attention is restricted to the subclass of Φ_α^{HR} of unbiased tests: There is no UMPU level-α rank test for H. It seems reasonable to restrict attention further to $\Phi_\alpha^{HRU^-}$, which is the subclass of unbiased level-α tests of Φ^{HR} for which $\beta_\phi(F, \Delta)$ is a monotonically decreasing function of Δ for any continuous F – intuitively, tests whose probabilities of an error of the second kind decrease with the "distance apart" of the two distributions.[11] But again, there is no $\phi \in \Phi_\alpha^{HRU^-}$ that is uniformly most powerful against all choices of (F, Δ).

So relaxing normality to continuity allows an element of arbitrariness to enter the choice of statistic, because the Neyman–Pearson approach fails to select an optimal test. Because we still have the power concept to rely on, the arbitrariness in choice of test statistic is not as extreme as in the tests of Section 2.3.4. Thus, the two most popular rank tests for our two-sample situation – the Wilcoxon rank sum and Fisher–Yates normal scores – often are defended on the grounds that their power for conventional α, if F is normal, is not much less than that of the classical t test (of the previous section).[12] For sufficiently small Δ, the rank-sum test is the most powerful test of Φ_α^{HR} if F is actually the logistic distribution, whereas the normal-scores test is the most powerful test of Φ_α^{HR} if F is actually normal (both these tests $\in \Phi_\alpha^{HRU^-}$).[13] Reintroducing assumptions about F, then, can make the choice of a level-α rank test for H less arbitrary.

Rank-based methods for our two-sample problem allow us to cap the probability of an error of the first kind assuming only continuous rather than normal probability distributions. But even with a rank-based approach, calculating the probability of an error of the *second* kind requires us to assume that F is normal, logistic, and so forth (i.e., to make a distributional assumption that is no less weak than the classical normality assumption). Approaches that *seem* to weaken the classical assumptions in power calculations, using the concept of asymptotic power and approximation theorems, do exist for *certain* tests. We shall now argue,

---

[11] Recall from Section 4.2.5 that the optimal test in the normal case had this property (see footnote 8).

[12] One may also employ the Wilcoxon rank-sum test when the observations themselves are ranks rather than from continuous distributions. See Neuberg (1978) for an example.

[13] They are still elements of $\Phi_\alpha^{HRU^-}$ if H is changed to $\Delta \leq 0$.

however, that such approaches really only alter rather than weaken the classical assumptions.

4.3.2 From a normal distribution to a distribution with finite variance

We begin by supposing that $t_1, \ldots, t_n$ are independent observations from the same normal probability distribution $N(\mu, \sigma)$, where $0 < \sigma^2 < \infty$ (and call this supposition *Axiom 4.5*). We wish to test the hypothesis $H: \mu = 0$ against the alternative $K: \mu > 0$.[14] Again let $\bar{t} = \sum_{i=1}^n t_i/n$ and $s_t^2 = \sum_{i=1}^n (t_i - \bar{t})/(n-1)$. We distinguish the two cases where σ^2 is either known or unknown and consider a different test for each. (Note that the σ^2-unknown case is the one-sample analogue of the classical two-sample situation of Section 4.2.5.)

Definition: For σ^2 known, ϕ_1 rejects H iff $[(n)^{1/2}\bar{t}/\sigma] \geq C_\alpha^1$, where C_α^1 is determined by $\int_{C_\alpha^1}^\infty g(z)\,dz = \alpha$, where g is the standard normal probability density function.

Definition: For σ^2 unknown, ϕ_2 rejects H iff $[(n)^{1/2}\bar{t}/s_t] \geq C_\alpha^2$, where C_α^2 is determined by $\int_{C_\alpha^2}^\infty t_{(n-1)}(y)\,dy = \alpha$, where $t_{(n-1)}$ is the probability density function for Student's t distribution with $(n-1)$ degrees of freedom.

The Neyman–Pearson theory results for σ^2 known and unknown are given in the following four theorems.

Theorem 4.3. *For σ^2 known, Axiom 4.5 implies that ϕ_1 is a UMP level-α test for H.*

Theorem 4.4. *For σ^2 known, let $\delta = (n)^{1/2}\mu/\sigma$. Then Axiom 4.5 implies that the power of ϕ_1 against any $\mu \in K$ is given by $1 - \beta_{\phi_1}(\delta) = \int_{-\infty}^{C_\alpha^1} g_\delta(z)\,dz$, where g_δ is the normal probability density function with mean δ and variance 1.*

Theorem 4.5. *For σ^2 unknown, Axiom 4.5 implies that ϕ_2 is a UMPU level-α test for H.*

---

[14] Again, the results we present all still hold if the hypothesis H is changed to $\mu \leq 0$.

Theorem 4.6. *For σ^2 unknown, let $\delta = (n)^{1/2}\mu/s_t$. Then Axiom 4.5 implies that the power of ϕ_2 against any $\mu \in K$ is given by $1 - \beta_{\phi_2}(\delta) = \int_{-\infty}^{C_\alpha^2} t_{(\delta,(n-1))}(y) \, dy$, where $t_{(\delta,(n-1))}$ is the probability density function for the noncentral t distribution with noncentrality parameter δ and $(n-1)$ degrees of freedom.*[15]

Next, suppose we drop the normality assumption from Axiom 4.5 and call the resulting supposition Axiom 4.6. Note first that the assumptions that remain still go beyond the observable, for LRFAs of "probability distribution" and "independent" assume the existence of and relations between unobservable limits (see Section 4.2.2). And because there exist probability distributions (e.g., Cauchy distributions) with infinite variances, the observations cannot assure $\sigma^2 < \infty$, nor can σ^2 be known solely on the basis of them. Now suppose that we are still interested in testing H: $\mu = 0$ against K: $\mu > 0$ (see footnote 13).

Consider again tests for the σ^2-known and -unknown cases. Let the test for the σ^2-known case again be ϕ_1, but define a new test ϕ_3 for the σ^2-unknown case.

Definition: For σ^2 unknown, ϕ_3 rejects H iff $[(n)^{1/2}\bar{t}/s_t] \geq C_\alpha^1$, where C_α^1 is as in the definition of ϕ_1.

What can be said about ϕ_1 (ϕ_3) and the probabilities of the two kinds of errors without the normality assumptions? Generally the significance level of ϕ_1 (ϕ_3) and the probabilities of the two kinds of errors depend on n and the actual underlying distribution F. But the Central Limit and Slutsky theorems provide two important results.

Theorem 4.7. *If $\alpha_{\phi_1}(n, F)$ $[\alpha_{\phi_3}(n, F)]$ is the significance level of ϕ_1 (ϕ_3), then Axiom 4.6 implies that $\alpha_{\phi_1}(n, F)$ $[\alpha_{\phi_3}(n, F)]$ approaches α as n approaches ∞ for any F such that $0 < \sigma^2 < \infty$.*

Theorem 4.8. *Let $\delta = \mu/\sigma$, and suppose that $\beta_{\phi_1}(\delta, n, F)$ $[\beta_{\phi_3}(\delta, n, F)]$ is the probability of an error of the second kind. Then Axiom 4.6 implies that*

---

[15] Proofs for Theorems 4.3 and 4.4 may be culled from Bickel and Doksum (1977, pp. 170–1, 194, 196), and for Theorems 4.5 and 4.6 from Lehmann (1959, pp. 165–6). Notice that the power of ϕ_1 (ϕ_2) again is a monotonically increasing function of δ.

$\beta_{\phi_1}(\delta, n, F)$ $[\beta_{\phi_3}(\delta, n, F)]$ *approaches* $\int_{C_\alpha^1}^\infty g_\delta(z)\, dz$ *as n approaches* ∞ *for any F such that* $0 < \sigma^2 < \infty$, *where* g_δ *is as in Theorem 4.4.*[16]

Notice that without normality, for our actual sample size n we know neither the significance level (or probability of an error of the first kind) of ϕ_1 (ϕ_3) nor the power (so probability of an error of the second kind) of ϕ_1 (ϕ_3) against any alternative. Certainly, then, we can't say that ϕ_1 (ϕ_3) is optimal, in any UMP-among-some-class-of-tests sense. What makes ϕ_1 (ϕ_3) a test of choice, with no normality, but σ^2 known (unknown) $< \infty$ assumed, is the fact that we can find the limit *as n approaches* ∞ of ϕ_1's (ϕ_3's) significance level against any alternative – something we can't do for just any test of H. Because for different F's these limits are approached at different rates, and because we only ever have a finite sample size n, knowing the asymptotic significance level and power of ϕ_1 (ϕ_3) seems of little value in itself. What makes such knowledge valuable, though, is that when we can find the *asymptotic* significance level and power, we can also often use an approximation theorem to approximate the true significance level and power of the test for our actual sample size n.

4.3.3   *An approximation theorem and an illustration of its use*

Suppose once again that $t_1, \ldots, t_n$ are independent observations from a probability distribution F with unknown mean μ and known variance $\sigma^2 < \infty$ (i.e., that Axiom 4.6 holds) and that we wish to test the hypothesis $H: \mu = 0$ against the alternative $K: \mu > 0$. In the previous section we proposed a test ϕ_1 which rejects H iff $[(n)^{1/2}\bar{t}/\sigma] \geq C_\alpha^1$, where C_α^1 is determined by $\int_{C_\alpha^1}^\infty g(z)\, dz = \alpha$, where g is the standard normal probability density function. Also, we stated what the asymptotic significance level and power of ϕ_1 were. Now, however, we wish to estimate the significance level and power of ϕ_1 for our actual finite number of observations n. We employ a famous approximation theorem in our estimation.

Theorem 4.9 (Berry-Esséen). *Let F be a probability distribution with* 0 *mean, variance* $\sigma^2 < \infty$, *and third moment* $\rho = \int_{-\infty}^\infty |x|^3 F\{dx\} < \infty$, *and*

---

[16] Lehmann's discussion (1959, pp. 167–8) essentially sketches proofs of Theorems 4.7 and 4.8. Similar two-sample results are proved by Bickel and Doksum (1977, pp. 350–1). We have shifted over to a one-sample situation in this section to make the mathematics of the next section's theorems less convoluted. We have added the σ^2-*known* case here because the approximation theorem of the next section deals only with that case.

suppose that $X_1, \ldots, X_n$ are independent observations from F. Then

$$\left| p[((X_1 + \cdots + X_n)/\sigma(n)^{1/2}) < x] - \int_{-\infty}^{x} g(z)\, dz \right| < (33/4)[\rho/\sigma^3(n)^{1/2}],$$

where g is as before.[17]

This truly remarkable result gives one a bound for how far off one is when employing a standard normal distribution to estimate the distribution of the standardized sum of independent observations from *F*.

The Berry-Esséen Theorem provides bounds for the significance level and power of our test ϕ_1 for *H*.

Theorem 4.10. *Let F be the distribution of the observations, ρ be as in Theorem 4.9, and $\delta = \mu/\sigma$. Then Axiom 4.6 implies*

(a) $|\alpha_{\phi_1}(n, F) - \alpha| < (33/4)[\rho/\sigma^3(n)^{1/2}]$ *and*

(b) $|[1 - \beta_{\phi_1}(F, \delta)] - \int_{C_\alpha^1}^{\infty} g_\delta(z)\, dz| < (33/4)[\rho/\sigma^3(n)^{1/2}],$

where C_α^1 is as before and g_δ is as in Theorem 4.4.[18]

Notice that the bounds for the significance level and power of ϕ_1 are in terms of ρ. Thus, to have actual bounds, we must know ρ. And knowledge of ρ is neither stronger nor weaker than knowing that *F* is normal (i.e., *F* normal does not imply $\rho = \rho_0$ for some ρ_0, and $\rho = \rho_0$ for some ρ_0 does not imply *F* normal). So our approximation-theorem approach merely alters, not weakens, the classical normality assumption. Note also that our approximation-theorem approach deals *only* with the σ^2-*known* case and that ϕ_1 with the dropped normality assumption and ρ known is still no longer optimal in any UMP-among-some-class-of-tests sense.

Now, in what sense is it possible to know ρ (σ^2)? Conceivably one might have a point estimate $\hat{\rho}\,(s_t^2)$ of ρ (σ^2) such that $\hat{\rho}$ approaches ρ in probability (s_t^2 approaches σ^2 in probability) as n approaches ∞. But the *rate* at which any such $\hat{\rho}\,(s_t^2)$ approaches ρ (σ^2) in probability depends on the actual underlying distribution *F*. Thus, using $\hat{\rho}\,(s_t^2)$ in place of ρ (σ^2) in

---

[17] See Feller (1966, pp. 515–17) for a proof of this theorem.

[18] The Berry-Esséen Theorem also provides bounds, when the normality assumption is dropped entirely, for the significance level and power of the standard test for the difference in two population (with common *known* variance σ^2) means. But the two-sample situation affords no new insights into what must be known to get the bounds, and it involves more convoluted mathematics.

calculating bounds for α_{ϕ_1} and β_{ϕ_1} by Theorem 4.10 introduces errors of unknown size into the stated bounds. Instead of a point estimate, one might choose to employ an interval estimate of $\rho\,(\sigma^2)$ in calculating the bounds for α_{ϕ_1} and β_{ϕ_1} by Theorem 4.10. But for such an interval estimate, one must assume something more about F. Thus, knowing $\rho\,(\sigma^2)$ goes beyond the observations, and the empiricist program of basing calculations of α_{ϕ_1} and β_{ϕ_1} solely on the observations appears unrealizable.

4.3.4   *From classical to consistent nonparametric regression*

To now we have considered either one- or two-sample situations in which each observation was a single real number and focused on testing a hypothesis about the probability distribution(s) presumed to have generated the observations. In social studies like the income maintenance experiments, however, the researchers usually choose to observe n-tuples of real numbers and employ some form of regression technique to analyze the observations. Accordingly, we turn to a one-sample situation (to keep things simple) in which the observations are *pairs* (again to keep things simple)[19] of real numbers, and we focus on the problem of point-estimating a feature of the probability distribution presumed to have generated the observations. We begin with Axiom 4.7 for both classical and consistent nonparametric regression approaches: $(Y_1, X_1),\ \ldots\ ,(Y_n, X_n)$ are independent observations from the same bivariate probability distribution $D(Y, X)$. Just as in earlier univariate situations, on a LRFA of "probability distribution," Axiom 4.7 goes beyond the observable. For any probability distribution $D(Y, X)$, the conditional expectation of Y given $X = x$, $E(Y|X = x)$, is some function of x, say $f(x)$. We wish to provide an estimate – say $\widehat{f(x)}$ – of $f(x)$ that is a function of the observations and has some desirable properties deducible from some assumptions on $D(Y, X)$.

Consider first the classical assumptions on $D(Y, X)$: $f(x)$ is linear [i.e., $E(Y|X = x) = f(x) = a + bx$ for some constants a and b], $Y - a - bX$ is normally distributed with mean 0 and constant (with respect to x) variance $0 < \sigma^2 < \infty$, the distribution of X doesn't depend on a, b, or σ^2, and X and $Y - a - bX$ are independent. These assumptions once again go beyond what is observable. Notice that under the classical assumptions,

---

[19] That is, all the results of this section are easily extended to two samples and/or n-tuples of observations, though the mathematics grows more intricate.

the problem of estimating $E(Y|X=x)=f(x)$ becomes the problem of estimating the constants a and b.

Now consider three desirable properties of estimates of constant parameters.

Definition: If the joint probability density function of the observations can be written as $\mathcal{L}(\theta)$, a function of the parameter θ called the likelihood function, then an estimate $\hat{\theta}$(observations) that maximizes $\mathcal{L}(\theta)$ is called a maximum-likelihood estimate (MLE) of θ.

Definition: $\hat{\theta}$(observations) is an unbiased estimate of θ iff $E(\hat{\theta}) = \theta$ (as in Chapter 2).

Definition: $\hat{\theta}$(observations) is a consistent estimate of θ iff $E(\hat{\theta} - \theta)^2$ approaches 0 as n approaches ∞.

Now consider the OLSQ regression estimates of a and b:

$$\hat{a} = \sum_{i=1}^{n} Y_i/n \quad \text{and} \quad \hat{b} = \sum_{i=1}^{n} Y_i(X_i - \overline{X})/ \sum_{i=1}^{n} (X_i - \overline{X})^2.$$

Then we have a theorem on desirable properties.

Theorem 4.11. *Axiom 4.7, with $E(Y|X=x)=f(x)$ and the classical assumptions, implies that $\hat{b}$ ($\hat{a}$) is an MLE, unbiased, and consistent estimate of b (a).*[20]

Next, notice that with $\hat{b}$ ($\hat{a}$) estimating b (a), $\widehat{f(x)} = \hat{a} + \hat{b}x$ is an estimate of $f(x)$ that may be written as

$$\widehat{f(x)} = \sum_{i=1}^{n} W_i^*(n, x, X_1, \ldots, X_n)Y_i,$$

where

$$W_i^*(n, x, X_1, \ldots, X_n) = \left[(1/n) + x(X_i - \overline{X})/ \sum_{i=1}^{n} (X_i - \overline{X})^2 \right]$$
$$\text{for } i = 1, \ldots, n.$$

This fact and the definition of consistency motivate a more general definition.

---

[20] Proof of Theorem 4.11 may be culled from Wonnacott and Wonnacott (1970, pp. 38–9, 151–2).

Definition: A set of weights $\{W_i(n, x, X_1, \ldots, X_n)\}_{i=1}^n$ is universally consistent for $f(x)$ iff $E[\Sigma_{i=1}^n W_i Y_i - f(x)]^2$ approaches 0 as n approaches ∞.

Notice that the consistency of $\hat{b}$ ($\hat{a}$) as an estimate of b (a) when $f(x)$ is linear implies that $\{W_i^*\}_{i=1}^n$ is a universally consistent set of weights for $f(x)$ when $f(x)$ is linear. It is natural, then, to ask how much the classical assumptions can be weakened and still result in the exhibitability of a universally consistent set of weights for $f(x)$. The statistician Charles Stone has recently provided an answer to (as well as formulated) this question.

Theorem 4.12. *Axiom 4.7, with $E(Y|X = x) = f(x)$ and $E(Y^2) < \infty$, implies that a universally consistent set of weights $\{W_i\}_{i=1}^n$ for $f(x)$ is exhibitable* (Stone 1977, p. 600).

Thus, we can replace the entire set of classical assumptions with $E(Y^2) < \infty$ and still exhibit a universally consistent set of weights for $f(x)$. $E(Y^2) < \infty$ is a *nonvacuous* assumption (e.g., Y-distributed Cauchy violates it), and it is also an assumption which goes beyond the observable. Notice, too, that by abandoning the classical assumptions, we also give up the MLE and unbiasedness results of Theorem 4.11, retaining only the (universal) consistency desirable-property result. The loss of the MLE property is especially serious, because under the classical assumptions, $\hat{a}$ ($\hat{b}$) is a *unique* MLE of a (b). With only $E(Y^2) < \infty$, we have universally consistent weights that are *not* unique. So an element of arbitrariness or nonobjectivity reenters the choice of point estimate with only the $E(Y^2) < \infty$ assumption.

Notice that $E(Y^2) < \infty$ is similar to the $\sigma^2 < \infty$ assumption of Section 4.3.2. Also, with the abandonment of the classical assumptions here, an element of nonobjectivity reenters the choice of point estimate just as it did the choice of test with the abandonment of the classical assumptions in Section 4.3.1. Another similarity between the test and point-estimate results is their asymptotic nature. In Section 4.3.2, under Axiom 4.6, we could say only that $\alpha_{\phi_1}(n, F)$ or $\alpha_{\phi_3}(n, F)(\beta_{\phi_1}(n, F, \delta)$ or $\beta_{\phi_3}(n, F, \delta))$ approaches $\alpha(\int_{C_\alpha^1}^\infty g_\delta(z)\, dz)$ as n approaches ∞ for any F such that $\alpha^2 < \infty$; we could not state the value of the significance level or power of ϕ_1 or ϕ_3 for our actual n. The *rate* at which $\alpha_{\phi_1}(n, F)$ or $\alpha_{\phi_3}(n, F)(\beta_{\phi_1}(n, F, \delta)$ or $\beta_{\phi_3}(n, F, \delta))$ approaches $\alpha(\int_{C_\alpha^1}^\infty g_\delta(z)\, dz)$ depends on the underlying unknown F. Here we can say only that $(\Sigma_{i=1}^n W_i Y_i - f(x))^2$ approaches 0 as n

approaches ∞, not what the value of $(\Sigma_{i=1}^{n} W_i Y_i - f(x))^2$ for our actual n is. Again, the *rate* at which $(\Sigma_{i=1}^{n} W_i Y_i - f(x))^2$ approaches 0 depends on the unknown underlying $f(x)$ and $D(Y, X)$.

4.4   Summary and conclusion

We began by arguing that in modern statistical inference, Hume's problem of induction becomes the problem of justifying the assumptions about the "population(s)" from which the observations are presumed to have come. The semantic maneuver of calling a "population" a "probability distribution" makes the problem no more tractable. A limiting relative-frequency account (LRFA) of "probability distribution" and "independent" supposes that certain limits of infinite sequences exist and that those limits have a certain relationship to each other. Also, a LRFA of "*normal* probability distribution" requires the observations to stem from an infinite set of possible observations. In those two ways, then, the assumptions at the base of modern statistical inference go beyond what is observable.

An account of "normal probability distribution" that does *not* rely on an infinite set of possible observations is possible, but such an account still relies on an assumption that goes beyond the observable in attaching what we write about to what we can observe. We defined a concept of test and two kinds of errors: rejecting the hypothesis when it is true, and accepting it when it is false. The Neyman–Pearson theory caps the probability of an error of the first kind by convention at level α and provides a test that minimizes the probability of an error of the second kind (β) for an allowed alternative, among all level-α tests. We offered an argument for choosing an unbiased test and examined the classical two-sample situation of independent normal observations. For this situation, the Neyman–Pearson theory provides a unique test for the difference in population means, which gives minimum β against *any* allowed alternative, among all unbiased level-α tests.

With the normality assumption weakened to continuity, several rank-based level-α tests are popular, but none maximizes β against *any* allowed (shift) alternative, among all unbiased level-α rank tests with power a monotonically increasing function of shift. So an element of nonobjectivity enters the choice of test. Further, to *calculate* β for the rank-based level-α test chosen, one must restore the normality assumption (or make one which is equally strong). Switching to a one-sample situation, when normality is weakened to $\sigma^2 < \infty$, we chose a test for which we can calculate what the α-level and β approach as n (= the number of observations) approaches infinity. To know the α-level and β of the test for any actual finite n requires us to know the value of the underlying distribution's σ^2 and absolute third moment.

Finally, in a situation where the observations are *pairs* presumed generated by a bivariate probability distribution $D(Y, X)$, the classical regression assumptions include a linear conditional mean and normal conditional distribution. These assumptions result in a unique point estimate for the conditional mean with three desirable properties, one of which is consistency [i.e., the square of the difference between the expected value of the estimate and the conditional mean approaches 0 as n (= the number of observations) approaches ∞]. When the classical assumptions are replaced by the assumption of a finite second moment for Y, a (universally) consistent estimate exists, but it is not in general unique. Table 4.1 depicts Chapter 4's conclusions.

In reconstructing Mill's methods, Mackie discovered that each method depended on an assumption and an observation. If the assumption was weakened, the observation had to be strengthened to get the method's result. From our reconstruction of modern statistical methods, we learn a parallel lesson: Each modern method requires observations and (classical) assumptions to get a result. But if the (classical) assumptions are weakened, rather than requiring stronger observations to get the same result, our modern methods simply provide weaker results. Mackie concluded: "By making explicit the need for assumptions we have abandoned any pretense that these methods themselves solve or remove the problem of induction" (1974, p. 316). We conclude more strongly: In the modern methods, the problem of induction becomes the problem of justifying the assumptions about the population, and this problem has no solution that does not go beyond the observable.

Table 4.1. *The logic of modern statistical inference: some examples*

	Neyman–Pearson hypothesis testing		Point estimation: one sample of pairs from $D(Y, X)$
	Two-sample situation (same variance)	One-sample situation	
Further assumptions	1. Normal distributions	3. $\sigma^2 < \infty$	5. $E(Y|X = x) = a + bx$; $Y - a - bX$ normal, with constant variance $\sigma^2 < \infty$; $p(X)$ doesn't depend on a, b, σ^2; X and $Y - a - bX$ independent
	2. Continuous distributions	4. $\sigma^2 < \infty$; third absolute moment both known	6. $E(Y^2) < \infty$
Hypothesis	1. No shift	3. Mean $= 0$	
	2. No shift	4. Mean $= 0$	
Alternative	1. Shift in one direction	3. Mean > 0	
	2. Shift in one direction	4. Mean > 0	

Result

1. Unique, uniformly most powerful unbiased level-α test

2. Nonunique, unbiased level-α rank tests with monotonically increasing functions; exact power calculation requires reintroduction of strong distributional assumption like normality

3. Test the limit of whose significance level and power, as number of observations approaches infinity, can be calculated

4. Test for which bounds on significance level and power for actual number of observations may be calculated

5. Unique maximum-likelihood estimators of a and b, which are unbiased and consistent

6. Nonunique, (universally) consistent estimator of $E(Y|X = x)$

Notes: (1) Assumptions for all examples: independent observations from probability distributions. (2) All assumptions, on limiting relative-frequency account of "probability distribution," go beyond the observations. (3) (Universal) consistency is only an asymptotic property of estimators, i.e. it is a feature of the limit of an estimator as the number of observations approaches infinity.

Summary and conclusion of Part I

Chapter 1 took classical notions of causality and probability and fused them into a concept of gambling-device law (GDL). We also argued that the proposition that a GDL governs a given situation cannot, in principle, be based solely on what is observable. On the basis of the GDL concept, some assumptions based on the physical act of randomization, and some further assumptions not baseable on the observations, Chapter 2 constructed a theory of causal point estimation for statistical analysis of data from controlled experiments. We also argued that extensions of the theory to tests of hypotheses of no causal effects and no mean causal effect entail conceptual difficulties.

Chapter 3 argued that special features of social experiments may make Chapter 2's theory of randomization-based causal point estimation inapplicable to them. We also argued that the choice of a complex experimental design and any (unnecessary) failure to account for the complexity in subsequent statistical analysis could prevent Chapter 2's theory of causal inference from applying to a social experiment's results. Chapter 4 developed a second concept of probability (not baseable solely on observation) and constructed a theory of statistical inference (in which randomization has no role to play) based on the second probability concept and other assumptions not baseable on the observations. A social experiment's statistical analysis may be based on the second theory, but if so, randomization becomes a superfluous procedure, and the relationship between the experiment's results and propositions on cause and effect becomes unclear.

The philosopher Hilary Putnam has recently argued: "If any problem has emerged as *the* problem for analytic philosophy in the twentieth century, it is the problem of how words 'hook onto' the world" (1984, p. 265). Put another way, the problem of induction and causality with which Hume and Mill grappled in the nineteenth century has become the problem of how we can attach what we write about to what we can observe (and to what is). Part I has examined this problem's manifestation in the

114

statistical foundations of the modern controlled experiment. If there were going to be a place in such experiments where observations alone were sufficient for making inferences or attaching our words to the world, we would expect it to be in the experiments' statistical foundations.

Chapters 1, 2, and 3 considered one set of words or assumptions connected with a controlled experiment, and Chapter 4 considered a second set. Though these two sets of words or assumptions often confusingly coexist in the same experiment description, we argued that they could be distinguished from each other. The first perspective (the counterfactual causal one) takes the observations as dependent, whereas the second perspective (modern statistical inference proper) takes them as independent. The counterfactual causal perspective leads to problematic hypothesis-testing, whereas if philosophers who claim that explaining causality requires counterfactual conditionals are right, the connection between modern statistical inference proper and causality is a problem.

The counterfactual causal perspective hooks its words "equiprobable distribution" onto the world through the physical act of randomization. But the perspective's inferential statements in point-estimating and testing hypotheses connect to the observations with words and assumptions (in addition to "equiprobable distribution") that have no physical-acts basis, and some of which, at least, go beyond the observable. For example, the perspective's inferential statements suppose that the observations are governed by a deterministic law of working – an ontological, not empirical, assumption. Thus, attachment of the perspective's words to the world remains problematic.

Modern statistical inference proper hooks its words "probability distribution" onto the world through the words of a limiting relative-frequency explanation. The words of such an explanation include some like these: "the limit as the number of repetitions approaches infinity of the proportion of outcomes falling below any value exists." The difficulty for such an explanation is that though we can write about an infinite number of repetitions, we can neither observe nor imagine, in the sense of "watch in the mind's eye," an infinite number of repetitions.[21] Thus, the connection of the words of modern statistical inference proper to the world, contrary to the doctrine of empiricism, goes beyond the observable.

By making explicit and carefully scrutinizing the assumptions of both statistical perspectives, we have arrived at the conclusion that the notion

---

[21] That is, the author has never been able to "see" an infinite number of repetitions in *his* "mind's eye." Readers will have to decide for themselves the capacity of the "mind's eye."

that modern statistics has somehow solved the problem of induction with which Hume, Mill, and other philosophers wrestled is false. By repeatedly seeking a pure observation world in the statistical foundations of controlled experimentation, we have been forced to the conclusion that none exists. We have found that inference requires knowledge, assumptions, perhaps theory, that are irreducible to the observations and that the weaker the assumed knowledge, the weaker the inferential conclusion.

In Part I, then, we have characterized empiricism as a program of knowing the world on the basis of the observable alone. Pointing to the statistical assumptions that go beyond the observable in recent controlled social experiments, we argued that an empiricist program is not realized in these (or other) controlled experiments. Our immanent critique of empiricism, then, suggests another program: knowing the world on the basis of observations *and* theory. We call such a program rationalism. Rationalism rather than empiricism has been the research program of those fields of inquiry (e.g., physics) that most agree are science. Further, though statistics (and statistical theory) play a role in such scientific inquiry, the theory that makes physics, for example, a science (and its program of inquiry, rationalism) is not statistical theory. We turn in Part II to an immanent critique of recent controlled social experiments as *science* experiments, realizing a rationalist program of social inquiry.

PART II

Economic logics

On difficulties with the confrontation of theory and observations in controlled economic experiments

CHAPTER 5

Problems with a rationalist account of classical mechanics

5.1 Introduction: Some philosophers on the logic of science and classical mechanics

5.1.1 *From Mill to Popper*

Our ultimate aim in Part II is to understand the sense in which microeconomic studies such as the income maintenance experiments may perhaps be science experiments, or realizations of a rationalist program of social inquiry. To achieve that aim, it will help to first develop a concept of physical science with which to compare our income maintenance examples. Just as in Chapters 2 and 4 we focused on the *logics* of statistical inference, here we shall consider the *logic* of classical mechanics experiments.[1] We develop our view of the logic of such experiments by distillation of the views of J. S. Mill, Karl Popper, and Thomas Kuhn on science generally and through close scrutiny of some examples from classical mechanics.

In explaining a feature of his Deductive Method, Mill offers an embryonic rationalist account of science:

. . . that of collating the conclusions of the ratiocination with the concrete phenomena themselves, or, when such are obtainable, with the empirical laws. The ground of confidence in any concrete deductive science is not the *a priori* reasoning itself, but the accordance between its results and those of observation *a posteriori*. (1973, pp. 896–7)

On Mill's account, then, one begins (at least on reconstruction) with a set of propositions, deduces some conclusions, and compares the deduced

---

[1] We choose to compare microeconomics with classical mechanics rather than with some other field because few dispute that classical mechanics is a scientific theory, and our training enables us to give a fairly detailed account of classical mechanics. Also, the most prominent philosophical literature on the nature of science often begins with classical mechanics, and microeconomics does have some important similarities to classical mechanics (as we shall demonstrate in Chapters 6 and 7).

conclusions with observations or perhaps with observed regularities (= "empirical laws"). So on Mill's view, a science experiment seems unproblematic.

Probably the dominant rationalist view of the logic of science in the twentieth century has been that of Sir Karl Popper. Popper summarizes his view in an account of testing a new theory against a current theory:

> Here too the procedure of testing turns out to be deductive. With the help of other statements, previously accepted, certain singular statements – which we may call "predictions"– are deduced from the theory; especially predictions that are easily testable or applicable. From among these statements, those are selected which are not derivable from the current theory, and more especially those which the current theory contradicts. Next we seek a decision as regards these (and other) derived statements by comparing them with the results of practical applications and experiments. (1968, p. 33)

Notice that if there is no current theory, Popper's account is close to Mill's. Both accounts stress the deductive element, and Popper's "comparing them with the results of . . . experiments" is close to Mill's "collating the conclusions of the ratiocination with the concrete phenomena themselves." But Popper's "with the help of other statements, previously accepted" begins to suggest where problems may lie, though Popper doesn't explicitly acknowledge any problems.

Popper argues that no "absolutely certain, irrevocably true statements" (1968, p. 37) can come out of science as he sees it, but that "if the conclusions have been *falsified,* then their falsification also falsifies the theory from which they were logically deduced" (1968, p. 33). Failure to falsify, though, merely calls for the temporary, conventional acceptance of the theory until the next test – it does *not* establish that the theory is "true." He notes: "Nothing resembling inductive logic appears in the procedure here outlined. I never assume that we can argue from the truth of singular statements to the truth of theories" (1968, p. 33). Finally, he makes a bold claim for his rationalist account of science: "all the problems can be dealt with that are usually called *'epistemological.'* Those problems, more especially, to which inductive logic gives rise, can be eliminated without creating new ones in their place" (1968, pp. 33–4). If Popper's claim is correct, then his account of science resolves the sort of problems that we argued in Chapter 4 plague the inductive logic of modern statistical inference.

5.1.2 *From Popper to Kuhn*

In the last quarter century, Popper's view of science has come under increasing attack by a wide range of philosophers of science. Thomas Kuhn, the man who started the tumult, offers some conditions for a mature science that are subtly at odds with the accounts of Mill and Popper:

First . . . for some range of natural phenomena concrete predictions must emerge from the practice of the field. Second, for some interesting sub-class of phenomena, whatever passes for predictive success must be consistently achieved. . . . Third, predictive techniques must have roots in a theory which, however metaphysical, simultaneously justifies them, explains their limited success, and suggests means for their improvement in both precision and scope. (1972b, p. 245)

Notice that Kuhn retains Popper's terms "theory" and "prediction." But whereas for Popper the predictions "are deduced from the theory," for Kuhn they "emerge from the practice of the field." Kuhn replaces Popper's (and Mill's) "deduced from" with the weaker "justifies" or "have roots in." Finally, note that "test" has disappeared from the lexicon in moving from Popper's account to Kuhn's account.

Kuhn speculates whether or not Popper's account of science may be salvaged as a demarcation criterion between a scientific theory and nonscientific theory. Sticking to *statements alone,* Kuhn ponders the following criterion: ". . . a theory is scientific if and only if *observation statements* – particularly the negations of singular existential statements – can be logically deduced from it, perhaps in conjunction with stated background knowledge. . . . To be scientific a theory need be falsifiable . . . by an observation statement" (1972a, p. 14). (Notice that Popper's "other statements, previously accepted" has become "stated background knowledge.") Kuhn is suggesting that Popper's account of science *conceivably* deserves a similar (but not quite identical) logical status of formal criterion or axiom as Mill's canon for the Method of Difference (Section 1.2.1) or the probability axioms (4.1 or 4.2).

But Kuhn is skeptical that all scientific theories can be put in strict deductive (or axiomatic) form: "I doubt that scientific theories can without decisive change be cast in a form which . . . this version of Sir Karl's criterion requires" (1972a, p. 15). But even if they can, Kuhn stresses that for *knowledge* to result from the scientific enterprise requires "that both the epistemological investigator and the research scientist be able to relate sentences derived from a theory not to other sentences but to actual observations and experiments" (1972a, p. 15). And he prescribes that "philosophers of science will need to follow other contemporary philosophers in examining, to a previously unprecedented depth, the manner in which language fits the world, asking how terms attach to nature" (1972a, p. 15). On Kuhn's view, then, contrary to Popper's claim, Popper's account of science not only fails to resolve but even fails to grapple with "*the problem for analytic philosophy in the twentieth century*" (see Section 4.4.2).

We consider in this chapter some examples from classical mechanics – a field that most observers agree is a science. By looking closely at these examples, we seek a more concrete understanding of Kuhn–Popper areas of disagreement and a partial (at least) synthesis of their views. Also, we aim to extend Kuhn's Popper criterion of demarcation of a scientific theory into an account of the logic of classical mechanics experiments to be compared (in Chapter 6) with the logic of microeconomic studies such as the income maintenance experiments. Before turning to our examples from classical mechanics, though, we must try to establish the logical or epistemological status of that theory's core: Newton's Second Law.

5.1.3   *Core of the theory: Newton's Second Law*

We shall formulate each of our examples from classical mechanics first in such a way as to see if the theory can be made to satisfy Kuhn's Popper criterion of a scientific theory (see Section 5.1.2). Our approach in the first formulation will be quasi-axiomatic – we shall state a theorem which links the theory to prediction statements. In this formulation, such terms as "particle," "force," "mass," "charge," "position," "velocity," and "time" will remain largely unexplained. Once a theorem has been formulated, we shall try to see how an observation statement containing previously unexplained terms might be attached to the observable in such a way that observations (not just the observation statement) can test (in some sense) the theory. Here we may need to reformulate the theorem and still may fail to get observations (rather than just an observation statement) to confront one of the theory's prediction statements. We begin our account with an axiom.

Axiom 5.1 (Superposition of forces).   *If* **F** *and* **F*** *are any forces acting on a particle, then* **F** *and* **F*** *satisfy the standard linear algebra axioms of a vector space.*

The axiom allows us to define the notion of the net force acting on a particle.

Definition: The *net force* acting on a particle is the vector sum of forces acting on the particle.

We regard the axiom's epistemological status as uncertain. All we can say is that satisfaction of the axiom is a necessary (but not sufficient) property of a force. The definition is, of course, an analytic truth.

We aim at an account of classical mechanics that both satisfies Kuhn's Popper criterion of a scientific theory *and* has Newton's Second Law (NSL) as its core (i.e., as a central proposition from which observation statements are deduced). But NSL is also a deductive *consequence* of a prior principle. Consider a mechanical system consisting of a single particle whose (mass, position, time) is $(m, \mathbf{r}, t)$, and let $\mathbf{F}$ be the force acting on the particle. Also, let a dot over a term indicate differentiation with respect to time, and let "$\nabla \cdot$" be $[(\partial/\partial x) + (\partial/\partial y) + (\partial/\partial z)]$, where (x, y, z) are the Cartesian coordinates of $\mathbf{r}$.

Definition: If there exists a function $V(\mathbf{r})$ (called the *potential energy* of the particle) such that $\mathbf{F} = -\nabla \cdot V(\mathbf{r})$, then $\mathbf{F}$ is a *conservative* force.

Definition: If $\mathbf{F}$ is a conservative force whose associated potential energy is $V(\mathbf{r})$, then $\mathscr{L} = V(\mathbf{r}) - [m|\dot{\mathbf{r}}|^2/2]$ is the system's *Lagrangian,* and $\mathscr{S} = \int \mathscr{L}(\dot{\mathbf{r}}, \mathbf{r}, t) \, dt$ is the system's *action.*

Suppose next that $\mathbf{r}(t)$ satisfies the following condition.

Hamilton's Principle (HP). $\mathbf{r}(t)$ minimizes $\mathscr{S}$.

We have, then, a theorem.

Theorem 5.1. *HP implies* $\mathbf{F} = m\ddot{\mathbf{r}}$ *(i.e., NSL).*[2]

So NSL becomes a deductive consequence of HP or the principle of least action. Notice that HP is no more intuitive than is NSL. We may say that HP means that nature (or perhaps God) minimizes the action. But in doing so, we discover no new reasons for believing NSL.

---

[2] Landau and Lifshitz (1976, pp. 2–5) provide a proof of Theorem 5.1.

So the problem of the epistemological status of NSL arises. Kuhn points out "that symbolic expression [$'f = ma'$] is . . . a law-sketch rather than a law. It must be written in a different symbolic form for each physical problem before logical and mathematical deduction are applied to it" (1972b, p. 272). So Kuhn sees $'f = ma'$ as a placeholder. In a mechanical system of a single particle, suppose that $\mathbf{r}(t)$, $m(\dot{\mathbf{r}})$, and $\mathbf{F}(\mathbf{r}, m, \dot{\mathbf{r}}, t)$ are placeholders for the particle's position, mass, and force functions, respectively. Then the "law-sketch" $"f = ma"$ is the placeholder $\mathbf{F}(\mathbf{r}, m, \dot{\mathbf{r}}, t) = m(\dot{\mathbf{r}})\ddot{\mathbf{r}}(t)$.

$\mathbf{F}(\mathbf{r}, m, \dot{\mathbf{r}}, t) = m(\dot{\mathbf{r}})\ddot{\mathbf{r}}(t)$ is a *differential-equation* placeholder. That is, in any physical problem, if we suppose $\mathbf{F}(\mathbf{r}, m, \dot{\mathbf{r}}, t)$ and $m(\dot{\mathbf{r}})$ to have *particular* forms, then $\mathbf{F}(\mathbf{r}, m, \dot{\mathbf{r}}, t) = m(\dot{\mathbf{r}})\ddot{\mathbf{r}}(t)$ becomes the differential equation of motion of the particle. [For example, in the case of a falling body near the earth's surface, $\mathbf{F}(\mathbf{r}, m, \dot{\mathbf{r}}, t)$ and $m(\dot{\mathbf{r}})$ might be $m\mathbf{g}$ and m, respectively – with m and $\mathbf{g}$ constants, so that $\mathbf{F}(\mathbf{r}, m, \dot{\mathbf{r}}, t) = m(\dot{\mathbf{r}})\ddot{\mathbf{r}}(t)$ would become the differential equation of motion $\mathbf{g} = \ddot{\mathbf{r}}(t)$.] The differential equation of motion is *deterministic*. That is, given a set of initial conditions [e.g., the values of $\mathbf{r}(t)$ and $\dot{\mathbf{r}}(t)$ at time t_1], the equation in principle "predicts" the value of $\mathbf{r}(t)$ at any time $t_2 > t_1$. In fact, though, only for *some* particular forms $\mathbf{F}(\mathbf{r}, m, \dot{\mathbf{r}}, t)$ and $m(\dot{\mathbf{r}})$ will the differential equation's solution be obtainable in closed form.

The philosopher W. Stegmüller has amplified Kuhn's insight on NSL as follows:

The law is represented by a simple formula containing three functions. The formula expresses a universal sentence. This representation immediately leads to the question: "Is it an empirical law or an *a priori* truth?" Numerous answers have been given, from "It is an elementary analytic truth, namely a mere definition of force" at one end of the spectrum to "It is an empirically falsifiable hypothesis" at the other. I maintain all of these answers are wrong. We can say even more: this whole discussion concerning the epistemological status of the second law is just nonsense. The sterile dispute has its roots in the misleading way of stating this law *as an isolated and "self-contained" universal sentence.* (1979, p. 53).

If Stegmüller is right, then we should attempt to formulate NSL as something that is *not "an isolated and 'self-contained' universal sentence."* Trying to respect the Kuhn–Stegmüller insights, we take NSL roughly as follows: Whatever the particular force ($\mathbf{F}(\mathbf{r}, m, \dot{\mathbf{r}}, t)$), mass ($m(\dot{\mathbf{r}})$), and position ($\mathbf{r}(t)$) functions of the particle in the particular example under consideration, $\mathbf{F}(\mathbf{r}, m, \dot{\mathbf{r}}, t) = m(\dot{\mathbf{r}})\ddot{\mathbf{r}}(t)$.

For a particular example under consideration, our NSL appears to be a statement that a priori could be true or false and hence not an analytic truth. Yet because of its non-"self-contained" nature, our NSL for the particular example is not an empirically falsifiable hypothesis (EFH) either, for a necessary (but perhaps not sufficient) condition for a statement to be an EFH is that the statement imply a "prediction" statement, and NSL *in isolation* does *not* imply one. But suppose in the particular example under consideration that "the particular mass function is such and such – e.g., *m*" and "the particular force function is so and so – e.g., *m*g" are *M** and *F**, respectively. Then the *conjunction* of NSL, *F**, and *M** implies a "prediction" statement. Thus, (NSL and *F** and *M**) *may* be an EFH.

When a mechanical system consists of *two* particles of (mass, position) $(m_1, \mathbf{r}_1)$ and $(m_2, \mathbf{r}_2)$, then the *i*th particle's force- and mass-function placeholders are $\mathbf{F}_i(\mathbf{r}_1, \mathbf{r}_2, m_1, m_2, \dot{\mathbf{r}}_1, \dot{\mathbf{r}}_2, t)$ (or $\mathbf{F}_i$ for short) and $m_i(\dot{\mathbf{r}}_i)$ (or m_i for short), respectively, for $i = 1, 2$. So, for such a system, NSL will be: Whatever the particular force $(\mathbf{F}_i)$, mass (m_i), and position $(\mathbf{r}_i(t))$ functions of the particles in the particular example under consideration, $\mathbf{F}_i = m_i\ddot{\mathbf{r}}_i(t)$ for $i = 1, 2$. The assumption M (= in the particular example under consideration, all mass functions are constant) will be made for each example in Sections 5.2 and 5.3. Thus, the account to which we now turn is of *nonrelativistic* (classical) mechanics.

5.2 Motion of earthly and heavenly bodies

5.2.1 *One particle, no force: Newton's First Law*[3]

The simplest case is that of a single particle with no force acting on it.

Theorem 5.2 (Newton's First Law). *If* $\mathbf{F}(\mathbf{r}, m, \dot{\mathbf{r}}, t) = \mathbf{0}$, *and M and NSL hold, then* $\dot{\mathbf{r}} = constant$.

Notice that *m* has vanished from the prediction statement (= theorem's conclusion) $\dot{\mathbf{r}}$ = constant. To test the theory (= theorem's hypothesis), we would want to compare the prediction statement $\dot{\mathbf{r}}$ = constant with what is observed. But we don't observe the instantaneous velocity $\dot{\mathbf{r}}(t)$ of the particle at any time *t*, but only its average velocity between two observed positions.

So let the positions of the particle at *t* and $t + \Delta t$ be $\mathbf{r}$ and $\mathbf{r} + \Delta\mathbf{r}$, respectively. Then we reformulate.

---

[3] Section 5.2.1 formalizes a portion of Goldstein (1959, pp. 1–4).

Theorem 5.3

> *(i)   If* $F(r, m, \dot{r}, t) = 0$, *and M and NSL hold, then* $\dot{r} = constant,$
>
> *(ii) $\dot{r} = constant$ implies* $(\Delta r / \Delta t) = constant.$

In fact, to keep things simple, suppose motion is constrained to a single dimension x (e.g., imagine a wheeled cart on a "frictionless" track). Let the cart be observed in positions x_1, $x_1 + \Delta x_1$, x_2, $x_2 + \Delta x_2$ at times t_1, $t_1 + \Delta t_1$, t_2, $t_2 + \Delta t_2$ (in temporal order), respectively. The new theorem is as follows.

Theorem 5.4

> *(i)   If* $y = y_0 = constant$, $z = z_0 = constant$, $F(r, m, \dot{r}, t) = 0$, *and M and NSL hold, then* $\dot{x} = constant,$
>
> *(ii) $\dot{x} = constant$ implies* $(\Delta x / \Delta t) = constant,$
>
> *(iii) $(\Delta x / \Delta t) = constant$ implies* $(\Delta x_1 / \Delta t_1) = (\Delta x_2 / \Delta t_2).$

Letting $\bar{v}_i = (\Delta x_i / \Delta t_i)$ for $i = 1, 2$, the theory's prediction statement becomes $\bar{v}_1 - \bar{v}_2 = 0$. With a perfect observation or reporting system, the theory's prediction statement becomes $r^*(\bar{v}_1) - r^*(\bar{v}_2) = 0$, where $r^*(\ )$ means "the observed or reported value of." Then if $r^*(\bar{v}_1) - r^*(\bar{v}_2)$ were not 0, we could conclude: the reporting system was *not* perfect, or y or $z \neq constant$, or $F \neq 0$, or M is false, or NSL is false. Alternatively, with an additive measurement or reporting error ϵ, with $E(\epsilon) = 0$ and some statistical distributional assumptions (of the Chapter 4 variety), the theory's prediction statement becomes: $r^*(\bar{v}_1) - r^*(\bar{v}_2)$ is distributed as ϵ. Repeating the experiment of sending the cart down the track and taking the four (position, time) pairs of measurements n times allows a "test" of the prediction statement in the Chapter 4 statistical sense. Rejecting the prediction statement would then mean concluding: y or $z \neq constant$, or $F \neq 0$, or M is false, or NSL is false, or $E(\epsilon) \neq 0$, or ϵ is not distributed as supposed, or an unusual event occurred.

Our first example suggests that the sort of deductive statement-level form of Kuhn's Popper criterion may be within reach in classical mechanics at least. It also suggests that with certain attachment assumptions (e.g., perfect reporting or a statistical model of the reporting system) we may even be able to preserve a stronger sense of "deductive testing" or "falsification" (in classical mechanics) than the mere statement-level sense. To do so, however, we need to regard what is potentially falsifiable not as what we would wish (i.e., NSL) but rather as the *conjunction* of NSL, a constraint, a particular force form, M, and the attachment assumptions.

It also must be admitted that making actual observations (not merely an observation statement) confront the theory's prediction statement requires brushing past certain previously unmentioned attachment problems. For example, to know whether or not a track is "frictionless" seems to require assuming the very theory allegedly being tested – an apparent circularity. Further, we don't really believe any system in the world is "frictionless"– this is an idealization. Taken literally, then, $F = 0$ would lead us to *always* reject the conjunction of the constraint, $F = 0$, M, NSL, and the attachment assumptions on the basis of an observation-world cart-on-a-track experiment. In fact, though, the physicist *accepts* the conjunction on the basis of such an experiment. Apparently, then, we need to modify $F = 0$ to $F \approx 0$ and (say in the perfect-reporting case) $r^*(\bar{v}_1) - r^*(\bar{v}_2) = 0$ to $r^*(\bar{v}_1) - r^*(\bar{v}_2) \approx 0$ if we are to succeed in making the theory's prediction statement confront the actual observations. And we appear to need an explicit or implicit convention on what constitutes "≈ 0." For now, though, we put aside these troubling attachment problems and turn to a system of two particles and no (net) force to see if there are any further such problems.

5.2.2 *Two particles, no (net) force: the law of conservation of linear momentum*[4]

In Section 5.1.3 we stated NSL for a two-particle system as $F_1 = m_1\ddot{r}_1$ and $F_2 = m_2\ddot{r}_2$. Now suppose $F_i = F_{i\text{ext}} + F_{ji}$ for $i = 1, 2$ and $j \neq i$, where $F_{i\text{ext}}$ is the external force acting on particle i and F_{ji} is the (internal) force on particle i due to particle j. The theorem then is the following.

Theorem 5.5 (Law of conservation of linear momentum). *If M and NSL hold, if $F_{1\text{ext}} = F_{2\text{ext}} = 0$ and $F_{12} = -F_{21}$ (Newton's Third Law), then $m_1\dot{r}_1 + m_2\dot{r}_2$ (the system's linear momentum) = constant.*

Notice that m_1 and m_2 *don't* vanish from the prediction statement (= theorem's conclusion) $m_1\dot{r}_1 + m_2\dot{r}_2$ = constant. To test the theory (= theorem's hypothesis), we would want to compare the prediction state-

---

[4] Section 5.2.2 formalizes a portion of Goldstein (1959, pp. 4–10).

ment $m_1\dot{r}_1 + m_2\dot{r}_2 =$ constant with what is observed. The first attachment problem is again that we don't observe the instantaneous velocity $\dot{r}_i$ of a particle at any time t, but only its average velocity between two observed positions. In fact, as the theorem presently stands, $\dot{r}_i$ need not be constant over *any* time interval.

Again imagine a "frictionless" track, in the x direction, with now *two* wheeled carts moving and colliding via a "massless" spring attached to one of them. Let cart i be observed before (after) the collision in positions x_i, $x_i + \Delta x_i$ (x_i', $x_i' + \Delta x_i'$) at times t_i, $t_i + \Delta t_i$ (t_i', $t_i' + \Delta t_i'$) in temporal order, respectively. Then we reformulate.

Theorem 5.6. *If $y = y_0 =$ constant, $z = z_0 =$ constant, M and NSL hold, $F_{1ext} = F_{2ext} = 0$, $F_{12} = -F_{21}$, and $F_{21} \approx 0$ (except during the collision), then $m_1(\Delta x_1/\Delta t_1) + m_2(\Delta x_2/\Delta t_2) \approx m_1(\Delta x_1'/\Delta t_1') + m_2(\Delta x_2'/\Delta t_2')$.*

Notice that dealing with the first attachment problem now requires a convention on what constitutes ≈ 0 (see the previous section).

If $\bar{v}_i = \Delta x_i/\Delta t_i$ and $\bar{v}_i' = \Delta x_i'/\Delta t_i'$ for $i = 1, 2$, the theory's prediction statement becomes: $m_1(\bar{v}_1 - \bar{v}_1') + m_2(\bar{v}_2 - \bar{v}_2') \approx 0$. As in the previous section, we could assume perfect reporting [or an additive measurement error ϵ, with $E(\epsilon) = 0$, and some statistical distributional assumptions], and rejection of the new prediction statement would still mean concluding that a conjunction of statements including NSL was false (or an unusual event occurred). So our second example also suggests that the deductive statements-level form of Kuhn's Popper criterion is realizable as long as one is satisfied with the a priori possibility of falsifying only a *conjunction* of statements including NSL. And our second example possesses all the attachment problems of our first (apparent circularity in assuming the track is frictionless, need for ≈ 0 convention, etc.).

There is, however, one attachment problem present in the second example, but absent from the first example: How can we attach the *masses* of the prediction statement to the world of observations? One approach is to use two "identical" carts (i.e., made of the same material, in the same mold, etc.) and reformulate the theorem.

Theorem 5.7. *If $y = y_0 =$ constant, $z = z_0 =$ constant, M and NSL hold, $F_{1ext} = F_{2ext} = 0$, $F_{12} = -F_{21}$, $F_{21} \approx 0$ (except during the collision), and $m_1 = m_2$, then $(\bar{v}_1 - \bar{v}_1') + (\bar{v}_2 - \bar{v}_2') \approx 0$.*

The disadvantage of this approach is simply that it fails to deal with the unequal-masses case.

Alternatively, we might tie the masses together at rest, with the fully compressed spring between them, burn the string through, and take the ratio of the two carts' resulting average velocities as the ratio of their masses. Such an approach, though, seems circular in the sense that it (at least implicitly) appears to assume the theorem's hypotheses that we seek to test. Finally, we could take relative *weights* (as measured by a balance scale) as relative masses. Again, though, there is an *almost* circular aura: assuming the theory ($= M$, NSL, and earth–cart gravity force) in another situation to test it in the situation of our theorem's hypotheses. Though its attachment problems remain troubling, we put aside our second example for now to explore for further such problems in an example from astronomy.

5.2.3   Newton's Law of Universal Gravitation and Kepler's laws[5]

Consider again a two-particle system, and let $\mathbf{r} = \mathbf{r}_2 - \mathbf{r}_1$, $|\mathbf{r}| = r$. As before, suppose that $\mathbf{F}_{21} = -\mathbf{F}_{12}$, but now also suppose that $|\mathbf{F}_{12}| = Gm_1m_2/r^2$ for some G constant over time and the same for any two particles, and that the *direction* of $\mathbf{F}_{12}$ is from particle 2 toward particle 1 along $\mathbf{r}$ (i.e., that $\mathbf{F}_{12}$ is attractive). With the convention that an attractive force has a minus sign, we can write $\mathbf{F}_{12} = [-(d/dr)(Gm_1m_2/r)](\mathbf{r}/r)$. (We call this particular force relation Newton's Law of Universal Gravitation, NLUG.) Also let θ be the angle made by $\mathbf{r}$ with respect to some arbitrary 0 line through m_1 (see Figure 5.1A). Next we require a definition.

Definition

(a) $\mu = m_1m_2/(m_1 + m_2)$ is the system's *reduced mass,*

(b) $V(r) = Gm_1m_2/r$ is the system's *potential energy,*

(c) $\lambda(t) = \mu r^2(t)\dot{\theta}(t)$ is the system's *angular momentum* (with respect to its center of mass),

(d) $T^+(t) = (\frac{1}{2})\mu(\dot{r}^2(t) + r^2(t)\dot{\theta}^2(t))$ is the system's *kinetic energy* (with respect to its center of mass),

[5] Section 5.2.3 formalizes a portion of Goldstein (1959, pp. 76–80).

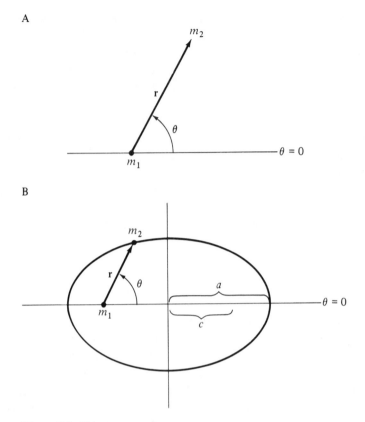

Figure 5.1. Planetary motion.

(e) $E^*(t) = V(r(t)) + T^+(t)$ is the system's *energy,* and

(f) $dA(t)/dt = (\tfrac{1}{2})r^2(t)\dot{\theta}(t)$ is the system's *areal velocity* or the area swept out by $\mathbf{r}(t)$ per unit time.

Next we require two lemmas, the first from analytic geometry.

Lemma 5.1. *The polar equation of an ellipse is* $r = (1 - e^2)/(1 + \cos(\theta))$, *where a* = *half major axis, c* = *distance from center to focus, and e* = *c/a (Figure 5.1B).*

Lemma 5.2. *If M and NSL hold,* $\mathbf{F}_{1\text{ext}} = \mathbf{F}_{2\text{ext}} = \mathbf{0}$, $\mathbf{F}_{12} = -\mathbf{F}_{21}$, $\mathbf{F}_{12} = -(Gm_1m_2/r^2)(\mathbf{r}/r)$, *then*

(i) $\lambda = \lambda_0 = $ *constant over time (conservation of angular momentum)*,

(ii) $E^* = E_0^* = $ *constant over time (conservation of energy)*.

The theorem then is the following.

Theorem 5.8. *If M and NSL hold, if* $\mathbf{F}_{1ext} = \mathbf{F}_{2ext} = 0$, $\mathbf{F}_{12} = -\mathbf{F}_{21} = -(Gm_1m_2/r^2)(\mathbf{r}/r)$, *and* $E^* < 0$, *then*

(i) $r = (\lambda_0^2/\mu Gm_1m_2)/(1 + (2E_0\lambda_0^2/\mu(Gm_1m_2)^2)\cos(\theta))^{1/2}$; *that is, particle 2 traces out an ellipse, with particle 1 at one focus and* $e = (1 + (2E_0\lambda_0^2/\mu(Gm_1m_2)^2)^{1/2}$, $a = Gm_1m_2/2E_0^*$ *(Kepler's First Law)*,

(ii) $dA(t)/dt = $ *constant over time (Kepler's Second Law)*,

(iii) *if* τ *is the time period for one circuit of the ellipse, then* $\tau = 2\pi a^{3/2}(1/G(m_1 + m_2))^{1/2}$ *(Kepler's Third Law)*.

Let m_1 be the sun, and m_2 a planet. An astronomer observing the planet orbiting the sun might be content to see a near ellipse and merely modify the theorem's hypothesis $\mathbf{F}_{iext} = 0$ to $\mathbf{F}_{iext} \approx 0$ and/or $\mathbf{F}_{12} = -(Gm_1m_2r^2)(\mathbf{r}/r)$ to $\mathbf{F}_{12} \approx -(Gm_1m_2/r^2)(\mathbf{r}/r)$ to get (i)'s shape-prediction statement to accord with the observed shape, within some "$\approx$" convention. Or one might try to account for the deviation from an ellipse that is observed; for example, observing a *precessing* ellipse, the astronomer might let $\mathbf{F}_{12} = -[(Gm_1m_2/r^2) + (c/r^3)](\mathbf{r}/r)$ to give a new shape-prediction statement of a precessing ellipse. In either case, the astronomer leaves the theory's core – M and NSL – intact, but tries by altering other theorem hypotheses or through an attachment convention for "$\approx$" to get the theory's orbital-shape-prediction statement to match the observed orbital shape. In neither case, though, are the predicted values of e and a in (i) tested.

Now suppose the planet is observed at $(r_1, \theta_1, t_1), \ldots, (r_4, \theta_4, t_4)$, where the t_i's are in temporal order and $t_2 - t_1 = t_4 - t_3$. Then the theorem's prediction statement (ii) implies $(r_1 + r_2)^2(\theta_2 - \theta_1) \approx (r_3 + r_4)^2(\theta_4 - \theta_3)$. Thus, we again get a deductive consequence of the theory to confront the observations through an attachment convention for "$\approx$." Next, suppose we observe *two* planets orbiting the sun, and assume that neither sun–planet system affects the other, so that the theorem's hypotheses hold for each system separately. Let m_1 be the sun's mass, m_2 and m_3 the planets' masses, and τ_{1j} and a_{1j} the period and semi-major axis of planet j's orbit for $j = 1, 2$. Then the theorem's (iii) implies $(\tau_{12}/\tau_{13})^2 = (a_{12}/a_{13})^3[(m_1 + m_3)/(m_1 + m_2)]$. So if we add to the theorem's hypothesis the condition m_2, $m_3 \ll m_1$, then $(\tau_{12}/\tau_{13})^2 \approx (a_{12}/a_{13})^3$ is a consequence of (iii), and so again we get the observations to confront a prediction statement.

Notice that in this example we have managed to get prediction statements to confront observations with some conventions on "$\approx$" in the prediction statement (and/or theorem hypothesis). Also, whereas in the previous section we needed (m_1/m_2) "known" from a situation where the theory was assumed to hold, here we merely suppose m_2, $m_3 \ll m_1$. As before, what is a priori falsifiable is merely the *conjunction* of NSL and the theorem's other hypotheses [e.g., in the case of (iii)'s consequence $(\tau_{12}/\tau_{13})^2 \approx (a_{12}/a_{13})^3$, rejection means that M is false, or NSL is false, or some external forces are not zero, or $\mathbf{F}_{12} \neq -(Gm_1m_2/r^2)(\mathbf{r}/r)$ for some universal constant G, or $\mathbf{F}_{12} \neq -\mathbf{F}_{21}$, or E_0^* for one of the two systems is not <0, or the sun–two-planet system cannot be regarded as two separate sun–planet systems, or m_2, m_3 is not $\ll m_1$]. Let us turn next to an example of gravity at the earth's surface.

5.2.4   *Falling bodies near the earth's surface*

We now wish to study another situation where the same gravitational force as in the previous section is at work: the motion of a falling body near the earth's surface. Let m_1 be the mass of the earth, and m_2 that of the falling body. Let r be the distance from the center of the earth to m_2, and we suppose that we can view m_1 as concentrated at its center for purposes of understanding its interaction with m_2. Also, let R be the radius of the earth, let z^* be the height of m_2 above the earth's surface, so that $r = R + z^*$, and let x^*–y^* be a plane perpendicular to $\mathbf{r}$ and tangent to the earth.

Theorem 5.9.   *If M and NSL hold, if $\mathbf{F}_{1\text{ext}} = \mathbf{F}_{2\text{ext}} = 0$, $\mathbf{F}_{12} = -(Gm_1m_2/r^2)(\mathbf{r}/r)$, $x^* = x_0^* = $ constant, $y^* = y_0^* = $ constant, $h = $ height from which m_2 is dropped, and z^*, $h \ll R$, then while m_2 is still above the earth, with $[Gm_1/(R + z^*)^2] \approx (Gm_1/R^2)$,*

(i) $E^* = m_2z^{*2}/2 + m_2gz^* \approx E_0^* = $ constant $= m_2gh$,

(ii) $h - z^* = (\frac{1}{2})gt^2$, where t is the time elapsed from the dropping of m_2.

It is prediction statement (ii) that we make confront the observations. Let m_2 be dropped from a tower of height h and observed twice en route to earth: at $(t_1, h - z_1^*)$ and $(t_2, h - z_2^*)$, where the t_i's are in temporal order and $h - z^*$ is measured from height h. Then prediction statement (ii) implies $[(h_1 - z_1^*)/(h - z_2^*)] \approx (t_1^2/t_2^2)$. So we can get (ii) to confront the observations with only a convention of "$\approx$." Of course, as always, we might also assume nonperfect measurement. For example, we might take the prediction statement to be tested (in the sense of Chapter 4) as $[(h - z_{1i}^*)/(h - z_{2i}^*)] - (t_{1i}/t_{2i})^2 \approx \epsilon_i$ for $i = 1, \ldots, n$ in n repetitions of the experiment, where ϵ_i is a measurement-error term, $E(\epsilon_i) = 0$, and ϵ_i has some assumed distribution. Then we are a priori capable of concluding only that one of the theorem's hypotheses (which include NSL) is false, or $E(\epsilon) \neq 0$, or the distributional assumptions are false, or an unusual event occurred. Consider finally an example where once again a gravity force is involved.

5.2.5   The motion of a pendulum

As in the previous section, we consider here the motion of a body near the earth's surface, but now the body will not be free to fall, but will be a pendulum bob (i.e., a point mass m_2 suspended from a "massless" string). Again let m_1 be the mass of the earth, and r the distance from the center of the earth to m_2, and suppose m_1 may be viewed as concentrated at its center for understanding the motion of m_2. Also, let R be the radius of the earth, let z^* be the height of m_2 above the earth (so that $r = z^* + R$), and let x^*–y^* be a plane perpendicular to $\mathbf{r}$ and tangent to the earth. Finally, let L be the length of the string, $\mathbf{L}$ its directed length, and β the angle it makes with the vertical (Figure 5.2). Let τ be the pendulum's period [i.e., the time until the (first) return to any point in its motion]. Let $\mathbf{F}_{12} = -(Gm_1m_2/r^2)(\mathbf{r}/r)$ be the gravity force on m_2 due to m_1, and suppose $\mathbf{T} = (Gm_1m_2/r^2)(\mathbf{L}/L)\cos(\beta)$ is the force on m_2 due to the tension in the string. In this situation, NSL for the pendulum becomes $m_2\ddot{\mathbf{r}} = \mathbf{F}_{12} + \mathbf{T}$. Finally, we suppose that if $\beta_{\max}$ is the largest that β gets, then $\sin(\beta_{\max}) \approx \beta_{\max}$. Our theorem then is the following.

Theorem 5.10.   *If M and NSL hold, if* $\mathbf{F}_{12} = -(Gm_1m_2/r^2)(\mathbf{r}/r)$, $\mathbf{T} = -\mathbf{F}_{12}(\mathbf{L}/L) \cos(\beta)$, $x^* = x_0^* = constant$, $y^* = y_0^* = constant$, $z^* \ll R$, *and* $\beta_{\max} \approx \sin(\beta_{\max})$, *then with* $[Gm_1/(R + z^*)^2] \approx (Gm_1/R^2) = g$, $\tau \approx 2\pi(L/g)^{1/2}$.

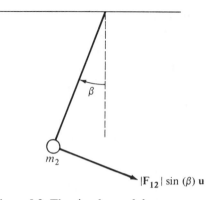

Figure 5.2. The simple pendulum.

Notice that the theorem's prediction statement is that the pendulum's period is proportional to (length of the string)$^{1/2}$, but does not depend on m_2.

Now consider a set of pendula with varying masses and string lengths. If (τ_1, L_1) and (τ_2, L_2) are the (period, string length) for any two pendula, then the theorem's prediction statement implies $(\tau_1/\tau_2) \approx (L_1/L_2)^{1/2}$. Just as in the previous section, we need a convention for "$\approx$" here and probably would want a probabilistic measurement model. What is new here is the approximation method. The hypotheses of the theorem [without $\beta_{max} \approx \sin(\beta_{max})$] imply that the vector sum of forces acting on m_2 is $|\mathbf{F}_{12}|\sin(\beta)\mathbf{u}$, where $\mathbf{u}$ is a unit vector perpendicular to $(\mathbf{L}/L)$ (Figure 5.2). With $|\mathbf{F}_{12}|\sin(\beta)\mathbf{u}$ as the force in placeholder NSL, the resulting differential equation's solution is rather complicated. But with $\beta_{max} \approx \sin(\beta_{max})$, we can approximate the conjectured force $|\mathbf{F}_{12}|\sin(\beta)\mathbf{u}$ with $|\mathbf{F}_{12}|\beta\mathbf{u}$, and then the differential equation's solution becomes relatively simple. So we use a mathematical fact [$\beta \approx \sin(\beta)$ for small β] and the assumption that β is small (which we can test *directly* by observation) to approximate a particular hypothesized force function.[6] Let us next consider some atomic and subatomic phenomena.

---

[6] With the small-angle approximation $\beta_{max} \approx \sin(\beta_{max})$, the theorem's prediction statement was $\tau \approx 2\pi(L/g)^{1/2}$. Without the approximation, it would have been $\tau = 2\pi(L/g)^{1/2}$ $[1 + (\frac{1}{2}^2)\sin^2(\beta_{max}/2) + (1 \cdot 3^2/2^2 4^2)\sin^4(\beta_{max}/2) + \ldots]$ (Landau and Lifshitz, 1976, pp. 26–7). The simple pendulum equation's approximate solution is an example of a class of small-oscillation techniques in classical mechanics (Landau and Lifshitz, 1976, Chapter V). Similar approximation methods are employed in certain wave mechanical problems (Dicke and Wittke, 1960, Chapter 14). These approximation techniques all involve the same basic approach. A *particular* conjectured force or potential is first approximated

5.3 Classical mechanical phenomena of atomic and subatomic origin

5.3.1 The ideal-gas law[7]

We suppose (i) that an "ideal" or "dilute" gas consists, on the microscopic level, of a collection of N identical particles, "moving bodies" (= molecules), or equal "point masses" m. In this situation, NSL becomes $m\ddot{\mathbf{r}}_i = \mathbf{F}_{ij} + \mathbf{F}_{i\text{ext}}$ for $i \neq j$ for $i = 1, \ldots, N$. At the macroscopic level, let there be R^* moles of a gas in a container of volume V^*, and suppose the temperature of the gas is T^* and the pressure it exerts on the container's walls is P. On the microscopic level, we suppose further that (ii) $\mathbf{F}_{i\text{ext}} = \mathbf{0}$ for all $i = 1, \ldots, N$, (iii) $\mathbf{F}_{ij} = \mathbf{0}$ in the absence of a collision, (iv) $\mathbf{F}_{ij} = -\mathbf{F}_{ij}$ (along the line between particles i and j) during a collision (Newton's Third Law), and (v) collisions between particles and between particles and the container's walls are perfectly elastic and of negligible duration.

Also, we assume (vi) that the average distance between molecules is large compared with their size (i.e., we indeed idealize "point masses" or a "dilute" gas), (vii) that at any given point in time the empirical distribution of the velocity $\dot{\mathbf{r}}$ of one of the molecules has the property $E(x^2) = E(|\dot{\mathbf{r}}|^2)/3$ for any direction $\mathbf{r}$, and (viii) that T^* is proportional to $NmE(|\dot{\mathbf{r}}|^2)$. We then have the following theorem.

Theorem 5.11. *If M, NSL, and (i)–(viii) hold for the identical particles of a gas, then $PV^* \approx KR^*T^*$ (the ideal-gas law), where K is a universal constant, which is the same no matter which "real" "ideal" or "dilute" gas we are considering.*

How can we make the theorem's prediction statement confront some observations here? Suppose we have two containers of variable volumes, each containing a different ideal gas. If we measure (P, V^*, R^*, T^*) once for each container, then the theorem's conclusion implies $P_1 V_1^* \approx P_2 V_2^*$. Notice that we already need a "$\approx$" attachment convention because of "$\approx$" conventions implicit in (iv) and (v). We might want to modify "perfect reporting" to a system with additive measurement-error terms ϵ,

---

with a mathematical relationship known to hold over the observed range of values of the force's or potential's argument. Then the approximate force or potential equation is substituted into the Newton's Second Law (or Schrödinger equation) placeholder, and a solution to the resulting (approximate) differential equation is deduced.

[7] Section 5.3.1 formalizes a portion of the material in Chapters 3 and 4 of Huang (1963).

η, with $E(\epsilon - \eta) = 0$ and $\mathrm{Var}(\epsilon) = \mathrm{Var}(\eta)$. For example, we might take a series of n (m) intentionally varied measurements $P_{1i}V^*_{1i}/R^*_1 T^*_{1i}$ $(P_{2j} V^*_{2j}/R^*_2 T^*_{2j})$ on the first (second) container, supposing, say, that $[P_{1i}V^*_{1i}/R^*_1 T^*_{1i}]$ $([P_{2j} V^*_{2j}/R^*_2 T^*_{2j}]) \sim N(K_1, \sigma^2) (N(K_2, \sigma^2))$. Then we would test the hypothesis (= prediction statement here) that $K_1 = K_2$ in the manner of Section 4.2.5. Rejection would mean that one or more of M or NSL or (i)–(viii) or the measurement model assumptions were false or that an unusual event occurred.

Notice that Theorem 5.11, as we have stated it, does not predict a particular value for K. Once we had tested in the manner prescribed the consequence of the theorem, though, if we failed to reject the prediction statement that $K_1 = K_2$, we could use, say, the n observed $(P_{1i}V^*_{1i}/R^*_1 T^*_{1i})$'s to interval-estimate $K_1 = K$ (= Boltzmann's constant $\times$ Avogadro's number) in standard statistical fashion. Such a step would be an empirical filling out of the theory. But it would not be empiricism, because it would respect, and be compatible with, tested theory. Having considered an example whose particles were the not directly observable molecules of a dilute gas, we turn now to some whose particles are those supposed to constitute the structure of a molecule's atom.[8]

5.3.2   The atom's nucleus and Rutherford scattering[9]

Suppose (i) that an atom consists of a point nucleus (mass, charge) (m_2, q_2) and (mass, charge) $(m, -q_2)$ (with $q_2 > 0$) of unspecified distribution in motion outside the nucleus. We are interested in predicting the motion of an α-particle (ii) supposed to be a point (mass, charge) (m_1, q_1) (with $q_1 > 0$) passing through such an atom. To deduce this motion, we suppose further (iii) that when the α-particle is outside the atom, the effect of the atom on its motion is negligible, and (iv) when the α-particle is inside the atom, the presence of $(m, -q_2)$ has a negligible effect on its motion. Our system then consists of two point (mass, charge)'s (m_1, q_1) and (m_2, q_2), with $q_1, q_2 > 0$.

Since the two particles are *charged,* they will have *electromagnetic* (as well as possibly gravitational) internal forces acting on them. On the

---

[8] Notice that in this section we have ignored some of the attachment difficulties involved in observing temperature and pressure. These difficulties appear to be of the same logical sort we discuss in connection with other examples.

[9] Section 5.3.2 formalizes a portion of Goldstein (1959, pp. 81–5).

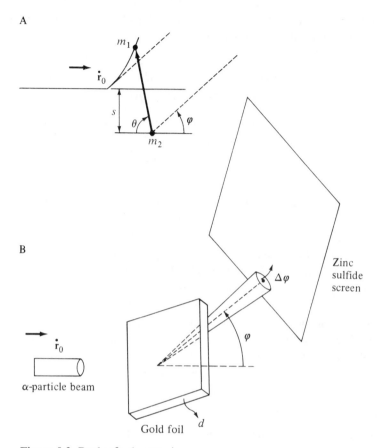

A

B

Zinc
sulfide
screen

α-particle beam

Gold foil d

Figure 5.3. Rutherford scattering.

classical view, *all* the forces acting on the particle must obey the place-holder differential equations of classical mechanics – NSL – and the electromagnetic forces must obey the laws of classical electrodynamics – Maxwell's Equations (ME). In particular, by Coulomb's Law (= the first of ME), $\mathbf{F}_{ji}$ (= the electromagnetic force on particle i due to particle j) will be $(q_1 q_2/r^3)\mathbf{r}$ (repulsive, because $q_1 q_2 > 0$), where $\mathbf{r}$ is the vector between the two particles, and $r = |\mathbf{r}|$. Also by ME, an accelerated charge will radiate energy.

Suppose further: (v) $\mathbf{F}_{exti} = \mathbf{0}$, (vi) $|Gm_1 m_2| \ll |q_1 q_2|$, $m_1 \ll m_2$ (so that m_2 will remain roughly stationary), (vii) m_1's initial velocity is $\dot{\mathbf{r}}_0$ along a line passing a distance s from m_2 (Figure 5.3A), and (viii) $(m_1 |\dot{\mathbf{r}}_0|^2/2) \gg (q_1 q_2/r_0)$, E^*_{Rad1} (= the radiation from charge 1 over the time of the motion). Also let φ of Figure 5.3A be the angle of scatter. Then we have the following theorem.

Theorem 5.12. *If M, NSL, and ME hold, and if (i)–(viii) hold for the two point (mass, charge)'s (m_1, q_1) and (m_2, q_2), then*

(a) $(1/r) \approx -[q_1 q_2/\mu|\dot{\mathbf{r}}_0|^2 s^2](1 + e \cos(\theta))$, *where* $e =$ $(1 + (\mu|\dot{\mathbf{r}}_0|^2 s/q_1 q_2)^2)^{1/2}$ *(i.e., m_1 traces out a hyperbola)*,

(b) $\varphi \approx 2 \cos^{-1}[(1 + (q_1 q_2/\mu|\dot{\mathbf{r}}_0|^2 s)^2)^{-1/2}]$.

In fact, we cannot observe a single α-particle passing a single nucleus; so we must alter the theorem's hypothesis to try to get its prediction statement to confront what we *can* observe. So consider a thin piece of gold foil of thickness d and a beam of α-particles (N_0 per unit cross section) incident perpendicular to the foil. Also suppose that a zinc sulfide screen that registers N_φ (the number of α-particles per unit area hitting a small element of solid angle of width $\Delta\varphi$) sits perpendicular to the direction of motion of the scattered particles at φ (Figure 5.3B). Finally, add three further assumptions: (ix) the empirical distribution of gold nuclei in the foil is uniform, (x) the α-particles (gold nuclei) are identical, and (xi) d is sufficiently small that a negligible number of α-particles pass through more than one atom. Then we have a revised theorem.

Theorem 5.13 (Rutherford scattering). *If M, NSL, and ME hold, and if (i)–(xi) hold for the α-particles and gold nuclei point (mass, charge)'s, then N_φ is approximately proportional to $N_0 d|\dot{\mathbf{r}}_0|^4 \sin^4(\varphi/2)$.*

The observations then can be made to confront the theorem's prediction statement by varying N_2, d, $|\dot{\mathbf{r}}_0|$, and/or the angular location φ of the screen. Notice that for attachment, we suppose that the screen registers N_φ, which assumes that the very ME of the theorem's hypothesis is true in another situation. Also, remember we cannot *see* the α-particles tracing their hyperbolas here as we could see the moving particles of Sections 5.2.1–5.2.5. So playing a more obvious role in attachment here than in the earlier examples (where it was also in fact present) is the picture of the motion in the mind's eye created by the theorem's (and its explanation's) words and the accompanying figure. Consider next the atom's heretofore neglected negative charge.

5.3.3 *The hydrogen atom: A classical model*[10]

Suppose (i) that the hydrogen atom consists of two point (mass, charge)'s – a nucleus (m_2, e^*) and an electron (m_1, $-e^*$) (with $e^* > 0$). Again suppose: (ii) $\mathbf{F}_{exti} = \mathbf{0}$, (iii) $|Gm_1 m_2| \ll e^{*2}$, and (iv) $m_1 \ll m_2$. Also, on the classical view, the particles' motion should be governed by NSL and ME. The Coulomb force is an attractive one – $\mathbf{F}_{ji} = -(e^{*2}\mathbf{r}/r^3)$ – and so has the same *form* as the gravity force of Section 5.2.4 (with e^{*2} replacing $Gm_1 m_2$). Thus, with the added assumption $E_0^* < 0$, if $|E_{Rad1}^*|$ were $\ll |E_0^*|$, the two particles would move in accordance with Kepler's laws of motion (see Theorem 5.8), with some changes in the values of the constants. In fact, we shall assume (v) that $E_0^* = -(\mu e^{*2}/2\lambda_0^2)$. So if $|E_{Rad1}^*|$ were $\ll |E_0^*|$, m_1 would trace out a circular orbit around m_2. But though $|E_{Rad\,1}^*| \ll |E_0^*|$ was plausible in the case of the hyperbolic motion of short time duration in the previous section, it is implausible in the case of extended-duration periodic motion.

So let ΔE_{Rad1}^* be the energy radiated during a single orbital cycle from time t to $t + \Delta t$, and suppose (vi) that $|\Delta E_{Rad\,1}^*| \ll |E^*(t)|$. Let $\tau = (2/3)(e^{*2}\mu c^3)$, where c is the speed of light. Then we have a theorem.

Theorem 5.14. *If M, NSL, and ME hold, and if (i)–(vi) hold for* (m_1, $-e^*$),

(a) $r(t') \approx$ *constant for any* $t' \in [t, t, + \Delta t]$,

(b) $r^3(t) \approx r^3(0) + 9[e^{*2}(c\tau^3)/\tau]t$, *so that if* $t \approx t^* = [r^3(0)\tau/9e^{*2}(c\tau^3)]$, $r^3(t^*) = 0$,

(c) *if* $v(t)$ *is the frequency of the radiation the system emits, then* $v(t)$ *is a monotonically increasing continuous function of* t *on* $[0, t^*]$.

So (a) predicts a nearly perfect circle in a single orbit, (b) predicts instability or eventual collapse of the atom, and (c) predicts emission of a continuous range of energy frequencies. The mind's-eye picture is of a micro sun–planet-like system where the planet performs roughly circular orbits while slowly spiraling into the sun – emitting a continuous range of energy frequencies in the process.

To get the theorem's prediction statement (c) to confront the observations, we dissociate hydrogen molecules in a discharge tube and analyze

---

[10] Section 5.3.3 formalizes a portion of Jackson (1962, pp. 581–4, problem 17.2 on p. 608).

the frequency of the light emitted with a spectroscope. With the attachment assumption of ME of the theorem's hypothesis in the situation of the spectroscope's analysis of the emitted light, what we observe is a *discrete* rather than the predicted *continuous* sequence of frequencies. Thus, some subset of the hypotheses of Theorem 5.14 appear *not* to characterize the hydrogen atom. Our classical model of the hydrogen atom seems inadequate in the sense that one of its prediction statements contradicts the observations.

5.3.4   *Bohr's model of the hydrogen atom: toward quantum wave mechanics*[11]

In the previous section we saw that a set of classical mechanical assumptions about the hydrogen atom's structure produced a prediction statement that contradicted the observations. So let us retain the previous section's assumptions (i)–(v), but scrap the last three of ME, retaining only Coulomb's Law (CL). And let us consider two further assumptions: (vi') $\lambda = nh^*/2\pi$ for positive integer n and Planck's constant h^*, and (vii') when the system radiates energy, $v = (W_1 - W_2)/h^*$ (the Einstein Frequency Condition, EFC), where W_1 is the initial and W_2 the final energy. Also, let $R^+ = 2\pi^2 \mu e^{*4}/ch^{*3}$. We then have a theorem.

Theorem 5.15 (Bohr hydrogen atom).

(a)   *If M, NSL, CL, (i)–(v), and (vi') hold, then*
   (1) $r = r_n = n^2 h^{*2}/4\pi e^{*2}$,
   (2) $E^* = E_n^* = -2\pi^2 \mu e^{*4}/h^{*2}n^2$.

(b)   *If M, (v), (vi'), and (vii') (or EFC) hold, then $v = R^+[(1/n_2^2) - (1/n_1^2)]$ as E^* goes from $E_{n_1}^*$ to $E_{n_2}^*$ ($n_1 > n_2$); in particular, $v_n \approx R^+[(\frac{1}{4}) - (1/n^2)]$ (Balmer's formula) as E^* goes from E_n^* to E_2^* for $n \geq 3$.*

The theorem suggests a certain mind's-eye attachment picture. By (a), the electron first orbits the nucleus in a circle characterized by n, in motion governed by NSL and CL. When the electron moves from a circle characterized by n to one characterized by $m < n$, NSL and CL are suspended, and the frequency of the radiation emitted is governed by EFC [see (b)]. Notice that the theorem leaves as mysteries when and why a transition from one orbit to another occurs, as well as the nature of

---

[11] Section 5.3.4 formalizes material from Dicke and Wittke (1960, pp. 10–13).

motion between orbits. Notice also that $n \geq 1$; so r of Bohr's hydrogen atom does *not* collapse to 0. Finally, (b) predicts *discrete* observed emission frequencies, which, with the previous section's spectroscopic attachment assumption, is what we got.

As a further refinement, we wish to know if the predicted frequencies of Balmer's formula are in accordance with the observations. First we refine our spectroscopic attachment assumption: The distance between any two observed spectral lines is proportional to the difference between the inverses of their frequencies. Then Balmer's formula implies that when we dissociate hydrogen molecules in a discharge tube and analyze the frequency of the light emitted with a spectroscope, there will be a series of spectral lines, and for the $U_1 < U_2 < U_3$th lines of the series,

$$[d(U_1, U_2)/d(U_2, U_3)]$$
$$\approx [[U_2(U_2 + 2)/U_1(U_1 + 2)] - 1]/[1 - [U_2(U_2 + 2)/U_3(U_3 + 2)]],$$

where $d(U_i, U_j)$ is the distance between the U_i and U_jth lines of the series. With a convention for "$\approx$," such a series is in fact observed.

Historically, Bohr's model of the hydrogen atom was one of a series of accounts that were developed in roughly the period 1900–25 to explain observations originating at the subatomic level that contradicted the predictions of classical mechanics. Each of these accounts entailed three sorts of assumptions foreign to classical mechanics: (i) energy comes in *discrete* quanta, (ii) EFC governs the system's radiation, and (iii) NSL is suspended during radiation. Planck, Einstein, and Bohr, for example, accounted for the photoelectric effect, black-body radiation, and hydrogen-atom spectral emissions, respectively, with such assumptions. Still, observations arose that physicists were unable to account for by merely suspending NSL. Bohr, for example, was unable, despite many attempts, to extend his model to account for spectral emissions of higher elements. In 1926, Schrödinger took a radical step: replacement of NSL with another placeholder differential equation in accounts of observations originating at the subatomic level, and modern wave mechanics was born. We turn now to some generalizations from our classical mechanical examples.

5.4   Summary and conclusion

5.4.1   *The formalization*

We began by defining what might be called Kuhn's challenge: to construct a scientific theory as statements from which "prediction" statements (falsifiable by observation statements) may be deduced *and* to show how the theory's statements can be attached to the world of actual observations. We took up Kuhn's challenge for the case of nonrelativistic classical (or Newtonian) mechanics. Our response to the first part of the challenge is the formalization of classical mechanics represented by the theorems of Sections 5.2 and 5.3. We take the theorems, or perhaps just the theorem hypotheses, as the theory statements of classical mechanics. Each theorem has as its referent a particular physical-world example. The theorem hypotheses all have two statements (which constitute the theory core) – M and NSL – in common. But each theorem hypothesis contains at least two other statements – one stating the example's force function and one specifying a reporting system. Most of the theorem hypotheses contain other statements as well. The theorems' conclusions are the theory's "prediction" statements. Figure 5.4A depicts the theory's structure.

Kuhn–Popper suggest a criterion, or necessary condition, for a scientific theory: "Prediction" statements (falsifiable by observation statements) must be logically deducible from the conjunction of the theory and "other statements previously accepted" (Popper) or "background knowledge" (Kuhn). Our account of classical mechanics exhibits the deductive element of the Kuhn–Popper criterion, but differs slightly from the criterion in two respects. First, the criterion fails to, whereas our account does, allow for a scientific theory as a collection of distinct theorems, each with a physical-world example referent.[12] Second, in place of the criterion's theory and "other statements previously accepted," our account has theory core and other statements of theorem hypothesis. But we do *not* take these other statements as previously accepted. Rather, within a given theorem, our other statements have precisely the same logical or epistemological status as the theory core. They differ from the theory core only in possibly not appearing in every one of the theory's theorem hypotheses.

Finally, our account's differences from the Kuhn–Popper criterion imply differences in *what* is understood as falsified when an observation statement contradicts a prediction statement. On the criterion, because the theory and the other statements jointly imply the prediction state-

---

[12] Our account shares the notion of a theory as a set of examples with the point of view that Stegmüller labels "structuralism." See, for example, Stegmüller (1979, p. 27). Of course, our account also seeks to formalize the formalizable elements of Kuhn's (1970) concept of "paradigm": "some accepted examples of actual scientific practice – examples which include law, theory, application, and instrumentation together" (p. 10).

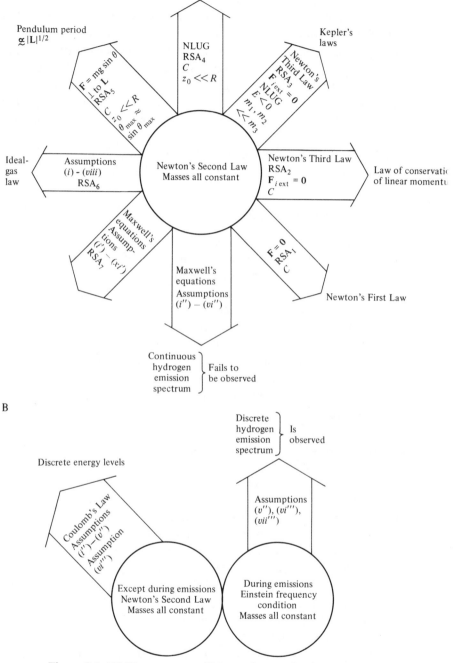

Figure 5.4. (A) The structure of Newtonian mechanics. (B) Bohr's hydrogen atom. RSA_i, reporting system assumptions; C, constraint that x, y are constants; NLUG, Newton's Law of Universal Gravitation; E, total energy; z_0, initial height; R, earth's radius; L, pendulum's directed length; θ_{max}, L's maximum angle with the vertical.

ment, *and the other statements are previously accepted,* the theory has been falsified. On our account, though, because the other statements in any theorem are *not* previously accepted, what has been falsified is only the theorem's hypothesis, not necessarily the theory's core. Further, because the theorems do not largely entail *each other,* then even if the theory's core were false for a particular example, it does not logically follow that the theory core is false for other examples. Thus, for example, the contradiction by an observation statement of a prediction statement implied by the classical model of the hydrogen atom (Section 5.3.3) does not imply that NSL is false in the theorems of Section 5.2.

5.4.2   *Attachment difficulties*

Our response to the second portion of Kuhn's challenge – to show how a theory's statements can be attached to the world of actual observations – has been to generate a catalogue of attachment difficulties. Popper himself cites our first difficulty: "[Newton's First] law, even though no body in this system moves in accordance with it, does not lose its significance within the solar system" (1964, p. 101). Thus, in *every* example, because of an idealization in the statement of the theorem's hypothesis and/or because of an assumption like $m \ll M$ in the theorem's hypothesis, we are forced to adopt a convention for when observations are "$\approx$" or "approximately proportional to" in order to make actual observations confront prediction statements.

Another sort of attachment difficulty was the apparent circularity of having to assume NSL in order to test it. This difficulty arose, for example, in the need to measure the ratio of two masses in the example of the law of conservation of linear momentum (Section 5.2.2). An apparent solution to the circularity difficulty is to add a preliminary calibration experiment or procedure, which *in itself* is not a test, to the example under consideration. Thus, for the example of conservation of linear momentum, we might first assume that NLUG acts between each cart and the earth and take the relative *weights* (as determined by a gravity balance) of the two carts as their relative masses. Then, finding the prediction statement of the theorem of Section 5.2.2 false would imply that either the theorem's hypothesis or the calibration assumption (that NLUG acts between each cart and the earth) is false.

It is not completely clear that our proposed solution to the circularity difficulty is adequate. *Some* physicists might object to it on the grounds that it confuses two concepts that they regard as distinct: inertial and

Table 5.1. *Catalogue of difficulties with the statements account of classical mechanics*

1. Newton's Second Law not logically falsifiable
2. Need for "$\approx$" or "approximately proportional to" convention in prediction statements
3. Circularity difficulties
4. Complicated (or even intractable) differential equation(s) of motion
5. Value of constants not implied
6. Cannot observe the atomic–subatomic realm
7. Scientist's reporting may be imperfect

gravitational mass. *Some* philosophers argue that avoiding circularity requires something considerably more sophisticated than our simple calibration-experiment notion. Stegmüller, for example, argues that to deal with circularity difficulties, we need to replace the usual observational–theoretical dichotomy: "For, by itself it [= the observational–theoretical dichotomy] is the confused amalgam of two different dichotomies, i.e. the *epistemological* dichotomy 'observational–non-observational' and . . . the '*quasi-semantical*' dichotomy '*T*-theoretical–non-*t*-theoretical'." [13] Thus, we make no *strong* claim to have *solved* the circularity problem. For our argument it suffices to have identified the problem as an attachment difficulty.

A third difficulty arose in the case of the pendulum (Section 5.2.5): The hypothesized net force leads to a differential equation whose solution is rather complicated. The difficulty's solution was to consider only small angles and approximate the hypothesized force, using a known mathematical relationship. In the case of the ideal-gas law, we found that the theorem's prediction statement involved the existence, not the value, of a constant. Once the prediction statement was corroborated, though, the constant could be estimated from the statement and the observations. Finally, in the atomic and subatomic realms, where we cannot see the moving bodies, we resorted to mind's-eye pictures created by words and diagrams to help connect our theory's statements with the observable. Table 5.1 lists the epistemological difficulties of our account of classical mechanics.

---

[13] Stegmüller (1979, p. 80). Stegmüller (1979, Chapter 3) gives a fuller account of the two-dichotomies view (which he attributes to the philosopher J. D. Sneed) and its relationship to various kinds of circularity in classical mechanics.

Recall Popper's bold claim that within his falsification-oriented account of science, "all the problems can be dealt with that we usually call *'epistemological'."* At least the problems of the immunity of the theory core to strictly logical falsification and the need for the "$\approx$" and "approximately proportional to" conventions for attaching statements to observations (as well as perhaps circularity difficulties)[14] belie Popper's claim. Still, we must not push epistemological difficulties too far. For example, when Rutherford offered his account of scattering, he was obviously worried about the problem of the instability of the atom that follows from considerations like those in Section 5.3.3: "The question of the stability of the atom proposed need not be considered at this stage, for this will obviously depend upon the minute structure of the atom, and on the notion of the constituent charged parts" (1911, p. 671). Yet he was able to ignore the instability deductive consequence of his model not because of any epistemological problems of the sort we catalogued but rather because another deductive consequence of his model – the scattering pattern – seemed in accordance with the observations.

Within two years, however, Bohr considered the "difficulties of a serious nature arising from the apparent instability of the system of electrons" (1913, pp. 1–2) and proposed the solution of Section 5.3.4 – at least for the case of the hydrogen atom. Thus, scientists may for a while ignore the apparent incompatibility of the observations and statements deduced from their theories, but eventually they seem to view such incompatibility as a problem – even if logically arriving at the incompatibility is itself fraught with epistemological problems. Thus, physicists seem to somehow respect the statements deduced from their theories, despite the philosophical difficulties of attaching those statements to the world of observations.[15]

---

[14] If our calibration-experiment-type solution holds philosophical water and can be generalized to cover other classical mechanical circularity difficulties, then perhaps we should exclude circularities from our list of attachment difficulties.

[15] Our account of classical mechanics takes a test as a (possibly problematic) *"two-cornered fight between theory and experiment."* The philosopher I. Lakatos offers a more sophisticated account (which is an elaboration of Popper's view – see Section 5.1.1) of science in which "tests are – at least – three-cornered fights between rival theories and experiment." [Both preceding quotations are from Lakatos (1972, p. 115)]. We have stuck primarily to what can be said about a *single* theory versus experiment, because our ultimate aim is to understand some microeconomic experiments that involved no second theory.

Note that in taking a single-theory view, we need not consider such epistemological difficulties as the possible incommensurability of rival theories [see, for example, Steg-

5.4.3   *Backward to Chapter 4 and forward to Chapters 6 and 7*

Popper's claim that "nothing resembling inductive logic appears in the procedure here outlined. I never assume that we can argue from the truth of singular statements to the truth of theories" is certainly borne out in our account of classical mechanics, for the theory core of classical mechanics appears not in the *conclusion* of each theorem but in its hypothesis, so that the reasoning in the theorem is from the theory to a singular statement, not vice versa. But we should not make too much of the absence of "inductive logic" (in Popper's sense) from our account, for we have made ample room in our account for the reasoning of modern statistical inference in the manner of Chapter 4.

A set of statistical assumptions (of the Chapter 4 sort) about the entities of a classical mechanics theorem's prediction statement is added to the theorem's hypothesis as a reporting-system model and converts the prediction statement into a statistical hypothesis. The statistical assumptions are *simultaneously* a hypothesis of a (Chapter-4-type) theorem in statistical theory whose conclusion is that the test of statistical hypothesis (= prediction statement) has certain optimality properties. The observations then are marshaled as the statistics required by the statistical theory, and (ignoring epistemological difficulties) if the statistical hypothesis is rejected, we can logically conclude that the classical mechanics theorem's hypothesis is false or that the statistical assumptions are false or that an unusual event has occurred.

Once a classical mechanics theorem's prediction statement has been accepted, statistical estimation theory may be employed to estimate a physical constant. Kuhn (1970, pp. 27–8) included such constant estimation among the activities he originally labeled "normal science." Figure 5.5 depicts the role we give to statistical inference in a classical mechanics experiment. Notice that the two theories (physical and statistical) fit together in such a way that the deductive logic of each is left completely intact. The statistical theory merely augments and aids in attaching, but fails to dominate and alter the basic structure of, the physical theory.

---

müller's discussion (1979, Chapter 11)] such as classical and wave mechanics. It may, however, be true that by taking a single-theory view of mechanics and microeconomics, our accounts forgo the dissolution (or at least lessening) of certain epistemological difficulties. For example, in experiments for deciding between classical and wave mechanics (or a classical atom and Bohr atom – see Sections 5.3.3 and 5.3.4) there may be little need for "approximately proportional to" or "$\approx$" conventions (or for any statistical measurement model).

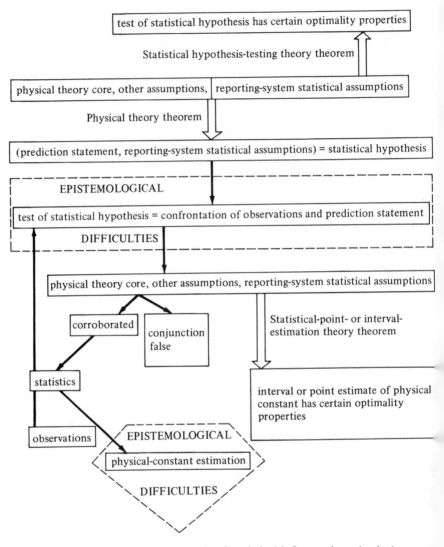

Figure 5.5. The logic and role of statistical inference in a classical mechanics experiment.

In Chapter 6 we shall seek to give an account of microeconomic theory and compare it with the present chapter's account of classical mechanics. We shall be looking for any differences between the two theories in theory structure, epistemological problems, and/or the role of statistical inference in the theory's experiments. We shall take it as a premise that classical mechanics is a science and seek in Chapter 7 to decide if any of the differences discovered in Chapter 6 mean that microeconomics is not a science, is not yet a science, or perhaps never will be able to be a science.

CHAPTER 6

Microeconomics striving to be a classical-mechanics–like science

6.1 Introduction: Addressing some possible objections

In the next two chapters we seek some tentative answers to questions about the nature of microeconomics as science. Is microeconomics science in the same sense as classical mechanics? If not, is it science in some other sense? Were the income maintenance experiments science experiments in some important sense? If microeconomics is not presently science, could it become science? What are the barriers to its becoming science? Does microeconomics face possibly insurmountable barriers to becoming science? Our approach to answering these questions will be first to attempt to reconstruct microeconomics, while noting differences between the developing reconstruction and the previous chapter's account of classical mechanics, and then try to decide if the noted differences demarcate a science from a nonscience. At least two sorts of objections can be raised to our questions and approach to answers.

The first sort of objection runs roughly as follows: Because microeconomics is in fact science like classical mechanics, our first question has been answered in the affirmative, and so our other questions don't apply. Consider, for example, a remark by Popper: "It must be admitted, however, that the success of mathematical economics shows that one social science at least has gone through its Newtonian revolution" (1964, p. 60). Yet Popper never details "the success of mathematical economics"; so his remark is merely a claim, without argument or supporting evidence, that mathematical economics is a science on a par with classical mechanics.

Mathematical microeconomics is, of course, a deductively formulated collection of propositions, but if such a collection were sufficient for a field to be a science, then mathematics itself would be a science (which it certainly isn't). On the previous chapter's Kuhn–Popper approach, to judge the possible scientific nature of a theory, one must first identify prediction statements deduced from it, then consider whether or not the prediction statements can be contradicted by observation statements, and finally consider how *actual* observations might be made to confront prediction statements. We take Popper's remark, then, not as an objection to our line of questioning but rather as a conjecture for which we seek evidence and argument, of a Kuhn–Popper criterion sort.

From the emergence of mathematical economics in the late 1940s until quite recently, many economists seemed to take it for granted that science was something like Popper's account and that economics was (or could at least aspire to be) science in that sense. Thus, for example, Milton Friedman argued in 1953 that

. . . the relevant question to ask about the "assumptions" of a theory is not whether they are descriptively "realistic," for they never are, but whether they are sufficiently good approximations for the purpose in hand. And this question can be answered only by seeing whether the theory works, which means whether it yields sufficiently accurate predictions. The two supposedly independent tests thus reduce to one test. (1953, p. 15)

Recall (from Section 5.1.1) that Popper saw some "assumptions" as "statements, previously accepted." So we need not suppose that he would concur with Friedman that one can *completely* ignore the question of whether or not a theory's "assumptions" are descriptively "realistic." Also, Friedman does not *explicitly* stress (as Popper does) the *deductive* element in going from theory to predictions. Still, Friedman understands science as involving predictions as a test of theory in Popper-like fashion.

Very recently, however, some economists have drawn a radical rejection of Popper's account of science from some post-Popper philosophy of science. For example, D. N. McCloskey argues that

The common claim that prediction is the defining feature of a real science, and that economics possesses the feature, is equally open to doubt. . . . Economics, in other words is not Science in the way we came to understand that term in high school. But neither, really, are other sciences. (1983, pp. 487, 491)

Notice that McCloskey not only questions Popper's view of science (see Section 5.1.1) but also questions the prediction-involving conditions (see Section 5.1.2) that Kuhn sees as "tantamount to the description of a good scientific theory." McCloskey also worries that economists might judge their work against "an ideal science" [1] (like the one of Popper's account) and end up with "the neurotic inhibitions of an artificial methodology of Science" (McCloskey 1983, p. 515). The second sort of objection to our approach runs roughly as follows: The notion that implicitly underlies the

---

[1] We employ the term "ideal science" here in place of McCloskey's "Science."

line of questioning – namely, that we have an adequate conception of science against which to measure microeconomics – is false.

It could be that prediction in no sense is the defining feature of a real science. But if our previous chapter's account captures the essence of classical mechanics, then there is at least one real science which involves prediction, albeit in a philosophically problematic sense. McCloskey is certainly right that accounts like Popper's are of "an ideal science" rather than of any real science. But it does not follow that an account of "an ideal science" cannot be of aid in judging whether or not a field is a real science. And McCloskey is sensible to worry that economists may be immobilized trying to get their work to measure up as "an ideal science" without realizing the limitations (which post-Popper philosophy of science has laid bare) of accounts of "an ideal science." But, again, it does not follow that economists should rule out questions like, In what sense is microeconomics science? as somehow illegitimate.

In the previous chapter, we found classical mechanics wanting as "an ideal science," but it did not follow that classical mechanics was not a real science. In the next two chapters we shall similarly find microeconomics not "an ideal science," but again it will not follow that microeconomics is not a real science. The logic of our argument, then, is not that microeconomics is not "an ideal science," and so microeconomics is not a real science. Rather, it is that classical mechanics, a real science, fails to be "an ideal science" in certain pinpointable ways (see Chapter 5). If microeconomics is *not* (a real) science, it might be because microeconomics fails to be "an ideal science" in ways other than those in which classical mechanics does.

Since World War II, microeconomic theory has actually undergone *two* successive mathematical revolutions. With the publication (originally in 1947) of Samuelson's 1965 book (actually completed before the war), the theory's propositions were expressed on the mathematical level of advanced calculus. Soon thereafter, Debreu's 1959 book weakened the assumptions in the advanced calculus approach and expressed the theory's propositions on the more abstract mathematical level of real analysis and topology. In this chapter, we develop microeconomic theory on the advanced calculus level, because both classical mechanics experiments and empirical economic studies such as the income maintenance experiments occur at the advanced calculus level. So as to avoid *forcing* microeconomics into a classical mechanics mold, Chapter 8 will contain an account of microeconomic theory at the mathematical level of real analysis and topology.

Section 6.2 is an account of microeconomic theory. We begin with the consumer theory core and then deduce some abstract features of individual and aggregate supply and demand functions. Then we introduce some particular but *hypothetical* utility and production functions which allow us to deduce some concrete prediction statements which can confront concrete observation statements. Section 6.3 considers the attachment difficulties which arise in trying to make the deduced prediction statements confront actual observations. We discover that each of the attachment difficulties of classical mechanics has a microeconomic counterpart, but also that there are some microeconomic difficulties without classical mechanics counterparts. Finally we examine an *actual* example from microeconometric practice that is not "ideal science," but may not be "real science" either. Section 6.4 is the usual summary and conclusion. The Appendix proves or sketches the proofs of the particular aggregate supply–demand and budget-share models that we employ.

6.2   Microeconomic theory

6.2.1   *The consumer theory core*

Our account of consumer theory relies on a series of words, including "consumer," "price," "income," "commodity," "tastes," "utility," and "satisfaction." We shall define some of these words (implicitly or explicitly) in terms of others. Perhaps still others have meanings we all share by virtue of our speaking the same ordinary language. Ultimately, though, the attachment of our words to the world is no less a philosophical problem in economics than it was in statistics and physics. Just as in our accounts of statistics (Chapters 2 and 4) and classical mechanics (Chapter 5), our interest is in the *logic* of consumer theory. Also as in the case of classical mechanics, we seek ultimately to put microeconomics in the form of theorems whose conclusions are prediction statements.

Consider a consumption sector consisting of two commodities and, for the moment, just one consumer. Let (q_1, q_2) be the quantities of the commodities the consumer consumes in a given time period, and let $\mathcal{Q} = [(q_1, q_2): q_1 \geq 0, q_2 \geq 0,$ but $(q_1, q_2) \neq (0, 0)$ are the consumer's possible consumptions of the two commodities]. Suppose also that (p_1, p_2), with $p_1, p_2 > 0$, are the prices of the two commodities, and $Y^+ > 0$ is the consumer's money income, so that $p_1 q_1 + p_2 q_2 = Y^+$ is the consumer's budget constraint. Finally, if $u: \mathcal{Q} \rightarrow R^1$ is any function, we say that u is a utility function if it in some sense (yet to be elaborated) measures the satisfaction the consumer receives from consuming quantities of the two commodities. We shall assume that a consumer's satisfaction is represented by a utility function.

With the concepts so far developed we have two related areas of attachment difficulties that appear to differ from those we encountered in our account of classical mechanics. First, the existence of a consumer's utility function is an assumption which appears to go beyond either the observations or introspection, though it does not seem to *contradict* the evidence of introspection. For now, then, a utility function represents a feature of the consumer's (in some sense) unconscious. In contrast to the masses of classical mechanics, then, here the consciousness of our object of inquiry (the consumer) is a source of attachment difficulties. More can be said on how to attach the utility concept to the world through introspection (i.e., on how to base a utility function's existence on assumptions accessible to introspection), but we reserve such further discussion for Chapter 8.

Second, we must somehow choose a time period over which to measure quantities of commodities consumed. Further, the time period needs to be one over which we can suppose that a particular utility function represents the consumer's satisfaction (i.e., one over which the consumer's tastes remain fixed). Unclear, here, is whether by "time period" we mean a particular *historical* period or a period of time more in the sense in which the phrase has meaning in classical mechanics. Much more needs to be said about the attachment problem of time-period choice, but we postpone most further discussion of it until Chapter 8 and the Conclusion.

Next we suppose that the following principle holds.

Principle of Utility Maximization (UM): If a consumer's utility function exists, then the consumer maximizes that utility function on $\mathcal{Q}$ subject to the budget constraint $p_1 q_1 + p_2 q_2 = Y^+$. Further, we suppose that in maximizing u, the consumer has no control over prices p_1 and p_2 – that is, the consumer takes p_1 and p_2 as given and chooses (q_1, q_2) from among those elements of $\mathcal{Q}$ satisfying the budget constraint so as to maximize $u(q_1, q_2)$.

Two arguments against UM's plausibility are often voiced: (i) UM supposes an unrealistically selfish consumer, and (ii) because the consumer is not consciously aware of $u(q_1, q_2)$, the consumer cannot choose (q_1, q_2) so as to maximize it. Argument (i) is less a case against UM's plausibility than a case for expanding the concept of consumer utility function to include the consumptions of (some) other individuals as arguments in an individual consumer's utility function – that is, to allow for consumption externalities (a question we shall take up in Chapter 8). For now, we respond to (ii) by saying that the consumer's choice of (q_1, q_2) [like what $u(q_1, q_2)$ represents] is in some sense unconscious. Once we have made $u(q_1, q_2)$ more accessible to introspection (see Chapter 8), UM may appear more plausible.

Next we have the consumer theory core's main theorem.

Theorem 6.1. *Suppose $p_1q_1 + p_2q_2 = Y^+$ holds. Let $u(q_1, q_2)$ be a consumer utility function defined on $\mathcal{Q}$, and suppose $\partial u/\partial q_1$, $\partial u/\partial q_2$ exist and are > 0 everywhere on the interior of $\mathcal{Q}$. Then, if $u(q_1, q_2)$'s constrained absolute maximum on $\mathcal{Q}$ is not on the boundary of $\mathcal{Q}$, UM implies (p_1/p_2) $= [\partial u(q_1, q_2)/\partial q_1]/[\partial u(q_1, q_2)/\partial q_2]$ (the marginal utility condition, MUC) at $u(q_1, q_2)$'s constrained absolute maximum on $\mathcal{Q}$.*[2]

The theorem's conclusion, MUC, is a partial-differential-equation placeholder [i.e., like Newton's Second Law (NSL) in our Chapter 5 account of classical mechanics]. In conjunction with a particular utility function, MUC becomes a statement that might (with some further assumptions) be made to confront the observations.

Notice the parallels and differences between the classical mechanical and consumer theory cores. In classical mechanics we had nature or possibly God minimizing the action (i.e., Hamilton's Principle, HP), possibly subject to a constraint leading to a second-order-in-time differential-equation law-sketch – NSL. In consumer theory we have the consumer maximizing utility (i.e., UM) subject to a budget constraint, leading to a first-order-in-quantities-consumed partial-differential-equation law-sketch – MUC. For the plausibility of HP, we have nowhere to appeal, whereas for the plausibility of UM, we hold out the hope for an appeal (in part) to introspection.

6.2.2   *Further demand-side theory*[3]

In classical mechanics, to get a prediction statement that would confront the observations, we needed, in addition to NSL, a particular force rela-

[2] For the theorem's proof, see Samuelson (1965, pp. 98, 362–5) or Henderson and Quandt (1971, pp. 14–15).

[3] For the proof of Theorem 6.2, see Samuelson (1965, pp. 100–7) or Henderson and Quandt (1971, pp. 14–19, 31–7). For the proof of Theorem 6.3, see Samuelson (1965, pp. 98–9) or Henderson and Quandt (1971, pp. 19–23). Theorem 6.4 follows immediately from the definition of market demand function and Theorem 6.2.

tionship. Can we, in consumer theory, get a prediction statement that will confront the observations *without* a particular utility relationship? Consider first a condition.

Axiom 6.1. *$u(q_1, q_2)$ is twice differentiable; $\partial u/\partial q_1$, $\partial u/\partial q_2 > 0$; and*

$$(\partial^2 u/\partial q_1^2)(\partial u/\partial q_2)^2 - 2(\partial^2 u/\partial q_1 \partial q_2)(\partial u/\partial q_1)(\partial u/\partial q_2)$$
$$+ (\partial^2 u/\partial q_2^2)(\partial u/\partial q_1)^2 < 0$$

everywhere on the interior of $\mathcal{Q}$.

Let UCAM mean unique constrained absolute maximum on $\mathcal{Q}$. We then have a theorem.

Theorem 6.2. *Suppose that Axiom 6.1 for $u(q_1, q_2)$ and $p_1 q_1 + p_2 q_2 = Y^+$ hold. Then for any $p_1, p_2, Y^+ > 0$,*

(i) *if MUC holds for some (q_{10}, q_{20}) in $\mathcal{Q}$'s interior, then (q_{10}, q_{20}) is a UCAM of $u(q_1, q_2)$, whereas if MUC holds for no (q_{10}, q_{20}) in $\mathcal{Q}$'s interior, there is a (q_{10}, q_{20}) on $\mathcal{Q}$'s boundary that is a UCAM of $u(q_1, q_2)$ [we call $q_{i0} = q_{i0}(p_1, p_2, Y^+)$, or $q_i = q_i(p_1, p_2, Y^+)$, for $i = 1, 2$, an individual consumer demand function],*

(ii) *$q_i(tp_1, tp_2, tY^+) = q_i(p_1, p_2, Y^+)$ for any $t > 0$,*

(iii) *if $q_1, q_2 > 0$, then $\partial q_i/\partial p_j$, $\partial q_i/\partial Y^+$ exist for $i \neq j = 1, 2$, and $(\partial q_1/\partial p_2) + q_2(\partial q_1/\partial Y^+) = (\partial q_2/\partial p_1) + q_1(\partial q_2/\partial Y^+)$ (Slutsky's Equation, SE).*

Can the theorem's conclusions (= prediction statements) be in some sense tested? On (iii), Samuelson points out that

[SE] reflect differential properties of our demand functions which are hard to visualize and hard to refute. For our empirical data consists of isolated points. These must be smoothed in some sense before our relation can be tested; the smoothing, even by the best known statistical methods, is to a degree arbitrary, and so refutation and verification are difficult.

I have tried, but thus far with no success, to deduce implications of . . . SE . . . which can be expressed in finite form, i.e., be conceivably refutable merely by a finite number of point observations. (1965, p. 107)

With (ii), the difficulty for testing against the observations is that only by chance do we observe multiples of any given (p_1, p_2, Y^+). With no *particular* demand functional form (a consequence of a *particular* utility functional form) involved, (ii) and (iii) seem too abstract to be made to confront the observations. But (ii), at least, seems testable in another sense: against introspection. That is, if the consumer's income and the prices of all available commodities are simultaneously multiplied by the same positive factor, then introspection suggests that the consumer will not alter the amount of any commodity desired. There seems, then, to be a role for introspection to play in testing one abstract prediction statement before a particular utility function is added to the theory.

Consider next another theorem.

Theorem 6.3. *If (q_{10}, q_{20}) satisfies $p_1 q_1 + p_2 q_2 = Y^+$ for given (p_1, p_2, Y^+) and is a UCAM of $u(q_1, q_2)$ that satisfies Axiom 6.1, then (q_{10}, q_{20}) is a UCAM of $F[u(q_1, q_2)]$ for any monotonic differentiable function $F: R^1 \rightarrow R^1$.*

Notice that any given particular individual consumer demand functional form corresponds not to a single utility form but to an equivalence class of such forms under differentiable monotonic transformation. Since it will be the demand function (or, more precisely, something constructable from it) which will be made to confront the observations, our test of a prediction statement will really test not a unique utility functional form but an equivalence class of such forms.

Next we expand our consumption sector to include N consumers. Let the superscript i refer to the ith consumer; for example, $Y = (Y^1, \ldots, Y^N)$ is the distribution of income (i.e., Y^i is the ith consumer's income).

Definition: If $q^i_j(p_1, p_2, Y^i)$ is the ith individual consumer's demand for the jth commodity, then $q_{Dj}(p_1, p_2, Y) = \Sigma_{i=1}^{N} q^i_j(p_1, p_2, Y^i)$ is the *market (consumer) demand function* for the jth commodity for $j = 1, 2$.

We then have the following theorem.

Theorem 6.4. *Suppose that $u^i(q^i_1, q^i_2)$ satisfies Axiom 6.1, that $p_1 q^i_1 + p_2 q^i_2 = Y^i$, and that UM holds for the ith consumer for $i = 1, \ldots, N$. Then*

(i)   *there exists a market (consumer) demand function $q_{Dj}(p_1, p_2, Y)$,*

(ii) *$q_{Dj}(tp_1, tp_2, tY) = q_{Dj}(p_1, p_2, Y)$ for any $t > 0$ for $j = 1, 2$.*

Conclusion (ii) of the theorem is difficult to test against the observations [just as is (ii) of Theorem 6.2 – the *individual* consumer demand case] because only by chance do we observe multiples of any given $(p_1, p_2, \mathbf{Y})$. So to get a market (consumer) demand function to confront the observations, we must specify particular utility functional forms so that the market (consumer) demand function takes a particular form. But we also must add an account of producer theory to our account of consumer theory, because microeconomics sees the observations as determined *jointly* by laws of supply and demand. In the next section, we offer an account of producer theory that largely parallels our account of consumer theory.

6.2.3   *The producer theory core and further supply-side theory*[4]

In producer theory, we again employ a series of terms [e.g., "producer" (or "firm"), "price," "commodity," "input," "output," and "profit"], some of which will be defined in terms of others. The ultimate problem, though, of attaching propositions involving the terms to the world of observations remains philosophically difficult. We consider a production sector consisting for now of a single firm producing an output with two inputs. Let (x_1, x_2) be the quantities of the inputs the firm utilizes in a given time period, and let $\mathscr{X} = \{(x_1, x_2): x_1 \geq 0 \text{ and } x_2 \geq 0 \text{ are the firm's input quantity possibilities}\}$. Also, let $r_1, r_2 > 0$ be the input prices and $q \geq 0, p > 0$ be the output quantity and price. And if $f: \mathscr{X} \rightarrow R^1$ is any function, we say that f is a firm's production function if it gives the firm's output quantity in terms of its input quantities. We shall assume that a firm's production process is represented by a production function.

Two attachment difficulties are again apparent. The first resembles those we have seen before in statistics and classical mechanics. It is possible – at least in principle – to conjecture a particular production function, add a statistical measurement-error model (of the Chapter 4 sort) as was done in our account of classical mechanics, and then rerun the firm's production process several times and test (in the Chapter 4 statistical sense) the particular production function. But as in Chapter 4 and in our account of classical mechanics, such a test's conclusion is not based *solely* on the observations. Thus, the particular production function assumed goes beyond the observations.

The second attachment difficulty resembles one we noted in consumer

---

4 For the proof of Theorem 6.5, see Samuelson (1965, pp. 362–5) or Henderson and Quandt (1971, pp. 67–9). For the proof of Theorem 6.6, see Henderson and Quandt (1971, pp. 67–9). Theorem 6.7 follows immediately from the definition of market supply function and Theorem 6.6.

theory (see Section 6.2.1), which differs from those we encountered in classical mechanics. We must somehow choose a time period over which to measure input quantities. Further, the time period needs to be one over which we can suppose that a particular production function represents the production process (i.e., one over which the firm's production technology remains fixed). Again unclear is whether "time period" refers to a classical mechanical or historical sense of time. And again we postpone further discussion of this question to Chapter 8 and the Conclusion.

Let $\Pi = pq - r_1 x_1 - r_2 x_2$ be the firm's profits. We next suppose that the following principle holds.

Principle of Profit Maximization (PM): For a given price vector (p, r_1, r_2), the firm chooses (x_1, x_2) from $\mathscr{X}$ subject to production function constraint $q = f(x_1, x_2)$ so as to maximize profits.

Notice that we suppose that in maximizing profits, the firm has no control over prices (p, r_1, r_2). That a privately owned firm *seeks* to maximize profits gains plausibility from the socially agreed upon convention that it *should* do so. What is possibly less plausible is that such a firm will *know* (and therefore be able to choose) the (x_1, x_2) that will maximize its profits. Friedman's response (1953, pp. 21–3) to such a critique is to deny that such knowledge has a role to play in "testing" PM and to claim that only tests of PM's predicted consequences are germane. The difficulty with Friedman's view, we are arguing, is that getting prediction statements to confront the observations is not without problems.[5]

Next we consider the producer theory's core theorem.

Theorem 6.5. *Suppose $q = f(x_1, x_2)$ is the firm's production function constraint, and suppose $\partial f/\partial x_1$, $\partial f/\partial x_2$ exist and are > 0 everywhere on $\mathscr{X}$'s interior, and suppose Π's constrained absolute maximum on $\mathscr{X}$ is*

---

[5] In fact, Friedman (1953) implicitly concedes the difficulty of testing PM by its predicted consequences. He explains that the "evidence for the maximization-of-returns hypothesis . . . is extremely hard to document" (p. 22). And he adds that "the evidence *for* a hypothesis always consists of its repeated failure to be contradicted, continues to accumulate so long as the hypothesis is used, and by its very nature is difficult to document at all comprehensively. It tends to become part of the tradition and following of a science revealed in the tenacity with which hypotheses are held rather than in any textbook list of instances in which the hypothesis has failed to be contradicted" (p. 23). Perhaps comprehensive documentation – whatever that means – is indeed difficult. But if it is true that there is a sense in which the predicted consequences of PM have *repeatedly* failed to be contradicted, one would think that Friedman could have cited *at least one* such instance – which he does not.

not on the boundary of $\mathscr{X}$. Then PM implies $p = (r_1/\partial f(x_1, x_2)/\partial x_1) = (r_2/\partial f(x_1, x_2)/\partial x_2)$ (the marginal product condition, MPC) at Π's constrained absolute maximum on $\mathscr{X}$.

The theorem's conclusion – MPC – is, like MUC and NSL in our accounts of consumer theory and classical mechanics, a partial-differential-equation *placeholder*. In conjunction with a particular production function form, MPC becomes a statement that might (with some further assumptions) be made to confront the observations.

Consider next an axiom.

Axiom 6.2. *$f(x_1, x_2)$ is twice differentiable, and $\partial f/\partial x_1$, $\partial f/\partial x_2 > 0$, $\partial^2 f/\partial x_1^2, \partial^2 f/\partial x_2^2 < 0$, and $(\partial^2 f/\partial x_1^2)(\partial^2 f/\partial x_2^2) - (\partial^2 f/\partial x_1 \partial x_2)^2 > 0$ everywhere on the interior of $\mathscr{X}$.*

Let UAM mean unique absolute maximum on $\mathscr{X}$. Further supply-side results are then possible without specification of a particular production functional form.

Theorem 6.6. *Suppose Axiom 6.2 holds for $f(x_1, x_2)$. Then for any p, r_1, $r_2 > 0$,*

  (i) *if MPC holds for some (x_{10}, x_{20}) in $\mathscr{X}$'s interior, then (x_{10}, x_{20}) is a UAM of $\Pi(x_1, x_2) = pf(x_1, x_2) - r_1 x_1 - r_2 x_2$ [with $q_0 = f(x_{10}, x_{20})$, we call $q_0 = q_0(p, r_1, r_2)$ or $q = q(p, r_1, r_2)$ the firm's supply function for the output],*

  (ii) *$q(tp, tr_1, tr_2) = q(p, r_1, r_2)$ for any $t > 0$.*

Notice that the theorem's conclusions (i) and (ii) are precisely parallel to the demand-side results (i) and (ii) of Theorem 6.2. Notice also that the difficulty here for testing (ii) against the observations is that only by chance do we observe multiples of any given (p, r_1, r_2). [Recall that a similar difficulty occurred with Theorem 6.2(ii).]

Next we expand our production sector to include M firms, all producing the same output with the same two inputs. Let the superscript i refer to the ith firm.

Definition: If $q^i(p, r_1, r_2)$ is the ith firm's supply function for the output, then $q_s = \Sigma_{i=1}^N q^i(p, r_1, r_2)$ is the *(firms') market supply function* for the output.

We then have the following theorem.

Theorem 6.7. *Suppose $q^i(x_1^i, x_2^i)$ satisfies Axiom 6.2 and PM holds for the ith firm for $i = 1, \ldots, M$. Then*

    (i)   *there exists a (firms') market supply function $q_s(p, r_1, r_2)$,*

    (ii) *$q_s(tp, tr_1, tr_2) = q_s(p, r_1, r_2)$ for any $t > 0$.*

Again, conclusion (ii) of the theorem is difficult to test against the observations because only by chance do we observe multiples of any given (p, r_1, r_2). So to get a (firms') market supply function to confront the observations, we must specify particular production functional forms so that the market supply function takes a particular form. We now have nonparticular accounts of both consumer and producer theories. In the next section we shall put the two accounts together, add particular utility and production functional forms, and examine a logical difficulty which arises in testing the MUC and MPC placeholder relationships for the consumers and firms.

6.2.4   *A particular concrete, hypothetical, artificial example and a logical difficulty*

To knit the two-commodity consumption and three-commodity production sectors into an economy, suppose that the first consumption commodity is the production output, so that $p_1 = p > 0$. Suppose also that the first consumption commodity is never a production input and that the second production input is never a consumption commodity, so that the first production input is the second consumption commodity. And suppose that the two inputs in the production of the first consumption commodity are themselves produced – one with the other, but make no further assumptions about the nature of that production process. We concentrate attention on the market for the first consumption commodity only. The amount of that commodity consumed (produced) by the ith (jth) consumer (producer) is q_{1i} (q^j), and the consumer (firm) market demand for (supply of) that commodity is q_D (q_S).

So far we have avoided specifying particular utility and production functions, but now we suppose $\mathscr{C}$: The N consumers' utility functions are $U^i = \alpha_i q_{1i}^{\beta_{1i}} q_{2i}^{\beta_{2i}}$ for constants $\alpha_i, \beta_{1i}, \beta_{2i} > 0$ for $i = 1, \ldots, N$, with $(\beta_{2i}/\beta_{1i}) = (\beta_{2j}/\beta_{1j})$ for $i \neq j$ during the period when the observations occur. And we also suppose $\mathscr{P}$: The M firms' production functions are

$q^j = \gamma_j x_{1j}^{\delta_{1j}} x_{2j}^{\delta_{2j}}$ for constants γ_j, δ_{1j}, $\delta_{2j} > 0$, $\delta_{1j} + \delta_{2j} < 1$ for $j = 1, \ldots, M$, with $(\delta_{2j}/\delta_{1j}) = (\delta_{2i}/\delta_{1i})$ for $j \neq i$ during the period when the observations occur.[6] Notice that $\mathcal{C}$ and $\mathcal{P}$ implicitly suppose that a period can be found during which consumer tastes and production technologies remain fixed. What makes $\mathcal{C}$ ($\mathcal{P}$) *artificial* is that it supposes the ratio of each consumer's (producer's) marginal utilities (products) is the same. We take up a less artificial example in Section 6.3.4. Unlike in Chapter 5, we do not conjecture that $\mathcal{C}$ and $\mathcal{P}$ characterize any *actual* situation in the world, but rather through Section 6.3.4 explore the consequences of $\mathcal{P}$ and $\mathcal{C}$ (or a modification of $\mathcal{C}$) as *hypothetical* examples.

Let $R_i = \ln(r_i)$ for $i = 1, 2$, $Y = \ln(\Sigma_{i=1}^N Y^i)$, $P = \ln(p)$, $Q_D = \ln(q_D)$, and $Q_S = \ln(q_s)$.

Let S_2 be

$$Q_D = a_{1D}P + a_{2D}Y + a_{3D}R_1 + a_{4D}R_2 + a_{5D}$$

and

$$Q_S = a_{1S}P + a_{2S}Y + a_{3S}R_1 + a_{4S}R_2 + a_{5S},$$

where the a_{iD}'s and a_{iS}'s are constants, with $a_{1D} = -1 \neq a_{1S} = -(a_{3S} + a_{4S})$, $a_{2D} = 1$, and $a_{3D} = a_{4D} = a_{2S} = 0$.

And let S_1 be, for given $(r_1, r_2, \mathbf{Y})$, with $r_1, r_2, Y^i > 0$,

$$Q_D = a_{1D}P + K_{2D} \text{ (called a } market\ demand\ curve),$$

and

$$Q_S = a_{1S}P + K_{2S} \text{ (called a } market\ supply\ curve),$$

where a_{1D}, K_{2D}, a_{1S}, and K_{2S} are constants, with $a_{1D} = -1 \neq a_{1S}$.

We can now state a theorem with particular concrete conclusions.

Theorem 6.8. *Let S_4 be: UM and $p_1 q_{1i} + p_2 q_{2i} = Y^i$ for each consumer, with $p_1, p_2, Y^i > 0$; PM for each producer, with $p = p_1, r_1, p_2 = r_2 > 0$; $\mathcal{C}$ and $\mathcal{P}$ hold. Let S_3 be: MUC for each consumer, MPC for each producer, $\mathcal{C}$ and $\mathcal{P}$ hold. Then S_4 implies S_3 implies S_2 implies S_1.*

Now compare the theorem's *logic* with the logic of the classical mechanics theorems of Chapter 5. In Chapter 5, if a theorem's conclusion

---

[6] If we had insisted that $\delta_{1j} + \delta_{2j} = 1$, rather than < 1, our production function would be of the sort referred to as Cobb-Douglas. Such a production function was first employed by Cobb and Douglas in 1928. With the $\delta_{1j} + \delta_{2j} = 1$ restriction, Axiom 6.2 doesn't hold, and, in fact, depending on the values of δ_{1j}, γ_j, p, r_1, and r_2, Π either has no maximum or has its maximum on the boundary of $\mathcal{X}$ at $(0, 0)$, where $\partial f/\partial x_1$, $\partial f/\partial x_2$ aren't even defined.

was false, then either the particular force equation or NSL (and hence the particular force equation and HP) was false. Here, if S_1 or S_2 is false, then one or more of the following is false: a particular utility or production function equation, MUC for some consumer or MPC for some producer (and hence a particular utility or production function equation, UM for some consumer or PM for some producer). So Chapter 5's theorems involved a single placeholder NSL and one particular force equation, whereas here we have N MUC placeholders, M MPC placeholders, N particular utility function equations, and M particular production function equations.

We discover, then, a logical difficulty in our account of microeconomics that is similar to one we unearthed in our account of classical mechanics. In both cases, the law-sketches (or placeholder relations) involved cannot be falsified, but only their conjunction with particular force or utility-production function equations can be. Thus, neither classical mechanics nor microeconomics is an "ideal science" in Popper's sense of consisting of laws *each* of which can in principle be *individually* falsified in a confrontation of its deduced consequences with the observations. Yet if classical mechanics is a "real science" despite the fact that experiments cannot conclusively falsify its law-sketch, then microeconomics cannot fail to be a "real science" for exhibiting precisely the same logical difficulty. In Section 6.3 we take up some attachment difficulties which arise in trying to get the deduced conclusions of microeconomic theory to confront the observations.

6.3   Attachment difficulties

6.3.1   *Equilibrium and the identification problem and its solution in the particular concrete, hypothetical, artificial example[7]*

Each conclusion of Theorem 6.8 involves a pair of equations: one for ln of market demand, Q_D, and one for ln of market supply, Q_S. To get these equations to confront the observations, we need an attachment assumption.

Market Equilibrium Condition (MEC): Market demand for the commodity $= q_D = q_S =$ market supply of the commodity $= q^* =$ the observed amount of the commodity bought–sold. Equivalently, $Q_D = Q_S = Q$, where $Q = \ln(q^*)$.

---

[7] The proofs of Theorems 6.9–6.11 follow directly from the theory of solving a system of linear equations by determinants. See, for example, Vance (1955, pp. 157–9).

In one sense, MEC is similar to certain attachment assumptions made implicitly, but not remarked upon, in our account of classical mechanics. For example, we supposed that a mass's observed position was a value of the position variable of our equation. Here, though, we assume that the ln of the observed aggregate amount of a commodity bought–sold is *simultaneously* a value of *two* variables — Q_S and Q_D — of our equations.

Notice also that we assume that equilibrium exists and is observed in the market only for the one commodity of interest to us. Though we made some assumptions about production and consumption of the other two commodities, we did not specify these processes completely and do not assume that equilibrium exists or is observed in these other two markets. Thus, our analysis is essentially a partial equilibrium analysis. We postpone further discussion of MEC until Chapter 8 and the Conclusion, for now noting only that it is an attachment assumption with no complete classical mechanical analogue, and employing it in the remaining part of the chapter.

Now how can we make the conclusions of the previous section's theorems confront the observations? Consider S_1 first. Assuming MEC and viewing P, Q as unknown variables, S_1 contains a system of two linear equations (with unequal slopes) in two unknowns. A theorem expresses the system's solution.

Theorem 6.9. *If* $Q_D = a_{1D}P + K_{2D}$, $Q_S = a_{1S}P + K_{2S}$, $a_{1D} \neq a_{1S}$, *and* $Q_D = Q_S = Q$, *then*

$$(P_0, Q_0) = [(K_{2D} - K_{2S})/(a_{1S} - a_{1D}), (a_{1S}K_{2D} - K_{2S}a_{1D})/(a_{1S} - a_{1D})]$$

solves the system.

Graphically, we have two nonparallel straight lines intersecting at the unique point (P_0, Q_0). For given (Y, r_1, r_2), (P_0, Q_0) is the predicted observation, and we have expressed (P_0, Q_0) in terms of a_{1D}, K_{2D}, a_{1S}, and K_{2S}. To test something like $a_{1D} = -1$, though, we need to be able to express a_{1D} in terms of predicted observations. We call the problem of expressing the a_{iD}'s and a_{iS}'s in terms of predicted observations the identification problem.

If we stick to a given (Y, r_1, r_2), we cannot hope to solve the identification problem, for we have only a *single* predicted observation point. With that point fixed, we have a system of two linear equations, $K_{2D} = -P_0 a_{1D} + Q_0$ and $K_{2S} = -P_0 a_{1S} + Q_0$, in the four unknowns a_{1D}, K_{2D}, a_{1S}, and K_{2S}. Such a system has no unique solution. Graphically, there is an *infinite* number of pairs of straight lines that intersect at (P_0, Q_0), and each of those pairs corresponds to a different $(a_{1D}, K_{2D}, a_{1S}, K_{2S})$. To solve the identification problem, we shall have to allow (Y, r_1, r_2) to vary and try to get S_2 to confront the observations.

With $a_{3D} = a_{4D} = a_{2S} = 0$, S_2's two equations become $Q_D = a_{1D}P + a_{2D}Y + a_{5D}$ and $Q_S = a_{1S}P + a_{3S}R_1 + a_{4S}R_2 + a_{5S}$. Assuming MEC and viewing P, Q, Y, R_1, and R_2 as unknown variables, S_2's system is two linear equations in five unknowns. Again, a theorem expresses the system's solution.

Theorem 6.10. *If* $Q_D = a_{1D}P + a_{2D}Y + a_{5D}$, $Q_S = a_{1S}P + a_{3S}R_1 + a_{4S}R_2 + a_{5S}$, $a_{1D} \neq a_{1S}$, *and* $Q_D = Q_S = Q$, *then*

$$(P_0, Q_0, Y_0, R_{10}, R_{20}) = [(a_{2D}Y_0 - a_{3S}R_{10} - a_{4S}R_{20} + a_{5D} - a_{5S})$$
$$/(a_{1S} - a_{1D}), (a_{2D}a_{1S}Y_0 - a_{1D}a_{3S}R_{10} - a_{1D}a_{4S}R_{20} + a_{5D}a_{1S} - a_{1D}a_{5S})$$
$$/(a_{1S} - a_{1D}), Y_0, R_{10}, R_{20}].$$

Notice that S_2's system doesn't determine a unique point solution as S_1's did. Rather, S_2's system determines (P_0, Q_0) in terms of (Y_0, R_{10}, R_{20}) and $(a_{1D}, a_{2D}, a_{5D}, a_{1S}, a_{3S}, a_{4S}, a_{5S})$. Also, the identification problem has now become to express the a_{1D}'s and a_{1S}'s in terms of the predicted observations.

Next we have a theorem which solves the identification problem and a corollary which allows S_2 to confront the observations.

Theorem 6.11. *Suppose* $Q_D = a_{1D}P + a_{2D}Y + a_{3D}R_1 + a_{4D}R_2 + a_{5D}$, $Q_S = a_{1S}P + a_{2S}Y + a_{3S}R_1 + a_{4S}R_2 + a_{5S}$, *with* $a_{3D} = a_{4D} = a_{2S} = 0$, $a_{1D} \neq a_{1S}$, *and* $Q_D = Q_S = Q$. *Let* $(P_i, Q_i, Y_i, R_{1i}, R_{2i})$ *for* $i = 1, \ldots, 4$ *be four observations that satisfy the two equations and for which*

$$D(\neq 0) = \begin{vmatrix} Y_1 & R_{11} & R_{21} & 1 \\ Y_2 & R_{12} & R_{22} & 1 \\ Y_3 & R_{13} & R_{23} & 1 \\ Y_4 & R_{14} & R_{24} & 1 \end{vmatrix},$$

and let $D_j(C_j)$ *be D, with the jth column replaced with*

$$\begin{pmatrix} P_1 \\ P_2 \\ P_3 \\ P_4 \end{pmatrix} \begin{pmatrix} Q_1 \\ Q_2 \\ Q_3 \\ Q_4 \end{pmatrix} \quad \text{for } j = 1, \ldots, 4.$$

Then D_1, $D_2 \neq 0$, and

(i) $a_{1D} = C_2/D_2$,

(ii) $a_{1S} = C_1/D_1$,

(iii) $a_{2D} = (D_1/D)[(C_1/D_1) - (C_2/D_2)]$,

(iv) $a_{3S} = (D_2/D)[(C_2/D_2) - (C_1/D_1)]$,

(v) $a_{4S} = (D_3/D)[(C_2/D_2) - (C_1/D_1)]$,

(vi) $a_{5D} - a_{5S} = (D_4/D)[(C_1/D_1) - (C_2/D_2)]$.

Corollary. *If the theorem's hypotheses hold, then*

(a) *if $a_{1D} = -1$, then $C_2 = -D_2$,*

(b) *if $a_{2D} = 1$, then $C_1 D_2 - C_2 D_1 = DD_2$,*

(c) *if $a_{3S} + a_{4S} = -a_{1S}$, then $D_2^2 C_1 - C_2 D_1 D_2 - C_2 D_3 D_2 - C_1 D_2 D$*
     $= 0$.

Notice that the theorem solves the identification problem by expressing the a_{iS}'s and a_{iD}'s in terms of the observations, while the corollary's conclusions are conditions on the observations that follow from MEC and S_2. Thus, if the observations fail to satisfy one of the conditions of the corollary's conclusions, then either MEC or S_2 (and hence S_3 and hence S_1) is false. Hence, solving the identification problem and supposing MEC allows us to confront Theorem 6.8's prediction statement S_2 with the observations. Why, though, have we succeeded in solving the identification problem in the case of S_2's equations, but failed to do so in the case of S_1's equations? To answer the question, consider first a definition.

Definition: In a system of equations, variables whose values the system fails to determine are called *exogenous,* whereas those whose values are determined by the system in terms of the exogenous variables (and the coefficients) are called *endogenous.*

Thus, in the case of S_1's linear system of two equations in two unknown variables, both variables Q and P were endogenous, and there were no exogenous variables, whereas in the case of S_2's linear system of two equations in five unknown variables, Q and P were endogenous, while Y, R_1, and R_2 were exogenous variables.

The following theorem provides the reason why we could solve the identification problem for the second system but not the first system.

Theorem 6.12. *Suppose that in a system of equations linear in variables and coefficients, some variables are excluded from some equations (i.e., some variables have zero coefficients in some equations). Then, that the coefficients of an equation are identifiable implies that the number of exogenous variables excluded from the equation must at least equal the number of endogenous variables included on the right-hand side of the equation.*[8]

Both our systems of equations were linear in variables and coefficients. In S_1's system there were no exogenous variables left out in either equation, and one endogenous variable – P – was included on the right-hand side of each equation. So neither equation satisfies the theorem's identifiability condition. In S_2's system, both the demand and supply equations had one endogenous variable – P – included on the right-hand side, while the demand equation had two exogenous variables – R_1, R_2 – and the supply equation had one endogenous variable – Y – excluded. Thus, both equations of the second system satisfy the theorem's identifiability condition.

6.3.2 *The identification problem as a philosophical attachment difficulty*

In the previous section we were able to identify the demand and supply function coefficients of Theorem 6.8(i) because certain variables were absent from each equation – R_1, R_2 from demand and Y from supply. Put another way, the identification was possible because we supposed $a_{3D} = a_{4D} = a_{2S} = 0$. What is the source of such a supposition? In a treatise on the identification problem, the economist F. Fisher argues that

. . . the nature of the theoretical model to be estimated itself implies that the parameters of a given equation cannot logically be inferred on the basis of empirical data alone. Structural estimation is impossible without the use of a priori information concerning the equation to be estimated. Such information must be provided from a source outside the real-world data at hand – either from economic theory or from the results of other studies of different types of data (1966, p. 1)

---

[8] See Wonnacott and Wonnacott (1970, pp. 182–3, 345–53) for the theorem's proof.

On our account of the identification problem, there are two difficulties with Fisher's argument. First, his claim that a supposition like $a_{3D} = a_{4D} = a_{2S} = 0$ may stem "from the results of other studies of different types of data" seems open to question. Second, his labeling such a supposition "a priori information concerning the equation to be estimated" and suggesting that "such information must be provided from a source outside the real-world data at hand" is belied by our account.

The claim that a supposition like $a_{3D} = a_{4D} = a_{2S} = 0$ may stem from other empirical studies resembles Mill's argument (see Section 1.2.2) that his Method of Difference could be based on *observations* in a setting other than the experiment at hand. Such a claim leads to logical difficulties, for suppose the supposition needed for identification in the first study comes from the result of a second empirical supply–demand study. For identification to be possible in that *second* study, a second supposition of variables missing in equations is required. But if the second supposition rests on a third empirical study, we have started through an infinite regress of empirical studies. If, instead, the second supposition results from the first empirical study, we have a logical circularity. So Fisher is correct (as our account illustrates) in saying that the supposition of missing variables stems from economic theory.

But on our account of the identification problem it is not quite correct to claim that the supposition of missing variables is "a priori information." Such a claim resembles Popper's "with the help of other statements, previously accepted." On our account,

$$Q_D = a_{1D}P + a_{2D}Y + a_{3D}R_1 + a_{4D}R_2 + a_{5D}$$

and

$$Q_S = a_{1S}P + a_{2S}Y + a_{3S}R_1 + a_{4S}R_2 + a_{5S},$$

where the a_{iD}'s and a_{iS}'s are constants, with $a_{3D} = a_{4D} = a_{2S} = 0$, $a_{1D} = -1 \neq a_{1S} = -a_{3S} - a_{4S}$, and $a_{2D} = 1$, follows deductively from the theory. Further, a conclusion like $C_1 D_2 - C_2 D_1 = DD_2$ with which we confront the observations follows deductively from $Q_D = Q_S = Q$ (which is not "previously accepted"),

$$Q_D = a_{1D}P + a_{2D}Y + a_{3D}R_1 + a_{4D}R_2 + a_{5D},$$

and

$$Q_S = a_{1S}P + a_{2S}Y + a_{3S}R_1 + a_{4S}R_2 + a_{5S},$$

where the a_{iD}'s and a_{iS}'s are constants, with $a_{3D} = a_{4D} = a_{2S} = 0$ and $a_{2D} = 1$ [Corollary (b) of Theorem 6.11]. So theory implies the supposition $a_{3D} = a_{4D} = a_{2S} = 0$, and we use "the real-world data at hand" to test the conjunction of the supposition and other theory-implied propositions.

With perfect reporting, then, the identification problem is an attachment difficulty which closely resembles the problem of attaching classical mechanics' term "mass" to the world of observations (see Sections 5.2.2 and 5.4.2). In classical mechanics, the problem was to use the observations in a noncircular way to distinguish mass and force, whereas here the problem is to use the observations in a noncircular way to distinguish supply and demand equations. Whether or not our proposed "solutions" to these two problems are philosophically adequate is only of secondary concern to us. What counts for us is that the problem of mass measurement and the identification problem are conceptually similar, and the existence of the problem of mass measurement does not imply that classical mechanics is not a "real science." Therefore, the existence of the identification problem does not imply that microeconomics is not a "real science."

6.3.3   *Three further attachment difficulties*

What if one of the particular utility or production function equations in C holds only approximately – for example, $U^i \approx \alpha_i q_{1i}^{\beta_{1i}} q_{2i}^{\beta_{2i}}$ for some i? Then Theorem 6.11 and its corollary require modification. In the theorem's hypothesis, the demand function equation must be modified to

$$Q_D \approx a_{1D}P + a_{2D}Y + a_{3D}R_1 + a_{4D}R_2 + a_{5D},$$

and *all* the "=" signs in the theorem's conclusion must be changed to "$\approx$" (e.g., $a_{1D} \approx C_2/D_2$). Similarly, all the "=" signs in the corollary must be changed to " $\approx$ " (e.g., if $a_{2D} \approx 1$, then $C_1 D_2 - C_2 D_1 \approx DD_2$). Thus, making a particular utility or production function equation into an approximate equality requires the same sort of " $\approx$ " convention in deduced relations among the observations as making a particular force equation into an approximate equality required in classical mechanics – see, for example, Section 5.2.1.

Notice next that Theorem 6.11 and its corollary implicitly suppose perfect reporting. We may, however, add a statistical reporting model of the Section 2.2.3 sort to the corollary's conclusions. For example, with an additive reporting-error term ϵ, with $E(\epsilon) = 0$, and some statistical distributional assumptions (of the Chapter 4 variety), corollary (b)'s prediction statement becomes that $r^*(C_1 D_2 - C_2 D_1 - DD_2)$ is distributed as ϵ. With repeated sets of four observations $(P_i, Q_i, Y_i, R_{1i}, R_{2i})$ for $i = 1, \ldots, 4$ then giving us repeated observed values of $r^*(C_1D_2 - C_2D_1 - DD_2)$, we could test (in a Chapter 4 sense) the prediction statement. Thus, a statistical reporting model may be employed in dealing with attachment, when the assumption of perfect reporting is dropped, in precisely the same way as it was employed in classical mechanics – again see, for example, Section 5.2.1.

Now suppose that one has tested the prediction statements of the corollary of Theorem 6.11 and concluded that they do not contradict the observations. Can one then find the coefficients involved in Theorem 6.11's demand and supply equations? From the theorem, we have immediately that $a_{3D} = a_{4D} = a_{2S} = 0$, $a_{1D} = -1$, $a_{2D} = 1$, and $a_{1S} = -(a_{3S} + a_{4S})$. So the two equations become $Q = -P + Y + a_{5D}$ and $Q = a_{3S}(R_1 - P) + a_{4S}(R_2 - P) + a_{5S}$, which, with $Q + P - Y = Q^+$, $(R_1 - P) = R_1^*$, and $(R_2 - P) = R_2^*$, become $Q^+ - a_{5D} = 0$ and $Q = a_{3S}R_1^* + a_{4S}R_2^* + a_{5S}$. Notice that there are two distinct equations, *not* a simultaneous system. Having successfully tested the theory, we can use the observations to find a_{5D}, a_{3S}, a_{4S}, and a_{5S}. Suppose, first, that there is perfect reporting. Let Q_0^+ be an observed value of Q^+. Then the first equation trivially implies $a_{5D} = Q_0^+$. The following theorem gives the rest of the coefficients.

Theorem 6.13. *Suppose $R_1^* a_{3S} + R_2^* a_{4S} + a_{5S} = Q$. Let $(Q_i, R_{1i}^*, R_{2i}^*)$ for $i = 1, 2, 3$ be three observations that satisfy the equation and for which*

$$\Delta \, (\neq 0) = \begin{vmatrix} R_{11}^* & R_{21}^* & 1 \\ R_{12}^* & R_{22}^* & 1 \\ R_{13}^* & R_{23}^* & 1 \end{vmatrix},$$

and let Δ_j be Δ with the jth column replaced by

$$\begin{matrix} Q_1 \\ Q_2 \quad \text{for } j = 1, 2, 3. \\ Q_3 \end{matrix}$$

Then $a_{3S} = \Delta_1/\Delta$, $a_{4S} = \Delta_2/\Delta$, and $a_{5S} = \Delta_3/\Delta$.[9]

Now suppose that instead of perfect reporting there are additive, independent error terms ϵ and ν for the two equations, where $E(\epsilon) = E(\nu) = 0$, so that from the first equation, $r^*(Q^+) - a_{5D}$ is distributed as ϵ, whereas

---

[9] The theorem follows from the theory of solving a system of linear equations by determinants, cited in footnote 7.

from the second equation, $r^*(Q) - a_{3S}R_1^* - a_{4S}R_2^* - a_{5S}$ is distributed as v. Then the average of the observed $r(Q^*)$'s is an unbiased estimate of a_{5D} in the first equation, and the OLSQ coefficients are unbiased estimates of a_{3S}, a_{4S}, and a_{5S} in the second equation. So with either perfect reporting or a statistical reporting model, we can use the observations to estimate those coefficients of the demand and supply equations not implied by theory. Such use of the observations here is precisely the same as their use in classical mechanics, once theory has been tested, to estimate a constant not implied by theory – see, for example, estimation of the constant in the ideal-gas law in Section 5.3.1.

6.3.4 *Differences in individual tastes and intractable supply and demand equations*

Until now we have artificially assumed that all consumers have the same tastes or the same individual consumer demand function for a commodity. We now wish to examine the difficulty that arises in getting prediction statements to confront the observations when we relax the same-tastes assumption. So let $\mathcal{P}$ be as in Section 6.2.4, and let $\mathcal{C}'$ be $\mathcal{C}$ with the restriction $N = 2$ and without the restriction $(\beta_{21}/\beta_{11}) = (\beta_{22}/\beta_{12})$ – that is, there are two consumers with utility functions as in Section 6.2.4, but whose marginal utility ratios may differ. As in Section 6.3.4, let $R_i = \ln(r_i)$ for $i = 1, 2$, $P = \ln(p)$, $Q_D = \ln(q_D)$, and $Q_S = \ln(q_S)$. Also, let $Y_1^* = \ln(Y^2)$, $Y_2^* = Y^2/Y^1$, and $\beta_i = \beta_{1i}/(\beta_{1i} + \beta_{2i})$ for $i = 1, 2$. Finally let S_2' be: $Q_S = a_{1S}'P + a_{2S}'Y_1^* + a_{3S}'Y_2^* + a_{4S}'R_1 + a_{5S}'R_2 + a_{6S}'$ for constant a_{iS}''s, where $a_{2S}' = a_{3S}' = 0$, $a_{1S}' = -(a_{4S}' + a_{5S}')$, and $Q_D = a_{1D}'P + \ln(\beta_1 Y^1 + \beta_2 Y^2) + a_{4D}'R_1 + a_{5D}'R_2 + K_{6D}$ for constant a_{iD}''s and K_{6D}, where $a_{1S}' \neq a_{1D}' = -1$, $a_{4D}' = a_{5D}' = K_{6D} = 0$, $0 < \beta_1, \beta_2 < 1$. We then have a new theorem.

Theorem 6.14. *Let S_4' be: UM and $p_1 q_{1i} + p_2 q_{2i} = Y^i$ for each consumer, with $p_1, p_2, Y^i > 0$; PM and for each producer, with $p_1 = p, r_1, p_2 = r_2 > 0$; $\mathcal{C}'$ and $\mathcal{P}$ hold. Let S_3' be: MUC for each consumer, MPC for each producer, $\mathcal{C}'$ and $\mathcal{P}$ hold. Then S_4' implies S_3' implies S_2'.*

Notice first that with the relaxation of the same-tastes assumption, we cannot, as we could earlier, say that market demand depends on *aggregate income*, $Y^1 + Y^2$. Now we can say only that market demand depends on the *distribution of income*, (Y^1, Y^2). Notice also that the presence of the $\ln(\beta_1 Y^1 + \beta_2 Y^2)$ term in the demand equation makes that equation not linear in variables and coefficients and so bars an appeal to Theorem 6.12 to assure identifiability. With not obviously solvable supply–demand equations, our situation is similar to the one of Section 5.2.5's pendulum, where we had an apparently intractable differential equation. There we appealed to a known mathematical fact – $\sin(\theta) \approx \theta$ for small θ – to give us an approximate differential equation which was solvable. Here, similarly, we appeal to another known mathematical fact – $\ln(1 + x) \approx x$ for $0 < x \ll 1$ – to give us an approximate demand equation which together with the supply equation forms an identifiable system.

Now let S_2'' be

$$Q_S = a_{1S}'P + a_{2S}'Y_1^* + a_{3S}'Y_2^* + a_{4S}'R_1 + a_{5S}'R_2 + a_{6S}'$$

for constant a_{iS}''s, where $a_{2S}' = a_{3S}' = 0$, $a_{1S}' = -(a_{4S}' + a_{5S}')$, and $Q_D \approx a_{1D}'P + a_{2D}'Y_1^* + a_{3D}'Y_2^* + a_{4D}'R_1 + a_{5D}'R_2 + a_{6D}'$, where $a_{1S}' \neq a_{1D}' = -1, a_{2D}' = 1, a_{4D}' = a_{5D}' = 0$, and $a_{3D}' = \beta_2/\beta_1$, where $(\beta_2/\beta_1)Y_2^* \ll 1$. And let S_4'' be: $(\beta_2/\beta_1)Y_2^* \ll 1$ and S_4'. And let S_3'' be: $(\beta_2/\beta_1)Y_2^* \ll 1$ and S_3'. Then $\ln(1 + x) \approx x$ for $0 < x \ll 1$ allows us to modify Theorem 6.14.

Theorem 6.15. S_4'' *implies* S_3'' *implies* S_2''.

Notice that from S_2'' and $Q_D = Q_S = Q$, the supply and demand equations may be written

$$Q = a_{1S}'P + a_{4S}'R_1 + a_{5S}'R_2 + a_{6S}'$$

and

$$Q \approx a_{1D}'P + a_{2D}'Y_1^* + a_{3D}'Y_2^* + a_{6D}'.$$

Thus, the system becomes linear in variables and coefficients, with Q and P endogenous, and Y_1^*, Y_2^*, R_1, and R_2 exogenous. In each equation there are two exogenous variables excluded and one endogenous variable included on the right-hand side. Hence, by Theorem 6.12, the system's coefficients are identifiable.

In fact, we have the following theorem.

Theorem 6.16. *Suppose* $Q_S = a_{1S}'P + a_{2S}'Y_1^* + a_{3S}'Y_2^* + a_{4S}'R_1 + a_{5S}'R_2 + a_{6S}'$, $Q_D \approx a_{1D}'P + a_{2D}'Y_1^* + a_{3D}'Y_2^* + a_{4D}'R_1 + a_{5D}'R_2 + a_{6D}'$, *with* $a_{2S}' = a_{3S}' = a_{4D}' = a_{5D}' = 0$, $a_{1D}' \neq a_{1S}'$, *and* $Q_S = Q_D = Q$. *Let* $(P, Q, Y_1^*, Y_2^*, R_1, R_2)$ *for* $i = 1, \ldots, 5$ *be five observations which satisfy the* *supply and demand equations and for which*

$$D' \, (\neq 0) = \begin{vmatrix} Y_{11}^* & Y_{21}^* & R_{11} & R_{21} & 1 \\ Y_{12}^* & Y_{22}^* & R_{12} & R_{22} & 1 \\ Y_{13}^* & Y_{23}^* & R_{13} & R_{23} & 1 \\ Y_{14}^* & Y_{24}^* & R_{14} & R_{24} & 1 \\ Y_{15}^* & Y_{25}^* & R_{15} & R_{25} & 1 \end{vmatrix},$$

and let D_j' (C_j') be D' with the jth column replaced by

$$\begin{matrix} P_1 \\ P_2 \\ P_3 \\ P_4 \\ P_5 \end{matrix} \begin{pmatrix} Q_1 \\ Q_2 \\ Q_3 \\ Q_4 \\ Q_5 \end{pmatrix} \quad \textit{for } j = 1, \ldots, 5.$$

Then D_1', $D_2' \neq 0$, and

(i) $\quad a_{1S}' \approx C_1'/D_1'$,

(ii) $\quad a_{1D}' \approx C_3'/D_3'$,

(iii) $\quad a_{2D}' \approx (D_1'/D')[(C_1'/D_1') - (C_3'/D_3')]$,

(iv) $\quad a_{3D}' \approx (D_2'/D')[(C_1'/D_1') - (C_3'/D_3')]$,

(v) $\quad a_{4S}' \approx (D_3'/D')[(C_3'/D_3') - (C_1'/D_1')]$,

(vi) $\quad a_{5S}' \approx (D_4'/D')[(C_3'/D_3') - (C_1'/D_1')]$,

(vii) $\quad a_{6D}' - a_{6S}' \approx (D_5'/D')[(C_1'/D_1') - (C_3'/D_3')]$.

Corollary. *If the theorem's hypotheses hold, then*

(a) *if $a_{1D}' = -1$, then $-C_3' \approx D_3'$,*

(b) *if $a_{2D}' = 1$, then $C_1'D_3' - C_3'D_1' \approx D'D_3'$,*

(c) *if $a_{1S}' = -(a_{4S}' + a_{5S}')$, then $C_1'D_3'D' \approx C_1'D_3'^2 - C_3'D_3'D_1' + C_1'D_4'D_3' - C_3'D_4'D_1'$,*

(d) *if $a_{3D}' = (\beta_2/\beta_1)$ and $a_{3D}'Y_{2i}^* = (\beta_2/\beta_1)Y_{2i}^* \ll 1$ for $i = 1, \ldots, 5$, then $(C_1'D_2'D_3' - C_3'D_2'D_1')Y_{2i}^* \ll D'D_1'D_3'$ for $i = 1, \ldots, 5$.[10]*

---

[10] The theorem follows from Theorem 6.15 and the theory of solving a system of linear equations by determinants, cited in footnote 7.

Notice that the theorem solves the identification problem by expressing the a'_{iS}'s and a'_{iD}'s in terms of the observations, while the corollary's conclusions are conditions on the observations that follow from MEC and S''_2. Thus, if the observations fail to satisfy one of the conditions of the corollary's conclusions, then either MEC or S''_2 (and hence S''_3 and hence S''_4) is false. Hence, solving the identification problem and supposing MEC allows us to confront the prediction statement S''_2 with the observations. Here our use of the approximation $\ln(1 + x) \approx x$ for $0 < x \ll 1$ leads the prediction statements to entail an "$\approx$" convention, just as our use of the approximation $\sin(\theta) \approx \theta$ for small θ did in the case of the pendulum's motion in classical mechanics. In the case of the pendulum, we could observe θ_{max} and so be sure it was in the range such that the approximation held. Here we are supposing $(\beta_2/\beta_1) Y^*_2 \ll 1$, but we cannot observe if this is so because we cannot observe (β_2/β_1). Still, approximation plays a role here similar to its role in classical mechanics. There we approximated a particular force equation that allowed us to find an approximate solution to an otherwise intractable differential equation. Here we approximate a particular demand equation that allows us to solve and identify approximately an otherwise intractable system of a supply equation and a demand equation.

The sort of approximation technique we have illustrated here is not usually employed in practice by economists to deal with differences in individual tastes. For one thing, our illustration is still artificial – there are only two consumers. In the market an economist usually conceives, there may be thousands or even millions of consumers. For the existence of practical solutions we would need the consumers to fall into a few same-individual-consumer-demand-function equivalence classes, and we would need to know the aggregate income of each equivalence class. Also, there is no reason that two equivalency classes would have to have the same demand functional *forms* (as they did in our example – $q = \beta_i Y^i/P$ for $i = 1, 2$), further complicating supply–demand equation solutions. Conceivably, approximations could be devised to deal with such further complications – that is, it isn't clear that differences in individual tastes pose a difficulty that *in principle* prevents empirical microeconomic studies from having a similar relationship to microeconomic theory as classical mechanics experiments have to classical mechanical theory. In practice, however, microeconomic studies often fail, whereas classical mechanics studies succeed, in confronting deductions from theory with the observations.

6.3.5   *The translog utility function: an actual example from microeconometric practice*

Let there be three consumption commodities rather than the two we have considered so far. The economists L. R. Christensen, D. W. Jorgenson, and L. J. Lau have proposed, as a basis for empirical microeconometric studies, the particular individual-consumer utility function

$$-\ln U(q_1, q_2, q_3) = \eta_0 + \sum_{i=1}^{3} \eta_i \ln(q_i) + \tfrac{1}{2} \sum_{i=1}^{3} \sum_{j=1}^{3} \rho_{ij} \ln(q_i) \ln(q_j),$$

where q_i is the quantity of the ith commodity the individual consumes, and the η_i's and ρ_{ij}'s are constants. They call the function the translog utility function and argue that such "utility functions provide a local second-order approximation to any utility function" (1975, p. 368). By the same argument, one could claim that the utility function we employed beginning in the hypothetical example in Section 6.2.4 provides a local first-order approximation to any utility function.

More precisely, Christensen, Jorgenson, and Lau should say that the translog utility function provides a local second-order approximation to any utility function for which $-\ln U$ has a convergent power series expansion in the $\ln(q_i)$'s. And they should add that without knowing the utility function, we can't say how good the approximation is. Approximation here, then, is an attempt to weaken assumptions – that is, to assume only that the utility function falls in a certain class of functions rather than to assume a particular utility function – just as it was in Section 4.3.3. In Section 5.2.5, in contrast, we approximated $\sin(\theta)$ with θ, in a conjectured particular force equation, to make a differential equation tractable. Here and in Section 4.3.3, approximation is an empiricist avoidance of conjecturing a particular function which has the cost that we don't know how good the approximation is. When we approximate $\sin(\theta)$ with θ, in contrast, we know how good the approximation is.[11]

Christensen et al. (1975, p. 370) deduce a theorem where p_i is the ith commodity's price, and the individual's income $Y^+ = \Sigma_{i=1}^{3} p_i q_i$.

Theorem 6.17. *MUC and*

$$-\ln U(q_1, q_2, q_3) \approx \eta_0 + \sum_{i=1}^{3} \eta_i \ln(q_i) + \tfrac{1}{2} \sum_{i=1}^{3} \sum_{j=1}^{3} \rho_{ij} \ln(q_i) \ln(q_j)$$

[11] For an earlier version of this paragraph's critique in the context of a translog *cost* function, see Neuberg (1977b, fn. 22).

imply

$$(p_j q_j / Y^+) \approx \left[a_{j0} + \sum_{i=1}^{3} a_{ji} \ln(q_i) \right] \Big/ \left[b_0 + \sum_{i=1}^{3} b_i \ln(q_i) \right]$$

for $j = 1, 2, 3$ *for constant* a_{ji}*'s and* b_i*'s, where*

$$a_{ij} \approx a_{ji} \quad \text{for } i \neq j \quad \text{and} \quad b_i \approx \sum_{j=1}^{3} a_{ji} \quad \text{for } i = 1, 2, 3.$$

Now suppose the consumption sector consists of N consumers, each governed by the same translog utility function. Let q_i^k be the amount of the ith commodity consumed by the kth consumer, $Q_i^* = \Sigma_{k=1}^{N} q_i^k$, Y^k be the kth consumer's income, $y = \Sigma_{k=1}^{N} Y^k$, $Y^k = \Sigma_{i=1}^{3} p_i q_i^k$, and $\mathbf{Y} = (Y^1, \ldots, Y^N)$. Then we have another theorem.

Theorem 6.18. *MUC holding for each consumer, and*

$$-\ln U(q_1^k, q_2^k, q_3^k) \approx \eta_0 + \sum_{i=1}^{3} \ln(q_i^k) + \tfrac{1}{2} \sum_{i=1}^{3} \sum_{j=1}^{3} p_{ij} \ln(q_i^k) \ln(q_j^k),$$

for $k = 1, \ldots, N$, *imply*

$$(p_j Q_j^*/y) \approx \sum_{k=1}^{N} (Y^k/y) \left[\left(a_{j0} + \sum_{i=1}^{3} a_{ij} \ln(q_i^k) \right) \Big/ \left(b_0 + \sum_{i=1}^{3} b_i \ln(q_i^k) \right) \right]$$

for $j = 1, 2, 3$ *for constant* a_{ji}*'s and* b_i*'s, where*

$$a_{ij} \approx a_{ji} \quad \text{for } i \neq j \quad \text{and} \quad b_i \approx \sum_{j=1}^{3} a_{ji} \quad \text{for } i = 1, 2, 3.$$

Notice that with the translog rather than with the utility function of our Section 6.2.4 hypothetical example, the two aggregate budget-share equations in Theorem 6.18's conclusion involve the distribution of income $\mathbf{Y}$ as well as aggregate income y, *even if we don't relax the same-tastes assumption*. Further, these two aggregate budget-share equations involve individual as well as aggregate quantities of commodities consumed.

Now consider the following statement.

Statement 6.1

$$(p_j Q_j^*/y) \approx \left(a_{j0} + \sum_{i=1}^{3} a_{ji} \ln(Q_i^*) \right) \Big/ \left(b_0 + \sum_{i=1}^{3} b_i \ln(Q_i^*) \right)$$

for j = 1, 2, 3 for constant a_{ji}'s and b_i's, where

$$a_{ij} \approx a_{ji} \quad for \ i \neq j \quad and \quad b_i \approx \sum_{j=1}^{3} a_{ji} \quad for \ i = 1, 2, 3.$$

Notice that the statement's two equations are aggregate budget-share equations, though by Theorem 6.18 they are *not* the aggregate budget-share equations which follow deductively from these economists' assumptions. The statement's two aggregate budget-share equations follow from substituting (Q_1^*, Q_2^*, Q_3^*, y) for (q_1, q_2, q_3, Y^+) in the two *individual* budget-share equations of Theorem 6.17's conclusion. Such substitution, however, is *not* a *deductive* step. Note also that the statement's two equations do not depend on the distribution of income **Y** or on individual quantities of commodities consumed.

Christensen, Jorgenson, and Lau suggest that by suitably adding error terms to the two individual budget-share equations in the conclusion of Theorem 6.17, they can use observations of an individual's (q_1, q_2, q_3, Y^+) and (p_1, p_2) to estimate the a_{ji}'s and b_i's and then test the relations among the a_{ji}'s and b_i's of the theorem's conclusion. What they aim to test, though, is not what they *actually* test. They explain what they do:

> Our empirical results are based on time-series data for U.S. personal consumption expenditures for 1929–72. The data include prices and quantities of the services of consumers' durables, nondurable goods, and other services. . . . We have fitted equations for the services of consumers' (durables) and for nondurable goods (nondurables). There are forty-four observations for each behavioral equation. (1975, p. 377)

So with Q_1^*, Q_2^*, Q_3^* the aggregate quantities of durables, nondurables, and other services, respectively, these economists fit the observations to the two aggregate budget-share equations of Statement 6.1 and then test the relations among the a_{ji}'s and b_i's of the statement. We can understand, then, these economists' goal: "Our objective has been to test the theory of demand" (1975, p. 381). But their conclusion – "we conclude that the theory of demand is inconsistent with the evidence" (1975, p. 381) – is fallacious because they never test a statement which follows deductively from the theory of demand.[12]

---

[12] The deductive fallacy involved here is a special case of a more general fallacy: supposing a relationship among aggregate variables follows from the relationship's holding among variables prior to aggregation. The fallacy is just a kind of inverse of the ecological fallacy much discussed in the econometric, epidemiological, and statistical literatures: supposing a relationship between aggregate variables also holds among the difficult-to-observe

6.4   Summary and conclusion

6.4.1   *The formalization*

We began by arguing, against the objections of some authors, that questions like, Is microeconomics science in the same sense classical mechanics is? or even, Is microeconomics science? are legitimate questions. Our Chapter 5 inquiry discovered certain philosophical attachment difficulties which prevented classical mechanics from being an "ideal science" in the sense of the philosopher Karl Popper's account of science. Nevertheless, classical mechanics is surely science. So we argued that reconstructing microeconomics along the lines of our Chapter 5 account of classical mechanics might reveal ways in which microeconomics fails to be an "ideal science" that differ from those ways in which classical mechanics fails to be an "ideal science." Among microeconomics' failures might conceivably be some which even prevent microeconomics from being a "real science."

Our account of classical mechanics assumed that masses do not vary with time, while our account of microeconomics supposed prices to be the same for all consumers and producers who regard them as constants in maximizing utility and profits. We saw that microeconomics has two principles – utility and profit maximization – that are akin to action minimization (or Hamilton's Principle) in classical mechanics. From these two principles and the existence of utility and profit equations – akin to classical mechanics' force equations – we deduced marginal utility and product conditions that are placeholder relations akin to Newton's Second Law.

We also deduced abstract features of both individual and aggregate supply and demand equations that were not sufficiently concrete to confront the observations. To get prediction statements that could be made to confront the observations, we had to assume particular utility and production functions – akin to the particular force equations assumed for

Footnote 12 (*cont.*)

variables prior to aggregation. Notice, then, that Christensen, Jorgenson, and Lau ignore a deductive fallacy known to economists and test a statement that does not follow from the theory articulated. In Section 5.3.2 we noted Rutherford ignoring one consequence of theory that seemed to be false (i.e., the hydrogen atom's collapse) and testing another consequence (i.e., the theory's predicted scattering pattern). In a way, these two situations are similar. Yet Rutherford at least tests a deductive consequence of theory, whereas Christensen, Jorgenson, and Lau do not. Thus, while Rutherford's work is science in crisis, the work of Christensen, Jorgenson, and Lau may not be science at all.

Table 6.1. *Theory parallels between classical mechanics and microeconomics*

Feature	Classical mechanics	Microeconomics
Constant	Masses	Prices
Optimization principles	Action minimization (Hamilton's Principle)	Utility, profit maximization
Placeholder relations	Newton's Second Law	Marginal utility, product conditions
Particular equations	Force equation	Utility, production functions

various situations in classical mechanics. Here, though, we studied the logic of testing with some *hypothetical* utility and production functions, making no claim that these functions correspond to any actual situation in the world. Table 6.1 gives some parallels between our accounts of classical mechanics and microeconomics.

Section 6.2's development of microeconomic theory discovered two kinds of difficulties which seem to have no parallels in Chapter 5's account of classical mechanics. First, the consciousness of our objects of inquiry – consumers and producers – played a role in the theory. For example, introspection (so far) failed to make the existence of a utility function plausible – for now we supposed such a function was a feature of the consumer's unconscious. Yet given the existence of a utility function, introspection made plausible the assumption that consumers *seek to* maximize utility, though not the assumption that they *succeed*. Second, time's meaning and role in microeconomics was uncertain. For example, a time period was required during which consumer tastes and production technologies remained fixed, but whether time meant historical time or time in the sense implicit in classical mechanics wasn't clear. Also, the marginal utility and production placeholders, in contrast to Newton's Second Law, did not involve time.

6.4.2 *Attachment difficulties*

Using two sets of hypothetical particular utility and production functions, we discovered attachment difficulties, each of which had a classical mechanical analogue. First, we could falsify the placeholder marginal utility and product conditions only in conjunction with particular utility and production functions, just as in classical mechanics we could falsify Newton's Second Law only in conjunction with a particular force equation. Second, the identification problem – which was how to use the observations in a noncircular way to distinguish supply and demand equations – had its classical mechanical counterpart in the problem of mass measurement – how to use the observations in a noncircular way to distinguish mass and force.

Also, both microeconomics and classical mechanics require a convention for "$\approx$" and can take similar statistical reporting-error models in place of perfect-reporting assumptions. And in both microeconomics and classical mechanics, once the theory has been tested and not falsified, the observations may be employed to estimate the values of certain constants that are not implied by the theory. In approximating otherwise intractable equations, we used mathematical facts [$\ln(x) \approx (1 + x)$ for $0 < x \ll 1$, and $\sin(\theta) \approx \theta$ for small θ] in microeconomics and classical mechanics, respectively. Finally, in classical mechanics, a mind's-eye picture of phenomena we couldn't see aided in attaching statements to the world of observations, while in microeconomics, introspection sometimes made plausible statements with no basis on observations. Table 6.2 presents the parallel attachment difficulties of classical mechanics and microeconomics.

Except for the difficulty of introspection, when it becomes the larger problem of the consciousness of the object of inquiry, the seven microeconomic attachment difficulties of Table 6.2 cannot prevent microeconomics from being a "real science," because their counterparts do not prevent classical mechanics from being a "real science." In studying microeconomic attachment difficulties, though, we discovered three without apparent classical mechanical parallels: the observation of market equilibrium, differences in individual consumer tastes and production technologies, and the practical problem of needing to observe the distribution of income (not just the aggregate income). If there are difficulties, then, that prevent microeconomics from being a science in the way classical mechanics is, they are found among the following: conscious objects of inquiry, the meaning and role of time in the theory, the existence of market equilibrium, differences in individual consumer tastes and production technologies, and the need to observe the distribution of income (not just the aggregate income).

We ended the chapter by examining an actual example from microeconometric practice. The example failed as "ideal science" in a funda-

Table 6.2. *Attachment difficulty parallels between classical mechanics and microeconomics*

Difficulty	Classical mechanics	Microeconomics
Law-sketch not falsifiable	Newton's Second Law falsifiable only in conjunction with particular force equation	Marginal utility condition (MUC) and marginal product condition (MPC) falsifiable only in conjunction with particular utility and production functions
Circularity	Mass measurement	Identification problem
Meaning of "≈"	Convention needed	Convention needed
Reporting not perfect	Statistical reporting model	Statistical reporting model
Theory doesn't imply value of all constants	Estimate with observations once theory successfully tested	Estimate with observations once theory successfully tested
Intractable equation	Approximate $\sin(\theta)$ with θ for small θ	Approximate $\ln(x)$ with x for $0 < x \ll 1$
Certain phenomena can't be seen	Mind's-eye picture	Introspection

mentally different way than did the classical mechanics experiments of Chapter 5 or the hypothetical microeconomic example of Sections 6.2.4 and 6.3.1–6.3.3. In the classical mechanics and hypothetical microeconomics examples we had a prediction statement *deduced* from the theory's statements confront an observation statement. The difficulty was in getting the deduced prediction statement to confront *actual* observations. In the actual example from microeconometric practice, though, we have a statement that has *not* been deduced from the theory's statements confront an observation statement. Thus, on the level of statements, ignoring attachment difficulties, an account of science as deductions from theory tested by observations fits the classical mechanics and hypothetical microeconomics examples, but not the actual example from microeconometric practice.

CHAPTER 7

The income maintenance experiments: microeconomic science or scientism?

7.1   Introduction: Science and scientism in microeconomics

Let "ideal science" be the reconstruction of science as predictions deduced from theory and tested by observations. In Chapter 5 we argued that classical mechanics, though science, failed as "ideal science" because of a set of attachment difficulties. In Chapter 6, with hypothetical particular utility and production functions, we argued that microeconomics failed as "ideal science" because of precisely the same set of attachment difficulties. Therefore, there exist utility and production functions which make microeconomics into a scientific research program in precisely the same way in which classical mechanics is a scientific research program.

Yet in Chapter 5 we also argued that classical mechanics is a *successful* scientific research program. That is, we exhibited particular force equations for various situations in the world. And we argued that in each such situation the conjunction of the particular force equation and Newton's Second Law implied a prediction statement which, ignoring the attachment difficulties, would confront and fail to be contradicted by actual observations. In Chapter 6, though, we made no claim that our hypothetical utility and production functions characterized any situation in the world or that their hypothetical prediction statements succeeded in confronting and failed to be contradicted by actual observations. Our account of microeconomics is so far one of an *undeveloped, nonmature,* or *not yet successful* scientific research program.

Thus, in classical mechanics we have a placeholder relation – Newton's Second Law – together with particular force equations (e.g., no force, Newton's Law of Universal Gravitation, Hooke's Law of spring forces) that experiments have shown to govern various situations. In microeconomics, so far, we have the analogue of Newton's Second Law in the marginal utility and product conditions. But we do not yet have the analogue of Newton's Law of Universal Gravitation or Hooke's Law – that is, particular utility and production functions – that experiments have shown to govern, in conjunction with MUCs and MPCs, various

182

situations. We need to examine some actual microeconometric studies in search of scientific research program successes.

Chapter 6 has also given us a sense of some possible barriers to becoming successful science that actual microeconometric work faces. These barriers include the following: conscious objects of inquiry, tastes which may vary between individuals, tastes and technologies which may not remain fixed over time, markets which may not be in equilibrium, and a difficult-to-observe distribution of income. In examining actual microeconometric studies, we shall be on the lookout to see how a study overcomes these barriers and becomes successful science. And we shall be alert to the possibility that a microeconometric study fails to overcome these barriers and so is not successful science but rather only appears to be successful science.

Borrowing from some French authors, the economist F. A. Hayek introduced the term "scientism"– by which he meant the "slavish imitation of the method and language of Science" (1979, p. 24) – into English political economic discourse in the early 1940s. Popper borrowed Hayek's term "as a name for the imitation of *what certain people mistake* for the method and language of science" (1964, p. 105). Both authors use the term "scientism" to critique historically oriented authors (e.g., Comte, Marx, Mill, Toynbee) who borrow metaphors from physics (e.g., social dynamics, forces, movements, and the directions and velocities of such forces and movements) to create the illusion that there is or can be a science of history.

Recall that our accounts of classical mechanics and microeconomics with the hypothetical examples involved theory statements implying prediction statements confronting observation statements. Difficulties arose in getting prediction statements to confront actual observations. But we concluded the last chapter by examining an actual econometric study in which the statement tested was not one deducible from the theory statements. Yet the authors of the actual microeconometric study "conclude that the theory of demand is inconsistent with the evidence." In doing so, though, they appear to move their study across a line between undeveloped science and scientism (i.e., faulty "imitation of the method and language of Science").

So we let Science be a reconstructed account of science that at the level of statements entails theory statements implying prediction statements confronting observation statements. Then science, at least in the sense in which classical mechanics or our hypothetical microeconomic examples are science, is an enterprise which is Science with the attachment difficulties which we have catalogued. In the spirit of Hayek and Popper, we take scientism to be work which at the level of statements fails to confront observation statements with prediction statements implied by theory statements, yet in imitation of science still speaks of tests of theory. In examining actual microeconometric studies, we shall try to distinguish successful science from undeveloped science from scientism.

Section 7.2.1 is an account of the theory of labor–leisure choice which underlies the income maintenance experiments. Section 7.2.2 adds taxes, including a negative income tax (NIT), to the theory, and Section 7.2.3 extends the theory from an individual to a family budget constraint. Section 7.3 examines the income maintenance experiments as tests of the theory of labor–leisure choice. For each of the experiments in turn, we study first its deductive structure and then any attachment difficulties with no apparent classical mechanical counterparts. We seek to distinguish successful science from underdeveloped science from scientism in the experiments. Section 7.4 is the usual summary and conclusion.

7.2   The theory of labor–leisure choice

7.2.1   The theory of labor–leisure choice without taxes

Consider a worker whose hours of leisure, disposable income, hours of work, wage rate, and nonwage income in a given time period of T hours are L, y_d, h, $w > 0$, and $y > 0$, respectively. Let $\mathcal{L}$ $(\subseteq R^2) = \{(L, y_d):$ $0 \leq L \leq T$ and $y_d \geq 0$ are the worker's possible hours of leisure and disposable income}. Suppose, for now, that the worker faces no taxes. Then the worker's budget constraints are $h + L = T$ and $wh + y = y_d$. Also, if $u: \mathcal{L} \to R^1$ is any function, we say that u is a utility function if it in some sense measures the satisfaction the worker receives from hours of leisure and disposable income. We assume that a worker's satisfaction is represented by a utility function. Finally, using the constraint $T = h + L$, we can write $u(L, y_d) = u(T - h, y_d) =$, say, $u^+(h, y_d)$.

Next we have a principle and axiom similar to UM (of Chapter 6) and Axiom 6.1 of consumer theory.

Principle of Utility Maximization (UM): If a worker's utility function exists, then the worker maximizes that utility function subject to the worker's budget constraints. Further, we suppose that in maximizing u, the worker has no control over the wage rate – that is, the worker takes w

as given and chooses (L, y_d) from among those elements of $\mathscr{L}$ satisfying the budget constraints, so as to maximize $u(L, y_d)$.

Axiom 7.1. $u(L, y_d)$ *is twice differentiable;* $(\partial u/\partial L)$, $(\partial u/\partial y_d) > 0$; *and*

$$(\partial^2 u/\partial L^2)(\partial u/\partial y_d)^2 - 2(\partial^2 u/\partial L \partial y_d)(\partial u/\partial L)(\partial u/\partial y_d)$$
$$+ (\partial^2 u/\partial y_d^2)(\partial u/\partial L)^2 < 0$$

everywhere on the interior of $\mathscr{L}$.

Thus far, the concepts of labor–leisure choice theory resemble those of consumer theory. So our critical comments on the existence of u and plausibility of UM in consumer theory (see Section 6.2.1) apply here as well. But an additional critical comment on the logical compatibility of consumer and labor–leisure choice theories seems germane. Call the two theories' utilities $u_1(q_1, q_2)$ and $u_2(L, y_d)$. From consumer theory, $y_d = p_1 q_1 + p_2 q_2$; so $u_2(L, y_d) = u_2^*(L, q_1, q_2)$ for some function u_2^*. So the consumer-worker *simultaneously* maximizes $u_1(q_1, q_2)$ and $u_2^*(L, q_1, q_2)$ subject to $w(T - L) + y = y_d = p_1 q_1 + p_2 q_2$. In general, a solution (L_0, q_{10}, q_{20}) to such a simultaneous-maximum problem doesn't exist. So one either must make no claim that particular versions of consumer and labor–leisure choice theories are logically compatible (our approach in Chapters 6 and 7) or must make assumptions on the consumer-worker's preferences among L, q_1, and q_2 that make the two theories logically compatible. For example, the economists G. Burtless and J. A. Hausman suggest "assuming homogeneous weak separability of preferences between labor-supply [h] and the other . . . goods [q_1, q_2]" (1978, p. 1111).

Another critical remark on the theory of labor–leisure choice as so far developed is that it treats leisure (hence labor) and disposable income (hence money) as consumer theory treats *commodities,* but labor and money are not *really* commodities. The political economist Karl Polanyi (1968), for example, argued that "commodities are defined . . . as objects produced for sale on the market" (p. 31), so that "labor . . . and money are obviously *not* commodities" (p. 32) because they are not "produced for sale." For example, "labor is only another name for a human activity which goes with life itself, which in its turn is not produced for sale but for entirely different reasons" (p. 32). Yet Polanyi also concedes that labor and money "are being actually bought and sold on the market; their demand and supply are real magnitudes" (p. 32) and so suggests our response to his critique: Our theory's treatment of labor and money merely mirrors our society's (deplorable, to be sure) historical tendency to organize these items *as if* they were "objects produced for sale on the market."

Let CAM mean constrained absolute maximum on $\mathscr{L}$, and let UCAM mean unique constrained absolute maximum on $\mathscr{L}$. Then we have a theorem whose analogue lies in Theorems 6.1 and 6.2 of consumer theory.

Theorem 7.1. *Let $u(L, y_d)$ be a worker utility function, where $(\partial u/\partial L)$, $(\partial u/\partial y_d)$ exist and are >0 everywhere on $\mathscr{L}$'s interior, and suppose $h + L = T$ and $wh + y = y_d$ hold.*

(i) *$u(L, y_d)$ $(= u^+(h, Y_d))$ has a (not necessarily unique) CAM, and if the CAM is not on the boundary of $\mathscr{L}$, UM implies*

$$[(\partial u/\partial L)/(\partial u/\partial y_d)] = w = -[(\partial u^+/\partial h)/(\partial u^+/\partial y_d)]$$

at $u(L, y_d)$'s CAM. If $u(L, y_d)$ satisfies Axiom 7.1, then

(ii) *$u(L, y_d)$ has a UCAM for any $w_0 > 0$, $y_0 \geq 0$. If (L_0, y_{d0}) is the UCAM and $h_0 = T - L_0$, we call $h_0 = h_0(w_0, y_0)$, or $h = h(w, y)$, the individual-worker labor-supply function and $v(w_0, y_0) = u(L_0, y_{d0})$, or $v(w, y)$, the individual-worker indirect utility function. Let*

$$(\partial h(w, y)/\partial w)|_{u=\text{constant}} = (\partial u/\partial y_d)^3/[2(\partial^2 u/\partial L \partial y_d)(\partial u/\partial L)$$
$$(\partial u/\partial y_d) - (\partial^2 u/\partial L^2)(\partial u/\partial y_d)^2 - (\partial^2 u/\partial y_d^2)(\partial u/\partial L)^2] \quad at\ (L_0, y_{d0})$$

if the UCAM is in $\mathscr{L}$'s interior. Then

(iii) *$(\partial h(w, y)/\partial w) = (\partial h(w, y)/\partial w)|_{u=\text{constant}} + h(w, y)(\partial h(w, y)/\partial y)$*
$$\textit{(Slutsky's Equation, SE)}$$

if $u(L, y_d)$'s UCAM is in $\mathscr{L}$'s interior for (w, y).[1]

Next consider the following result from calculus and analytic geometry.

Proposition 7.1. *If $R^+ = f(r_1, r_2)$ is a real-valued function on $\mathscr{R}$ $(\subseteq R^2)$, if $(\partial f/\partial r_i)$ exist and are >0 everywhere on $\mathscr{R}$'s interior, and if $f(r_1, r_2) = c$ (constant), then dr_2/dr_1 $[= \text{the slope of the curve } f(r_1, r_2) = c] = -(\partial f/\partial r_1)/(\partial f/\partial r_2)$ at any point (r_{10}, r_{20}) on the curve in $\mathscr{R}$'s interior.*[2]

---

[1] The proofs of Theorems 6.1 and 6.2 constitute a proof of Theorem 7.1.
[2] See Hildebrand (1962, p. 339) for the proposition's proof.

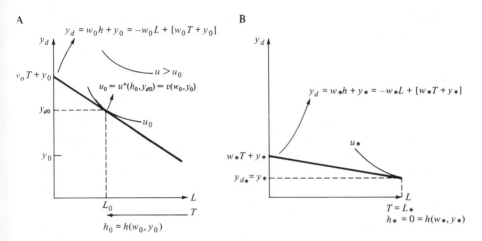

Figure 7.1. Some of Theorem 7.1's results.

The proposition allows us to depict some of the theorem's results graphically. The same utility function underlies parts A and B of Figure 7.1. Each of the curved lines is a locus of equal utility, called an indifference curve. The bowing of the curves toward the origin depicts the second-order condition of Axiom 7.1, and $(\partial u/\partial L)$, $(\partial u/\partial y_d) > 0$ means that utility rises as one moves outward from the origin.

In Figure 7.1A, $(w, y) = (w_0, y_0)$, and u's UCAM [see (ii) of the theorem] is at (L_0, y_{d0}), where u's value is $v(w_0, y_0)$, while the value of the individual-worker labor-supply function is $h_0 = h(w_0, y_0)$. Also, since u's UCAM is not on the boundary of $\mathcal{L}$, the slope of the constraint, which is $-w_0$, is equal to the slope of u_0 at the UCAM, which is $-(\partial u/\partial L)/(\partial u/\partial y_d)$ by the proposition [see (i) of the theorem]. In Figure 7.1B, $(w, y) = (w_*, y_*)$, and u's UCAM [see (ii) of the theorem] is at (L_*, y_{d*}) – on the boundary of $\mathcal{L}$ – where u's value is $v(w_*, y_*)$, while the value of the individual labor-supply function is $h_* = h(w_*, y_*) = 0$. Also note that $[(\partial u/\partial L)/(\partial u/\partial y_d)] \neq w$ at (L_*, y_{d*}); yet (i)'s conclusion is not contradicted because (L_*, y_{d*}) is on the boundary of $\mathcal{L}$.

Section 7.2.2 extends the theory of labor–leisure choice to the sort of budget constraints that include taxes faced by income maintenance experimental subjects before and during the experiments. Section 7.2.3 is a further extension from the sort of *individual* budget constraints of Sections 7.2.1 and 7.2.2 to the *family* budget constraints faced by income maintenance experimental subjects. Finally, Theorem 7.1's results (and those in the next two sections to follow) are still all of a placeholder sort. So

Section 7.3 turns to the actual experiments themselves and their (explicit or implicit) particular utility functions.

7.2.2   Adding taxes to the theory of labor–leisure choice

Suppose that a worker has no source of nontaxable, nonwage income (i.e., ignore welfare and NIT guarantees for now). Also suppose that a worker faces only a federal income tax on gross income (i.e., ignore such taxes as state and social security). And suppose the federal income tax is zero on gross income up to an exemption level E, and a fixed rate t_F on income above E (i.e., ignore higher tax brackets, because wage rates and taxable nonwage income of the low-income workers of interest to us bar higher tax brackets with a physically reasonable number of hours worked). Let y_d, h, L, and T be as in the previous section, let w_g, $y_g > 0$ be the gross (i.e., before taxes) wage rate and taxable nonwage income (e.g., interest on savings), and suppose that $y_g < E$ – a plausible assumption for our low-income workers. Then our worker faces the following.

Budget Constraint 7.1

$$y_d = \begin{cases} -w_g(1 - t_F)L + [(w_g T + y_g)(1 - t_F) + t_F E] \\ \quad \text{if } 0 \leq L \leq T + (y_g/w_g) - (E/w_g) \\ -w_g L + (w_g T + y_g) \\ \quad \text{if } T + (y_g/w_g) - (E/w_g) \leq L \leq T. \end{cases}$$

Figure 7.2A depicts such a budget constraint. In the period before the income maintenance experiments, all subjects (and during the experiments, control-group subjects) faced such a budget constraint.

Now consider a worker who may choose between an NIT with (negative tax rate, guarantee level) (t_n, G) and Budget Constraint 7.1. Suppose also that $t_n > t_F$, $G > t_n E$, and $G - t_F E < w_g T(t_n - t_F)$. Such a worker faces the following.

Budget Constraint 7.2

$$y_d = \begin{cases} -w_g(1 - t_F)L + [(w_g T + y_g)(1 - t_F) + t_F E) \\ \quad \text{if } 0 \leq L \leq T + (y_g/w_g) - [(G - t_F E)/(t_n - t_F)w_g] \\ -w_g(1 - t_n)L + [(w_g T + y_g)(1 - t_n) + G] \\ \quad \text{if } T + (y_g/w_g) - [(G - t_F E)/(t_n - t_F)w_g] \leq L \leq T. \end{cases}$$

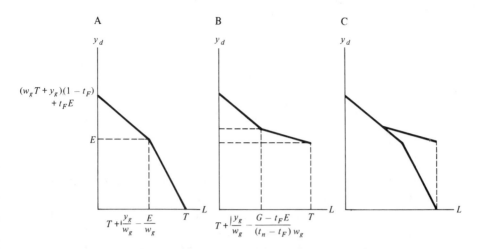

Figure 7.2. Income maintenance experiments' control and treatment-group budget constraints. (A) Control group; (B) treatment group; (C) control and treatment groups.

Figure 7.2B depicts such a budget constraint: $t_n > t_F$ and $G > t_n E$ for all NIT plans of the New Jersey, Seattle-Denver, and Gary income maintenance experiments. Also, $G - t_F E < w_g T(t_n - t_F)$ seems to be true for *nearly* all male heads of two-parent families on each of the experiments' NIT plans.[3] Thus, nearly all treatment-group subjects of the three experi-

---

[3] For each individual worker, $T = 24 \times 365 = 8{,}760$ hours (per year), and $t = .15$. Consider as an example a family of 4. For a New Jersey family of 4, E (in 1968) was $3,000. For six of New Jersey's eight (G, t_n) NIT plans, $G - t_F E < w_g T(t_n - t_F)$ for all $w_g \geq \$1.15$ per hour, the 1968 minimum wage. For the last two plans, though, $G - t_F E < w_g T(t_n - t_F)$ only for all $w_g \geq \$1.20$ and $1.50 per hour. Thus for these two plans, there may have been some male family heads receiving very low wages whose budget constraint when they were regarded as individual workers was the second piece only of Budget Constraint 7.2. But, as we shall see in Section 7.3.1, the New Jersey researchers correctly regarded individuals as facing a family budget constraint with a condition analogous to $G - t_F E < w_g T(t_n - t_F)$ which *all eight* NIT plans satisfied (see Section 7.2.2).

For a Gary or Seattle-Denver family of 4, E (in 1971) was $3,750. For three of Gary's four (G, t_n) plans, $G - t_F E < w_g T(t_n - t_F)$ for all $w_g \geq \$1.60$ per hour, the 1971 minimum wage. For the fourth plan, though, $G - t_F E < w_g T(t_n - t_F)$ only for all $w_g \geq \$1.71$ per hour. So for this plan there may have been some male heads receiving very low wages whose budget constraint when they were regarded as individual workers was the second piece only of Budget Constraint 7.2. Yet Gary researcher Kehrer (1979, p. 432) explains: "The families with a male head of household present (almost all of which were intact husband-wife families) usually had low income but generally were not extremely poor. The husbands were typically full-time workers who were able to earn enough to keep their families out of poverty (only 10 percent of these families had incomes below the poverty

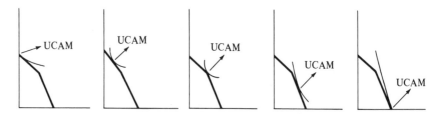

Figure 7.3. Control-group maxima for various shapes of indifference curves.

ments who were male heads of two-parent families faced Budget Constraint 7.2. Figure 7.2C superimposes B on A.

Notice that while the previous section's budget constraint was linear, Budget Constraints 7.1 and 7.2 are only *piecewise* linear. Now consider first a worker whose budget constraint is that depicted in Figure 7.2A. Suppose the worker's utility function satisfies Axiom 7.1, so that the worker's indifference curves are bowed toward the origin. Then, depending on the precise shape of the utility function's indifference curves, it achieves a UCAM either at one of the three corner points or on the interior of one of the two segments of the budget constraint. Figure 7.3 depicts the five possibilities. For each of the five possibilities we can define

$$h[T, w_g(1 - t_F), (w_g T + y_g)(1 - t_F) + t_F E, w_g, (w_g T + y_g)]$$
$$(= \text{the value of } h \text{ at the UCAM})$$

as the individual-worker labor-supply function.

Footnote 3 *(cont.)*

level)." Thus, few, if any, Gary experiment husbands could have been making $<\$1.71$ per hour.

For 10 of Seattle-Denver's 11 (G, t_n) NIT plans, $G - t_F E < w_g T(t_n - t_F)$ for all $w_g \geq \$1.60$ (in 1971). For the 11th plan, though, $G - t_F E < w_g T(t_n - t_F)$ only for all $w_g \geq \$1.64$ per hour. So for this plan there may have been a few male family heads receiving very low wages whose budget constraint when they were regarded as individual workers was the second piece only of Budget Constraint 7.2. Yet Keeley et al. (1977, p. 16) reveal that the participant husbands' average wage rate was $3.30 per hour; so there probably weren't many husbands with wage rates $<\$1.64$ per hour.

For the two experiments (Gary and Seattle-Denver) that, as we shall see in Section 7.3, failed to model the *family* nature of the budget constraint, there may have been a few heads whose wage rates were such that as individuals they faced a strictly linear during-experiment budget constraint. Sections 7.3.3, 7.3.5, and 7.3.7 will note how these heads could have been dealt with. Figures for E are from Commerce Clearing House (1969, p. 17; 1971, p. 23), while those for minimum wages are from U.S. Bureau of the Census (1984, p. 422). The 8 New Jersey, 4 Gary, and 11 Seattle-Denver (G, t_n) NIT plans for various family sizes are detailed in Kershaw and Fair (1976, pp. 6–11), Kehrer et al. (1975, pp. 3–6), and Keeley, Spiegelman, and West (1980, pp. 5–10), respectively.

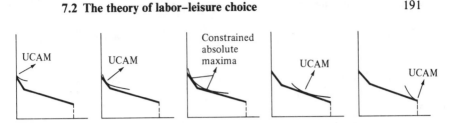

Figure 7.4. Treatment-group maxima for various shapes of indifference curves.

Next consider a worker whose budget constraint is that depicted in Figure 7.2B. Suppose the worker's utility function satisfies Axiom 7.1, so that the indifference curves are bowed toward the origin. Then, depending on the precise shape of the utility function's indifference curves, either it achieves a UCAM at one of the two corner points or on the interior of one of the two segments of the budget constraint or it achieves a CAM at *two* points – one on the interior of each of the two segments of the budget constraint. Figure 7.4 depicts the five possibilities. For each of the five possibilities, we can define

$$H[T, w_g(1 - t_F), (w_gT + y_g)(1 - t_F) + t_FE, w_g(1 - t_n),$$
$$(w_gT + y_g)(1 - t_n) + G]$$
$$= \{\text{the values of } h \text{ at the constrained absolute maximum of } u(L, y_d) \text{ on } \mathscr{L}\}$$

as the individual-worker labor-supply correspondence.

Now consider two definitions.

Definition: A subset of R^2 is convex iff the straight line between any two points in the set is in the set.

Definition: The area between the worker's budget constraint, the portion of the two axes intersecting at the origin, and the vertical line at $L = T$ is the worker's *budget set*.

Notice that control-group workers have convex budget sets, and treatment-group workers do not. In fact, it is the convexity of the control-group worker's budget set which allows us to define an individual-worker labor-supply function and the nonconvexity of the treatment-group worker's budget constraint which forces us to settle for the weaker notion of an individual-worker labor-supply correspondence.

Finally, we may summarize this section's discussion as a theorem.

Theorem 7.2. *Let $u(L, y_d)$ be a worker's utility function that satisfies Axiom 7.1.*

(i) *If Budget Constraint 7.1 is the worker's budget constraint, then $u(L, y_d)$ has a UCAM, say (L_0, y_{d0}), and*

$$h[T, w_g(1 - t_F), (w_g T + y_g)(1 - t_F) \\ + t_F E, w_g, (w_g T + y_g)] = T - L_0$$

is the individual worker labor-supply function, while

$$v[w_g(1 - t_F), (w_g T + y_g)(1 - t_F) \\ + t_F E, w_g, (w_g T + y_g)] = u(L_0, y_{d0})$$

is the individual-worker indirect utility function.

(ii) *If Budget Constraint 7.2 is the worker's budget constraint, then $u(L, y_d)$ has one, say (L_0, y_{d0}), or two, say (L_0, y_{d0}) and (L_*, y_{d*}), CAMs, and*

$$H[T, w_g(1 - t_F), (w_g T + y_g)(1 - t_F) + t_F E, w_g(1 - t_n), \\ (w_g T + y_g)(1 - t_n) + G] = \{T - L_0\} \quad \text{or} \quad \{T - L_0, T - L_*\}$$

is the individual worker labor-supply correspondence, while

$$v[w_g(1 - t_F), (w_g T + y_g)(1 - t_F) + t_F E, w_g(1 - t_n), \\ (w_g T + y_g)(1 - t_n) + G] = u(L_0, y_{d0}) \quad (= u(L_*, y_{d*}))$$

is the individual-worker indirect utility function.

Notice that the theorem's results are still of a placeholder sort. To get prediction statements which can confront the observations, we must, as in the previous chapter, conjecture a particular utility function – a task which we reserve for Section 7.3. First, however, we take up the issue of *family* versus individual utility function and budget constraints.

7.2.3   *Family utility function and budget constraint in the theory of labor–leisure choice*

In our development of the theory of labor–leisure choice, we have so far supposed that an *individual* worker faces an *individual* budget constraint. But in the income maintenance experiments, we shall be interested in two-parent *families* of workers. Hence, it seems more appropriate to take the viewpoint of a *family*, with a family utility function facing a budget constraint involving family earnings. Without loss of generality, suppose the family contains three potential wage earners – wife, husband, and child designated by subscripts 1, 2, and 3, respectively. Let L_i, h_i, and T_i be the hours of leisure, work, and total available for family member i, where $L_i + h_i = T_i$ for $i = 1, 2, 3$. Also suppose that y_d is family disposable income, and let

$$\mathcal{L} = \{(L_1, L_2, L_3, y_d): 0 \leq L_i \leq T_i \text{ and } y_d \geq 0\}.$$

Then we suppose that the family's utility function is of the form $u(L_1, L_2, L_3, y_d)$ and that the family maximizes this utility function subject to its budget constraint.

Suppose the family has no source of nontaxable, nonwage income (i.e., again ignore welfare and NIT guarantees for now). Also suppose the family faces only a federal income tax on its gross income (i.e., again ignore such taxes as state income and social security). And suppose the federal income tax is zero on family gross income up to an exemption E, and a fixed rate t_F on income above E (i.e., ignore higher tax brackets, because wage rates and taxable nonwage income of the low-income families of interest to us bar higher tax brackets with a physically reasonable number of hours worked). Let $w_{gi} > 0$ be family member i's gross wage rate for $i = 1, 2, 3$, let $y_g > 0$ be gross (i.e., before taxes) taxable nonwage income (e.g., interest on savings), and suppose that $y_g < E$ – a plausible assumption for our low-income families. Then our family faces the following.

Budget Constraint 7.3

$$y_d = \begin{cases} -w_{g1}L_1 + w_{g2}L_2 + w_{g3}L_3)(1 - t_F) + (w_{g1}T_1 + w_{g2}T_2 + w_{g3}T_3 + y_g)(1 - t_F) + t_F E \\ \quad \text{if } 0 \leq w_{g1}L_1 + w_{g2}L_2 + w_{g3}L_3 \leq (w_{g1}T_1 + w_{g2}T_2 + w_{g3}T_3) + y_g - E \\ -w_{g1}L_1 + w_{g2}L_2 + w_{g3}L_3) + (w_{g1}T_1 + w_{g2}T_2 + w_{g3}T_3) + y_g \\ \quad \text{if } (w_{g1}T_1 + w_{g2}T_2 + w_{g3}T_3) + y_g - E \leq w_{g1}L_1 + w_{g2}L_2 + w_{g3}L_3 \\ \quad \leq (w_{g1}T_1 + w_{g2}T_2 + w_{g3}T_3). \end{cases}$$

In the period before the income maintenance experiments, *all* subject families (and during the experiments, control-group families) faced such a budget constraint.

Now consider a family which may choose between an NIT with (negative tax rate, guarantee level) (t_n, G) and Budget Constraint 7.3. Suppose also that $t_n > t_F$, $G > t_n E$, and $G - t_F E < (w_{g1}T_1 + w_{g2}T_2 + w_{g3}T_3)(t_n - t_F)$. Such a family faces the following.

Budget Constraint 7.4

$$
y_d = \begin{cases}
-(w_{g1}L_1 + w_{g2}L_2 + w_{g3}L_3)(1 - t_F) + (w_{g1}T_1 + w_{g2}T_2 + w_{g3}T_3 + y_g)(1 - t_F) + t_F E \\
\quad \text{if } 0 \le w_{g1}L_1 + w_{g2}L_2 + w_{g3}L_3 \le w_{g1}T_1 + w_{g2}T_2 + w_{g3}T_3 + y_g - (G - t_F E)/(t_n - t_F) \\
-(w_{g1}L_1 + w_{g2}L_2 + w_{g3}L_3)(1 - t_n) + (w_{g1}T_1 + w_{g2}T_2 + w_{g3}T_3 + y_g)(1 - t_n) + G \\
\quad \text{if } w_{g1}L_1 + w_{g2}L_2 + w_{g3}L_3 + y_g - (G - t_F E)/(t_n - t_F) \\
\quad \le w_{g1}L_1 + w_{g2}L_2 + w_{g3}L_3 \le w_{g1}T_1 + w_{g2}T_2 + w_{g3}T_3.
\end{cases}
$$

With $t_n > t_F$, $G > t_n E$, and $G - t_F E < (w_{g1}T_1 + w_{g2}T_2 + w_{g3}T_3)(t_n - t_F)$ for all NIT plans of the New Jersey, Seattle-Denver, and Gary income maintenance experiments, all two-parent treatment-group families with three potential workers in the three experiments faced Budget Constraint 7.4.

We also need to generalize Axiom 7.1. Consider first a definition.

Definition: If $u(L_1, L_2, L_3, y_d)$ is twice-differentiable at a point, let

(i) $$D_1 = \begin{vmatrix} \partial^2 u/\partial L_1^2 & \partial^2 u/\partial L_1 \partial L_2 & \partial u/\partial L_1 \\ \partial^2 u/\partial L_2 \partial L_1 & \partial^2 u/\partial L_2^2 & \partial u/\partial L_2 \\ \partial u/\partial L_1 & \partial u/\partial L_2 & 0 \end{vmatrix},$$

(ii) $$D_2 = \begin{vmatrix} \partial^2 u/\partial L_1^2 & \partial^2 u/\partial L_1 \partial L_2 & \partial^2 u/\partial L_1 \partial L_3 & \partial u/\partial L_1 \\ \partial^2 u/\partial L_2 \partial L_1 & \partial^2 u/\partial L_2^2 & \partial^2 u/\partial L_2 \partial L_3 & \partial u/\partial L_2 \\ \partial^2 u/\partial L_3 \partial L_1 & \partial^2 u/\partial L_3 \partial L_2 & \partial^2 u/\partial L_3^2 & \partial u/\partial L_3 \\ \partial u/\partial L_1 & \partial u/\partial L_2 & \partial u/\partial L_3 & 0 \end{vmatrix},$$

(iii) $$D_3 = \begin{vmatrix} \partial^2 u/\partial L_1^2 & \partial^2 u/\partial L_1 \partial L_2 & \partial^2 u/\partial L_1 \partial L_3 & \partial^2 u/\partial L_1 \partial y_d & \partial u/\partial L_1 \\ \partial^2 u/\partial L_2 \partial L_1 & \partial^2 u/\partial L_2^2 & \partial^2 u/\partial L_2 \partial L_3 & \partial^2 u/\partial L_2 \partial y_d & \partial u/\partial L_2 \\ \partial^2 u/\partial L_3 \partial L_1 & \partial^2 u/\partial L_3 \partial L_2 & \partial^2 u/\partial L_3^2 & \partial^2 u/\partial L_3 \partial y_d & \partial u/\partial L_3 \\ \partial^2 u/\partial y_d \partial L_1 & \partial^2 u/\partial y_d \partial L_2 & \partial^2 u/\partial y_d \partial L_3 & \partial^2 u/\partial y_d^2 & \partial u/\partial y_d \\ \partial u/\partial L_1 & \partial u/\partial L_2 & \partial u/\partial L_3 & \partial u/\partial y_d & 0 \end{vmatrix}.$$

Then the generalization of Axiom 7.1 is as follows.

Axiom 7.2. $u(L_1, L_2, L_3, y_d)$ *is twice-differentiable,* $(\partial u/\partial L_i)$, $(\partial u/\partial y_d) > 0$ *for* $i = 1, 2, 3,$ *and* $D_1 > 0$, $D_2 < 0$, $D_3 > 0$ *everywhere on the interior of* $\mathcal{L}$.

We can now state the generalization of Theorem 7.2 from the individual to the family.

Theorem 7.3. *Let* $u(L_1, L_2, L_3, y_d)$ *be a family utility function which satisfies Axiom 7.2.*

(i) *If the family faces Budget Constraint 7.3, then u has a UCAM, say* $(L_{10}, L_{20}, L_{30}, y_{d0})$, *and*

$$h_i[T_i, w_{g1}, w_{g2}, w_{g3}, t_F, (w_{g1}T_1 + w_{g2}T_2 + w_{g3}T_3 + y_g), E] = T_i - L_{i0}$$

is family member i's individual labor-supply function, while

$$v[w_{g1}, w_{g2}, w_{g3}, t_F, (w_{g1}T + w_{g2}T + w_{g3}T + y_g), E] = u(L_{10}, L_{20}, L_{30}, y_{d0})$$

is the family's indirect utility function.

(ii) *If the family faces Budget Constraint 7.4, then*

$$\mathcal{C} = \{(L_{10}, L_{20}, L_{30}, y_{d0}): (L_{10}, L_{20}, L_{30}, y_{d0}) \text{ is a CAM of } u\} \neq \varnothing$$

and

$$H_i[T_i, w_{g1}, w_{g2}, w_{g3}, t_F, t_n, (w_{g1}T_1 + w_{g2}T_2 + w_{g3}T_3 + y_g), E, G]$$
$$= \{T_i - L_{i0}: (L_{10}, L_{20}, L_{30}, y_{d0})$$
$$\text{is in } \mathcal{C} \text{ for some } L_{j0}\text{'s}, j \neq i, \text{ and } y_{d0}\}$$

is family member i's individual labor-supply correspondence, while

$$v[w_{g1}, w_{g2}, w_{g3}, t_F, t_n, (w_{g1}T_1 + w_{g2}T_2 + w_{g3}T_3 + y_g), E, G] = u(L_{10}, L_{20}, L_{30}, y_{d0}),$$

where $(L_{10}, L_{20}, L_{30}, y_{d0}) \in \mathcal{C}$ *is the family's indirect utility function.*

Notice that when we go over from individual to family utility function budget constraints we get a family indirect utility function, but individual labor-supply functions and correspondences. Notice also that once again the theorem's results are of an abstract nature. To get prediction statements which have a chance of confronting the observations, we turn to the actual income maintenance experiments themselves and their sometimes conjectured particular utility functions.

7.3   The income maintenance experiments as tests of the theory of labor–leisure choice

7.3.1   *The New Jersey experiment: "deduction" of a prediction statement*

Let L_i, h_i, T_i, w_{gi}, for $i = 1, 2, 3$, y_g, y_d, t_F, t_n, E, G be as in the previous section. Also let $w_{g1}T_1 + w_{g2}T_2 + w_{g3}T_3 = T^*$; let Δh_i = family member i's hours worked after minus hours worked before an NIT is introduced for $i = 1, 2, 3$; and recall from the previous chapter that MEC is the assumption that market demand = market supply = observed aggregate amount bought–sold. Let S be: (w_{gi}, T_i) for $i = 1, 2, 3$ doesn't change with the introduction of an NIT; $y_g = 0$; and market demand for low-income labor is perfectly elastic. And let $\mathscr{P}_1$ be: $u(L_1, L_2, L_3, y_d) = KL_1^{\beta_1}L_2^{\beta_2}L_3^{\beta_3}y_d^{1-\beta}$, with K, $\beta_i > 0$ for $i = 1, 2, 3$, and $\beta_1 + \beta_2 + \beta_3 < 1$. The last part of S allows us to ignore the demand side of the low-income labor market and study only individual labor supply, while $\mathscr{P}_1$ conjectures a particular family utility function.

Now let k index observations from different families, and let

$$Q_{ik} = \begin{cases} G_k/(1 - t_n)w_{gik} \text{ if } ik \text{ pays no federal taxes before and} \\ \quad \text{receives a payment during the NIT} \\ [G_k/(1 - t_n)w_{gik}] - [t_FE_k/(1 - t_F)w_{gik}] \text{ if } ik \text{ pays federal} \\ \quad \text{taxes before and receives a payment during the NIT} \\ 0 \text{ if } ik \text{ pays federal taxes before and during (and} \\ \quad \text{receives no payment during) the NIT.} \end{cases}$$

We then have a theorem.

Theorem 7.4. *Suppose UM, S, $\mathscr{P}_1$, and MEC hold; $G > t_nE$, $t_n > t_F$, and $(G - t_FE)/(t_n - t_F) < T^*$; in the period before (after) an NIT is introduced, Budget Constraint 7.3 (7.4) is the family budget constraint; y_d, w_{gi}, $L_i > 0$ for $i = 1, 2, 3$; and $w_{g1}L_1 + w_{g2}L_2 + w_{g3}L_3 \neq T^* - E + y_g$, T^* before and $\neq T^* - (G - t_FE)/(t_n - t_F) + y_g$, T^* after an NIT is introduced. Then $\Delta h_{ik} = -\beta_iQ_{ik} + c_i$, where $0 < \beta_i < 1$ and $c_i = 0$ for $i = 1, 2, 3$.*

Notice that the restrictions on w_{gi}, L_i for $i = 1$, 2, 3 and $w_{g1}L_1 + w_{g2}L_2 + w_{g3}L_3$ in the theorem's hypothesis eliminate any prediction statement for corner solutions.

One way to allow for variation in tastes and preferences between families is to suppose $\beta_i = \beta_i \cdot \mathbf{x}_i$ where $\beta_i = (\beta_{i0}, \beta_{i1}, \ldots, \beta_{im})$ and $= \mathbf{x}_i = (1, x_{i1}, x_{i2}, \ldots, x_{im})$ where x_{ij} is the value of a jth demographic variable (e.g., age, health, education, or location-indicator) for individual i. Such an assumption constitutes an ad hoc (in senses elaborated in Section 7.3.2) theory of taste and preference determination. We can also allow for measurement error by assuming an additive measurement-error term $\epsilon_i \sim N(0, \sigma_i)$ for Δh_{ik}. With these two added assumptions, Theorem 7.4's prediction statement becomes $\Delta h_{ik} = -(\beta_i \cdot \mathbf{x}_i)Q_{ik} + c_i + \epsilon_{ik}$, where $0 < \beta_i \cdot \mathbf{x}_i < 1$ (for each k) and $c_i = 0$.

Now let the subscripts f, m, o correspond to our 1, 2, 3, and let (Y, t, W_m) correspond to our (y_d, t_n, w_{g2}). Then, in their analysis related to the theory of labor–leisure choice, the New Jersey researchers Watts and Horner (1977) conjectured that "there is associated with each family a utility function $u = u[z(Y, L_m, L_f, L_o)]$" (p. 64) and "that z is produced by a simple Cobb-Douglas production function" (p. 65). Suppose there is "no change in wage rates or time availabilities for any family member" (p. 65) when an NIT with (guarantee level, tax rate) (G, t) is introduced. Then, if $\Delta H_m = $ our Δh_2, the researchers claim that their assumptions imply that for the treatment group, "$\Delta H_m = -\beta_m[G/(1 - t)W_m]$" (p. 65). And they also say: "By its derivation, β_m should be positive if the nonmarket time input has positive productivity for z" (p. 65). What makes formalizing these theoretical musings in a theorem difficult is the fact that the New Jersey researchers are not *explicit* about the budget constraint assumption behind them.

However, reverting back to *our* notation, the New Jersey theory appears to formalize as follows.

Theorem 7.5. *Let UM, S, $\mathscr{P}_1$, and MEC hold;*

$$-(w_{g1}L_1 + w_{g2}L_2 + w_{g3}L_3) + (T^* + y_g)$$
$$= y_d(-(w_{g1}L_1 + w_{g2}L_2 + w_{g3}L_3)(1 - t_n) + (T^* + y_g)(1 - t_n) + G = y_d)$$

be the budget constraint in the period before (after) an NIT is introduced; w_{gi}, $L_i > 0$ for $i = 1$, 2, 3; and $w_{g1}L_1 + w_{g2}L_2 + w_{g3}L_3 \neq T^*$ before and after an NIT is introduced. Then $\Delta h_i = -(\beta_i G)/[w_{gi}(1 - t_n)]$, where $0 < \beta_i < 1$ for $i = 1$, 2, 3.

Notice again that the restrictions on w_{gi}, L_i for $i = 1, 2, 3$ and $w_{g1}L_1 + w_{g2}L_2 + w_{g3}L_3$ in the theorem's hypothesis eliminate any prediction statement for corner solutions. Also note that the theorem's conclusion contains the Δh_i relation of the researchers' musings and that Δh_i is the same as the first of the three "pieces" of the Δh_i relation in Theorem 7.4's conclusion. Notice also that the researchers fail to mention their implicit assumption that $y_g = 0$ in the period just before and after an NIT is introduced, while we include it in S. Note, finally, that the researchers suggest only $0 < \beta_i$, while the theorem's conclusion contains the fuller $0 < \beta_i < 1$.

The theorem's Δh_i relation loses the other two "pieces" of the Δh_i relation of Theorem 7.4 because the theorem supposes that the budget constraints before and during an NIT are linear, whereas Theorem 7.4 supposed only that they were *piecewise* linear. Linearity during the NIT means that treatment-group families will accept payments and pay the tax rate t_n even if their gross (pre-payment-tax) income is large enough that it would be in their interest not to. Such an assumption seems implausible and at odds with the observation that some treatment-group families choose not to receive payments. Linearity before the NIT means that families face no federal income tax even if their gross taxable income is above E – an assumption which is patently false. The budget constraint assumptions which imply the New Jersey researchers' Δh_i relation then appear to be completely false.

Following Friedman (see Section 6.1), one might argue that "unrealistic assumptions" are not a difficulty here because what counts is whether or not prediction statements deduced from those assumptions are approximately in accordance with the observations. Yet as we are about to see, tests of prediction statements face severe attachment difficulties. Also, the researchers' unrealistic assumptions here appear different in principle from those we encountered in classical mechanics. In classical mechanics we assumed point masses or perhaps an approximation when a differential equation would otherwise have been intractable. Without such assumptions we would have been unable to deduce any prediction statement at all. Here, though, from more realistic assumptions, the researchers *could* have deduced a prediction statement (see Theorem 7.4). Thus, in both cases the unrealistic assumptions concern a failure to model the complexity of the real-world situation, but in classical mechanics they seem necessary for stating a testable theory, whereas here they do not.

Now let Q be $G/(1 - t_n)w_{g2}$ and $(\Delta H, \mathbf{B}_1', \overline{\mathbf{X}}_1, \mathbf{B}_1'\overline{\mathbf{X}}_1, c_m, e)$ be $(\Delta h_2,$

$-\beta_2, x_2, -\beta_2 \cdot x_2, c_2, \epsilon_2$). The New Jersey researchers Watts and Horner (1977) recognize that "the response parameter, β_m, may be a function of the characteristics of a family or its male breadwinner" (p. 65). Given their ΔH_m relation and the fact that they drop the upper limit on β_m, one might have guessed they would add a measurement-error term e and come up with a relation like $\Delta H = (\mathbf{B}_1'\overline{\mathbf{X}}_1)Q + c_m + e$, where $0 < \mathbf{B}_1'\overline{\mathbf{X}}_1$ and $c_m = 0$. Instead, the relation they employ is "$\Delta H = \mathbf{B}_0'\overline{\mathbf{X}}_0 + \mathbf{B}_1'\overline{\mathbf{X}}_1 Q + e$" (p. 67), where $\mathbf{B}_0'$ is a vector of parameters, $\overline{\mathbf{X}}_0$ is a vector of demographic variables, and $\mathbf{B}_0'\overline{\mathbf{X}}_0$ is the dot product of $\mathbf{B}_0'$ and $\overline{\mathbf{X}}_0$. The researchers abandon entirely the hypothesis which follows from their assumptions [i.e., $0 < \mathbf{B}_1'\overline{\mathbf{X}}_1 (< 1)$] and consider instead standard statistical hypotheses [e.g., that a component of $\mathbf{B}_0'$ ($\mathbf{B}_1'$) is < 0 or > 0]. Notice that the term $\mathbf{B}_0'\overline{\mathbf{X}}_0$ does *not* follow from Theorem 7.5's assumptions. The researchers explain that "the control variables [$\overline{\mathbf{X}}_0$] introduced in the regression – age, education, health status, prior work experience, and experimental site – are necessary both to reduce variation and to allow for possible nonrandom association between these variables and experimental status" (p. 65). Even if one concurred with the researchers' claim that inclusion of $\overline{\mathbf{X}}_0$ is necessary, $\overline{\mathbf{X}}_0$'s inclusion still shatters the deductive logic of the experiment's theory.

In summary, UM, S, $\mathcal{P}_1$, and MEC, budget constraints that characterize the situation, and an assumption of no corner solutions imply a prediction statement that – with the addition of an ad hoc theory of taste and preference determination and a measurement-error term – becomes $\Delta H_k = (\mathbf{B}_1'\overline{\mathbf{X}}_{1k})Q_{2k} + c_m + e_k$, where $0 < \mathbf{B}_1'\overline{\mathbf{X}}_{1k} < 1$ for each k, and $c_m = 0$. The New Jersey researchers, though, test another statement: $\Delta H_k = (\mathbf{B}_1'\overline{\mathbf{X}}_{1k})Q_k + (\mathbf{B}_0'\overline{\mathbf{X}}_{0k}) + e_k$, where the components of $\mathbf{B}_1'$ and $\mathbf{B}_0'$ are > 0 (or < 0) ($Q_k \neq Q_{2k}$, $\overline{\mathbf{X}}_{1k}$'s components are a subset of $\overline{\mathbf{X}}_{0k}$'s). UM, S, $\mathcal{P}_1$, and MEC, budget constraints that patently do not characterize the situation, and an assumption of no corner solutions imply a prediction statement that – with the addition of an ad hoc theory of taste and preference determination and a measurement-error term *and* the inclusion of $\overline{\mathbf{X}}_{0k}$ in the regression *and* the substitution of standard statistical hypotheses for those that follow from the assumptions – becomes the statement the New Jersey researchers test. So in contrast to the classical mechanics and hypothetical microeconomics experiments of Chapters 5 and 6, the statement the New Jersey experiment tests does *not* follow deductively from the researchers' assumptions. The New Jersey experiment violates the Kuhn–Popper scientific theory demarcation criterion (see Section 5.1.2), whereas the other experiments do not.

7.3.2   *The New Jersey experiment: attachment and induction difficulties*

In Chapters 5 and 6 we examined experiments which satisfied the Kuhn–Popper criterion, and the hypothetical microeconomic experiments of Chapter 6 also exhibited attachment difficulties not present in the classical mechanics experiments of Chapter 5. In the previous section, we argued that the New Jersey experiment violated the Kuhn–Popper criterion. The New Jersey experiment does exhibit most of the attachment difficulties of the hypothetical microeconomics experiments that were not present in the classical mechanics experiments. For example, for the experiment's results to apply to any period other than the one during which the data were gathered, tastes and preferences for laboring must be the same in both periods; yet tastes and preferences for laboring may evolve historically. This is the difficulty of time's role and meaning in the theory (of labor–leisure choice). In the case of the other Chapter 6 (but not Chapter 5) difficulties, the fact that the New Jersey experiment was an actual experiment rather than a hypothetical experiment allows us to see what those things we have identified rather abstractly as "attachment difficulties" become in practice.

For those who worked at all during the experiment, the researchers calculated observed wage rate W as an average of reported gross earnings (adjusted for inflation) divided by hours worked from the individual's quarterly interviews.[4] They noticed a systematic differential in treatment- and control-group observed wage rates that "is consistent with the idea that much of the differentials is an illusion brought about by systematic reporting errors" (Watts and Mamer 1977, p. 350) by participants, with the experimental group faster to learn the gross/net distinction because of more frequent (monthly) interviews. Thus an attachment difficulty is introduced by employing observations reported by conscious objects of inquiry. (See Section 6.2.1 on the problem of conscious objects of inquiry.)

Another attachment difficulty arises from the question whether or not what one observes is actually an equilibrium (see Section 6.3.1 on this issue). In an influential work, Friedman argues that observed income is the sum of a "permanent" (or equilibrium) component and a "transitory" component and that it is the "permanent" component that demand theory needs to measure. He suggests:

---

[4] See Watts and Horner (1977, p. 67, fn. 10) for details.

One indirect means is to use evidence for . . . other consumer units to interpret data for one consumer unit. . . . Suppose Mr. A's measured income in any period is decidedly lower than the average measured income of a group of individuals who are similar to him in characteristics that we have reason to believe affect potential earnings significantly – for example, age, occupation, race, and location. It then seems reasonable to suppose that Mr. A's measured income understates his permanent income. (1953, p. 21)

Friedman calls his view that observed income is a sum of equilibrium and transient components, and that the equilibrium component for an individual may be measured as roughly the average observed income of demographically similar individuals, "the permanent income hypothesis."

Friedman's "permanent income hypothesis" differs in several respects from the hypotheses – of classical mechanics, microeconomics, and the theory of labor–leisure choice – that we have thus far considered. First, there is no placeholder relation and particular law that imply the hypothesis. Second, Friedman's list of variables determining permanent income is somewhat off-the-cuff – he could have excluded some, and he meant not necessarily to exclude others. Third, Friedman fails to suggest *how* his variables determine (permanent) income (i.e., he fails to suggest a functional form). Friedman's "permanent income hypothesis," then, inserts an entire ad hoc (in the three specified senses) theory of income determination between the consumer theory and the data used to test it. (The meaning of "ad hoc" in the ad hoc theory of taste and preference determination discussed earlier in this and the previous sections encompasses precisely the three senses specified here.)

In Chapter 6 we saw that an attachment assumption in a classical mechanics experiment sometimes supposes that a certain theory holds. For example, to test the law of momentum conservation, which followed from NSL and no external forces, we had to add an assumption that NSL and NLUG hold in measuring masses. Or to test Kepler's Laws, which follow from NSL and NLUG, astronomers implicitly add an assumption that the classical theory of optics holds, because they really do not see or observe the planets themselves, but only their telescopic images. So use of Friedman's "permanent income hypothesis" as an attachment assumption in a test of consumer theory creates no *logical* difficulties not present in a classical mechanics experiment. Rather, it is the ad hoc (in the three specified senses) nature of Friedman's hypothesis that distinguishes it as an attachment assumption from the theories similarly employed in classical mechanics experiments.

For individuals who worked a lot during the experiment, the New Jersey researchers Watts and Horner (1977) employed observed wage rate $\overline{W}$ for W_m in estimating their ΔH relation of the previous section. But for such individuals they also regressed $\overline{W}$ "on selected characteristics" (p. 67): "age, education, average earnings in the year prior to the experiment, dummy variables [including some for race?], a dummy health variable, and two race–education interaction terms" (p. 67). Each individual who worked only a little or not at all during the experiment "and for whom the characteristic variables were available was then assigned the $\overline{W}$ [for W_m] which was predicted from the regression" (p. 67). Thus, the researchers employed a version of Friedman's "permanent income hypothesis" in measuring wage rate for some individuals, and their ad hoc theory of wage rate determination for those individuals becomes an attachment assumption when the New Jersey researchers test their statement.

In one of the three regressions that Watts and Horner (1977) report, they employ a second ad hoc theory – this time of taste and preference determination – to account for variation in β_m (i.e., in tastes and preferences for laboring) between families. Specifically, they let $\beta_{mk} = \mathbf{B}_1' \overline{\mathbf{X}}_{1k} = B_{11}' E_k + B_{12}' A_k^* + B_{13}' A_k^{*2}$, "where $E = 1$ if the husband completed high school, 0 otherwise" (p. 77), and A^* and A^{*2} are "the age and age squared of the husband" (p. 77). But then the hypothesis $0 < \beta_m < 1$ becomes $0 < \mathbf{B}_1' \overline{\mathbf{X}}_{1k} < 1$ for each k – a hypothesis for which statistical theory seems to provide no test. So the researchers are compelled by the attachment difficulty of variation in tastes and preferences for laboring to seek to test standard statistical null hypotheses (i.e., $B_{11}' = 0$, $B_{12}' = 0$, $B_{13}' = 0$) rather than the hypothesis which follows from their version of the theory of labor–leisure choice.

Watts and Horner (1977) employ yet a third ad hoc theory in adding $\mathbf{B}_0' \overline{\mathbf{X}}_0$ to their ΔH model – a step which, recall from the previous section, does not follow from the assumptions of their version of labor–leisure choice theory. $\overline{\mathbf{X}}_0$'s components are "the individual's normal wage rate, $1/\overline{W}$," (p. 77) A^* and A^{*2}, "L (where $L = 1$ if the husband had a health problem which currently limited his work effort, 0 otherwise) and L_m ($L_m = 1$ if information on the husband's health was missing, 0 otherwise)" (p. 77), E, "and E_m (where $E_m = 1$ if information on the husband's education was missing, 0 otherwise); the weeks worked by the husband in the year prior to the experiment, P; and three site dummy variables" (p. 77). Again, the researchers seek to test standard statistical null hypotheses (e.g., $B_{0i}' = 0$ for some i). But now, since the $\mathbf{B}_0' \overline{\mathbf{X}}_0$ term does not follow from their labor–leisure choice theory, we can say that theory implies $B'_{0i} = 0$ for *each i*. So when, for some i, the researchers reject the null hypothesis $B'_{0i} = 0$ (e.g., in the case of all three regressions for the coefficient of P), they are also rejecting (completely unawares) their version of labor–leisure choice theory!

In Chapters 5 and 6 we identified an attachment difficulty of needing to add a reporting model to a theory's prediction statement before testing it. This was a *difficulty* in the sense that the reporting model assumption did not follow from the theory (either classical mechanics or microeconomics). Once made, however, the assumption with the theory's prediction statement implied a hypothesis and its test. For example, adding (independent) measurement-error terms $\sim N(0, \sigma^{+2})$ for each of n observations of a prediction statement's relation $X = 0$ implies X_i (independent) $\sim N(\mu, \sigma^{+2})$, with $H_0: \mu = 0$, which implies that under H_0 (in Chapter 4 lexicon), $[\overline{X}/s(n)^{1/2}] \sim$ Student's t with $(n - 1)$ degrees of freedom (where n is the number of observations). So classical mechanical or microeconomic theory implies a prediction statement which, with a reporting model, implies H_0 and a statistic's distribution under H_0 (i.e., a reporting model links two deductive chains). Then the statistic's observed value can test H_0 in the Chapter 4 sense.

In the previous section we argued that the first deductive chain – labor–leisure choice theory to prediction statement – is not intact in the New Jersey experiment. And we argued a moment ago that the link between the two deductive chains fails in the New Jersey experiment because the hypotheses $B'_{1j} = 0$ for $j = 1, 2, 3$ that the researchers seek to test do not follow from their labor–leisure choice theory. Now we wish to argue that the second deductive chain – prediction statement + added assumptions (including hypotheses) to statistical tests of hypotheses – is also not intact in the New Jersey experiment. Recall that with the addition of a measurement-error term, the "prediction" statement's ΔH relation became "$\Delta H = \mathbf{B}'_0 \overline{\mathbf{X}}_0 + \mathbf{B}'_1 \overline{\mathbf{X}}_1 Q + e$," but the researchers also added an ad hoc theory of wage rate determination – say $W_m = \mathbf{C}_W \cdot \mathbf{Y} + \zeta$, where ζ is another measurement-error term, and some of the variables of the vector $\mathbf{Y}$ (e.g., age, and health and education dummies) are the same as some of those of the vector $\overline{\mathbf{X}}_0$. The researchers suppose that "$e =$ an error term with assumed distribution $(0, \sigma^2)$" (Watts and Horner 1977, p. 67). About their implicit ζ they say nothing.

Now the researchers report that their estimate $\hat{B}'_{0j}$ of the coefficient B'_{0j} of $X_{0j} (= A^* =$ age$)$ in their ΔH relation is "significant at the 5 percent level, one-tailed [t] test." [5] So they implicitly suppose that if $B'_{0j} = 0$, then $[\hat{B}'_{0j}/(\widehat{\text{Var}}(\hat{B}'_{0j}))^{1/2}] \sim$ Student's t-distribution with $(n - 1)$ degrees of freedom. Yet such a supposition does *not* seem to follow from their assumptions: Let independence and normality of e_k's and independent ζ_k's $\sim N(0, \sigma^{*2})$ be supposed. Had the researchers used an OLSQ estimate $\hat{B}'_{0j}$ of B'_{0j} *with observed values of* $\overline{W}$ *for each individual,* then if $B'_{0j} = 0$, their distributional assumption for $[\hat{B}'_{0j}/(\widehat{\text{Var}}(\hat{B}'_{0j}))^{1/2}]$ *would* follow.[6] But had they employed two-stage least squares (i.e., used OLSQ estimates of coefficients in the ad hoc theory and then employed $\overline{W}$ for *each individual* in OLSQ estimates of B'_{0j} in the ΔH relation), then though $\hat{B}'_{0j}$ *might* be a consistent estimate of B'_{0j}, if $B'_{0j} = 0$ it would not follow that $[\hat{B}'_{0j}/(\widehat{\text{Var}}(\hat{B}'_{0j}))^{1/2}]$ is asymptotically $N(0, 1)$, so that not even the asymptotic validity of the test results the researchers report would follow.[7] In any case, recall that the researchers employed neither OLSQ nor two-stage least-squares estimates of B'_{0j}. Instead, they used a two-stage procedure where the result of the first stage was *either* an OSLQ estimate of $\overline{W}$ (observed wage rate) *or* $\overline{W}$ itself, and the second stage was OLSQ. So even if we strengthen their stated assumptions on e and ζ, given their estimating procedure, the distribution the researchers suppose for $[\hat{B}'_{0j}/(\widehat{\text{Var}}(\hat{B}'_{0j}))^{1/2}]$ doesn't seem to follow. The second deductive chain is not intact in the New Jersey experiment.

A final attachment difficulty in the New Jersey experiment is one we didn't note in Chapter 6, but did in the previous section: unnecessary failure to model the complexity of the real-world situation's budget constraint. In Section 7.3.1 we saw that the researchers assumed a linear budget constraint in the period before and during the experiment, when in fact both those budget constraints were piecewise linear. However, the real-world situation was actually even more complex than we have so far supposed. In the period before the experiment, no welfare was available as

[5] Watts and Horner (1977, p. 68). Similarly, Watts et al. (1974) report: "The coefficient of Q turns out to have the expected negative sign and to be significant at the 5 percent level for a one-tailed [t] test."

[6] See Wonnacott and Wonnacott (1970, pp. 22–4) for proof of this fact. Here, as in Section 4.3.2, we can weaken the normality assumptions to $\sigma^{*2} < \infty$ and still get the weak inferential conclusion that $[\hat{B}'_{0j}/(\text{Var}(\hat{B}'_{0j}))^{1/2}]$ is asymptotically $N(0, 1)$. On this point, see Wonnacott and Wonnacott (1970, p. 22).

[7] Even when the first stage estimates one of the second stage's independent variables, we have only that $\hat{B}'_{0j}$ is a consistent estimate of B'_{0j}, not that $[\hat{B}'_{0j}/(\widehat{\text{Var}}(\hat{B}'_{0j}))^{1/2}]$ is asymptotically $N(0, 1)$. [On consistency and absence of asymptotic standard normality results, see Wonnacott and Wonnacott (1970, pp. 191, 397–400).] In the New Jersey research, since $\overline{W}$ is not even an independent variable, but $G/\overline{W}(1 - t_n)$ is, even the very weak result that $\hat{B}'_{0j}$ is a consistent estimate of B'_{0j} may not hold.

we have supposed. But soon after the experiment began, New Jersey introduced a fairly generous welfare program for two-parent families with an unemployed father (AFDC-UP). Treatment-group families were allowed to go back and forth from AFDC-UP to experimental payments at will, but never to receive both payments in a given month (Garfinkel 1977, p. 280).

Because AFDC-UP began only after the start of the experiment, the prior experiment budget constraint is unaffected by its availability. AFDC-UP may be characterized by an implicit tax rate and guarantee level. Reported earnings and hours worked during the experiment were from a single week during which a treatment-group member chose either welfare or NIT payments or nothing. Thus, the researchers, though they did not, *could have* included the welfare option in the experimental group's budget constraint. For the most generous NIT plan, no change in the two pieces of the budget constraint would have been needed; for the least generous NIT plan, a welfare piece would have replaced the NIT piece; for NIT plans in between, a third welfare piece would have been added to the NIT and federal tax pieces. The researchers' unnecessary failure to model the complexity of the real-world situation's budget constraint included a failure to include the welfare option in their model.[8]

7.3.3 The Seattle-Denver experiment: "deduction" of a prediction statement

We begin with the framework of Section 7.2.1 and let T, h, L, w, y, and y_d be as they were there. Rather than begin with a particular utility function (as we did in Section 7.3.1), we begin with some abstract knowledge of a utility function and some particular knowledge of its consequent labor-supply function and then *deduce* a particular utility function.

Theorem 7.6. *Let $u(L, y_d)$ be an individual utility function of a worker facing a budget constraint $wh + y = y_d$, $h + L = T$. Suppose that $u(L, y_d)$ satisfies Axiom 7.1; UM holds, $L \neq 0$, T, and $\partial h(w, y)/\partial y = \beta$ and*

---

[8] For somewhat contradictory speculation on the effect on the New Jersey experiment's results of the researchers' failures to include a welfare "piece" in their implicit budget constraint model, see Avery (1977) and Garfinkel (1977).

$[\partial h(w, y)/\partial w]|_{u=\text{constant}} = \alpha$ *for constants* α *and* β *for* $w, y > 0$. *Then*

(i) $u(L, y) = y_d - [(L^2/2) - TL]/\alpha$,

(ii) $h(w, y) = \alpha w$,

(iii) $\beta = 0, \alpha > 0$.

Now we shift to the framework of Section 7.2.2 and let T, h, L, w_g, y_g, t_n, t_F, E, and G be as they were there and let $\Delta h =$ worker's hours worked after minus hours worked before an NIT is introduced. Also, let S^* be: w_g and T are the same in the periods before and after an NIT is introduced; market demand for low-income labor is perfectly elastic. And let $\mathcal{P}_2$ be: $u(L, y_d) = y_d - [(L^2/2) - TL]/\alpha$. Finally, suppose MEC is as in Section 7.3.1.

Now let k index observations from different individuals, and let

$$Q_k^+ = \begin{cases} w_{gk}t_n \text{ if } k \text{ pays no federal tax before and receives a} \\ \quad \text{payment during the NIT} \\ w_{gk}(t_n - t_F) \text{ if } k \text{ pays federal tax before and receives a} \\ \quad \text{payment during the NIT} \\ 0 \text{ if } k \text{ pays federal tax before and during (and receives} \\ \quad \text{no payment during) the NIT.}^9 \end{cases}$$

We then have a theorem.

Theorem 7.7. *Suppose that UM, $S^*, \mathcal{P}_2$, and MEC hold; $G > t_n E, t_n > t_F$, and $(G - t_F E)/(t_n - t_F) < Tw_g$; in the period before (after) an NIT is introduced, Budget Constraint 7.1 (7.2) is the individual's budget constraint; $w_g, y_d > 0$; and $L \neq 0, T - (E/w_g) + (y_g/w_g), T (0, T - [(G - t_F E)/w_g t_n - t_F)] + (y_g/w_g), T)$ before (after) an NIT is introduced. Then $\Delta h_k = -\alpha Q_k^+ + c$, where $\alpha > 0$ and $c = 0$.*

Again notice that the restrictions on L in the theorem's hypothesis eliminate any prediction statement for corner solutions.

As in the New Jersey case, we could allow for variation in tastes and preferences between individuals by supposing $\alpha = \alpha \cdot \mathbf{x}$, where $x_1 = 1, x_i$ for $i \geq 2$ is the value of some demographic variable for the individual.

---

[9] For the few individuals who face the second piece only of Budget Constraint 7.2 during the experiment, i.e., those for whom $G - t_F E > Tw_g(t_n - t_F)$ (see footnote 3), only the first two pieces of Q_k^+ are possible, but these individuals may still be included as observations like all others.

Such an assumption again constitutes an ad hoc (in senses elaborated in the previous section) theory of taste and preference determination. Again we can also allow for measurement error by assuming an additive measurement-error term $\epsilon' \sim N(0, \sigma')$ for Δh. With these two added assumptions, Theorem 7.7's prediction statement becomes $\Delta h_k = -(\alpha \cdot x_k)Q_k^+ + c + \epsilon'_k$, where $\alpha_k = (\alpha \cdot x_k) > 0$ (for each k) and $c = 0$.

The Seattle-Denver researchers (Keeley et al. 1977) explain: "It is assumed that each individual maximizes a well-behaved utility function $U(L, Y_d)$, where L is leisure and Y_d is consumption of market goods (or disposable income) subject to the budget constraint $F = wT + Y_n = wL + Y_d$ where F is full income, w is net wage rate, T is total time available, and Y_n is net nonwage income" (p. 3). The same Seattle-Denver work supposes "α is the substitution effect, and β is the income effect" (p. 3). Thus, with their $(U, L, H, Y_d, w, T, Y_n, \alpha, \beta)$ equal to our $(u, L, h, y_d, w, T, y, \alpha, \beta)$, the Seattle-Denver researchers' assumptions seem the same as those of our initial Section 7.2.1 framework. Yet the researchers fail to deduce the conclusions of our Theorem 7.6 and also fail to take account (in our Theorem 7.7 fashion) of the *piecewise* budget constraint. (Both the Seattle-Denver work and our foregoing theorem exhibit another attachment difficulty of failing to model the complexity of the actual situation – i.e., they suppose an *individual*, when in the actual situation a *family*, budget constraint prevailed.) Their deductions follow instead a different course.

Consider first the following proposition.

Proposition 7.2. *Let the function $R^+ = f(r_1, r_2)$ be continuous and possess partial derivatives $\partial f/\partial r_i$ throughout a region $\mathcal{R}$: $|r_1 - r_{10}| < k_1$, $|r_2 - r_{20}| < k_2$ of the r_1–r_2 plane. Let $\partial f/\partial r_i$ be continuous at (r_{10}, r_{20}) in R^2. Let $\Delta R^+ = f(r_{10} + \Delta r_1, r_{20} + \Delta r_2) - f(r_{10}, r_{20})$. Then $\Delta R^+ = [\partial f(r_{10}, r_{20})/\partial r_1]\Delta r_1 + [\partial f(r_{10}, r_{20})/\partial r_2]\Delta r_2 + \eta_1 \Delta r_1 + \eta_2 \Delta r_2$, where η_1 and η_2 approach 0 when Δr_1 and Δr_2 approach 0.*[10]

Thus, we can say that if the propositions' hypotheses are satisfied, then

$$\Delta R^+ \approx [\partial f(r_{10}, r_{20})/\partial r_1]\Delta r_1 + [\partial f(r_{10}, r_{20})/\partial r_2]\Delta r_2,$$

where the sense of the "$\approx$" is in the proposition's conclusion. The Seattle-Denver researchers use this sense of "$\approx$" in their deductions.

Let (w_{np}, Y_{np}, H_p) and $(w_{np} + \Delta w, Y_{np} + \Delta Y_n, H_p + \Delta H)$ be [net

---

[10] See Thomas (1960, pp. 668–70) for the proposition's proof.

(after-tax) wage rate, net (after-tax) nonwage income, hours worked] before and after an NIT is introduced, respectively. We may formalize the Seattle-Denver deduction in a theorem.

Theorem 7.8. *Let $u(L, y_d)$ be an individual worker's utility function that satisfies Axiom 7.1. Suppose that UM, S^*, and MEC hold; $L \neq 0$, T before and after the NIT is introduced; and if $L \neq 0$, T and budget constraints $wh + y = y_d$, $h + L = T$ (of Section 7.2.1) prevail, then $\partial h(w, y)/\partial y = \beta$, $[\partial h(w, y)/\partial w]|_{u=\text{constant}} = \alpha$ for constants α and β for w, $y > 0$. If the worker's budget constraint is*

$$-w_g(1 - t_F)L + Tw_g(1 - t_F) + (1 - t_F)y_g = y_d$$
$$(-w_g(1 - t_n)L + Tw_g(1 - t_n) + (1 - t_n)y_g + G = y_d)$$

before (after) an NIT is introduced, then "$\Delta H \approx \alpha\Delta w + \beta(H_p\Delta w + \Delta Y_n)$." [11]

Notice that the researchers simply fail to draw the conclusion $\alpha > 0$, $\beta = 0$ that follows from their assumption.

Notice also that the researchers' theorem assumes *linear* budget constraints before and after the NIT is introduced. One *could*, however, account for the piecewise linearity of the actual budget constraints and still follow the Seattle-Denver researchers' approximation approach. Again, let k index observations from different individuals, Q_k^+ be as before, and

$$R_k = \begin{cases} H_{pk}w_{gk}t_n - G_k \text{ if } k \text{ pays no federal tax before and} \\ \quad \text{receives a payment during the NIT} \\ H_{pk}w_{gk}(t_n - t_F) - (G_k - t_F E_k) \text{ if } k \text{ pays federal tax} \\ \quad \text{before and receives a payment during the NIT} \\ 0 \text{ if } k \text{ pays federal tax before and during (and receives} \\ \quad \text{no payment during) the NIT.}^{12} \end{cases}$$

Then a theorem that accounts for the piecewise linearity of the actual budget constraints is the following.

Theorem 7.9. *Let $u(L, y_d)$ be an individual worker's utility function that satisfies Axiom 7.1. Suppose that UM, S^*, and MEC hold; $G > t_n E$, $t_n > t_F$, and $(G - t_F E)/(t_n - t_F) < Tw_g$; $L \neq 0$, $T - (E/w_g) + (y_g/w_g)$, T*

---

[11] Keeley et al. (1977, p. 4). See Keeley et al. (1977, pp. 3–4) and Keeley et al. (1978, pp. 8–9) for the deductions (which implicitly draw on Proposition 7.2).

[12] What was said in footnote 9 for Q_k^+ holds here for R_k as well.

before and $\neq 0$, $T - [(G - t_F E)/(t_n - t_F)w_g] + (y_g/w_g)$, *T after an NIT is introduced; and if budget constraint* $wh + y = y_d$, $h + L = T$ *(of Section 7.2.1) prevails, then* $[\partial h(w, y)/\partial w]|_{u=\text{constant}} = \alpha$, $\partial h(w, y)/\partial y = \beta$ *for constants* α *and* β *for* $w, y > 0$. *If the worker faces Budget Constraint 7.1 (7.2) in the period before (after) an NIT is introduced, then* $\Delta H_k \approx -\alpha Q_k^+ + \beta R_k + c$, *where* $c = 0$.

As before, we could modify the theorem's conclusion to

$$\Delta H_k \approx -(\alpha \cdot \mathbf{x}_k)Q_k^+ + (\beta \cdot \mathbf{x}_k)R_k + c + \epsilon'_k,$$

where $c = 0$, with addition of an ad hoc theory of tastes and preferences and a measurement-error term $\epsilon' \sim N(0, \sigma'^2)$.

The Seattle-Denver researchers, though, fail to account for piecewise-linear budget constraints in the fashion we have just suggested. They are certainly *aware* of the piecewise-linearity difficulty. For example: "the preexperimental budget constraint is also nonlinear because of progressivity of the positive income tax system" (Keeley et al. 1977, p. 15). Yet Keeley et al. (1978) claim that "many of the enrolled families are initially above the tax breakeven level. Even though the calculated values of Δw and $\Delta Y_d(H_p)$ $[= H_p \Delta w + \Delta Y_n]$ are zero for families with initial equilibriums above the breakeven level, some of those families will respond to the experiment" (p. 9). In fact, however, those who are initially above the tax breakeven level fall into two categories: those for which the calculated values of Δw and $\Delta Y_d(H_p)$ are indeed 0, and those for which the calculated value of Δw is $-w_g(t_n - t_F)$ and that of $\Delta Y_d(H_p)$ is $-[H_p w_g(t_n - t_F) - (G - t_F E)]$ (see Theorem 7.9).

The Seattle-Denver researchers, then, are confused about those initially above the tax breakeven level because they fail to lay out careful deductions. Yet Keeley et al. (1978) recognize that such subjects may respond to the experimental treatment and seek to account for such response:

We measure response for families above the breakeven level by defining a set of three explanatory variables that capture the location of the family relative to the breakeven level: FABOVE – a dummy variable equals 1 if a family is eligible for payments and is above the tax breakeven level, and 0 otherwise; BREAK – breakeven level of family earnings; and EARNABV – a distance, in family earnings, above the breakeven level. (p. 9)

Still, they are not satisfied with their approach: "This specification of the above-breakeven response is not entirely satisfactory because it is not consistent with the specification for below-breakeven response" (p. 9).

Elsewhere, the researchers employ a separate equation with a dummy variable for "participation" regressed on a set of demographic variables in a structural equations model with the ΔH relation of Theorem 7.8 as the second equation (Robins and West 1980, pp. 505–8). Such an approach avoids the inconsistent treatment of those below and above the tax break-even level of the three-variable-set approach. But there is no deductive basis for either of these approaches. The only approach which preserves statements-level deductive logic employs the variable Q_k^+ (and R_k) as in our Theorem 7.7 (7.9).

Another complication arises from the fact that some workers in the treatment group can choose between Budget Constraint 7.1 and an NIT with a declining,[13] rather than fixed, tax rate – that is, one for which for fixed r and t_e the negative tax rate

$$t_n = \begin{cases} t_e - r(w_g h + y_g) & \text{if } 0 \le w_g h + y_g \le t_e/2r \\ (t_e/2) & \text{if } (t_e/2r) < w_g h + y_g. \end{cases}$$

Suppose also that $(t_e/2) > t_F$, $G > (t_e/2)E$, and $2r(G - t_F E) > t_e[(t_e/2) - t_F]$ – conditions that hold for all Seattle-Denver declining tax rate plans. If $w_g T[(t_e/2) - t_F] > G - t_F E$, then during the NIT the worker faces the following.

Budget Contraint 7.5

$$y_d = \begin{cases} -w_g(1 - t_F)L + [(w_g T + y_g)(1 - t_F) + t_F E] \\ \quad \text{if } 0 \le T + (y_g/w_g) - (G - t_F E)/[(t_e/2) - t_F]w_g \\ -w_g[1 - (t_e/2)]L + [(w_g T + y_g)(1 - (t_e/2)) + G] \\ \quad \text{if } T + (y_g/w_g) - [(G - t_F E)((t_e/2) - t_F)w_g] \le L \le T + (y_g/w_g) - (t_e/2rw_g) \\ w_g^2 r L^2 - w_g[(1 - t_e) + 2r(w_g T + y_g)]L + [(w_g T + y_g)(1 - t_e + r(w_g T + y_g)) + G] \\ \quad \text{if } T + (y_g/w_g) - (t_e/2rw_g) \le L \le T. \end{cases}$$

If, however, $w_g T[(t_e/2) - t_F] \le G - t_F E$, then during the NIT the worker faces the following.

Budget Constraint 7.6

$$y_d = \begin{cases} -w_g[1 - (t_e/2)]L + [(w_g T + y_g)(1 - (t_e/2)) + G] \\ \quad \text{if } 0 \le L \le T + (y_g/w_g) - (t_e/2rw_g) \\ w_g^2 r L^2 - w[(1 - t_e) + 2r(w_g T + y_g)]L + [(w_g T + y_g)(1 - t_e + r(w_g T + y_g)) + G] \\ \quad \text{if } T + (y_g/w_g) - (t_e/2rw_g) \le L \le T. \end{cases}$$

---

[13] See Kurz, Spiegelman, and Brewster (1973, pp. 4–5, 16) on the declining tax rate program.

Thus, all Seattle-Denver workers on a declining tax rate plan faced either Budget Constraint 7.5 or 7.6 during the NIT.

Rather than studying, as previously, $\Delta h = h_{\text{NIT}} - h_b =$ hours worked during minus hours worked before the NIT, we now consider $(h_b/h_{\text{NIT}}) =$ ratio of hours worked before to hours worked during the NIT. Again let k index observations from different individuals, and let Q'_k, R'_k, Q''_k, and R''_k be defined by Table 7.1. We then have a theorem covering the declining tax rate plans.

Theorem 7.10. *Suppose UM, S*, $\mathscr{P}_2$, and MEC hold; $(t_e/2) > t_F$, $G > (t_e/2)E$, and $2r(G - t_F E) > t_e[(t_e/2) - t_F]$; w_g, $y_g > 0$; and the worker faces Budget Constraint 7.1 and $L \neq 0$, $T - (E/w_g) + (y_g/w_g)$, T in the period before the NIT is introduced.*

(i) *$w_g T[(t_e/2) - t_F] > G - t_F E$ and that the worker faces Budget Constraint 7.5 and*

$$L \neq 0, \; T + (y_g/w_g)$$
$$- [(G - t_F E)/((t_e/2) - t_F)w_g], \; T + (y_g/w_g) - (t_e/2rw_g), \; T$$

in the period after the NIT is introduced imply $(h_b/h_{\text{NIT}})_k = -\alpha Q'_k + c^* R'_k + c$, where $\alpha > 0$, $c^* = 1$, and $c = 0$.

(ii) *$w_g T[(t_e/2) - t_F] \leq G - t_F E$ and that the worker faces Budget Constraint 7.6 and $L \neq 0$, $T + (y_g/w_g) - (t_e/2rw_g)$, T in the period after the NIT is introduced imply $(h_b/h_{\text{NIT}})_k = -\alpha Q''_k + c^* R''_k + c$, where $\alpha > 0$, $c^* = 1$, and $c = 0$.*

Again notice that the restrictions on L in the theorem's hypotheses eliminate any prediction statement for corner solutions.

As before, with the addition of an ad hoc theory of tastes and preferences and an additive measurement-error term $\epsilon' \sim N(0, \sigma'^2)$, we could modify the theorem's conclusions to

$$(h_b/h_{\text{NIT}})_k = -(\alpha \cdot x_k)Q'_k + (c_1 \cdot x_k)R'_k + c + \epsilon'_k$$

and/or

$$(h_b/h_{\text{NIT}})_k = -(\alpha \cdot x_k)Q''_k + (c_1 \cdot x_k)R''_k + c + \epsilon'_k,$$

where c^* is a vector of parameters and $\alpha \cdot x_k > 0$, $c^* \cdot x_k = 1$ for each k and $c = 0$. Instead of employing our Theorem 7.10 approach of rededuction with the new budget constraints, in Keeley et al. (1978) "a dummy variable, labeled DECLINE, is included for families on the declining tax

Table 7.1. *Definitions of Q'_k, R'_k, Q''_k, and R''_k*

$Q'_k =$	$R'_k =$	Conditions
$-2w_{gk}^2 r/[(1 - t_e) + 2ry_{gk}]$	$1/[(1 - t_e) + 2ry_{gk}]$	if k pays no federal tax before and $0 <$ gross income $< t_e/2r$ during the NIT
0	$1/[(1 - (t_e/2)]$	if k pays no federal tax before and $(t_e/2r) <$ gross income $< (G_k - t_F E_k)/[(t_e/2) - t_F]$ during the NIT
0	$1/(1 - t_F)$	if k pays no federal tax before and $(G_k - t_F E_k)/[(t_e/2) - t_F] <$ gross income $< w_{gk}T + y_{gk}$ during the NIT
$-2w_{gk}^2 r(1 - t_F)/[(1 - t_F) + 2ry_{gk}]$	$(1 - t_F)/[(1 - t_e) + 2ry_{gk}]$	if k pays federal tax before and $0 <$ gross income $< (t_e/2r)$ during the NIT
0	$(1 - t_F)/[1 - (t_e/2)]$	if k pays federal tax before and $(t_e/2r) <$ gross income $< (G_k - t_F E_k)/[(t_e/2) - t_F]$ during the NIT
0	1	if k pays federal tax before and during (and receives no payment during) the NIT
$-2w_{gk}^2 r/[(1 - t_e) + 2ry_{gk}]$	$1/[(1 - t_e) + 2ry_{gk}]$	if k pays no federal tax before and $0 <$ gross income $< (t_e/2r)$ during the NIT
0	$1/[1 - (t_e/2)]$	if k pays no federal tax before and $(t_e/2r) <$ gross income $< w_{gk}T + y_{gk}$ during the NIT
$-2w_{gk}^2 r(1 - t_F)/[(1 - t_e) + 2ry_{gk}]$	$1/[(1 - t_e) + 2ry_{gk}]$	if k pays federal tax before and $0 <$ gross income $< (t_e/2r)$ during the NIT
0	$(1 - t_F)/[1 - (t_e/2)]$	if k pays federal tax before and $(t_e/2r) <$ gross income $< w_{gk}T + y_{gk}$ during the NIT

rate program" (p. 10) in the researchers' approximation relation (Theorem 7.8's conclusion) – a step for which there is simply no deductive basis.

Keeley et al. (1978), just like the New Jersey researchers, introduce a set of so-called control variables into their regression, explaining: "Finally, a set of assignment and other control variables, C, is included to account for nonexperimental effects, and variables are included to measure the effects of the manpower component of the experiment, M" (p. 10). The set C is similar to, but not the same as, the set of "control variables" employed in New Jersey: There the set included variables for age, education, health status, prior work experience, experimental site, and race (?), whereas here it includes age, experimental site, and race, as well as number of family members and number of children ages 0–5, but *not* health status and prior work experience. Again, there is no deductive basis for including any such variables in the regression. And again the researchers substitute standard statistical hypotheses (i.e., that a coefficient is >0 or <0) for any that follow by deduction from their assumptions.

To summarize the logic, let A_2 be: $u(L, y_d)$ satisfies Axiom 7.1; UM, S^*, and MEC hold; and if $L \neq 0$, T, $wh + y = y_d$, $h + L = T$, then $\partial h(w, y)/\partial y = \beta$ and $[\partial h(w, y)/\partial w|_{u=\text{constant}}] = \alpha$. Let B_2 be: Budget Constraint 7.1 (7.2) holds before (during) the NIT, and no corner solutions are allowed; and let B_2' (B_2'') be: Budget Constraint 7.1 holds before and 7.5 (7.6) holds during the NIT, and no corner solutions are allowed. Then for those who don't face a declining tax rate NIT program, A_2 and B_2 imply a prediction statement that, with the addition of a measurement-error term, becomes $\Delta h_k = -\alpha Q_k^+ + c + \epsilon_k'$, where $c = 0$, $\alpha > 0$ ($\beta = 0$), and allowing for variation in tastes and preferences, becomes $\Delta h_k = -(\alpha \cdot x_k)Q_k^+ + c + \epsilon_k'$, where $c = 0$, $\alpha \cdot x_k > 0$ for each k ($\beta = 0$). Or A_2 and B_2 imply an approximate prediction statement that, with the addition of a measurement-error term, becomes

$$\Delta h_k \approx -\alpha Q_k^+ - \beta R_k + c + \epsilon_k',$$

where $c = \beta = 0$, $\alpha > 0$. For those facing a declining tax rate NIT program, A_2 and B_2' (B_2'') imply an exact prediction statement that, with the addition of a measurement error term, becomes

$$(h_b/h_{\text{NIT}}) = -\alpha Q_k' + c^* R_k' + c + \epsilon_k' (-\alpha Q_k'' + c^* R_k'' + c + \epsilon_k'),$$

where $\alpha > 0$, $c^* = 1$, $c = 0$.

The Seattle-Denver researchers, though, assume that A_2 and linear budget constraints hold before and after an NIT is introduced, deduce an approximate prediction statement, and modify it in ways that do not follow from any assumptions. They add three variables for piecewise linearity of budget constraints, a dummy for declining tax rate program, and a set of "control variables," and they substitute standard statistical hypotheses for those that follow from their assumptions. Finally, with $H_e = H_p + \Delta H$, they add a measurement-error term and end up with the statement

"$H_e - H_p = \Delta H = b_0 + b_1 C + b_2 M + b_3 \Delta w + b_4 \Delta Y_d(H_p)$
$\quad + b_5 \text{FABOVE} + b_6 \text{BREAK} + b_7 \text{EARNABV} + b_8 \text{DECLINE} + \epsilon,$

where ϵ is an error term" (Keeley et al. 1978, p. 11), and $b_i > 0$ (or < 0) for $i = 0, 1, \ldots, 8$. So in contrast to the classical mechanics and hypothetical microeconomics experiments of Chapters 5 and 6, the statement the Seattle-Denver experiment tests does *not* follow deductively from the researchers' assumptions. Like the New Jersey experiment, the Seattle-Denver experiment violates the Kuhn–Popper scientific theory demarcation criterion (see Section 5.1.2), whereas the classical mechanics and hypothetical microeconomics experiments do not.

7.3.4   The Seattle-Denver experiment: attachment and induction difficulties

Just as in New Jersey, in addition to violating the Kuhn–Popper criterion, the Seattle-Denver experiment exhibits attachment difficulties of the sort we mentioned in the hypothetical microeconomics experiments of Chapter 6, but not in the classical mechanics experiments of Chapter 5. Again, for the experiment's results to apply to any period other than the one during which the data were gathered, tastes and preferences for laboring must be the same in both periods; yet tastes and preferences for laboring may evolve historically. This is the difficulty of time's role and meaning in the theory.

Also as in New Jersey, the Seattle-Denver researchers employed figures for income, gross earnings, and hours worked from interviews and de-fined gross wage rate as gross earnings/hours worked. From comparing their figures with IRS figures for a subset of the experiment's participants, the Seattle-Denver researcher Halsey (1980) concludes: "The wide var-iance in reported [gross earnings] is large enough to cause behaviorally significant error in parameter estimates in ordinary least squares regres-sions" (p. 47), and "nonwage income was reported with a higher var-iance . . . than earned income . . . in agreement with the discussion

[of] differential problems of recall" (p. 54). So again, employing observations reported by conscious objects of inquiry introduces an attachment difficulty.

For $\Delta Y_d(H_p)$, the Seattle-Denver researchers fail to follow Friedman's "permanent income hypothesis," but instead take interview-reported income as observed income. For observed Δw, they use "$-W(t_e - t_p)$, where W is gross wage rate [calculated from interview reports of gross earnings and hours worked]" (Keeley et al. 1977, p. 14), and t_e and t_p are the experimental and preexperimental tax rates, respectively. But "because [gross] wage rates are not observed for nonworkers, a wage equation is estimated for workers based on personal characteristics, and is used to predict wage rates for the entire sample. . . . The wage equation we estimate is a simple linear formulation. . . . Years of schooling, experience (defined as age minus years of schooling minus 5) and experience squared [and a dummy for race] are the explanatory variables" (Keeley et al. 1978, pp. 15–16). Thus, as in New Jersey, the Seattle-Denver researchers employ an ad hoc theory of wage rate determination to cope with the attachment difficulty of whether or not the observed wage rate is an equilibrium wage rate.

In order to cope with the attachment difficulty of variation in tastes and preferences for laboring between individuals, the Seattle-Denver researchers could have allowed the coefficients of Δw and $\Delta Y_d(H_p)$ in their ΔH relation to be determined by an ad hoc theory. That is, the researchers could have replaced α and β with $(\alpha, \boldsymbol{\alpha}) \cdot (1, \mathbf{x})$ and $(\beta, \boldsymbol{\beta}) \cdot (1, \mathbf{x})$, where $\mathbf{x}$ is a vector of individual characteristics, so that their ΔH relation would have become $\Delta H = \alpha \, \Delta w + (\boldsymbol{\alpha} \cdot \mathbf{x}) \, \Delta w + \beta \, \Delta Y_d(H_p) + (\boldsymbol{\beta} \cdot \mathbf{x}) \, \Delta Y_d(H_p)$ instead of $\Delta H = \alpha \, \Delta w + \beta \, \Delta Y_d(H_p)$. But the Seattle-Denver researchers failed to follow the New Jersey lead in employing such an ad hoc theory.

The Seattle-Denver study did, though, follow the New Jersey lead by employing an ad hoc theory in adding a term, say $\mathbf{b}_1^* \cdot \mathbf{C}$, where $\mathbf{b}_1^*$ is a vector of parameters and $\mathbf{C}$ is a "control vector" of individual characteristics, to their ΔH relation – as in the New Jersey case, a step which (recall from the previous section) does not follow from the assumptions of their version of labor–leisure choice theory. $\mathbf{C}$'s components were: a system of nine dummy variables for income in the year prior to enrollment, "hours worked in year prior to enrollment," dummies for site (Seattle or Denver) and race, "age (in years)," "number of family members," "number of children aged 0–5," and "AFDC benefits in year prior to enrollment ($1,000s)" (Keeley et al. 1978, p. 30). As in New Jersey, the researchers sought to test standard statistical null hypotheses (e.g., $b_{1i}^* = 0$ for some i).

But again, since the $b_1^* \cdot C$ term does not follow from their labor–leisure choice theory, we can say that theory implies $b_{1i}^* = 0$ for *each i*. So when for some *i* the researchers rejected the null hypothesis $b_{1i}^* = 0$ (e.g., in the case of the coefficient of the "age in years" variable), they were also rejecting (unawares) their version of labor–leisure choice theory!

The Seattle-Denver researchers, like their New Jersey counterparts, failed to seek to test hypotheses which followed from their prediction statement plus additional assumptions. But whereas the New Jersey researchers' use of an ad hoc theory of tastes and preferences for laboring led to a hypothesis for which statistical theory seems to provide no test, Seattle-Denver's failure to employ such an ad hoc theory means their prediction statement's hypothesis is of a sort which might conceivably have been tested. That is, the Seattle-Denver researchers should have sought to test a hypothesis like H: $\alpha > 0$, $\beta = 0$, $b_{1i}^* = 0$ for each *i*, which follows from their prediction statement.

Also, Seattle-Denver, like New Jersey, suffered from the induction difficulty of a second deductive chain – prediction statement + added assumptions (including hypotheses) to statistical tests of hypothesis – that was not intact. Recall that with the addition of a measurement error term, the "prediction" statement's ΔH relation became

"$\Delta H = b_0 + b_1 C + b_2 M + b_3 \Delta w + b_4 \Delta Y_d(H_p)$
$+ b_5 \text{FABOVE} + b_6 \text{BREAK} + b_7 \text{EARNABV} + b_8 \text{DECLINE} + \epsilon,$

where ϵ is an error term." But the researchers also added an ad hoc theory of (gross) wage rate determination – say $W = C_W \cdot Y + \zeta$, with ζ another measurement error term, and some of the variables of the vector Y (e.g., dummy for race) are the same as those of C. The researchers suppose that ϵ "is an error term with variance σ^2" (Keeley et al. 1978, p. 31), and they say nothing about their implicit ζ.

Now the researchers report that their estimate $\hat{b}_3$ of b_3 is significant "at the 5 percent level" (p. 12) in a one-tailed t test, implicitly supposing that if $b_3 = 0$, $[\hat{b}_3/(\widehat{\text{Var}}(\hat{b}_3))^{1/2}] \sim$ Student's t distribution with $(n - 1)$ degrees of freedom (where n is the number of observations). Yet, as in New Jersey, such a supposition does *not* seem to follow from their assumptions, for let the ϵ_k (ζ_k) be independent $\sim N(0, \sigma^2)$ ($N(0, \sigma^{*2})$). Then had the researchers used an OLSQ estimate $\hat{b}_3$ of b_3 *with observed values of W for each individual,* then if $b_3 = 0$ their distributional assumption for $[\hat{b}_3/(\widehat{\text{Var}}(\hat{b}_3))^{1/2}]$ *would* follow.[14] But the researchers don't employ an

[14] See footnote 6.

OLSQ estimate of b_3, but rather "the tobit estimation procedure is used" (Keeley et al. 1978, p. 11). Given their estimation procedure for b_3, it is not clear that if $b_3 = 0$ their distributional assumption for [or even asymptotically $\sim N(0, 1)$] $[\hat{b}_3/(\widehat{\text{Var}}(\hat{b}_3))^{1/2}]$ would follow, even *had* the researchers used observed values of W for each individual. But, in any case, the researchers do *not* employ such observed values, but use W for each individual (workers and nonworkers), where W's coefficients are OLSQ estimates in the ad hoc theory from the sample of workers only. So even if we strengthen their stated assumptions on ϵ and ζ, given their complicated two-stage estimation procedure for α, the distribution the researchers suppose for $[(\hat{b}_3/\widehat{\text{Var}}(\hat{b}_3))^{1/2}]$ doesn't seem to follow. The Seattle-Denver experiment's second deductive chain is not intact.

Next, as in New Jersey, we see in the Seattle-Denver experiment a failure to model, or inadequate models of, the complexity of the real-world budget constraint. First, rather than the actual situation's *family* utility function and budget constraint, the researchers employed an *individual* utility function and budget constraint. They explain:

For a family with more than one potential earner, the equation can be generalized to include cross-substitution effects [– e.g., one could add a term $b_3' \Delta w_s$ to the husband's ΔH relation where s designates wife.] In our empirical formulation of this model, it is assumed the cross-substitution effects are zero, partly because the net changes of both spouses are highly correlated and their effects are difficult to distinguish empirically. (Keeley et al. 1977, p. 3)

So one can extend the researchers' implicit individual to a family utility function in a completely natural fashion and deduce a new ΔH relation. But with actual available data, coefficient estimates for the new ΔH relation seem to have large standard error estimates due to the statistical difficulty of multicollinearity.

Second, the researchers used inadequate models of the budget constraint's *piecewise* linearity and the declining tax rate program (see previous section for details). Finally, in Seattle-Denver, welfare was available to all subjects in the period before the experiment, and during the experiment the researchers "required that persons receiving payments from the experiments give up welfare payments" (Keeley, Spiegelman, and West 1980, pp. 22–3). These circumstances should have been modeled by adding linear welfare pieces to budget constraints where appropriate and redefining our Q_k^+ (from the previous section) to take account of the possibility of a subject's choosing welfare before or after the experiment began. Instead, as we saw earlier, the Seattle-Denver researchers included the amount of any before-experiment welfare payments as a variable in **C**, the "control" variables.

7.3.5   *The Gary experiment: "deduction" of a prediction statement in a first study*

As in Section 7.3.3, we begin with the framework of Section 7.2.1 and let T, h, L, w, y, and y_d be as they were there. In the first study, the Gary experimenters make no explicit reference to an underlying utility function – either abstract or particular. Thus, the researchers themselves do not see their study as testing a prediction statement deduced from an assumed particular utility function, budget constraints, UM, and other assumptions. They begin with a model of hours worked: "$H = \gamma + \delta(1 - r - t)W + \eta(N + B_0)$ where $H =$ hours of work per month; $W =$ gross hourly wage rate; $r =$ average non-NIT tax rate on earnings; $t =$ average NIT tax rate on earnings; $B_0 =$ NIT benefit per month at zero hours ["equal to $G - tN$"]; and $N =$ non-NIT nonwage income" (Moffitt 1979, pp. 478–9). Supposing the Section 7.2.1 framework to hold, so that $r = t = 0 = B_0 = 0$, in the lexicon of that section, the researchers' hours-worked model becomes the "labor–supply function" $h = \gamma + \delta w + \eta y$, and the implicit budget constraint becomes $wh + y = y_d$, $h + L = T$.

We may then deduce the following theorem.

Theorem 7.11. *Suppose a worker faces a budget constraint $wh + y = y_d$, $h + L = T$. Then*

(i) *there is no twice-differentiable $u(L, y_d)$ for which UM implies $h = \gamma + \delta w + \eta y$,*

(ii) *$h = \gamma + \delta w + \eta y$, with $\delta > 0$, $\eta < 0$, and utility minimization imply*

$$\mathscr{P}_3 : u(L, y_d) = [(\eta y_d - (\delta/\eta) + \gamma)/(T - L - (\delta/\eta))] + ln[T - L - (\delta/\eta)].$$

The theorem's (i) means that there is no particular utility function for which the researchers' study may be understood as a test of a version of labor–leisure choice theory that includes the core proposition UM as one of its assumptions. The theorem's (ii) provides a particular utility function for which the researchers' study may be understood as a test of a version of labor–leisure choice theory that includes utility *minimization* as one of its

Table 7.2. *Definitions of* w_{nbk}, y_{nbk}, w_{nNITk}, *and* y_{nNITk}

$w_{nbk} = w_{gk}$	$y_{nbk} = y_{gbk}$	if k pays no federal income tax before the NIT
$= w_{gk}(1 - t_F)$	$= y_{gbk}(1 - t_F) + t_F E_k$	if k pays federal income tax before the NIT
$w_{nNITk} = w_{gk}(1 - t_n)$	$y_{nNITk} = y_{gNITk}(1 - t_n) + G_k$	if k participates in the NIT
$= w_{gk}(1 - t_F)$	$= y_{gNITk}(1 - t_F) + t_F E_k$	if k does not participate in the NIT

assumptions. Yet utility *minimization* is a rather bizarre and implausible assumption.

Next, as in Section 7.3.3, let us shift to the framework of Section 7.2.2, and let T, h, L, w_g, y_g, t_n, t_F, E, and G be as they were there. Also, let S^* be: w_g and T are the same in the periods before and after an NIT is introduced; market demand for low-income labor is perfectly elastic. And suppose MEC is as in Section 7.3.1. Let k index observations from different individuals, and let b and NIT indicate before and during the NIT, respectively. Finally, let w_{nbk}, y_{nbk}, w_{nNITk}, and y_{nNITk} be defined by Table 7.2.[15] We then have a theorem.

Theorem 7.12. *Let utility minimization, S^*, $\mathcal{P}_3$ (with $\delta > 0$, $\eta < 0$), and MEC hold; and suppose $G > t_n E$, $t_n > t_F$, $(G - t_F E)/(t_n - t_F) < T w_g$. If $L \neq 0$, $T - (E/w_g) + (y_g/w_g)$ $(0$, $T - [(G - t_F E)/(w_g(t_n - t_F)] + (y_g/w_g))$ and Budget Constraint 7.1 (7.2) holds before (during) the NIT, then*

$$h_{bk} = \gamma + \delta w_{nbk} + \eta y_{nbk} \ (h_{NITk} = \gamma + \delta w_{nNITk} + \eta y_{nNITk})$$
$$\text{with } \delta > 0, \ \eta < 0.$$

Notice that though we can model the actual NIT budget constraints exactly, without any approximation, such modeling does nothing for the problem of needing to replace the core proposition UM with the bizarre

---

[15] For the few individuals who face the second piece only of Budget Constraint 7.2 during the experiment, i.e., those for whom $G - t_F E > T w_g(t_n - t_F)$ (see footnote 3), only the first pieces of w_{nNITk} and y_{nNITk} are possible, but these individuals may still be included as observations like all others.

assumption of utility *minimization* to be able to understand any version at all of labor–leisure choice theory as behind the first Gary experiment study.

As in the New Jersey and Seattle-Denver cases, we could allow for variation in tastes and preferences between individuals with an ad hoc (in senses elaborated in Section 7.3.2) theory of taste and preference determination. Specifically, we might suppose $\gamma = \gamma \cdot x$, where $x_1 = 1, x_i$ for $i \geq 2$ is the value of some demographic variable for each individual. We could also allow for measurement error by assuming an additive measurement-error term ϵ_b (ϵ_{NIT}) for the h_b (h_{NIT}) relation. With these two added assumptions, Theorem 7.12's prediction statement becomes

$$h_{bk} = \delta w_{n\text{NIT}k} + \eta y_{n\text{NIT}k} + (\gamma \cdot x_{bk}) + \epsilon_{bk}$$
$$(h_{\text{NIT}k} = \delta w_{n\text{NIT}k} + \eta y_{n\text{NIT}k} + (\gamma \cdot x_{\text{NIT}k}) + \epsilon_{\text{NIT}k}),$$

with $\delta > 0, \eta < 0$. [Notice that if $x_{bk} = x_{\text{NIT}k}$, then the $(\gamma \cdot x)$ terms cancel in $\Delta h_k = h_{\text{NIT}k} - h_{bk}$.]

Next consider the budget constraint relation $y_d = wH(1 - r - t) + (1 - t)N + G$, where r and t are left vague for the moment. We can write this relation formally as follows.

Budget Constraint 7.7

$$y_d = -w(1 - r - t)L + [wT(1 - r - t) + N(1 - t) + G]$$
$$\text{for } 0 \leq L \leq T.$$

We then have a theorem.

Theorem 7.13. *Let utility minimization, S^*, $\mathcal{P}_3$ (with $\delta > 0$, $\eta < 0$), Budget Constraint 7.7, and $L \neq 0$, T hold. Then "$H = \gamma + \delta(1 - r - t)w + \eta(N + B_0)$," with $\delta > 0$, $\eta < 0$.*

Evidently, then, Budget Constraint 7.7 is the implicit budget constraint of the Gary experiment first study, though the researchers never explicitly state their budget constraint. But what is Budget Constraint 7.7's relation to the actual before- and during-NIT Budget Constraints 7.1 and 7.2? And what precisely are r and t?

The researchers are aware of "the piecewise-linear nature of most [actual] budget constraints" (Moffitt 1979, p. 479). To deal with the "kinked-budget-line problem" (Kehrer, McDonald, and Moffitt 1979, p. 15), they use "tax rate r and t [which] are averages of the marginal tax rates over the entire length of each individual's budget constraint. . . . The

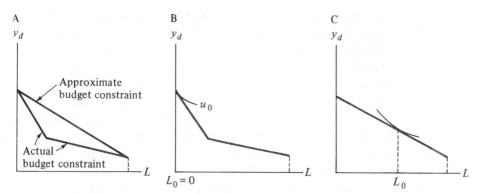

Figure 7.5. Linear approximation of a piecewise-linear budget constraint. (A) Approximation; (B) actual solution; (C) approximate solution.

averages are calculated by taking the two endpoints on the budget constraint and deriving the slope of the line between them" (Moffitt 1979, p. 479). Thus, the researchers suggest that Budget Constraint 7.7 approximates the actual underlying budget constraint with the line segment joining its two endpoints (Figure 7.5A). But such an approximation procedure departs in two ways from the approximation techniques of classical mechanics (Chapter 5) and hypothetical microeconomics (Chapter 6) experiments. First, the meaning of approximation here is vaguer than earlier, for $\sin(\theta) \approx \theta$ for small θ (Section 5.2.5) meant the limit, as θ approaches 1, of $(\sin(\theta)/\theta) = 1$, and $e^x \approx 1 + x$ for $0 \ll x < 1$ (Section 6.3.4) meant the limit, as x approaches 0, of $\ln[(1 + x)/x] = 1$. Yet the meaning of $y_d \approx -w(1 - r - t)L + [wT(1 - r - t) + G]$ is less clear, because a "very nearly" linear actual budget constraint might have a sharply different L_0 from its approximation (Figure 7.5B and C). Second, the Chapter 5 and 6 approximation techniques were employed to render *intractable* equations solvable, whereas here we can, as in Sections 7.3.1 and 7.3.3, deduce prediction statements assuming the actual piecewise-linear constraints (see Theorem 7.12).

So, had the researchers distinguished before- and during-NIT budget constraints and approximated each in turn, their approximation would have suffered from only the two difficulties just described. But they failed to distinguish before- and during-NIT constraints. Instead, t is the slope of a constraint that approximates one whose two pieces' y_d relations are the second-piece y_d relations of Budget Constraints 7.1 and 7.2, and r is the slope of a constraint that approximates Budget Constraint 7.1, and then r

and t are placed, together with the y_d intercept of the approximation whose slope is t, in Budget Constraint 7.7. By jumbling before- and during-NIT constraints together, Budget Constraint 7.7 seems a vaguer "approximation" of either before- or during-NIT than would have been an approximation of one of them by the previous paragraph's technique.

The researchers next modify the H relation of the prediction statement of Theorem 7.13 in two ways which have no deductive basis. First, the prediction statement's H relation "is modified slightly, but in an important way, to allow NIT and non-NIT variables to have different coefficients: $H = \gamma + \delta'w(1 - r) + \delta''(-wt) + \eta'N + \eta''B_0$" (Moffitt 1979, p. 480). Then "one pre-enrollment observation and up to seven during-experiment observations are available from interview data . . . and are pooled into one regression and the pre-enrollment observation is 'dummied out'" (Moffitt 1979, p. 480); that is, the H relation is modified a second time – to "$H = a + b(-wt)D + cB_0D + dw(1 - r) + eN + fD + g(-wt) + hB_0$ where D is a dummy variable equal to 1 if the observation is during the experiment and 0 if pre-enrollment" (Moffitt 1979, p. 480). Finally, $\mathbf{a} \cdot \mathbf{x}$, where $x_1 = 1$ and x_i for $i \geq 2$ is the value of a demographic variable, replaces a, and a measurement-error term is added to the H relation. The researchers test standard statistical hypotheses that, given their nondeductive modification of the H relation, have no clear relationship to the $\delta > 0$, $\eta < 0$ hypothesis that follows from Theorem 7.13's assumptions.

To summarize the "logic," the Gary researchers' first study appears to suppose H^* ("labor-supply function" $h = \gamma + \delta w + \eta y$ holds when A: $L \neq 0, T$), $wh + y = y_d$, $h + L = T$. But there is no utility function $u(L, y_d)$ for which UM and A imply H^*. However, for the utility function of $\mathcal{P}_3$ (with $\eta < 0, \delta > 0$), utility *minimization* and A imply H^*. So let B be: Budget Constraint 7.1 (7.2) holds before (during) the NIT, and no corner solutions are allowed. Then the researchers *could* have followed the following implicit logic: S^*, MEC, $\mathcal{P}_3$ (with $\eta < 0, \delta > 0$), B, and utility minimization imply a prediction statement – which with the substitution of $\gamma \cdot \mathbf{x}$ for γ and the addition of measurement-error terms becomes

$$h_b = \delta w_{nb} + \eta y_{nb} + (\gamma \cdot \mathbf{x}) + \epsilon_b \ (h_{\text{NIT}} = \delta w_{\text{NIT}} + \eta y_{n\text{NIT}} + (\gamma \cdot \mathbf{x}) + \epsilon_{\text{NIT}}),$$

with $\eta < 0, \delta > 0$. But the researchers fail to follow such an implicit logic. Let B^* be: Budget Constraint 7.7, which is a dubious amalgamation-approximation of the actual before- and during-NIT constraints, holds, and no corner solutions are allowed. Then the researchers follow the implicit logic: S^*, MEC, utility *minimization* and $\mathscr{P}_3$ (with $\delta > 0$, $\eta < 0$) imply a prediction statement that (after allowing NIT and non-NIT variables to have different coefficients, "dummying out" preenrollment observations, adding "control variables" and an additive measurement-error term) becomes

$$H = b(-Wt) + cB_0D + dW(1 - r) + eN + fD$$
$$+ g(-Wt) + hB_0 + \mathbf{a} \cdot \mathbf{x} + \epsilon.$$

So in contrast to the classical mechanics and hypothetical microeconomics experiments of Chapters 5 and 6, the statement that the first study of the Gary experiment tests does *not* follow deductively from the researchers' assumptions. Like the New Jersey and Seattle-Denver experiments, the Gary experiment first study violates the Kuhn–Popper scientific theory demarcation criterion (see Section 5.1.2), whereas the classical mechanics and hypothetical microeconomics experiments do not. But note that with the supposed "labor-supply function" of the Gary experiment first study, to understand the study as a test of some version of the theory of labor–leisure choice requires the implausible substitution of a utility-minimization assumption for UM. Thus, the Gary experiment first study seems an even worse logical shambles then were the New Jersey and Seattle-Denver experiments.

7.3.6   *The Gary experiment: attachment and induction difficulties in a first study*

As in New Jersey and Seattle-Denver, in addition to violating the Kuhn–Popper criterion, the Gary experiment first study exhibits attachment difficulties of the sort we noticed in the hypothetical microeconomics experiments of Chapter 6, but not in the classical mechanics experiments of Chapter 5. Again, for the experiment's results to apply to any period other than the one during which the data were gathered, tastes and preferences for laboring must be the same in both periods; yet tastes and preferences for laboring may evolve historically. This is the difficulty of time's role and meaning in the theory. As in New Jersey and Seattle-Denver, the researchers in the Gary experiment first study employed figures for income, gross earnings, and hours worked from interviews, and they defined gross wage rate as gross earnings divided by hours worked. They concluded that "the effect of underreporting on the response is unclear"

(Kehrer, McDonald, and Moffitt 1979, p. 41). So again, employing observations reported by conscious objects of inquiry introduces an attachment difficulty.

For gross wage rate, the researchers follow Friedman's "permanent hypothesis" in employing an ad hoc theory of wage rate determination to cope with the attachment difficulty of whether or not the observed wage rate is an equilibrium wage rate. The researchers took the observed gross wage rates, W, as those calculated from interview reports of *preenrollment* gross earnings and hours worked. They estimated an equation for W from the sample of workers only. Their independent variables were "years of Education (Dummies) [system of 3]," "years of Age," "years of Age Squared," and "Family Size" (Kehrer, McDonald, and Moffitt 1979, p. 68). As checks on each other, two approaches were employed in defining the wage variable in estimating hours: First, following New Jersey, W $(\hat{W})$ was used for workers (nonworkers) (see Section 7.3.2), and then, following Seattle-Denver, W was used for both workers and nonworkers (see Section 7.3.4).

To cope with the attachment difficulty of variation in tastes and preferences for laboring between individuals, the Gary experiment first study employed a second ad hoc theory in $\gamma \cdot \mathbf{x}$, where $\mathbf{x}$ is a vector of individual characteristics, in its hours relation. The components of $\mathbf{x}$ were "the number adults in the family, the total number of children, and the presence of children in various ranges as dummy variables," "the local SMSA unemployment rate and the season (one if summer, zero if not) at the time of the interview," and "a variable for . . . the presence of a separate family in the household which may share income, rent, expenses, etc." (Kehrer, McDonald, and Moffitt 1979, p. 26).

Had the Gary experiment first study not allowed NIT and non-NIT variables to have different coefficients and dealt with before–after experimental budget constraint differences by "dummying out" preenrollment observations, two hypotheses they sought to test (i.e., $\delta > 0$, $\eta < 0$) would have followed from their version of labor–leisure choice theory (a version with a utility-minimization assumption). As it was, though, these two hypotheses failed to link the study's two implicit deductive chains. Also, like New Jersey and Seattle-Denver, the Gary experiment first study suffered from the induction difficulty of a second deductive chain – prediction statement + added assumptions to statistical tests of hypotheses – that was not intact. Recall that with the addition of a measurement-error term, the "prediction" statement's H relation becomes

$$H = b(-Wt)D + cB_0D + dW(1 - r) + eN + fD + g(-Wt)$$
$$+ hB_0 + \mathbf{a} \cdot \mathbf{x} + \epsilon.$$

But the researchers also added an ad hoc theory of (gross) wage rate determination – say $W = \mathbf{C}_w \cdot \mathbf{Y} + \zeta$, with ζ another measurement-error term. About ϵ, they say only "ϵ is a randomly distributed error term" (Moffitt 1979, p. 478), and about their implicit ζ they say nothing.

As checks on each other, the researchers employed *two* approaches to estimating coefficients of their H relation. But in each approach they report "unsigned t-statistics" with their estimates and whether the estimate is "significant at the 90-percent [or "95-percent"] level,"[16] implicitly supposing that for an estimate, say $\hat{d}$ of coefficient d, if $d = 0$, then $[\hat{d}/(\widehat{\text{Var}}(\hat{d}))^{1/2}] \sim$ Student's t distribution with $(n - 1)$ degrees of freedom (where n is the number of observations). Yet, as in New Jersey and Seattle-Denver, such a supposition does *not* seem to follow from their assumptions: Let $\epsilon_k (\zeta_k)$ be independent $\sim N(0, \sigma^2) (N(0, \sigma^{*2}))$. Then had the researchers used an OLSQ estimated $\hat{d}$ of d *with observed values of W for each individual,* if $d = 0$ their distributional assumption for $[\hat{d}/(\widehat{\text{Var}}(\hat{d}))^{1/2}]$ *would* follow.[17]

In fact, the researchers employ OLSQ and TOBIT estimates of d, but in neither case do they employ observed values of W. They either use observed W for workers and $\hat{W}$ for nonworkers or use $\hat{W}$ for both workers and nonworkers, where $\hat{W}$'s coefficients are OLSQ estimates in the ad hoc theory from the sample of workers only. With none of the 4 (= 2 × 2) estimating approaches the researchers employed, then, is it clear that $\hat{d}$'s consistency as an estimate of d, let alone that if $d = 0$, then $[\hat{d}/(\widehat{\text{Var}}(\hat{d}))^{1/2}]$ is distributed as they suppose [or even asymptotically $\sim N(0, 1)$], follows from their assumptions. So even if we strengthen their stated assumptions on ϵ and ζ to standard ones, given their complicated two-stage estimation procedures for d, the distribution the researchers suppose for $[\hat{d}/(\widehat{\text{Var}}(\hat{d}))^{1/2}]$ doesn't seem to follow. The Gary experiment first study's second deductive chain is not intact.

Finally, as in New Jersey and Seattle-Denver, we had in the Gary first study failure to model, or an inadequate model of, the complexity of the real-world budget constraint. First, Kehrer, McDonald, and Moffitt

---

[16] The two quotations are from Kehrer, McDonald, and Moffitt (1979, p. 40). Similar reports may be found on p. 76 of the same source and in Moffitt (1979, p. 484).

[17] See footnote 6.

(1979) argued correctly but vaguely that since actual utility functions and budget constraints were *family* rather than individual, "the wage rates of all employable household members should appear as exogenous variables in the labor supply equation of each employable person" (p. 24). (Once the individual labor-supply function was modified in a particular way to include the wage rates of other family members, one would need to deduce the family utility function from the new labor-supply function and the family budget constraint.) Rather than include such wage rates, however, "in the husband's equations, we add to his nonwage income the value of his wife's wage rate times 95 hours per month, the mean in the wives' sample" (p. 25) – a step from which "two types of bias may arise" (p. 24). Second, the researchers used an inadequate model of the budget constraint's *piecewise* linearity (see previous section for details). (Because in Gary during and before the experiment there were no AFDC-UP programs for intact families, there was no need – as there was in New Jersey and Seattle-Denver – to model a welfare option in the budget constraint.)

7.3.7   The Gary experiment: "deduction" of a prediction statement in a second study[18]

As in the previous section, we begin with the framework of Section 7.2.1 and let T, h, L, w, y, and y_d be as they were there. Rather than begin with a particular utility function (as we did in Section 7.3.1), we begin with some abstract knowledge of a utility function and a particular consequent labor-supply function and *deduce* particular indirect (see Theorem 7.1) and direct utility functions.

Theorem 7.14. *Let $u(L, y_d)$ be an individual utility function of a worker facing budget constraint $wh + y = y_d$, $T = h + L$. Suppose that $u(L, y_d)$ satisfies Axiom 7.1, UM holds, $L \neq 0$, T, and $h = kw^\alpha y^\beta$, with $k > 0$, $\alpha \geq 0$, $\beta < 0$. Then*

(i)   *we cannot specify a closed-form $u(L, y_d)$ for each α, β,*

(ii) $V: v(w, y) = k[w^{(1+\alpha)}/(1 + \alpha)] + [y^{(1-\beta)}/(1 - \beta)]$, *where v is the worker's indirect utility function,*

(iii) *if $\alpha = 0$, then*

$$u(L, y_d) = [y_d/(T - L)] + [(T - L)^{(1/\beta)-1}/[(1/\beta) - 1]k^{1/\beta}].$$

---

[18] Notice that what we label a "second" study here was actually published the year *before* the materials that we label a "first" study.

As in the previous section, we next shift to the framework of Section 7.2.2 and let T, h, L, w_g, y_g, t_n, t_F, E, and G be as they were there. Also, let S^* be: w_g and T are the same in the periods before and after an NIT is introduced; market demand for low-income labor is perfectly elastic. Finally, suppose MEC is as in Section 7.3.1, and let h_{bk}, $h_{\text{NIT}k}$, w_{nbk}, y_{nbk}, $w_{n\text{NIT}k}$, and $y_{n\text{NIT}k}$ be as in the previous section.[19] We then have a theorem.

Theorem 7.15. *Let UM, S^*, V (with $k > 0$, $\alpha \geq 0$, $\beta < 0$), and MEC hold, and suppose $G > t_n E$, $t_n > t_F$, $(G - t_F E)/(t_n - t_F) < T w_g$. If $L \neq 0$, $T - (E/w_g) + (y_g/w_g)$ $(0,\ T - (G - t_F E)/w_g(t_n - t_F) + (y_g/w_g))$ and Budget Constraint 7.1 (7.2) hold before (during) the NIT, then $\ln(h_{bk}) = \alpha$ $\ln(w_{nbk}) + \beta \ln(y_{nbk}) + \ln(k)$ $(\ln(h_{\text{NIT}k}) = \alpha \ln(w_{n\text{NIT}k}) + \beta \ln(y_{n\text{NIT}k}) + \ln(k))$, with $\alpha \geq 0$, $\beta < 0$.*

As in the New Jersey, Seattle-Denver, and first Gary experiment studies, we could allow for variation in tastes and preferences between individuals with an ad hoc (in senses elaborated in Section 7.3.2) theory of taste and preference determination. Letting $k = e^{z \cdot \delta}$, where $z_1 = 1$, z_j for $j \geq 2$ is the value of some demographic variable for each individual, would be one possibility. With such an assumption, though, wage and income elasticity of labor supply $((\partial h/\partial w)/(h/w)$ and $(\partial h/\partial y)/(h/y))$ would be α and β, respectively (i.e., the same for each individual). [Also notice that if $y_{bk} = y_{\text{NIT}k}$, then the $e^{z \cdot \delta}$ terms cancel in $\ln(h_{\text{NIT}}/h_b)$.] So we could also suppose $\alpha = \alpha^+ \cdot z$, $\beta = \beta^+ \cdot z$, where z is as before. Finally, we could allow for measurement error by assuming an additive measurement-error term ϵ_{bk} $(\epsilon_{\text{NIT}k}) \sim N(0, \sigma^2)$ for the $\ln(h_b)$ $(\ln(h_{\text{NIT}}))$ relation. With these additional assumptions, Theorem 7.15's prediction statement becomes

$$\ln(h_{bk}) = (\alpha^+ \cdot z_k) \ln(w_{nbk}) + (\beta^+ \cdot z_k) \ln(y_{nbk}) + (z_{bk} \cdot \delta) + \epsilon_{bk}$$
$$(\ln(h_{\text{NIT}k}) = (\alpha^+ \cdot z_k) \ln(w_{n\text{NIT}k}) + (\beta^+ \cdot z_k)$$
$$\ln(y_{n\text{NIT}k}) + (z_{\text{NIT}k} \cdot \delta) + \epsilon_{\text{NIT}k}),$$

with $(\alpha^+ \cdot z_k) \geq 0$, $(\beta^+ \cdot z_k) < 0$ for each k.

Now let the subscript i replace the subscript k. The Gary experiment second study begins by deducing Theorem 7.14(i) and (ii) (Burtless and Hausman 1978, pp. 1111–13). Also, the study is the only one of the four that employs the actual before and during piecewise-linear budget constraint that the worker faces. Burtless and Hausman (1978) note that "two individuals with the same personal characteristics who face the same

---

[19] See footnote 15.

budget sets may prefer to work substantially different amounts" (p. 1115). To cope with this difficulty, they suggest "in estimating the unknown parameters k, α, and β, all may be specified to be functions of measureable and unmeasureable [i.e., stochastic error term] individual differences. However, this very general specification leads to an intractable estimation problem" (p. 1115). For example, if we let $\alpha = \alpha^+ \cdot z + \epsilon_\alpha$, $\beta = \beta^+ \cdot z + \epsilon_\beta$, and $\delta^* = \delta \cdot z + \epsilon_\delta$ for some stochastic error terms $\epsilon_\alpha, \epsilon_\beta, \epsilon_\delta$, then even if the error term distributions are relatively simple, we have an intractable estimation problem.

Probably the simplest way out of this difficulty would be to suppose that α and β are functions only of "measureable individual differences" and that δ^* is a function of "measureable and unmeasureable individual differences." For example, if we let $\alpha = \alpha^+ \cdot z$, $\beta = \beta^+ \cdot z$, and $\delta^* = \delta \cdot z + \epsilon_\delta$, with $\epsilon_\delta \sim N(0, \sigma_\delta^2)$, and suppose that ϵ is the measurement-error term of two paragraphs ago, with ϵ and ϵ_δ independent, then we get the prediction statement of two paragraphs ago, with $\epsilon + \epsilon_\delta = \epsilon^* \sim N(0, \sigma^2 + \sigma_\delta^2)$ replacing ϵ. Burtless and Hausman (1978) do suppose K^*: "$k_i = \exp(z_i\delta + \epsilon_{2i})$, where z_i is a vector of individual characteristics and ϵ_{2i} is assumed to be distributed $\mathcal{N}(0, \sigma_2^2)$" (p. 1116). And they specify A, "$\alpha_i = \bar{\alpha}$ [i.e. α] a constant" (p. 1116), which leads to no difficulty. But they also suppose B^+: "the individual parameter β_i . . . can be written as $\beta_i = \mu_\beta + \epsilon_{1i}$ where $\epsilon_{1i} \sim \mathcal{TN}(0, \sigma_1^2)$ with a truncation point from above of $-\mu_\beta$ [where] ϵ_{1i} and ϵ_{2i} are independent" (p. 1116) (note that truncation assures $\beta < 0$). In the next section we shall argue that B^+ leads to an estimation problem which, though perhaps tractable, the researchers fail to solve. For now, we concentrate on an approximation difficulty to which B^+ leads.

Let

$$w_{1i} = w_{gi}(1 - t_n), \; y_{1i} = y_{gi}(1 - t_n) + G, \; w_{2i} = w_{gi}(1 - t_F)$$

and

$$y_{2i} = y_{gi}(1 - t_F) + t_F E.$$

Then, for given values of $w_{1i}, y_{1i}, w_{2i}, y_{2i}, z_i, \alpha$, and δ for a worker facing Budget Constraint 7.2, there exists a unique β_i^* such that the worker chooses to participate in the NIT for β_i in $(-\infty, \beta_i^*)$, and not to participate for $\beta_i \in (\beta_i^*, 0)$. For $\beta_i = \beta_i^*$ the worker is indifferent between participating or not (Figure 7.6). So for $\beta_i = \beta_i^*$, the value of the indirect utility function is the same when calculated for either segment of the budget

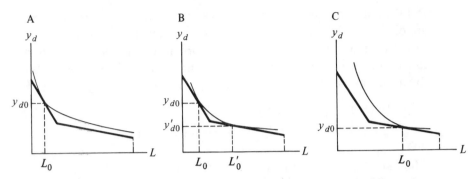

Figure 7.6. The existence of β^* for which the worker is indifferent between the budget constraint's two segments. (A) $\beta \in (-\infty, \beta^*)$; (B) $\beta = \beta^*$; (C) $\beta \in (\beta^*, 0)$.

constraint; that is, β_i^* is a solution of the equation

$$\text{``}[e^{z_i\delta}w_{1i}^{1+\overline{\alpha}}/(1+\overline{\alpha})] + [y_{1i}^{1-\beta_i^*}/(1-\beta_i^*)] = [e^{z_i\delta}w_{2i}^{1+\overline{\alpha}}/(1+\overline{\alpha})]$$
$$+ [y_{2i}^{1-\beta_i^*}/(1-\beta_i^*)]\text{''} \qquad \text{(p. 1117).} \quad (7.1)$$

The researchers argue that "for every experimental observation, this equation must be solved for β_i^* *each time the parameters change;* however, the solution may be cheaply obtained on a computer" (p. 1117, emphasis added); yet it is difficult to fathom what they could mean. For though z_i, w_{1i}, y_{1i}, w_{2i}, and y_{2i} differ for each individual i, $\overline{\alpha}$ and δ are the *unknown parameters,* which are the same for all individuals, so that β_i^* for any i is a function of $\overline{\alpha}$ and δ [i.e., $\beta_i^* = \beta_i^*(\overline{\alpha}, \delta)$], and the equation is not "solved" (cheaply or otherwise) unless β_i^* is so expressed. In fact, if for any i we let

$$e^{z_i\delta}(w_{1i}^{1+\overline{\alpha}} - w_{2i}^{1+\overline{\alpha}}) = C_i(\delta, \overline{\alpha}),$$
$$1 - \beta_i^* = x_i, \qquad y_{1i} = a_{1i}, \qquad \text{and} \qquad y_{2i} = a_{2i},$$

then equation (7.1) may be written as $a_{2i}^{x_i} - a_{1i}^{x_i} = C_i(\delta, \alpha)x_i$ – an equation with no closed-form solution x_i^*. One can, however, obtain an approximate solution of equation (7.1) by using second-order Maclaurin-series approximations of $a_{2i}^{x_i}$ and $a_{1i}^{x_i}$.

Proposition 7.3. *If $a_{ji}^{x_i} \approx 1 + x_i \ln(a_{ji}) + (x_i \ln(a_{ji}))^2/2!$ for $j = 1, 2$, then*

$$A^*: \beta_i^*(\overline{\alpha}, \delta) \approx 1 + [2/\ln(y_{1i}y_{2i})] - [2e^{z_i\delta}(w_{2i}^{1+\overline{\alpha}} - w_{1i}^{1+\overline{\alpha}})]/$$
$$(1+\overline{\alpha})[(\ln(y_{1i}))^2 - (\ln(y_{2i}))^2].$$

But then one employs not just one but as many approximations as there are observations in deducing the researchers' prediction statement.

Let $\varphi(\ )$ and $\Phi(\ )$ be the standard normal density and distribution functions, respectively, $PC_i(PNC_i)$ be the probability of observing H_i before (during) the NIT, $w_{3i} = w_{gi}$, $y_{3i} = y_{gi}$, $\tilde{H}_i = (E_i - y_{gi})/w_{gi}$,

$$BL_i = (\ln(\overset{\approx}{H}_i) - z_i\delta - \alpha \ln(w_{3i}))/\ln(y_{3i}),$$

$$BU_i = (\ln(\overset{\approx}{H}_i) - z_i\delta - \alpha \ln(w_{2i}))/\ln(y_{2i}),$$

$$\tilde{w}_{ji} = \ln(w_{ji}), \quad \tilde{y}_{ji} = \ln(y_{ji}) \quad \text{for } j = 1, 2, 3,$$

$$R(a, b, c) = [1/(1 - \Phi(\mu_\beta/\sigma_\beta)(\sigma_\beta^2 c^2 + \sigma_2^2)^{1/2})]$$
$$\varphi[(\ln(H_i) - z_i\delta - \alpha b - \mu_\beta c)/$$
$$(\sigma_\beta^2 c^2 + \sigma_2^2)^{1/2}] \times \Phi[((a - \mu_\beta)(\sigma_\beta^2 c^2 + \sigma_2^2)^{1/2}/\sigma_\beta\sigma_2)$$
$$- (\sigma_\beta c(\ln(H_i) - z_i\delta - \alpha b - \mu_\beta c))/\sigma_2^2(\sigma_\beta^2 c^2 + \sigma_2^2)^{1/2}],$$

and

$$X(BU_i, BL_i) = [1/\sigma_2^2[1 - \Phi(\mu_\beta/\sigma_\beta)]]\varphi((\ln(H_i)$$
$$- \ln(\overset{\approx}{H}_i))/\sigma_2)[\Phi((BU_i - \mu_\beta)/\sigma_\beta) - \Phi((BL_i - \mu_\beta)/\sigma_\beta)].$$

Then we may formalize deduction of the Gary experiment second study's prediction statement with a theorem.

Theorem 7.16. *Let UM, S*, V (with $\alpha \geq 0$, $\beta < 0$, $k > 0$), MEC, K*, A, B+, and A* hold; $G > t_n E$, $t_n > t_F$, $(G - t_F E)/(t_n - t_F) < Tw_g$. If Budget Constraint 7.1 (7.2) holds before (during) the NIT, then*

(i) . $PC_i \approx R(BL_i, \tilde{w}_{3i}, \tilde{y}_{3i}) + X(BU_i, BL_i) + 1 - R(BU_i, \tilde{w}_{2i}, \tilde{y}_{2i})$,

(ii) $PNC_i \approx R(\beta_i^*, \tilde{w}_{1i}, \tilde{y}_{1i}) + 1 - R(\beta_i^*, \tilde{w}_{2i}, \tilde{y}_{2i})$, *with $\alpha \geq 0$.*[20]

Notice that the theorem's prediction statement or conclusion is considerably more complicated than those we have heretofore encountered. Nevertheless, one could proceed to test it as in earlier cases (i.e., in Section 4.3 fashion). For example, to test $\alpha \geq 0$, we would need estimates $\hat{\alpha}$ and $\widehat{\text{Var}}(\hat{\alpha})$ of α and $\text{Var}(\hat{\alpha})$, respectively, for which the theorem's assumptions imply that if $\alpha = 0$, then $(\hat{\alpha}/(\widehat{\text{Var}}(\hat{\alpha}))^{1/2}) \sim$ (perhaps asymptotically) some known distribution. But here a serious logical difficulty (whose details we consider in the next section) arises: Though we can find esti-

---

[20] The Gary experiment second-study researchers prove the theorem (Burtless and Hausman 1978, pp. 1127–8). Here one would have to discard individuals who face the second piece only of Budget Constraint 7.2 during the experiment, i.e., those for whom $(G - t_F E) > Tw_g(t_n - t_F)$ (see footnote 3).

mates $\hat{\alpha}$ and $\widehat{\mathrm{Var}}(\hat{\alpha})$ of α and $\mathrm{Var}(\hat{\alpha})$, no (asymptotic) distribution of $(\hat{\alpha}/(\widehat{\mathrm{Var}}(\hat{\alpha}))^{1/2})$ seems to follow from the theorem's assumptions. Burtless and Hausman (1978) are aware of this difficulty and cautiously say only that "the estimated wage elasticity $[\hat{\alpha}]$ is .00003 [with 'asymptotic standard error . . . .01632']" [21] or "our estimate of α is zero" (p. 1127). If in fact α is zero, though, we can deduce the underlying direct (not just indirect) utility function (see Theorem 7.14(iii)) – a step the researchers do *not* take.

To summarize the logic, let H^+ be $h = kw^\alpha y^\beta$, with $k > 0, \alpha \geq 0, \beta < 0$, and let A be $L \neq 0, T; wh + y = y_d, h + L = T$. Then the Gary researchers' second study first shows that one cannot specify a particular closed-form *direct* utility function that with A and UM implies H^+, but that UM, A, and V (a particular indirect utility function with $k > 0, \alpha \geq 0, \beta < 0$) imply H^+. Then let J be: UM, S^*, V (with $\alpha \geq 0, \beta < 0, k > 0$), MEC, and K^* hold; $G > t_n E, t_n > t_F, (G - t_F E)/(t_n - t_F) < Tw_g$. And let A_1^* be $\alpha = \alpha^+ \cdot z$, and B_1^* be $\beta = \beta^+ \cdot z$. Then the researchers could have followed the following implicit logic: That J, A_1^*, B_1^*, and Budget Constraint 7.1 (7.2) hold before (during) the NIT implies

$$\ln(h_{bk}) = (\alpha^+ \cdot z_{bk}) \ln(w_{nbk}) + (\beta^+ \cdot z_{bk}) \ln(y_{nbk}) + z_{bk} \cdot \delta + \epsilon_{2bk}$$
$$[\ln(h_{\mathrm{NIT}}) = (\alpha^+ \cdot z_{\mathrm{NIT}k}) \ln(w_{n\mathrm{NIT}k}) + (\beta^+ \cdot z_{\mathrm{NIT}k}) \ln(y_{n\mathrm{NIT}k})$$
$$+ (z_{\mathrm{NIT}k} \cdot \delta) + \epsilon_{2\mathrm{NIT}k}],$$

with $(\alpha^+ \cdot z_{bk}), (\alpha^+ \cdot z_{\mathrm{NIT}k}) \geq 0, (\beta^+ \cdot z_{bk}), (\beta^+ \cdot z_{\mathrm{NIT}k}) < 0$, for each k. Instead, the researchers followed the following implicit logic: That J, A, B^+, A^*, and Budget Constraint 7.1 (7.2) hold before (during) the NIT implies

$$PC_i(PNC_i) \approx R(BL_i, w_{3i}, y_{3i}) + X(BU_i, BL_i) + 1 - R(BU_i, w_{2i}, y_{2i})$$
$$(R(\beta_i^*, w_{1i}, y_{1i}) + 1 - R(\beta_i^*, w_{2i}, y_{2i})),$$

with $\alpha \geq 0$.

---

[21] The quotations are from Burtless and Hausman (1978, p. 1123). The researchers' use of "asymptotic standard error" is at odds with standard statistical lexicon and therefore possibly misleading. Let $\hat{\alpha}$ be an estimate (= function of the observations) of parameter α. By "the asymptotic standard error of $\hat{\alpha}$," a statistician means the limit, as sample size approaches infinity, of $\hat{\alpha}$'s standard error. So $\hat{\alpha}$'s asymptotic standard error is *not* a function of the observations of any (finite) sample. In contrast, the researchers' "asymptotic standard error" seems to refer to an estimate (= function of the observations) of $\hat{\alpha}$'s standard error, where $\hat{\alpha}$ is the estimate of α from the last iteration that their estimating procedure attempts. (Presumably, the iterative estimating procedure terminates once $\hat{\alpha}$ and the estimate of $\hat{\alpha}$'s standard error stabilize, i.e., remain at the values .00003 and .01632, respectively, for several iterations.)

The conclusion the Gary experiment second-study researchers draw, then, is more complicated than the one they could have drawn using the stochastically simpler assumptions we suggested. And they rely on an approximation repeated as many times as there are observations without accounting for the error thereby introduced into the model. Still, on the level of statements, the logic of the study seems to be very similar to that of the classical mechanics and microeconomics experiments of Chapters 5 and 6. So the Gary experiment second study does not seem to violate the Kuhn–Popper scientific theory demarcation criterion (see Section 5.1.2) as did the other three income maintenance experimental studies. Yet the study still suffers from serious attachment difficulties – some of which conceivably make it not science – to which we now turn.

7.3.8   *The Gary experiment: attachment and induction difficulties in a second study*

We argued in the previous section that the Gary experiment second study was the only one of the four income maintenance experimental studies we have examined that apparently satisfied the Kuhn–Popper scientific theory demarcation criterion of our Chapter 5 and 6 experiments. That is, one could discern in the Gary experiment second study's statements an intact deductive chain from a version of labor–leisure choice theory to a prediction statement, whereas no similar chain is discernible in the other three income maintenance experimental studies. The Gary experiment second study, though, exhibits the same sort of attachment difficulty of the meaning and role of time in the theory as did the other three income maintenance experimental studies.

Burtless and Hausman (1978) explain that "data on workers' hours, wages, nonwage income . . . were taken from . . . periodic interviews administered to participants in the experiment during the period of NIT payments" (p. 1121). Thus, the Gary experiment second study exhibits the same attachment difficulty of conscious objects of inquiry as do the other three income maintenance experimental studies. The researchers do not follow Friedman's "permanent hypothesis" approach in supposing that gross wage rates and/or nonwage income depend on personal characteristics. Instead, "individuals not observed to be working are omitted from the current sample" (p. 1121), and gross wage rate and nonwage income are calculated for all remaining individuals as interview data averages. Thus, the researchers deal with the attachment difficulty of whether or not observed gross wage rates and nonwage income are equilibria by dropping individuals for whom they think observed values may not be equilibria (or are undefined in the case of wage rates) – that is, nonworkers – and by supposing that observed values are equilibria for all others.

The Gary experiment second-study researchers take two steps to cope with the attachment difficulty of variations in tastes and preferences for laboring between individuals. First, Burtless and Hausman (1978) replace their parameter k with $e^{\delta z + \epsilon_2}$, where ϵ_2 is an error term, δ is a parameter, and z is a vector of individual characteristics "chosen by reference to earlier research on the NIT experiments" (p. 1122). The z components were an education dummy ($= 1$ for <9 years of schooling), "number of persons age 16 or more residing in the household" (p. 1122), a healthy dummy ($= 1$ "if the individual reported his health to be 'poor in relation to others'") (p. 1122), and "age . . . minus 45 . . . if age exceeded 45 years, otherwise . . . zero" (p. 1122). In addition to this ad hoc theory of taste and preference determination, the researchers allowed β to vary with a (prior) probability distribution over $\beta < 0$ (which follows from their version of labor–leisure choice theory).

The standard statistical hypotheses Burtless and Hausman (1978) seek to test about δ neither follow from nor contradict their version of labor–leisure choice theory, while the hypothesis $\alpha \geq 0$ they also seek to test does follow from their version of labor–leisure choice theory and so provides a link between their study's two deductive chains. Recall from the previous section that the researchers' explicit stochastic assumptions are that their ϵ_{2i}'s are independent $\sim N(0, \sigma_2^2)$, "$\beta_i = \mu_\beta + \epsilon_{1i}$ where ϵ_{1i} [are independent and] $\sim \mathcal{T}\mathcal{N}(0, \sigma_1^2)$ with a truncation point from above of $-\mu_\beta$" (p. 1116), and that "ϵ_{1i} and ϵ_{2i} are independent sources of random variation" (p. 1116). Their parameters then are $(\delta, \alpha, \beta, \sigma_1^2, \sigma_2^2)$, "where β is the mean of the truncated distribution" (p. 1116).

The researchers seek maximum-likelihood estimates (MLEs) of their parameters with two iterative procedures and explain:

Two techniques were used to maximize the likelihood function. Convergence was obtained using the modified scoring method proposed by Berndt et al. (1974). Only first derivatives are required for this algorithm. However, as a check to make certain that the global maximum was achieved, the no-derivative conjugate gradient routine of Powell (1964) was also used to verify the parameter estimates. The reason for this caution is that, while the log-likelihood function . . . is everywhere differentiable in the parameters, the derivatives are not everywhere continuous because of the kink point. While proofs of the usual large sample properties of maximum likelihood were not attempted in this nonregular case, consistency of the estimates would follow from the usual type of proof. However, proof of asymptotic normality of the estimates is complicated by the lack of continuous derivatives, and reported asymptotic standard errors should be interpreted with this problem in mind. (pp. 1128–9)

In other words, the researchers are *pretty* sure their estimates are MLEs. But if $\hat{\delta}_j$ *is* an MLE estimate here for the jth component of δ, they only think, but haven't proved, that their stochastic assumptions imply that $\hat{\delta}_j$ is a consistent estimate of δ_j. And they evidently *don't* think that their stochastic assumptions imply that if $\delta_j = 0$, then $[\hat{\delta}_j/(\widehat{\text{Var}}(\hat{\delta}_j))^{1/2}]$ is asymptotically $\sim N(0, 1)$.

Yet despite their admirable cautioning that the "reported asymptotic standard errors should be interpreted with this problem in mind," Burtless and Hausman (1978) report that "all [components of δ] except the coefficient of age are significantly different from zero at the 5 percent level" (p. 1123) on the basis of a one-tailed t test. Such a report implicitly assumes that if $\delta_j = 0$, then $[\hat{\delta}_j/(\widehat{\text{Var}}(\hat{\delta}_j))^{1/2}] \sim$ Student's t distribution with $(n - 1)$ degrees of freedom (where n is the number of observations), ignoring their own cautionary advice! So the Gary experiment second study's second deductive chain – prediction statement + added assumptions (including hypotheses about δ's components) to statistical tests of hypotheses – is not intact.

Finally, recall from Section 7.3.7 that the Gary experiment second study was the only one of the four we have examined that correctly modeled the *piecewise*-linear nature of the actual budget constraint, and recall from Section 7.3.6 that no AFDC-UP was available for intact families in Gary, so that there was no need to include welfare in modeling the budget constraint. So the Gary experiment second study's only failure in modeling the complexity of the real-world budget constraint was to employ an individual rather than family budget constraint – a step about which the researchers make no remark. In fact, they *could* have replaced their h relation ($h = kw^{\alpha}y^{\beta}$) with $h = kw^{\alpha}y^{\beta}w_s^{\alpha'}$, where w_s is spouse's wage rate, and deduced a family (indirect) utility function from the new h relation and employed the new h relation in their empirical work.

7.4   Summary and conclusion

7.4.1   *Weaknesses of deduction, induction, and particular utility function choice*

In the previous two chapters we examined experiments behind which were two linked deductive systems of statements. That is, from a theory –

either classical mechanics or microeconomics – a prediction statement including hypothesis was deduced. Then from the prediction statement and a reporting model with stochastic assumptions and estimating procedure, the probability distribution of some statistic under the hypothesis was deduced, so that the hypothesis could be tested by the observed value of the statistic in Chapter 4 statistical theory fashion. We label the first chain of statements the experiment's deductive system, and the second its inductive system, even though the second chain was also deductive. In the present chapter we examined four experiments behind which were also deductive/inductive systems of statements, but in these experiments, in contrast to those of the earlier chapters, one or the other or both the deductive chains exhibited a gap in logic, and/or the link between the two systems was not intact.

Gaps in the New Jersey and Seattle-Denver deductive systems were the addition of ad hoc systems of individual characteristics to the change in hours-worked relations which followed deductively from their version of labor–leisure choice theory. The presence of these ad hoc systems meant that the hypotheses the researchers sought to test were not those which followed from their versions of labor–leisure choice theory (i.e., the hypotheses failed to serve as links between the experiments' deductive and inductive systems). Finally, gaps in the two experiments' inductive systems were the distributions given hypotheses supposed for the test statistics behind their reported test results not following from their prediction statements and reporting models with stochastic assumptions and estimating procedures.

The Gary experiment first study's implicit logic was the most bizarre of the four income maintenance experimental studies. First, gaps in the study's deductive logic occurred with the modification (by allowing NIT and non-NIT variables to have different coefficients and dummying out preenrollment observations) of the statement which follows from the researchers' implicit version of labor–leisure choice theory. Second, included in that version of labor–leisure choice theory was a utility-*minimization* rather than utility-maximization assumption! Thus, the Gary experiment first-study deductive system's implicit logic was an utter shambles. And the modification of the deduced prediction statement in two ways meant that the hypotheses the researchers sought to test were not those which followed from their version of labor–leisure choice theory; that is, the hypotheses failed to link the study's deductive and inductive systems. Finally, as in New Jersey and Seattle-Denver, the researchers reported test results which supposed test-statistic distributions (under hypotheses) that did not follow from their prediction statement and reporting model with stochastic assumptions and estimating procedure. So there was a gap in the logic of the study's inductive system.

The Gary experiment second study's logic was the closest of the four studies' logics to those of the experiments of the two previous chapters. From a version of labor–leisure choice theory including a particular indirect utility function, a prediction statement including a hypothesis follows. Further, the researchers seek to test that hypothesis. Thus, there are no gaps in the study's deductive system, and the link between deductive and inductive systems is intact. As in the other three studies, though, the researchers reported test results which supposed a test-statistic distribution (under the hypothesis) that did not follow from their prediction statement and reporting model with stochastic assumptions and estimating procedure. So there was a gap in the logic of the study's inductive system. Finally, the study's test-result conclusion implies a particular *direct* utility function behind their work, though the researchers never actually identify that function.

Table 7.3 compares the statements-level deductive logic of the previous two chapters' experiments with those of the income maintenance experiments. Notice that for the previous chapters' experiments, but for *none* of the income maintenance experiments, the two linked deductive chains are intact. So if intact, statements-level, linked deductive chains constitute a necessary (but not sufficient) condition for a developed science experiment, then the income maintenance experiments were not developed science experiments. Rather, they were logically immature or underdeveloped science experiments. What is unclear is whether or not the researchers could have formulated their deductive–inductive systems so that the experiments would have been both logically mature and successful (in the sense of corroborating deduced hypotheses) science experiments.

Because the income maintenance researchers' formulations were logically immature, we cannot say whether or not their experiments *could* have been logically mature and successful. When we view the four experiments *together,* and in contrast to any of the classical mechanics experiments, though, we see a second sense in which they were immature science experiments: The four studies are of essentially the same situation; yet each employs a different particular utility functional form, and none ends up rejecting its form. Hence, the experiments taken together are immature in the sense that they fail to decide on a particular utility

Table 7.3. *Deductive logics of classical mechanics, hypothetical microeconomics, and income maintenance experiments*

System	Deductive system	Link between systems	Inductive system
Characteristics	Theory version to prediction statement intact	Seek to test hypotheses implied by theory version	Prediction statement, and reporting model with stochastic assumptions and estimating procedures, to test statistic's distribution given hypothesis intact
Experiment			
Classical mechanical (Chapter 5)	Yes	Yes	Yes
Hypothetical microeconomics (Chapter 6)	Yes	Yes	Yes
New Jersey	No	No	No
Seattle-Denver	No	No	No
Gary first study	No	No	No
Gary second study	Yes	Yes	No

function for their situation. Table 7.4 depicts the second sense in which the income maintenance studies were immature or underdeveloped science experiments.

7.4.2 Attachment difficulties, ad hoc theories, and the necessity of immaturity

Each experiment's inductive system exhibits the sort of attachment difficulties that, as we saw in Chapter 4, are those of the Neyman–Pearson theory of statistical inference – that is, a meaning for probability which goes beyond the observations, and weaker stochastic assumptions leading

Table 7.4. *Four distinct versions of labor–leisure choice theory*

Experiment	Optimization principle	Particular utility function
New Jersey	Utility maximization	$u(L_1, L_2, L_3, L_4) = kL_1^{\beta_1}L_2^{\beta_2}L_3^{\beta_3}y_d^{1-\beta}$, with $k, \beta_i > 0$ for $i = 1, 2, 3$, and $\beta_1 + \beta_2 + \beta_3 = \beta < 1$ (family)
Seattle-Denver	Utility maximization	$u(L, y_d) = y_d - [(L^2/2) - TL]/\alpha$, with $\alpha > 0$ (individual)
Gary first study	Utility minimization (implicit)	$u(L, y_d) = [\eta y_d - (\delta/\eta) + \gamma]/[T - L - (\delta/\eta)] + \ln[T - L - (\delta/\eta)]$, with $\delta > 0$, $\eta < 0$ (individual)
Gary second study	Utility maximization	$u(L, y_d) = [y_d/(T - L)] + [(T - L)^{(1/\beta)-1}/[(1/\beta) - 1]k^{1/\beta}]$, with $k > 0$, $\beta < 0$ (individual; form follows from conjectured labor-supply function and empirical conclusion that parameter $\alpha = 0$)

to a weaker (i.e., asymptotic) sense of valid test. Each experiment's deductive system possesses attachment difficulties of the sort that, as we saw in Chapters 5 and 6, the classical mechanics and hypothetical microeconomics experiments shared. In this chapter, though, we explored the attachment difficulties shared by the hypothetical microeconomics (but not classical mechanics) experiments – that is, those which conceivably demarcate the income maintenance experiments from science experiments. And we also discovered some attachment difficulties exhibited by the income maintenance (but not hypothetical microeconomics or classical mechanics) experiments that might also demarcate the income maintenance experiments from science experiments.

All four studies exhibited the attachment difficulty of the meaning and role of time in their theories, that is, the problem of whether or not historically evolving tastes and preferences for laboring prevent results from applying to any period other than the one during which the data were gathered. None of the studies sought to deal with this difficulty. All four experiments also exhibited the difficulty of relying on wage rate observations, reported by conscious objects of inquiry, that may have differed systematically from actual wage rates. The first three studies noted, on the basis of treatment/control or IRS (for some individuals) comparisons, the problem, but none sought to deal with it. All four studies

Table 7.5. *Income maintenance experiment attachment difficulties shared by the hypothetical microeconomics (but not classical mechanics) experiments*

	Experiment				
Difficulty	New Jersey	Seattle-Denver	Gary first study	Gary second study
Historical evolution of tastes and preferences for laboring	Not dealt with	Not dealt with	Not dealt with	Not dealt with
Conscious objects of inquiry	No remedy attempted for presumed systematic reporting error	No remedy attempted for presumed systematic reporting error	No remedy attempted for presumed systematic reporting error	Not dealt with
Observed values may not be equilibria	Ad hoc theory of wage rate determination	Ad hoc theory of wage rate determination	Ad hoc theory of wage rate determination	Dropped nonworkers; supposed observed values = equilibria for workers
Variations in tastes and preferences for laboring between individuals	Two ad hoc theories based on individual characteristics	Ad hoc theory based on individual characteristics	Ad hoc theory based on individual characteristics	Ad hoc theory based on individual characteristics; (prior) probability distribution for a "parameter"

faced the difficulty of whether or not observed wage rates were equilibria. The first three followed Friedman's "permanent hypothesis" (i.e., used ad hoc theories of wage rate determination, which led to two-stage estimating procedures), whereas the fourth study simply dropped nonworkers and took observed wage rates as equilibria for all workers.

Finally, the four studies faced the difficulty of variations in tastes and preferences for laboring between individuals. New Jersey dealt with the problem with two ad hoc theories – one substituted a term dependent on a set of individual characteristics for a fixed parameter, and the second added a term dependent on a larger set of individual characteristics to the change-in-hours-worked relation. Seattle-Denver employed only a term added to the hours-worked relation, whereas the first Gary study used only a parameter-substitution ad hoc theory. The second Gary study employed a parameter-substitution ad hoc theory, but also allowed another "parameter" to have a (prior) probability distribution. Table 7.5 lists the four attachment difficulties shared with the hypothetical microeconomics (but not classical mechanics) experiments and indicates how the income maintenance experiments dealt with them.

The attachment difficulties shared by neither the hypothetical microeconomics nor classical mechanics experiments were all failures to model the complexities of the actual budget constraints. Seattle-Denver and the two Gary studies failed to model the budget constraint's *family* nature, and all but the second Gary study failed to model its *piecewise* linearity. Seattle-Denver also failed to model the quadratic nonlinearity of one piece of some participants' constraint. All but the New Jersey study employed approximations of a sort foreign to the Chapters 5 and 6 experiments. Finally, though welfare was available to families in Seattle-Denver and, after the experiment began, in New Jersey, neither experiment's researchers modeled it with the addition of a linear welfare "piece" in the budget constraint where appropriate. Table 7.6 depicts the experiments' failures in modeling the actual budget constraint's complexity – a third sense in which the experiments were immature or underdeveloped science.

Table 7.7 shows that when the four studies are considered together, they reveal a fourth sense – chaotic ad hoc theories sense – in which they were underdeveloped or immature science. For example, in the wage rate determination theory, family size appears in the first Gary study, but not in New Jersey or Seattle-Denver, whereas education, though appearing in all three theories, appears differently in each one (i.e., two dummies, years, or three dummies). Seattle-Denver was the only study that failed to

Table 7.6. *Income maintenance experiment attachment difficulties shared by neither hypothetical microeconomics nor classical mechanics experiments: modeled complexity of actual budget constraint*

	Experiment			
Complexity	New Jersey	Seattle-Denver	Gary first study	Gary second study
Family	Yes	No	No	No
Piecewise linearity	No	No	No	Yes
Quadratic linearity of one piece	—	No	—	—
Avoided dubious approximation	Yes	No	No	No
Welfare	No	No	—	—

Note: Dash indicates "not applicable."

substitute an ad hoc theory for a parameter, and the second Gary study was the only one that allowed a "parameter" to have a (prior) probability distribution.

So we have identified four senses – logical, unchosen utility function, unmodeled budget constraint complexity, and chaotic ad hoc theories – in which the income maintenance experiments were immature science experiments. But we also argued that such immaturity wasn't always necessary. We demonstrated with a series of theorems how the three experiments which failed to model their budget constraint's piecewise linearity, and the one which failed to model the quadratic nonlinearity of one piece of some participants' constraint, could have done so, as well as how the three which employed approximations need not have done so. And we suggested how the three experiments which failed to model the budget constraint's family nature and the two which failed to model welfare could have done so. So all four experiments could have retained their particular utility functions and still modeled the complexity of the actual budget constraint.

Had New Jersey and Seattle-Denver not added terms to their change-in-hours-worked relation and had the first Gary study not modified its labor-supply relation, then these three deductive systems would have been intact (though the first Gary study still would have sported a utility-

Table 7.7. *The ad hoc theories of the four studies and their variables*

	Experiment			
Theory	New Jersey	Seattle–Denver	Gary first study	Gary second study
Wage rate determination	Age, 2 (apparently) education dummies, earnings in year prior to experiment, health dummy, 2 race–education interaction terms	Education (years), experience (= age – education – 5), experience squared, race dummy	Age (years), age squared, 3 education dummies, family size	
Parameter substitute	Age, age squared, education dummy	No	No. adults, no. children, dummies for children in various age ranges, season dummy and unemployment rate at time of interview, dummy for second family in household	Education dummy, no. adults, health dummy, age

Parameter distribution	No	No	No	Yes
Term added to change-in-hours-worked relation	Wage rate^{-1} (estimate for nonworkers), age, age squared, 2 health dummies, 2 education dummies, weeks worked in year prior to experiment, 3 site dummies	Nine dummies for income prior to enrollment, hours worked in year prior to enrollment, dummies for site and race, age (in years), no. family members, no. children aged 0–5, AFDC benefits in year prior to enrollment	No	No

minimization assumption). Had the New Jersey, Seattle-Denver, and first Gary studies discarded nonworkers, supposed that workers' observed wage rates were equilibria, assumed independent $\sim N(0, \sigma^2)$ measurement-error terms, and used simple OLSQ estimating procedures, then those three inductive systems would have been intact. Had the second Gary study kept the "parameter" with the (prior) probability distribution fixed instead, then its inductive system would have been intact. Had none of the studies substituted ad hoc theories for parameters, all their links would have been intact too. So all four experiments could have retained their particular utility functions, been free of any ad hoc theories, and satisfied the Kuhn–Popper statements-level logical criterion (with the first Gary study still exhibiting the utility-minimization anomaly).

We have argued that logical, unmodeled budget constraint complexity, and chaotic ad hoc theories senses of immaturity were not necessary features of the income maintenance experiments. But had the researchers employed mature (in these three senses) formulations, they might have failed to corroborate their deduced hypotheses. Multicollinearity, for example, might have led to failure to corroborate a hypothesis following from a model of a family budget constraint, or an estimate's large variance might have led to failure to corroborate a hypothesis following from the assumption of no variation of tastes and preferences for laboring among individuals. So the experiments' real immaturity is this: Their theory of labor–leisure choice makes assumptions (such as that observed wage rates are equilibria, or that all individuals have the same tastes and preferences for laboring) that are sufficiently implausible that researchers are sensibly driven to ad hoc modifications of the assumptions, which in turn destroy the theory's logic. So though the experiments need not have been immature science, that they could have been mature, *successful* science remains unknown.

How, then, do we answer the question raised at the beginning of the chapter: Were the income maintenance experiments science or scientism? Like the example of scientism at the end of the previous chapter, our first three income maintenance experimental studies sought to test statements that did not follow by deduction from their version of theory. But these three did not conclude they had rejected or corroborated their version of theory, as the earlier example concluded "that the theory of demand is inconsistent with the evidence." Yet in reporting the "statistical significance" of their findings, all four income maintenance experimental studies *seemed* to suggest they were confronting the deductive consequences of a statement and a reporting model (with stochastic assumptions and an estimating procedure) with evidence, which they were not. Thus, we conclude that the income maintenance experiments were immature science, with some partially articulated scientistic overtones.

Of the barriers between the income maintenance experiments and developed science, the four attachment difficulties shared by the hypothetical microeconomics experiments – the meaning and role of time in the theory, conscious objects of inquiry, observed values may not be equilibria, and variation in tastes and preferences between individuals – seem the most serious. We shall therefore conclude this book with a deeper look at these difficulties. But first we must examine another extension of microeconomics: *welfare* microeconomics. If the income maintenance experiments are examples of the most advanced microeconomic science available, then microeconomics, if this chapter's argument is correct, is an immature science. Yet, though economists have failed to offer a developed scientific account of how the microeconomic world works, they are nonetheless audacious enough to offer us an account of what sort of microeconomic arrangements we *should* have (i.e., a deontology). We turn next to a critical account of microeconomic deontology.

CHAPTER 8

Microeconomics striving to be deontology

8.1   Introduction: Microeconomics as "normative science"

Chapter 5 catalogued the ways in which a standard set of classical mechanics experiments failed to measure up to Popper's rationalist ideal notion of science. Chapter 6 argued that hypothetical microeconomic experiments of necessity fail in the same ways as the classical mechanics experiments to be science in Popper's sense. But Chapter 7 argued that a set of actual microeconomic studies – the income maintenance experiments – fail in a whole additional set of ways to be science in Popper's sense. So if the income maintenance experiments are the best scientific studies microeconomists can offer, then microeconomics appears to be at best an undeveloped or underdeveloped science.

The nineteenth-century political economist John N. Keynes distinguished two senses of the term "science" in economics:

> The first belongs to positive science, the second to normative or regulated science (along with ethics, if indeed it be not a branch of ethics or of what may be called applied ethics). . . . As the terms are here used, a *positive science* may be defined as a body of systematized knowledge concerning what is; a *normative* or *regulative science* as a body of systematized knowledge discussing criteria of what ought to be. (Keynes 1891, p. 34)

On Keynes's distinction, our argument in Part II so far has been that microeconomics is an undeveloped or underdeveloped *positive* science. In Chapter 8 we argue that microeconomics is also an undeveloped or underdeveloped *normative* science.

Microeconomics as normative science begins as early as the eighteenth-century political economist Adam Smith's famous laissez-faire doctrine that a market economy leads profit-seeking individuals to further the interests of society:

> . . . it is only for the sake of profit that any man employs a capital in the support of industry. . . . As every individual, therefore, endeavors as much as he can to employ his capital . . . He generally, indeed, neither intends to promote the public interest, nor knows how much he is promoting it . . . he intends only his own gain, and he is . . . led by an invisible hand to promote an end which was no

part of his intention. Nor is it always the worse for society that it was no part of it. By pursuing his own interest he frequently promotes that of the society more effectively than when he really intends to promote it. (Smith 1961, p. 166)

Notice that Smith's claim is normative in the sense that it justifies a market economy of profit-seeking individuals on the grounds that such social arrangements promote the public interest. A skeptic might naively doubt Smith's claim and suppose that such arrangements merely promote the successful over the unsuccessful profit-seekers. We shall argue in this chapter that there is truth in both Smith's claim and the skeptic's view.

Beginning in the early 1940s, the economists Gerard Debreu, Kenneth Arrow, and others succeeded in developing a mathematical formalization of the laissez-faire doctrine which had been evolving in political economy and economics at least since Adam Smith. Chapter 8 examines the logic of the Debreu–Arrow theory. Section 8.2 presents the abstract version of microeconomic theory which is the basis of Debreu–Arrow welfare microeconomics. Sections 8.2.1–8.2.3 discuss consumer, producer, and equilibrium theories, respectively, and Section 8.2.4 considers the logical status of abstract microeconomic theory. Section 8.3 extends the abstract theory to Debreu–Arrow welfare microeconomics and then develops a critique of the theory. Sections 8.3.1–8.3.5 consider, respectively: Pareto efficiency; an example illustrating remaining difficulties; the problems of the distribution of income; the term "leads to" in Adam Smith's claim; and weakening the theory's assumptions. Section 8.4 is a summary and conclusion.

8.2 Microeconomic theory: an abstract version

8.2.1 *Consumer theory again*

In Chapter 6 we developed an account of consumer theory that supposed a consumer was characterized by a twice-differentiable utility function. To keep the theory parallel with classical mechanics, we assumed some particular form for the function. But we argued that no *direct* test of this utility function assumption by either observations or introspection seemed possible. The best one could do was to employ calculus to deduce some consequences of utility-production functional form and utility- and profit-maximization assumptions, and then submit those consequences to philosophically problematic tests against the observations.

In this chapter we begin with weaker assumptions than the utility-production function assumptions in Chapter 6. The weaker assumptions force the theory's development to a more abstract mathematical level than calculus. The existence of a utility function becomes a deductive consequence of the weaker consumer assumptions, which in turn become accessible to introspection, though not without problems. Abstract consequences of a positive economic sort follow from the weaker consumer–producer theory assumptions, but testing of these consequences is at least as problematic as in Chapter 6. Finally, we extend the positive microeconomic theory into a normative one that exhibits further problems.

We preface our account of consumer theory with some definitions from real analysis and point set topology.

Definition

   (i)   If X, $Y \le R^n$, then $X + Y = \{x + y: x \in X, y \in Y\}$

   (ii) If x, $y \in R^n$, then $x \le y$ iff each component of $x \le$ the corresponding component of y.

   (iii) $\Omega = \{x \in R^p: 0 \le x\}$.

Definition: $X \le R^n$ is

   (i)   *closed* iff the limit of x_n (as n approaches ∞) $\in X$ for any convergent sequence $\{x_n\}_{n=1}^{\infty}$,

   (ii) *bounded* iff it is contained in some closed "cube,"

   (iii) *compact* iff it is closed and bounded,

   (iv) *convex* iff x_1, $x_2 \in X$ implies $\lambda x_1 + (1 - \lambda)x_2 \in X$ for any $0 \le \lambda \le 1$.

Definition: If $X \le R^m$, $Y \le R^n$, then a mapping from X to $\mathscr{P}(Y) - \{\phi\}$ (where $\mathscr{P}(Y)$ is the power set, or set of subsets, of Y) is called a *correspondence* from X to Y.

Definition: Let r be a binary relation in which any two elements x, $y \in X \le R^n$ (the order of which is essential) either stand or do not stand. If $x \, r \, x$ *(reflexivity)*, and $x \, r \, y$ and $y \, r \, z$ imply $x \, r \, z$ *(transitivity)* for any x, y, $z \in X$, then r is called a *preordering* of X. If for any x, $y \in X$, $x \, r \, y$ or $y \, r \, x$, the preordering is *complete;* otherwise it is *partial.* And $x_g \in X$ is a *greatest element of X for r* if $x \, r \, x_g$ for all $x \in X$.

Next, let $Q_i \le \Omega$ be the possible consumptions of consumer i in a given time period. Let "$\le_i$" be a binary relation on Q_i that characterizes consumer i's preferences during this time period (i.e., for x, $y \in Q_i$, $x \leq_i y$ means that x is not preferred to y. Also, we write $x \sim_i y$ iff $x \leq_i y$ and $y \leq_i x$, and $x <_i y$ iff $x \leq_i y$ but y is not $\leq_i x$. Two immediate attachment difficulties are familiar from Chapter 6: How to stipulate a time period for which preferences hold is unclear, and "the present analysis does not cover the case where the consumption set of a consumer and/or his preferences depend on the consumptions of the other consumers (and/or on the productions of producers)" (Debreu 1959, p. 73). The economist Gerard Debreu suggests a third: "The preferences considered here take no account of the resale value of commodities; the ith consumer is interested in these only for the sake of the personal use he is going to make of them" (p. 54). But the preferences most immediately accessible to introspection may take account of resale, because resale is in fact possible. So introspective evidence of preferences (in the theory's sense) calls for an answer to a "what if" question: What would I prefer if resale, which is possible, were not?

Let $(\mathbf{p}, \mathbf{Y})$ be the consumers' (prices, income distribution). Consumer theory results follow from parts of two axioms and a principle.

Axiom 8.1. *Q_i for each i is*

   (i)   compact,   (ii)   convex.

Axiom 8.2. *For each i,*

   (i) $(Q_i, \leq_i)$ is a completely preordered set,

   (ii)   for every $\mathbf{q}_i' \in Q_i$, $\{\mathbf{q}_i \in Q_i\colon \mathbf{q}_i \leq_i \mathbf{q}_i'\}$ and $\{\mathbf{q}_i \in Q_i\colon \mathbf{q}_i \geq_i \mathbf{q}_i'\}$ are closed in Q_i (continuity).

If $\mathbf{q}_i^1$, $\mathbf{q}_i^2 \in Q_i$ and $t \in (0, 1)$, then

   (iii)   if $\mathbf{q}_i^2 >_i \mathbf{q}_i^2$, then $t\mathbf{q}_i^2 + (1 - t)\mathbf{q}_i^1 >_i \mathbf{q}_i^1$ (convexity),

   (iv)   if $\mathbf{q}_i^2 \sim_i \mathbf{q}_i^1$, then $t\mathbf{q}_i^2 + (1 - t)\mathbf{q}_i^1 >_i \mathbf{q}_i^1$ (strong convexity),

Principles of preference satisfaction and "parametric" consumer prices: For each i, consumer i

   (i)   seeks a greatest element with respect to $\leq_i$ of $\{\mathbf{q}_i \in Q_i\colon \mathbf{p} \cdot \mathbf{q}_i \leq Y_i\}$ (PS),

   (ii)   takes $\mathbf{p}$ as given (CPP).

Because preferences are accessible (albeit problematically) to intropsection, Chapter 6's argument – that the consumer can't maximize a utility function the consumer doesn't know – loses much of its force against PS. But the two axioms are not without attachment difficulties.

For some commodities (e.g., trucks) of which only integer numbers are consumed, Q_i's convexity [Axiom 8.1(ii)] is questionable. Debreu explains:

> A quantity of well-defined trucks is an integer; but it will be assumed instead that this quantity can be any real number. This assumption of perfect divisibility is imposed by the present stage of development of economics; it is quite acceptable for an economic agent producing or consuming a large number of trucks. Similar goods are machine tools, linotypes, cranes, Bessemer converters, houses, refrigerators, trees, sheep, shoes, turbines, etc. (1959, p. 30)

So in consumer theory, Q_i's convexity depends on its "perfect divisibility," which is like normality in classical Neyman–Pearson theory (see Chapter 4) – that is, not obviously plausible, but required for the theory's results to follow. But it is also an idealization like point masses in classical mechanical theory (see Chapter 5) – that is, as a mass may be regarded as a point when it is far from and/or much smaller than other masses, so may a consumption possibilities set that includes large quantities be regarded as "perfectly divisible."

Introspection questions whether or not $(Q_i, \leq_i)$ is a completely preordered set [Axiom 8.2(i)]. For quantities of paint thinner and mustard, one may make no preference judgment; so completeness is dubious. Transitivity is questionable too. Let there be three commodities; suppose that more of each is preferred to less, and let preferences between bundles be decided by a majority of the three commodities. Then if the three bundles are **A** (10, 6, 3), **B** (9, 5, 5), and **C** (11, 4, 4), $A \geq_i B$, $B \geq_i C$, but **A** is not $\geq_i$ **C**. One can construct a similar example with three commodities, each with three distinct attributes. Anyone who has tried to choose between three neckties – the first one's favorite color, another one's favorite material, and the third one's favorite style – will soon come to doubt transitivity.[1] Preference continuity [Axiom 8.2(ii)] is also questionable. Let there be two commodities, and suppose (q_{11}, q_{12}) is preferred to (q_{21}, q_{22}) if either $(q_{11} > q_{21})$ or $(q_{11} = q_{21}$ and $q_{12} > q_{22})$. Such lexicographic preferences are not continuous.[2] Again, one can construct a similar exam-

---

[1] The examples of nontransitivity of preferences merely adapt, to our individual-consumer preferences context, an example of a "paradox of voting" (Arrow 1975, p. 467).

[2] See Quirk and Saposnik (1968, pp. 17–18) for proof.

ple with two commodities, each with two distinct attributes. For example, one might prefer an auto with a higher safety rating to one with a lower safety rating, or, if the two ratings are the same, the less expensive to the more expensive. Debreu admits that preference strong convexity [Axiom 8.2(iv)] is not "intuitively justified," but claims that preference convexity [Axiom 8.2(iii)] is (Debreu 1959, p. 71). But suppose one prefers red to yellow. Is it clear that one will prefer 2.5 gallons of red and 2.5 gallons of yellow to 5 gallons of yellow to paint a (5-gallon) room? Preference convexity isn't always obvious.

The counterexamples of the previous paragraph do not disprove consumer theory in any formal mathematical sense. Here we do well to heed in our context the warning of the philosopher John Rawls on counterexamples of his theory of justice (a theory we take up in Section 8.3): "Objections by way of counterexamples are to be made with care, since these may tell us only what we know already, namely that our theory is wrong somewhere. The important thing is to find out how often and how far it is wrong. All theories are presumably mistaken in places" (1971, p. 52). Friedman's gambit (see Section 6.1) seems reasonable here: Let us reserve judgment on sometimes implausible assumptions until we have had a chance to assess their deductive consequences.

A definition and two theorems contain our consumer theory results.

Definition: u_i: $Q_i \to R^1$ is a *utility function* iff for $\mathbf{q}_1, \mathbf{q}_2 \in Q_i$,

  (i) $\mathbf{q}_1 <_i \mathbf{q}_2$ implies $u_i(\mathbf{q}_1) < u_i(\mathbf{q}_2)$,

  (ii) $\mathbf{q}_1 \sim_i \mathbf{q}_2$ implies $u_i(\mathbf{q}_1) = u_i(\mathbf{q}_2)$.

Theorem 8.1.[3] *If Q_i is convex [Axiom 8.1(ii)] and preferences are completely preordered and continuous [Axiom 8.2(i)–(ii)], then there exists a continuous utility function u_i: $Q_i \to R$.*

Theorem 8.2.[4] *Let $S_i = \{(\mathbf{p}, Y): \{\mathbf{q}_i \in Q_i: \mathbf{p} \cdot \mathbf{q}_i \leq Y_i\} = T_i \neq \varnothing\}$, and suppose Axioms 8.1 and 8.2(i)–(ii) and CPP hold.*

  (i) *There exists a correspondence $\boldsymbol{\epsilon}_i(\mathbf{p}, Y)$ from S_i to Q_i (called consumer i's individual-consumer demand correspondence) and a correspondence $\boldsymbol{\epsilon}(\mathbf{p}, Y) = \Sigma_{i=1}^n \boldsymbol{\epsilon}_i(\mathbf{p}, Y)$ from $\cap_{i=1}^n S_i$ to $\Sigma_{i=1}^n Q_i$*

---

[3] See Debreu (1959, pp. 56–9) for proof.
[4] See Debreu (1959, pp. 62–72) for proof.

> *(called the market demand correspondence), where $n = number$ of consumers, $\epsilon_i(\mathbf{p}, \mathbf{Y}) = \{\mathbf{q}_i \in Q_1: \mathbf{q}_i$ is a greatest element of T_i for "$\leq_i$"\}, and $\epsilon_i(t\mathbf{p}, t\mathbf{Y}) = \epsilon_i(\mathbf{p}, \mathbf{Y})$ and $\epsilon(t\mathbf{p}, t\mathbf{Y}) = \epsilon(\mathbf{p}, \mathbf{Y})$ for any $t > 0$.*

> (ii)   *If Axiom 8.2(iv) holds, then $\epsilon_i(\mathbf{p}, \mathbf{Y})$ and $\epsilon(\mathbf{p}, \mathbf{Y})$ are continuous functions.*

> (iii) *If PS and Axiom 8.2(iii) or (iv) hold, then for any $(\mathbf{p}, \mathbf{Y}) \in S_i$, consumer i will want to choose an element of $\epsilon_i(\mathbf{p}, \mathbf{Y})$.*

Theorem 8.1's result – the existence of a continuous utility function – is essentially the assumption with which we began Section 6.2.1, where we argued that it could not be directly tested against either observation or introspection. The results of Theorem 8.2(i)–(ii) also cannot be tested against either observation or introspection, because they are all the same as or weaker than those of Theorems 6.2 and 6.3, which could not be. If one supposes that consumers can know their own preference structures through introspection, Theorem 8.2(iii)'s result could conceivably be tested (at least in principle) by polling consumers, but we shall delay further discussion of this point until Section 8.3.1. So our new consumer theory variant's conclusions [with the possible exception of Theorem 8.2(iii)'s] are too abstract to be tested against either observations or introspection.

From Theorem 8.1 we also see that "the use of correspondences in the study of consumers could be avoided only by making the strong-convexity assumption on preferences . . . for which there is little intuitive justification" (Debreu 1959, p. 66). But strong convexity is not merely an assumption to which a few counterexamples can be exhibited. It means that for any two commodity bundles between which one is indifferent, one prefers a weighted average (with $0 <$ weights < 1, and sum of weights $= 1$) of the two bundles to either of them. Intuition, introspection, and an apparent consensus of consumer theorists suggest that commodities satisfying preference strong convexity are rather rare.

So in consumer theory we discover an anomaly not unlike one from the historical period of classical mechanics' crisis. In Section 5.3.2 we had Rutherford seeking to test a consequence of a classical model of the hydrogen atom, while ignoring another consequence (i.e., the atom's collapse) contradicted by observation. In consumer economic empirical research, econometricians frequently seek to fit a market demand function to (quantity, price) observations when the consensus of theorists is that no such market demand functions exists. So our new variant of consumer theory, coupled with much consumer econometric empirical research, seems to be undeveloped or incoherent positive science – or perhaps positive science in crisis. We turn next to a new variant of producer theory.

8.2.2 Producer theory again

For our account of producer theory we must add two definitions to those of the previous section.

Definition: If $x \in R^n$, then $x > 0$ iff each component of $x \geq 0$ and at least one component of $x > 0$.

Definition: $X \leq R^n$ is *strictly convex* iff x_1, $x_2 \in X$ implies $\lambda x_1 + (1 - \lambda)x_2 \in$ interior X for any $0 < \lambda < 1$.

Next let $Q_i^* \leq R^p$ be the set of production possibilities for producer i for $i = 1, \ldots, m$, where each component of $q_i^* \in Q_i^*$ is either an output or input quantity. [By convention, the inputs (outputs) have negative (positive) signs.] So $Q^* = \Sigma_{i=1}^m Q_i^*$ is the entire production sector's possibilities. Also, let $S^p = \{p \in R^p: p > 0\}$. Producer theory results then follow from parts of an axiom and a principle.

Axiom 8.3. *For each i, (i) $0 \in Q_i^*$ (possibility of inaction), and Q_i^* is (ii) compact, (iii) convex, and (iv) strictly convex.*

Principles of profit maximization and "parametric" prices: For each i, $p \in S^p$; producer i

(i) seeks $q_i^* \in Q_i^*$ so as to maximize $p \cdot q_i^*$ (PM),

(ii) takes p as given (PPP).

Notice first that as with consumers, we implicitly suppose no "*external economies and diseconomies,* that is, the case where the production set of a producer depends on the productions of the other producers (and/or on the consumption of consumers)" (Debreu 1959, p. 49). Given, say, the historical existence of chemical plant pollution of lakes, assuming that a fishing boat's catch on a lake doesn't depend on the production of a chemical plant on the lake's shore seems questionable. Also, with the

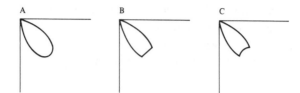

Figure 8.1. A simple one-input, one-output production process. (A) Strict convexity; (B) (nonstrict) convexity; (C) nonconvexity.

historical existence of monopolies, cartels, and price collusion, PPP seems questionable. For the moment, though, we employ Friedman's gambit to see where these two assumptions will lead us. The possibility of inaction means simply that any producer may choose to produce nothing, which seems plausible enough. Finally, Section 6.2.3's justification of PM applies here as well.

Q_i^*'s convexity, like Q_i's, needs the idealization that all real values "between" integers are possible. Also, consider the example of a simple one-input, one-output production process. Then Figure 8.1 (parts A, B, and C) depicts strictly convex, convex (but not strictly convex), and nonconvex Q_i's, respectively. Convexity, the possibility of inaction, and upper-boundedness are very restrictive assumptions because they imply *decreasing returns to scale* (i.e., if $\mathbf{q}_i^* \in Q_i^*$, then $t\mathbf{q}_i^* \in Q_i^*$ for any $0 < t < 1$, and there exists $\mathbf{q}_{i0}^* \in Q_i^*$, $t_0 > 1$ such that $t_0\mathbf{q}_{i0}^*$ is not $\in Q_i^*$) (Debreu 1959, p. 41). For the plausibility of Q_i^*'s (strict) convexity, one might seek to appeal directly to observation (i.e., rerun the production process many times and see what emerges). But whether or not the set of *actual* productions that would emerge from such repetitions would be generated by a (strictly) convex set of *possible* productions seems to be a question that is unanswerable on the basis of observations alone.

Theorem 8.3.[5] *Suppose Axiom 8.3(i)–(ii) and PPP hold.*

   (i)   *If Axiom 8.3(iii) holds,* $\max \mathbf{p} \cdot \mathbf{q}_i^* \geq 0$ *for any* $\mathbf{p} \in S^p$, *and there exists a correspondence* $\eta_i(\mathbf{p})$ *from* S^p *to* Q_i^* *(called producer i's*

---

[5] See Arrow and Hahn (1971, Theorems 3, 4, and 5, pp. 69–71) for proof. Arrow and Hahn include an assumption – free disposal (which we shall consider in the next section) – in the hypotheses of their Theorems 4 and 5, but they do not seem to use this assumption in their proofs of these two theorems. Further, Arrow and Hahn *do* employ the assumption that Q_i^* for each i is compact (and hence bounded) in their proof of Theorem 5, and Q_i^* bounded for each i *contradicts* free disposal. Hence, the only way to make the theorem statements and proofs of Arrow and Hahn coherent here is to delete their free-disposal assumption, which is what we have done in our statement of Theorem 8.3.

firm supply correspondence) and a correspondence $\eta(\mathbf{p}) = \sum_{i=1}^{m} \eta_i(\mathbf{p})$ *from* S^p *to* Q^* *(called the market supply correspondence) where* $\eta_i(\mathbf{p}) = \{\mathbf{q}_i^* \in Q_i^* : \mathbf{q}_i^*$ *is a maximum of* $\mathbf{p} \cdot \mathbf{q}_i^*\}$, *and* $\eta_i(t\mathbf{p}) = \eta_i(\mathbf{p})$ *and* $\eta(\mathbf{p}) = \eta(t\mathbf{p})$ *for any* $t > 0$.

(ii) *If Axiom 8.3(iv) holds, then* $\eta_i(\mathbf{p})$ *and* $\eta(\mathbf{p})$ *are continuous functions.*

(iii) *If PM and Axiom 8.3(iii) or (iv) hold, then for any* $\mathbf{p} \in S^p$, *producer i will want to provide an element of* $\eta_i(\mathbf{p})$.

The conclusions of Theorem 8.3(i)–(ii) cannot be tested against the observations because they are all the same as or weaker than those of Theorems 6.5 and 6.6, which could not be. But whereas in the previous section the existence of a market demand function depended on a-priori-implausible preference strong convexity, here the existence of a market supply function depends on not-a-priori-implausible production possibilities strict convexity. So, in contrast to the previous section's consumer economics, when producer economic empirical researchers try to fit a market supply function to (quantity, price) observations, there is no anomaly. Theorem 8.3(iii)'s conclusion could conceivably be tested by polling producers, but we again delay further consideration of this point until Section 8.3.1. We turn next to the question of equilibrium.

8.2.3   *Equilibrium*

To extend the results of the previous two sections to a theorem on equilibrium, we must first fill out our description of an economy a little further. We shall suppose that the economy is one in which the consumers own all the resources. Let $\omega_i \in R^p$ be consumer i's resources or initial endowment of commodities at the beginning of the period, so that the economy's total resources are $\omega = \sum_{i=1}^{n} \omega_i$. (Included in consumer i's initial endowment are i's capacities for various kinds of labor during the period.) Also, θ_{ij} is consumer i's share of profits in producer j for the period $[\theta_{ij} \geq 0, \sum_i \theta_{ij} = 1$ (for every j)]. Hence, if $\mathbf{p}$ is the price vector and $\mathbf{q}_j^*$ the production vector of the jth producer, then $Y_i = \mathbf{p} \cdot (\omega_i + \sum_j \theta_{ij}\mathbf{q}_j^*)$.

Next we require four definitions, the first of which formalizes the previous paragraph's remarks.

Definition: $E = ((Q_i, \leq_i), (Q_j^*), (\omega_i), (\theta_{ij}))$ is a *private-ownership economy* (POE).

Definition: **y** is a *lower bound* for $X \leq R^n$ iff $\mathbf{y} \leq \mathbf{x}$ for all $\mathbf{x} \in X$.

Definition: $((\mathbf{q}_i^+), (\mathbf{q}_j^{*+}), (\mathbf{p}^+))$ is

(i)   a *consumer equilibrium of a POE* if for every i, $\mathbf{q}_i^+$ is a greatest element of $\{\mathbf{q}_i \in Q_i: \mathbf{p}^+ \cdot \mathbf{q}_i \leq \mathbf{p}^+ \cdot \omega_i + \Sigma_j \, \theta_{ij}(\mathbf{p}^+ \cdot \mathbf{q}_j^{*+})\}$,

(ii) a *producer equilibrium of a POE* if for every j, $\mathbf{q}_j^{*+}$ maximizes $\mathbf{p}^+ \cdot \mathbf{q}_j^*$ for $\mathbf{q}_j^* \in Q_j^*$,

(iii) a *market equilibrium of a POE* if $\Sigma_{i=1}^n \mathbf{q}_i^+ - \Sigma_{j=1}^m \mathbf{q}_j^{*+} = \omega$,

(iv) an *equilibrium of a POE* if it is a consumer, producer, and market equilibrium of a POE.

Definition: $\mathbf{q}_i^0$ is a

(i)   *satiation consumption of* Q_i if $\mathbf{q}_i \leq_i \mathbf{q}_i^0$ for all $\mathbf{q}_i \in Q_i$,

(ii) *resource-diminishing consumption of* Q_i if a component of $\mathbf{q}_i^0 <$ (=) the corresponding component of ω_i when the component of $\omega_i > (=) 0$.

Our result on equilibrium is based on the following axiom.

Axiom 8.4.   *For every consumer i,*

(ci) Q_i *is closed and convex and has a lower bound for* "$\leq$,"

(cii) *preferences constitute a complete preordering of* Q_i *and are continuous and convex,*

(ciii) Q_i *contains a resource-diminishing consumption, but no satiation consumption for* Q_i.

For each producer j,

(pi) $\mathbf{0} \in Q_j^*$,

(pii) Q_j^* *is closed and convex,*

(piii) $Q^* \geq (-\Omega)$ *(free disposal), and* $Q^* \cap (-Q^*) \leq \{\mathbf{0}\}$ *(irreversibility).*

We have previously discussed the meaning and plausibility of all but (ciii) and (piii).

Table 8.1. *Two incompatible production possibilities under irreversibility*

	C1 C2	C1 C2	Total production C1 C2
First production	P1 (2, −3)	P1 (−5, 3)	(−3, 0)
Second production	P1 (9, −5)	P2 (−6, 5)	(3, 0)

Note: P1 and P2 are two producers, and C1 and C2 are two commodities.

The existence of a resource-diminishing consumption of Q_i means simply that a consumer can choose to consume less than that consumer's initial endowment of a commodity (if that initial endowment > 0), which seems plausible enough. Notice that satiation is with respect to "$\leq_i$" (*not* "$\leq$"), so that no satiation is not intuitively implausible. Free disposal means that "if a total production has all its outputs null, it is possible" (Debreu 1959, p. 42). That is, *any* total input set with no total outputs is possible. Given that a commodity's total production is the net of all producers' inputs of it, $Q^* \cap (-\Omega) \neq \phi$ is certainly plausible. But $Q^* \geq (-\Omega)$ implies that Q^* is not bounded below (i.e., that total production can use up *any* amount of any net input). Such an implicit assumption certainly is not based on what is observable.

Debreu (1959) says that irreversibility means "if the total production y, whose inputs and outputs are not all null, is possible, then the total production $-y$ is not possible," (p. 40), and he defends the assumption by arguing that "the production process cannot be reversed since, in particular, production takes time and commodities are dated" (p. 40). But though his meaning is accurate, his argument seems flawed. Suppose there are two producers (P1 and P2) and two commodities (C1 and C2) and that P1 produces C1 with C2, while P2 produces C2 with C1. Then irreversibility implies that the two productions of Table 8.1 are not both possibilities in a given period, for which Debreu offers no a priori justification. His claim that "production takes time and commodities are dated," while true, has no bearing on the irreversibility assumption. Arrow and Hahn do, however, make a good case for irreversibility: "For the validity of this irreversibility postulate, it suffices that there exists at least one non-produced input that is needed, directly or indirectly, for all production; labor provides an obvious example" (1971, p. 64). With labor (viewed as nonproduced) required (at least indirectly) for all production, which seems plausible enough, the labor component of any $q^* \in Q^*$ will be < 0, so that $Q^* \cap (-Q^*) \leq \{0\}$.

Our results on equilibrium are contained in the following theorem.

Theorem 8.4.[6] *Suppose E is a private-ownership economy for which Axiom 8.4, CPP, and PPP hold. Then*

(i) *an equilibrium for E exists,*

(ii) *if PS and PM also hold, consumers and producers will want to choose consumer and producer equilibria, respectively, for E.*

Theorem 8.4(i)'s conclusion – that an equilibrium for E exists – cannot be made to confront the observations because, for example, we cannot test whether or not the observed aggregate vector of commodities bought–sold is a market equilibrium (see Section 6.3.1). Because, for example, the set of consumer–producer equilibria may *properly* contain the set of market equilibria, the words "want to" cannot be removed from Theorem 8.4(ii)'s conclusion. These remarks again raise the issue of the logical status of the Section 8.2.1–8.2.3 abstract version of microeconomic theory.

8.2.4 *The logical status of the abstract microeconomic theory*

We have argued that the theorems of Sections 8.2.1–8.2.3 seem to contain no conclusions which can be made to confront the observations, for the theorems' hypotheses are all weakened versions of the hypotheses of the Section 6.2 theorems, whose conclusions also could not be made to confront the observations. The hypotheses of the Section 6.2 theorems were all placeholder relations and so could lead only to placeholder relations under deduction. And weakening the placeholder relations of the Section 6.2 theorems' hypotheses still gives placeholder relations which can lead under deduction only to placeholder relations (i.e., those of the conclusions of the theorems of Sections 8.2.1–8.2.3). So, that the conclusions of the theorems of Sections 8.2.1–8.2.3 cannot be made to confront the observations is no more surprising than our Chapter 5 finding that Newton's Second Law alone cannot be made to confront the observations.

If the conclusions of the theorems of Sections 8.2.1–8.2.3 cannot themselves be tested against the observations, the question arises as to

---

[6] See Debreu (1959, pp. 83–8) for proof.

what if any role such theorems have to play within microeconomics. Popper suggests another line "along which the testing of a theory could be carried out. First there is the logical comparison of the conclusions among themselves, by which the internal consistency of the system is tested" (1968, p. 32). The economist E. R. Weintraub has argued that the theory of Sections 8.2.1–8.2.3, especially Theorem 8.4, is something like such a test of internal consistency:

> . . . one must accept the hard core proposition "agents optimize" to work in the neo-Walrasian program. The question then arises: are the hard core propositions of the neo-Walrasian research program consistent? . . . In other words, consistency requires the production of a model in which a competitive equilibrium exists. The entire line of papers which culminated in those of McKenzie and Arrow-Debreu are exactly of this form. . . . It is nonsense to test a theory about the demand for electricity in the state of North Carolina . . . if there is no model containing . . . an equilibrium notion. . . . Such derived theories, like demand theory . . . must indeed be tested and corroborated. . . . *It is a category mistake to ask about the falsifiability of the Arrow–Debreu–McKenzie model.* (1985, pp. 30, 34–5)

Weintraub's account, because it fails to distinguish placeholder and particular relations, leaves some important points vague: What is "demand theory"? From precisely what and how is it "derived"? Does "demand theory" include a *particular* demand function? If not, how can it be "tested and corroborated"? But he *seems* to suppose that tests which focus primarily on a single market (e.g., those of Chapter 6) are "nonsense" without a *general* competitive equilibrium model behind them.

In Sections 6.2.4 and 6.3.1, though, we postulated a three-commodity economy and assumed (1) particular utility functions which consumers maximize subject to budget constraints, (2) profit-maximizing producers of one commodity in terms of the other two, with particular production functions, (3) equilibrium in the market for the one commodity, and (4) no consumer-producer control of prices. From these assumptions we deduced a relation between the (market quantity, price) of the one commodity and the distribution of income and the prices of the other two commodities. We then proposed a test (subject to certain philosophical difficulties) of the deduced relation against the observations. We could instead have added two further assumptions – (5) profit-maximizing producers of the other two commodities, with particular production functions, and (6) equilibria in the markets for those two commodities – and deduced that a set $P = \{(p_1, p_2, p_3)\}$ of price vectors satisfied all six assumptions. If we were guided in our choice of particular utility-production function relations by Theorem 8.4, we would find $P \neq \phi$; otherwise, P might $= \phi$. If $P \neq \phi$, we would have a particular general competitive equilibrium model and could test the six assumptions by testing (subject as always to the usual philosophical difficulties) our predicted $P \neq \phi$ against observed prices. It is unclear, however, that our stopping shy of a model containing a general competitive equilibrium (for our three-commodity economy) turns our proposed test of the deductive consequences of the first four assumptions against the observations into nonsense.

Suppose, though, that one *can* view Theorem 8.4 as a test of consistency of a set of hard-core propositions of the foundations of a *positive* empirical microeconomics. Such a view fails to explain why general equilibrium theory continues onward to questions of Pareto optimality or efficiency. Arrow and Hahn open their presentation of general equilibrium theory by referring it to the claim of Adam Smith with which we began the chapter:

Adam Smith's "invisible hand" is a positive expression of the most fundamental of economic balance relations . . . the notion that a social system moved by independent actions in pursuit of different values is consistent with a final coherent state of balance, and one in which the outcomes may be quite different from those intended by the agents, is surely the most important intellectual contribution that economic thought has made to the general understanding of social processes.

Smith also perceived the most important implication of general equilibrium theory, the ability of a competitive system to achieve an allocation of resources that is efficient in some sense. Nothing resembling a rigorous argument for, or even a careful statement of the efficiency proposition can be found in Smith, however. (Arrow and Hahn 1971, pp. 1–2)

So we may also regard the theory of Sections 8.2.1–8.2.3, especially Theorem 8.4, as formalizing Adam Smith's *normative* claim for a market economy of gain-seeking individuals into an argument.

But notice the subtle shifts from Adam Smith's original claim to Arrow and Hahn's translation of it and Theorem 8.4's formalization of it into an argument. First, the modern accounts (in contrast to Smith's) include consumers as well as producers. Second, Smith's "the public interest" and "interest of society" become in Theorem 8.4 "an equilibrium of a private-ownership economy" and in Arrow and Hahn's translation "a final coherent state of balance," though there is promise that under further formalization they will become "an allocation of resources that is efficient in some sense." Third, certain restrictions (i.e., Axiom 8.4) are introduced in the argument of Theorem 8.4 which were not present in Smith's original claim.

Most notably, however, while Smith's claim has an invisible hand *leading* people pursuing their own gain *to* the public interest, the modern accounts merely have people pursuing their own gain *consistent with* an equilibrium or final coherent state of balance. Remember that on Theorem 8.4, consumers and producers must take prices as given (CPP and PPP), so that though the theorem establishes that an equilibrium price vector exists, it fails to establish that there is an invisible hand which will lead the markets from their initially given prices to the equilibrium price vector. Still, the theorem establishes something, for "the immediate 'commonsense' answer to the question 'What will an economy motivated by individual greed and controlled by a very large number of different agents look like?' is probably: There will be chaos" (Arrow and Hahn 1971, p. vii). Though the theorem fails to establish that there will not be chaos, it does establish that there need not be.

So we view the theory of Sections 8.2.1–8.2.3, especially Theorem 8.4, as formalizing Adam Smith's essentially normative claim – that a private-ownership market economy is in the public interest – into an argument. But it is an argument with gaps and problems. First we must show that an equilibrium is an efficient allocation in some sense. Then we must see if there is a sense in which such an allocation is in the public interest. Also, we should again examine the restrictions (i.e., Axiom 8.4) which are in fact attachment difficulties which must be dealt with to convert the referent of the argument from a possible to the actual economy. Finally, we must see if the invisible hand is formalizable, so that "leads to" can replace "is consistent with" in the argument. Section 8.3 takes up these remaining tasks.

8.3 Welfare microeconomic theory

8.3.1 *Pareto efficiency*

This section extends the theory of Sections 8.2.1–8.2.3. We first require three definitions.

Definition: $((\mathbf{q}_i), (\mathbf{q}_j^*))$ for $\mathbf{q}_i \in Q_i$, $\mathbf{q}_j^* \in Q_j^*$ is an *attainable state* of POE $E = ((Q_i, \leq_i), (Q_j^*), (\omega_i), (\theta_{ij}))$ if $\sum_{i=1}^n \mathbf{q}_i - \sum_{j=1}^m \mathbf{q}_j^* = \omega$.

Definition: If A is the set of attainable states of POE E, and $((\mathbf{q}_i), (\mathbf{q}_j^*))$, $((\mathbf{q}_i'), (\mathbf{q}_j^{*\prime})) \in A$, then $((\mathbf{q}_i), (\mathbf{q}_j^*)) \leq ((\mathbf{q}_i'), (\mathbf{q}_j^{*\prime}))$ or $((\mathbf{q}_i'), (\mathbf{q}_j^{*\prime}))$ is *Pareto-superior* to $((\mathbf{q}_i), (\mathbf{q}_j^*))$ if for every i, $\mathbf{q}_i \leq_i \mathbf{q}_i'$.

Notice that "$\leq$," or Pareto superiority, is defined only in terms of consumer preferences (not producer states) and is a partial preordering of A.

Definition: A *Pareto-efficient* allocation is a maximal element of $(A, \leq)$.

The following theorem gives sufficient conditions for an equilibrium of a POE to be a Pareto-efficient allocation, and vice versa.

Theorem 8.5.[7] *Let E be a POE where for each i, Q_i and $(Q_i, \leq_i)$ are convex and Q_i contains no satiation consumption.*

  (i) *If $((\mathbf{q}_i), (\mathbf{q}_j^*), \mathbf{p})$ is an equilibrium of E, $((\mathbf{q}_i), (\mathbf{q}_j^*))$ is a Pareto-efficient allocation.*

Suppose also that for each i, Q_i is continuous and contains a resource-diminishing consumption, and that for each j, Q_j is convex.

  (ii) *If $((\mathbf{q}_i), (\mathbf{q}_j^*)$ is a Pareto-efficient allocation, then there exists $\mathbf{p} \neq \mathbf{0}$ such that $((\mathbf{q}_i), (\mathbf{q}_j^*), \mathbf{p})$ is an equilibrium of E.*

So the following theorem cumulates the formalization, so far, of Adam Smith's original claim.

Theorem 8.6.[8] *Let E be a POE for which Axiom 8.4, CPP, and PPP hold.*

  (i) *There exists an equilibrium of E.*

  (ii) *If $((\mathbf{q}_i), (\mathbf{q}_j^*), \mathbf{p})$ is an equilibrium of E, then $((\mathbf{q}_i), (\mathbf{q}_j^*))$ is a Pareto-efficient allocation.*

  (iii) *If $((\mathbf{q}_i), (\mathbf{q}_j^*))$ is a Pareto-efficient allocation, then there exists $\mathbf{p} \neq \mathbf{0}$ such that $((\mathbf{q}_i), (\mathbf{q}_j^*))$ is an equilibrium of E.*

The economist Oskar Morganstern suggests one reason why the Pareto-superiority relation interests economists so much:

> If one considers an ensemble of economic individuals, say consumers all, and asserts that when each is in a state of equilibrium all must be better off when nobody's position is deteriorated but one single person's position is improved – according to his own testimony – one seems indeed, to be confronted with an unchallengeable statement. This, then is perhaps a way to get rid of the unpleasant

---

[7] See Debreu (1959, pp. 94–6) for proof.
[8] Follows from Theorems 8.4 and 8.5.

dilemma of economics which is that on the one hand we cannot compare utilities of different individuals while still staying with the realm of strictly scientific observations involving no value judgments, but on the other hand economists must make assertions about economic welfare. (1976, p. 253)

But Morganstern suggests that determining a Pareto-efficient allocation, without comparing utilities of different individuals, may be problematic. We shall add that identifying "economic welfare" with (i.e., translating Adam Smith's "public interest" or "interest of society" as) Pareto efficiency is certainly problematic.

A second reason, however, that Pareto efficiency interests economists, more than other possible welfare criterion candidates, is that they suppose, in line with Adam Smith's original claim, that an invisible hand leads a private-ownership, perfectly competitive economy of interest-maximizing individuals to a Pareto-efficient allocation. But we have already emphasized that the theorems so far entail merely "is consistent with," rather than "leads to," and we shall argue that existing extensions of the theory to "leads to" are problematic. Finally, we have already shown (see Theorem 8.6) that under *certain conditions* (i.e., Axiom 8.4), a private-ownership, perfectly competitive economy of interest-maximizing individuals is consistent with a Pareto-efficient allocation. We shall argue that the sufficient conditions constitute attachment difficulties of the Chapter 6 sort.

On our account, the conclusions of the section's two theorems are of a placeholder sort (e.g., state that a point with certain properties exists, but not what that point is) and so cannot themselves be made to confront the observations. But Morganstern (1976) suggests: "The maximum [i.e., the Pareto-efficient allocation] is determined by reliance either (a) on the statements to be obtained by participants or (b) must be recognizable objectively by an outside observer" (p. 254). He proposes a thought experimental search for a Pareto-efficient allocation. An outside observer places the economy in one state and then moves it to another. If the outside observer judges whether an individual I_0 prefers one state to the other, "the outside observer . . . then has to make a statement about I_0's utility, thereby introducing an interpersonal utility comparison between himself and I_0" (p. 257). If, on the other hand, the individual's testimony is relied on, "he may not tell the truth. He may deny the existence of a benefit in order to obtain a still larger one by stating that the benefit to him occurs only when a certain minimum quantity – larger than the one offered – has been reached" (p. 257).

Morganstern (1976) concedes that though relying on the individual's testimony for Pareto superiority is a problem, "there is no apparent difficulty" (p. 257) in relying on his testimony for Pareto efficiency (i.e., if one of the two states is a Pareto-efficient allocation, the individual should cease any lying). But his thought experiment suggests two difficulties which Morganstern fails to explore: (1) If the search for a Pareto-efficient allocation is over an infinite (countable or uncountable) consumption possibility set, it need not terminate. (2) An individual need not know if a consumption is a maximum. In Section 8.2.1 we argued that basing consumer theory on preferences makes the theory's assumptions more accessible to introspection. But that a consumer knows that she prefers bundle A to bundle B doesn't imply that she knows that she prefers A to any of a possibly (countably or uncountably) infinite set of consumption possibility bundles. The first proposition seems introspectively plausible, the second less so. Difficulty (2) is merely the attachment problem (discussed in Chapter 6) of whether or not one knows one's entire preference structure or utility function.

Difficulty (1) – the possibly nonterminating nature of the search for a Pareto-efficient allocation – arises because of the theoretically incomplete nature of the framework of Morganstern's thought experiment. If particular utility and production relations are specified, then particular equilibrium-point candidates follow deductively from those particular and the placeholder relations. Testing deduced equilibria against the observations has the attachment difficulties of Chapter 6. If the observations confirm, with Chapter 6 philosophical caveats, equilibria, then Pareto efficiency of the equilibria follows deductively from Theorem 8.6(i) (if the particular utility relations satisfy the theorem's sufficient conditions). Thus, Morganstern's difficulty of determining a Pareto-efficient allocation appears to reduce to Chapter-6-type attachment difficulties of whether or not an individual knows his or her utility functions and/or the difficulties of a test of particular deduced supply–demand relations against the observations. We turn next to an example which illustrates the three remaining problems with formalizing Adam Smith's claim into an argument.

8.3.2   *An example*

Before examining the three remaining problems for formalizing Adam Smith's claim into an argument, we consider a very simple example which, without loss of generality, illustrates the three problems. Let E be a POE with two commodities, a single producer, for whom $Q_1^* = \{0\}$ (so

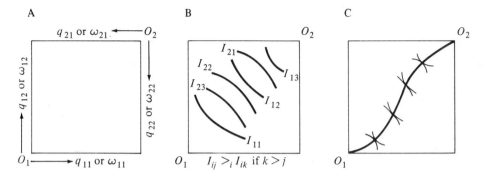

Figure 8.2. A two-commodity, two-consumer economy. (A) Attainable
states; (B) indifference curves; (C) Pareto-efficient allocations.

that share ownership – i.e., the θ_{ij}'s – is irrelevant), and two consumers.
Such a POE is often labeled a pure exchange economy, but such an
appellation obscures the role of prices in E. The attainable states of E, that
is, those for which $\mathbf{q}_1 + \mathbf{q}_2 = \omega_1 + \omega_2$, with $\mathbf{q}_i = (q_{i1}, q_{i2})$, $\omega_i = (\omega_{i1}, \omega_{i2})$,
may be depicted as the points of a standard Edgeworth box diagram
(Figure 8.2A), where the first (second) consumer's origin is the lower left
(upper right) corner of the box.

Figure 8.2B includes a few indifference curves for each consumer; each
consumer's curves are convex to his or her origin, and movement away
from one's origin is in the direction of increasing preferences. In Figure
8.2C, the locus of points of tangency of the two consumers' indifference
curves – O_1O_2 (sometimes labeled the contract curve, an appellation
which again obscures the role of prices in E) – is the set of Pareto-efficient
allocations (P). The first remaining problem for formalizing Smith's
claim concerns the assumptions implicit in our drawing of the indifference curves: What is the effect on the conclusion $P \neq \phi$ of altering those
assumptions? How can one know that the assumptions hold for an actual
economy? That is, how are the assumptions' words attached to the world?

Figure 8.3A depicts a distribution of resources or initial endowments,
the two indifference curves (one for each consumer) on which the initial
endowments lie, and the set of Pareto-efficient allocations. Notice that BC
is a subset of O_1O_2 preferred by each consumer to his or her initial
endowments ω and that prices $\mathbf{p} = (p_1, p_2)$. Notice that $M_1 \neq M_2$ means
that $\mathbf{p}$ is not an equilibrium price vector. Figure 8.3C depicts the budget
constraints and preferred attainable states ($M_1^* = M_2^*$) for each consumer
corresponding to the initial endowments ω and the prices $\mathbf{p}^* = (p_1^*, p_2^*)$.

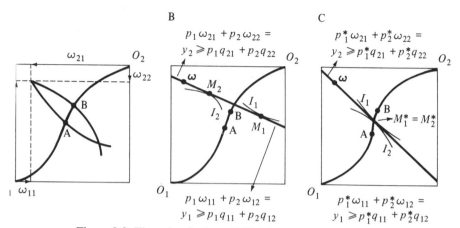

Figure 8.3. The role of prices. (A) Initial endowments; (B) nonequilibrium price vector; (C) equilibrium price vector.

Notice that $M_1^* = M_2^*$ means that $\mathbf{p}^*$ is an equilibrium price vector (and so $M_1^* = M_2^*$ is a Pareto-efficient allocation). So the second remaining problem for formalizing Smith's claim is specifying a mechanism that might, when both consumers take prices as given, move prices from a possibly initial nonequilibrium price vector like $\mathbf{p}$ to an equilibrium price vector like $\mathbf{p}^*$.

Figure 8.4 depicts an equilibrium price vector $\mathbf{p}^*$ with three different initial endowments, hence income distributions, and three different consequent Pareto-efficient allocations. Figure 8.4A's (B's) distribution gives much more to the second (first) consumer, while C's is more even. Recall from Sections 8.2.1, 8.2.3, and 8.3.1 that the individual-consumer preference maxima which determine a Pareto-efficient allocation are actually constrained maxima, where the constraints are the individuals' incomes. Thus, a Pareto-efficient allocation is really a Pareto-efficient allocation relative to a given distribution of income.

The economist Amartya K. Sen reminds us:

> But there is a danger in being exclusively concerned with Pareto-optimality. An economy can be optimal in this sense even when some people are rolling in luxury and others are near starvation as long as the starvers cannot be made better off without cutting into the pleasures of the rich. If preventing the burning of Rome would have made Emperor Nero worse off, then letting him burn Rome would have been Pareto-optimal. In short, a society or an economy can be Pareto-optimal and still be perfectly disgusting. (1970, p. 22)

So the third remaining problem for formalizing Smith's claim is that "Pareto-efficient allocation" is not a very good translation of his "public

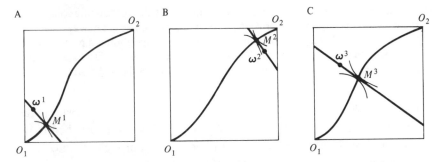

Figure 8.4. Three income distributions. (A) Second consumer's income much larger; (B) first consumer's income much larger; (C) similar incomes.

interest" or "interest of society." To improve the translation, one must provide an argument for both what the distribution of income should be and how that distribution of income should be related to the notion of a Pareto-efficient allocation. In taking up the three remaining problems for formalizing Smith's claim, we reverse the order in which we introduced them.

8.3.3   *The problem of the distribution of income*

The philosopher John Rawls's theory of justice suggests one answer to the question what should be the distribution of income. To decide on a conception of justice to govern society, Rawls argues, we should imagine ourselves in a hypothetical "original position" under a "veil of ignorance" (i.e., with no individual knowing his or her social position, natural abilities, conception of good life, etc.). Since, he argues, rational individuals in such a position would choose the following principle of justice to govern their society, we should decide on it too: "All social primary goods – liberty and opportunity, income and wealth, and the bases of self-respect – are to be distributed equally unless an unequal distribution of any or all of these goods is to the advantage of the least favored" (1971, p. 303).

Rawls's theory also suggests an answer to the question how the distribution of income should be related to the notion of a Pareto-efficient allocation. Rawls summarizes the formalization of Smith's claim into an argument (see Sections 8.2.1–8.2.3 and 8.3.1), noting the attachment difficulties and supposing "leads to" (rather than the "is consistent with" we have so far established) holds:

The theory of general equilibrium explains how, given the appropriate conditions, the information supplied by prices leads economic agents to act in ways that sum to achieve this [Pareto-efficient] outcome. Perfect competition is a perfect procedure with respect to efficiency. Of course, the requisite conditions are highly special ones and they are seldom if ever fully satisfied in the real world. (1971, p. 272)

But on his theory of justice, the "principle of justice is lexically prior to the principle of efficiency" (1971, p. 302). So, on Rawls's theory, before we can translate Smith's "public interest" or "interest of society" into "Pareto-efficient allocation" we must first be certain that extra-market background institutions governed by his principle of justice exist (e.g., that an income-wealth transfer system, which assures income and wealth are distributed equally unless unequal distribution is to the advantage of the least favored, is in place).

Rawls's principle of justice may be used to choose between various forms of negative income tax (NIT). For example, in proposing an NIT, Milton Friedman offered no coherent suggestion on what the size of the guarantee level should be: "I see no way of deciding 'how much' except in terms of the amount of taxes we – by which I mean the great bulk of us – are willing to impose on ourselves for the purpose. . . . The precise floor set would depend on what the community could afford" (Friedman 1962, pp. 191–2). Besides being vague, Friedman's suggestion is ambiguous, since "what we are willing to impose on ourselves" need not equal "what the community could afford." But on Rawls's principle of justice, of two guarantee levels, the one which assures greater equality in the distribution of income is preferred, unless the level with lesser equality is "to the advantage of the least favored."

Difficulties in applying Rawls's principle of justice center on the rider: "unless an unequal distribution of any or all of these goods is to the advantage of the least favored." Consider an example he raises:

To illustrate the difference principle, consider the distribution of income among social classes. . . . Now those starting out as members of the entrepreneurial class in property-owning democracy, say, have a better prospect than those who begin in the class of unskilled laborers. . . . What then, can possibly justify this kind of initial inequality in life prospects? According to the difference principle . . . the inequality in expectation is permissible only if lowering it would make the working class even worse off. Supposedly . . . the greater expectations allowed to entrepreneurs encourage them to do things which raise the longterm prospects of [the] laboring class. Their better prospects act as incentives so that the economic process is more efficient, innovation proceeds at a faster pace, and so on. Eventually the resulting material benefits spread throughout the system and to the least advantaged. I shall not consider how far these things are true. The point is that something of this kind must be argued if these inequalities are to be just by the difference principle. (1971, p. 78)

The difficulties arise, however, when one does try to "consider how far these things are true." If this book's argument so far is correct, the science for deciding whether or not an inequality is to the advantage of the least favored simply doesn't exist yet. Thus, whether one places the burden of proof on those who favor or those who oppose an inequality, that burden is heavy indeed. One conceivable way to deal with the difficulty would be to replace the rider as: "unless an unequal distribution of any or all of these goods is to the advantage of, or acceptable to, the least favored."

Some have cited other difficulties with Rawls's theory, and some have offered different theories with arguments for different distributions of income and relationships with Pareto efficiency.[9] Our focus was on Rawls's theory as a well-developed example of such an argument. Our point remains: *Some* argument for *some* distribution of income and its relationship to Pareto efficiency is necessary (but not sufficient) to translate, into a formal argument, Adam Smith's claim that a perfectly competitive, private-ownership economy leads to commodity allocations which are in the public interest. We turn next to the problem of strengthening the "is consistent with" of Sections 8.2.3 and 8.3.1 to the "leads to" of Smith's claim.

8.3.4   *Strengthening "is consistent with" to "leads to"*

In this section we take up the problem of strengthening the "is consistent with" of the theorem of Section 8.2.3 into the "leads to" of Adam Smith's original claim. But in the previous section we saw that as penetrating a thinker as John Rawls supposes that general equilibrium theory provides

---

[9] Nozick (1974), for example, sketches a theory of distributive justice whose key feature he describes: "The entitlement theory of justice in distribution is *historical;* whether a distribution is just depends on how it came about" (p. 153). He sees Rawls's theory, in contrast, as consisting of "*end-results principles* or *end-state principles*" (p. 155) and offers an extensive critique of Rawls's view (pp. 183–231). Rather than justify a distribution of income on the grounds that it would be chosen in a hypothetical "original position," Nozick would justify it on the grounds that principles of justice in acquisition and transfer (p. 151) were satisfied in its actual historical coming about. But whether one accepts Rawls's, Nozick's, or some other view (see, for example, Roemer 1982, especially pp. 305–10, for critiques of both Rawls and Nozick), our point is the same: General equilibrium theory formalizations of Adam Smith's claim for the public interest of an economy of profit-utility-seeking individuals are incomplete without *some* justification of a distribution of income.

results of a "leads to" sort. So it is worth pausing to review, from another angle, our general equilibrium theory results so far. We wish to be certain that our "is consistent with"/"leads to" distinction is meaningful and that we so far have only "is consistent with" rather than "leads to" formalizations. We reconsider the issue with another comparison of classical mechanical theory and microeconomic theory.

For simplicity, consider a two-particle classical mechanical system and a two-commodity, two-consumer microeconomic system. For the classical mechanical system, we suppose that God or nature selects $r_1(t^*)$, $r_2(t^*)$ so as to minimize the action $\mathcal{S}(r_1(t^*), r_2(t^*), \dot{r}_1(t^*), \dot{r}_2(t^*), t^*, m_1, m_2)$. Mathematically, the problem is one of minimizing a *single* function of related arguments, with m_1 and m_2 held constant at any values in the minimization. For the microeconomic system, we suppose that the first (second) individual selects q_1 (q_2) so as to maximize $u_1(q_1)$ ($u_2(q_2)$) subject to $p \cdot q_1 \le Y_1$ ($p \cdot q_2 \le Y_2$). Mathematically, the problem *for either individual* (as for God or nature in classical mechanics) is one of maximizing a *single* function of an argument , with p held constant at any value in the maximization. But *the system* must somehow select a p^* so that both $u_i(q_i)$'s are *simultaneously* maximized; so far, we have merely specified conditions under which a p^* exists. Selecting such a p^* has no classical mechanical analogue. If Rawls is right, and general equilibrium theory provides "leads to" results, we have yet to see them.

Seeking to strengthen "is consistent with" to "leads to," we first recast slightly and then extend some previous results. For the remainder of this section, the POE $E = ((Q_i, \le_i), (Q_j^*), (\omega_i), (\theta_{ij}))$, with $S_*^p = \{p \in R^p : \Sigma p_i = 1, p > 0\}$ the set of allowed prices, and consumers and producers take $p \in S_*^p$ as given. We begin with an axiom similar to those of Sections 8.2.1–8.2.2 and a theorem which recasts the results of those sections.

Axiom 8.5. *For each i, Q_i is compact and convex and has a resource-diminishing consumption; $(Q_i, \le_i)$ is a completely preordered, continuous, and strongly convex set; and for each j, $0 \in Q_j^*$, which is compact and strictly convex.*

Theorem 8.7.[10] *Let Axiom 8.5 hold for POE E. For fixed Y,*

(i)   *there exists a function $z(p) = \Sigma_{i=1}^n \epsilon_i(p) - \Sigma_{j=1}^m \eta_j(p) - \omega$ from S_*^p to R^p (called an excess demand function), where $\epsilon_i(p) = a$ unique*

---

[10] With the assumption of a resource-diminishing consumption for each i, the set S_i (in Theorem 8.2) $\ne \varnothing$, and the theorem follows from Theorems 8.2 and 8.3. We don't claim

greatest element for "$\leq_i$" of $\{q_i \in Q_i : p \cdot q_i \leq Y_i\}$, and $\eta_j(p) = a$ unique maximum of $\{p \cdot q_j^* : q_j^* \in Q_j^*\}$,

(ii) $z(p)$ *is continuous,*

(iii) $z(kp) = z(p)$ *for every* $k > 0$, $p > 0$ *(homogeneity of degree 0 in* p*).*

Since our effort at strengthening "is consistent with" to "leads to" is based on this theorem, it is based on preference strong convexity – an assumption which is sharply at odds with introspection.[11]

that the theorem's hypothesis is the weakest possible that will imply its conclusion. Quirk and Saposnik (1968, p. 176) suggest that the theorem's conclusions follow from a weaker hypothesis: For each i, Q_i has a resource-diminishing consumption; $(Q_i, \leq_i)$ is a completely preordered, continuous, and strongly convex set; and for each j, Q_j^* is closed, bounded above, and strictly convex.

[11] Recall (from Section 8.2.1) that Debreu (1959) noted that the existence of demand *functions* depends on the anti-intuitive strong-convexity assumption. This fact posed no problem for Debreu's 1959 results because they were of an "is consistent with" sort, requiring only the existence of demand *correspondences*. But to get "leads to" or stability analysis results requires at least the existence of demand *functions:* "The need to restrict any stability analysis to cases involving single-valued and continuous excess demand functions arises because, in general solutions to differential equations systems in which functions are discontinuous cannot be guaranteed and, as Arrow and Hurwicz remark with respect to correspondences, '[in this case] the precise meaning of the adjustment equations becomes problematical'" (Quirk and Saposnik 1968, pp. 175–6).

Arrow and Hahn (1971) acknowledge that the existence of demand *functions* depends on a preference strong-convexity assumption (p. 100). But they never remark on the anti-intuitive nature of the strong-convexity assumption and fail to stress the dependency of all their uniqueness and stability analysis results (Chapter 9–14) on that assumption. Quirk and Saposnik (1968) also never remark on the anti-intuitive nature of the strong-convexity assumption, but do concede, perhaps a bit reluctantly, that "leads to" results depend on preference strong convexity: "strong convexity [of preferences] is not needed in most of the modern theory (except possibly in stability and comparative static analysis)" (p. 27). In a recent (1985) formal reconstruction of general equilibrium theory for pure exchange economies (PEE), the economist D. W. Hands argues: "In PEE both individual utilities and individual demands are one-step-removed from what 'really' matters. What really matters is market demand and market supply. It is these market functions which should be the fundamental concept in any reconstruction which is true to the economic literature" (p. 261). Hands never even alludes to any individual consumer preference structures behind market demand functions, let alone to needing to suppose that the preferences are strongly convex to get the market demand functions from them!

So general equilibrium theorists seem, on the whole, to be rather reluctant to admit the anti-intuitive foundations of their "leads to" results. Because we shall see in a moment that those results are of a placeholder sort with an end-of-days qualifier and are not easily made to confront the observations, tests against introspection or intuition are the only sort that can bridge the gap between what is logically possible and what is empirically plausible. Given that their results fail such tests miserably, it is neither surprising nor admirable that general equilibrium theorists would be reluctant to admit it.

To recast and extend the results of Section 8.2.3 requires a definition, two more axioms, and another definition.

Definition: If $z(p)$ is an excess demand function on S_*^p, then

(i) $p^* \in S_*^p$ is an *equilibrium* if $z(p^*) \leq 0$ ($E^* = \{p \in S_*^p : z(p) \leq 0\}$),

(ii) *Walras's "law"* holds if $p \cdot z(p) = 0$ for all $p \in S_*^p$,

(iii) (GS) commodities $i \neq j$ are *gross substitutes* at $p \in S_*^p$ if $\partial z_i(p)/\partial p_j$ exists and is > 0.

Axiom 8.6

(i)   *Walras's "law" holds,*

(ii) *all commodities are GS for any $p \in S_*^p$.*

Axiom 8.7 ("Law" of supply and demand).[12]   *For $i = 1, \ldots, p$,*

$$p_i = \begin{cases} 0 \text{ if } p_i < 0 \text{ and } z_i(p) < 0 \\ G_i[z_i(p)] \text{ otherwise,} \end{cases}$$

where $G_i(z_i)$ is a sign-preserving function of z_i, $G_i(0) = 0$, and $dG(z_i)/dz_i$ exists and is > 0.

Definition: Axiom 8.7 is *globally stable* if $p(t^*, p(0))$ approaches p^* as t^* approaches ∞ for some $p^* \in E^*$ for any initial price vector $p(0) \in S_*^p$ and price time path $p(t^*, p(0))$ satisfying Axiom 8.7.

Walras's "law" is equivalent to $p \cdot q_i =$ (not just $\leq$) Y_i for each i (Quirk and Saposnik 1968, pp. 154–5). Supposing that consumers prefer more to less of each commodity (as we did in Chapter 6) justifies Walras's "law," but is plausible for many commodities only if one drops the implicit assumption (see Section 8.2.1) that preferences are as if resale were not allowed. Certainly empirical econometricians often conjecture a $z_i(p)$ for which i and any other commodity are GS for any $p \in S_*^p$ (see, for example, our extended example in Section 6.3). But Arrow and Hahn (1971) argue convincingly that the contention that *all* commodities are GS for any $p \in S_*^p$ is rather implausible (pp. 225–7). Thus, Axiom 8.6 is implausible and seems to require a justification which contradicts the meaning we have given to preferences.

---

[12] Notice here the implicit assumption that price is a *differentiable* function of time.

Axiom 8.7 formalizes the actions of the metaphorical person with Adam Smith's metaphorical invisible hand, an imaginary "superauctioneer who calls a given set of prices **p** and reviews transaction offers from agents in the economy. If these do not match, he calls another set of prices, following [Axiom 8.7]" (Arrow and Hahn 1971, p. 264). The axiom's logical status then seems similar to that of the differential-equation placeholder in physics which formalizes sound's travel as if it were a wave (see Section 1.2.3), or even to that of Newton's Second Law (see Chapter 5). The axiom says that if its market is out of equilibrium, then a commodity's price goes up (down) when consumers want more (less) than producers want to provide, which *seems* intuitive. But Arrow and Hahn warn us:

. . . while it is reasonable to postulate that, for instance, an actual attempt to buy more shoes at a given price than are supplied at that price will drive up the price of shoes, it is quite another matter to say that the *plan* to buy more shoes than are planned for supply will drive up their price. It might just happen that when several markets are out of equilibrium and I have already found myself unable to buy the socks I had planned to buy, my actual pressure to have shoes supplied to me also will be less than I had planned it to be before I discovered my inability to fulfill my plans. Therefore, is it sensible to deduce from people's plans what the actual pressure on individual markets is going to be when these plans are inconsistent? (1971, p. 267)

In our lexicon, Arrow and Hahn note that there is a difficulty in how the axiom's "$z(\mathbf{p})$," or the words "excess demand at prices **p**," attach to the worlds both of our observations and of introspection when **p** is not an equilibrium.

The following theorem recasts and extends the results of Section 8.2.3.

Theorem 8.8.[13] *Let E be a POE for which a continuous excess demand function which is homogeneous of degree 0 in* **p** *exists.*

(i) *If Axiom 8.6(i) holds, then there exists an equilibrium* **p***.

(ii) *If Axiom 8.6 holds,* **p*** *is unique.*

(iii) *If Axioms 8.6 and 8.7 hold, then Axiom 8.7 is globally stable.*

Then a definition and a corollary complete our effort at strengthening "is consistent with" to "leads to."

---

[13] For proof, see Arrow and Hahn (1971, Corollary 7, p. 223, Theorem 4, p. 288).

Definition: If Axiom 8.7 is globally stable and *POE* has a unique equilibrium, then that equilibrium is *globally stable.*

Corollary.[14] *If E is a POE for which Axioms 8.5–8.7 hold, then E has a unique, globally stable equilibrium* **p***.

The corollary, which is the strongest sort of conclusion that so-called stability analysis provides, introduces a kind of dynamics into microeconomics, but is not a very satisfactory solution to the "is consistent with"/ "leads to" problem, for it merely tells us that a perfectly competitive POE satisfying a more restrictive list of assumptions than in Sections 8.2.3 and 8.3.1 will be led, *by the end of days,* to an equilibrium which is a Pareto-efficient allocation. Given that preference strong convexity – which is sharply at odds with introspection – is on the more restrictive list of assumptions, one is inclined to agree with the recent remark of Frank Hahn (1986), one of the founders of stability analysis:

> Adam Smith started it with the metaphor of "the invisible hand." He was soon credited with the demonstration that "the price mechanism" or "law of demand and supply" acted as if it had been designed to ensure a satisfactory or perhaps optimal state of the economy. After two hundred years of pondering Smith's insight however we have only succeeded in demonstrating that under certain very restrictive assumptions a state of the economy in which the invisible hand is at rest is possible and Pareto efficient. (p. 1)

Notice that the referents of Hahn's metaphorical discussion are our results in Sections 8.2.3 and 8.3.1. So he implicitly deems the present section's results (with which he is certainly familiar, because they are largely his results) so unsatisfactory that they can be left out of a summary of our present state of knowledge. We turn finally to the question of weakening the restrictive assumptions.

8.3.5   *Weakening assumptions*

In previous sections we focused attention on the plausibility of the various assumptions and/or axioms we employed. Here we consider the relations of some conclusions to each other and to Smith's original claim, the relations between the systems of assumptions needed to draw the conclusions, and the effect on the conclusions of weakening certain assumptions. The three conclusions of interest are that an equilibrium exists, a

---

[14] Follows from Theorem 8.8 and the definition of a globally stable equilibrium.

unique globally stable equilibrium exists, and an equilibrium is a Pareto-efficient allocation. Table 8.2 depicts these three conclusions and the systems of assumptions from which we draw them. In saying that D is weaker than D', we mean that D' implies D.

Conclusion I is weaker than II, and if either of their assumption systems holds, conclusion I follows. The assumptions needed to deduce conclusion I are generally weaker on an individual basis than those needed to deduce conclusion II. But conclusion I's assumption system is *not* weaker than conclusion II's, because II's doesn't imply free disposal and irreversibility.[15] The three conclusions piece together as two formalizations – a weak version and a strong version – of Adam Smith's original claim: Given a set of assumptions A_w (A_s), a perfectly competitive private-ownership economy is consistent with (leads to, by the end of days) an equilibrium that is a Pareto-efficient allocation, which is in the public interest if the distribution of income is in the public interest. Notice that both formalizations are weaker than Smith's original claim (e.g., because of the income distribution qualifier) and even weaker than the result Rawls mistakenly supposed to follow from general equilibrium theory (e.g., because of the strong version's "end of days" qualifier).

Consider the weak formalization first. Suppose perfect competition (PC) is weakened to this: Monopolies (i.e., firms that do not take some p_i's as given) exist in some markets. Then, if A_w remains the same as before for the perfectly competitive firms and some further assumptions are made for the monopolies, equilibrium can be redefined so that one still exists, but may not be a Pareto-efficient allocation. So the weak formalization collapses.

Alternatively, let "no externalities" be replaced by an assumption that for each i, the ith individual's utility is of the form $u_i(\mathbf{q}_i, \omega)$. Then A_w may be modified (without changing PC), and equilibrium can again be redefined so that it still exists, but again may not be a Pareto-efficient allocation. So the weak formalization again collapses. Thus, weakening PC to

---

[15] Note, however, footnote 5 suggests that conclusion II's assumption set can evidently be weakened considerably. In particular, upper-boundedness could replace boundedness of Q_j^* for each j without losing conclusion II. With only upper-boundedness in conclusion II's assumption system, the addition of free disposal to that system, though not needed for conclusion II, would not seem to contradict the other assumptions of the system. Then only irreversibility of conclusion I's set would not seem to follow from conclusion II's set. Probably some further slight tinkering (e.g., just adding irreversibility to conclusion I's set) with the two assumption sets could get conclusion I's set weaker than II's set and keep each conclusion as is.

Table 8.2. *The theory results*

	Conclusions		
Assumptions	I Existence of equilibrium	II Existence of unique globally stable equilibrium	III Pareto efficiency of equilibrium
Assumption set	A_w	A_s	
Private ownership	Yes	Yes	Yes
Preferences & production technologies constant in time	Yes	Yes	Yes
PC, all consumers & producers take **p** as given	Yes	Yes	Yes
No externalities	Yes	Yes	Yes
Consumption			
Preferences constitute complete preordering	Yes	Yes	Yes
Q_i closed	Yes	Yes	No
Q_i boundedness	Lower bound for "$\leq$"	Bounded for "$\leq$"	No
Q_i convexity	Yes	Yes	Yes
$(Q_i, \leq_i)$ continuity	Yes	Yes	No
$(Q_i, \leq_i)$ convexity	Convex	Strongly convex	No
Q_i contains resource-diminishing consumption of Q_i	Yes	Yes	No
Q_i contains no satiation consumption of Q_i	Yes	Implicit in Walras's "law"	Yes

Production			
$0 \in Q_j^*$	Yes	Yes	No
Q_j^* closed	Yes	Yes	No
Q_j^* boundedness	No	Yes	No
Q_j^* convexity	Convex	Strictly convex	No
Free disposal	Yes	No	No
Irreversibility	Yes	No	No
Stability			
Walras's "law"	No	Yes	No
All commodities GS for any $\mathbf{p} \in S_*^p$	No	Yes	No
"Law" of supply and demand	No	Yes	No

Note: PC means perfect competition; No means no assumption made.

allow for some monopolies, or relaxing "no externalities," collapses the weak formalization.[16]

The size of the economy and bounded nonconvexities (of consumer preferences and production possibility sets) can be defined, and equilibrium redefined, so that if "bounded nonconvexities" replaces the assumptions of convexity (of consumer preferences and production possibility sets), then equilibrium exists, but again may not be Pareto-efficient, *as the size of the economy approaches infinity* (Arrow and Hahn 1971, pp. 169–82). So again the weak formalization collapses. Next, let *weak convexity* of preferences be: For every $\mathbf{q}_i' \in Q_i$, $\{\mathbf{q}_i \in Q_i : \mathbf{q}_i \gtrsim_i \mathbf{q}_i'\}$ is convex. Then we can replace A_w with a weaker $A_{w'}$ – which excludes Q_i lower-boundedness for $\leq$, $(Q_i, \leq_i)$ convex, the (non) existence of a resource-diminishing (satiation) consumption of Q_i, Q_j^* closed, bounded, and convex, and free disposal and irreversibility, but includes weak convexity of $(Q_i, \leq_i)$ – *without making any other alteration in the weak formalization* (Debreu 1962). Thus, weakening convexity in one way collapses, and in another (which includes weakening "assumptions [such as free disposal and the existence of a resource-diminishing consumption which] have not been readily accepted because of their strength, and this in spite of the simplicity that they give to the analysis" (Debreu 1962, p. 257)) leaves intact, the weak formalization.

Consider the strong formalization next. Weakening its preference strong-convexity assumption (which is sharply at odds with introspection) to convexity collapses the formalization. There may be assumptions that are weaker than preference strong convexity (and perhaps, but not necessarily, stronger than preference convexity), that are not sharply at odds with introspection, and that imply the existence of demand functions and so do not collapse the strong formalization. If so, though, the literature fails to develop them. Next, let WGS be: Two commodities $i \neq j$ are *weak gross substitutes* at $\mathbf{p} \in S_*^p$ if $\partial z_i(\mathbf{p})/\partial p_j$ exists and is ≥ 0. Also, let the economy be *connected at* $\mathbf{p}$ if there is no set of commodities I such that $(\partial z_i(\mathbf{p})/\partial p_j) = 0$ for all $i \in I$, j not $\in I$. Then weakening GS to WGS (and connectedness at all $\mathbf{p} \in S_*^p$) collapses (leaves intact) the strong formalization (Arrow and Hahn 1971, pp. 289–92). And since "the increase in generality obtained by replacing GS by WGS is by no means great" (Arrow and Hahn 1971, p. 227), there seems little prospect of an important weakening of the strong formalization's GS assumption.

---

[16] See Arrow and Hahn (1971, pp. 151–68, 132–6) for the results stated on monopolies and externalities, respectively.

Table 8.3A. *Weakening assumptions: weak formalization*

Assumption	Assumption weakened to	Equilibrium still exists	Equilibrium still a Pareto-efficient allocation
Perfect competition	Some monopolies	Yes	Maybe not
No externalities	Utilities a function of distribution of initial endowments	Yes	Maybe not
Convexities	Bounded nonconvexities	Yes	Maybe not
Preference convexity, free disposal, existence of resource-diminishing consumption, etc.	Preference weak convexity	Yes	Yes

Table 8.3B. *Weakening assumptions: strong formalization*

Assumption	Assumption weakened to	(Unique) globally stable equilibrium still exists	Equilibrium still a Pareto-efficient allocation
Preference strong convexity	Preference convexity	Maybe not	Yes
GS	WGS, connectedness at all $\mathbf{p} \in S^p_*$	Yes	Yes
GS	WGS	Maybe not	Yes

Tables 8.3 summarize our account of weakening assumptions. Since the strong formalization is closer than the weak to Adam Smith's even stronger original claim, the most important result of the assumption-weakening analysis is that there is no obvious way to make the strong

formalization less at odds with introspection without collapsing it. If we take introspection seriously, then, the strong formalization of Adam Smith's original claim seems not to refer to our economic world.

8.4   Summary and conclusion

We began the chapter with the economist J. N. Keynes's distinction between a positive science and a normative science. Recently, the philosopher E. W. Händler (1982) offered a related distinction among empirical, pure, and normative theories in economics:

> A leading feature of almost all present work in theoretical microeconomics is the fact that no explicit statements are made about the ontological status of the objects these theories refer to. . . . The mathematical structures of microeconomic theories can be and have in practice been applied in three different ways. . . . An empirical theory is characterized by the claim that the entities forming its model are part of the ontological inventory of the actually existing world. A pure theory does not intend to speak about reality. A pure theory is just a (sometimes very complex) picture of a possible world which does not actually exist . . . a possible world is not discovered but stipulated by the descriptive conditions one associates with it. Normative theories correspond to the situation where a group of persons establishes a norm. (pp. 74–5)

Händler (1982) argues that general equilibrium theory is a pure rather than empirical theory on the grounds that the theory's "books and papers are never concerned with the measurement of the economic variables their theories deal with" (p. 75). "The economists gain knowledge about their possible worlds by inquiring into the logical relationships between several different stipulations for these possible worlds. This entails that contemporary general equilibrium theory has to be reconstructed as a bundle of pure theories" (p. 76). In the words of two of general equilibrium theory's most prominent practitioners: "The proposition having been put forward and very seriously entertained, it is important to know not only whether it *is* true, but also whether it *could* be true. A good deal of what follows is concerned with this last question, which seems to us to have considerable claim on the attention of economists" (Arrow and Hahn 1971, p. vii).

We noted that, against philosophers like Händler, the economist E. R. Weintraub claims that (a portion at least of) general equilibrium theory, rather than being a pure theory, tests the logical consistency of the core propositions of an empirical theory. Weintraub (1985) claims that accounts like Händler's err in only telling part of the story: "What may appear to be nonchalance about the empirical referents of the competitive equilibrium models now appears to be a sensible division of labor. Some theorists [i.e., general equilibrium theorists] develop models that can frame the basic facts about our daily economic life. . . . Other theorists and applied economists test those frameworks [against the observations, in Chapter 6 and 7 fashion]" (p. 33). We suggested that because he fails to distinguish placeholder and particular relations, Weintraub's conception of the tests performed and theories tested by applied economists, and the relationship of those tests and theories to the work of general equilibrium theorists, remains vague. Citing our examples of Sections 6.2.4 and 6.3.1, we also argued that Weintraub may be wrong to suppose that the development of empirical microeconomic theory makes little sense without a general equilibrium theory foundation.

Neither Händler's view that general equilibrium theory (GET) is a *pure* theory nor Weintraub's view that it is the foundation of an *empirical* theory accounts for why GET continues onward to welfare microeconomics. To explain this fact, one must view GET as a *normative* theory (i.e., as an argument for a certain set of social arrangements). Over two hundred years ago, Adam Smith argued that an invisible hand would lead a certain kind of economy, in which people pursued their own interests, to the public interest. We reconstructed GET as a formalization – two formalizations actually – of Smith's original argument. First we developed consumer and producer theories that specified assumptions under which market supply and demand functions exist. Later we added sufficient assumptions to allow our consumer–producer theory to be extended into a strong formalization of Smith's argument. Meanwhile, we also constructed a weak formalization of Smith's argument based on another set of assumptions. We also demonstrated that both formalizations require the addition of a philosophical argument for what the distribution of income should be before they become normative, and we explored weakening the formalizations' assumptions. Figure 8.5 depicts the logical structure of our efforts.

The strong (weak) formalization of Smith's argument reads: Under a set of restrictive assumptions A_s (A_w), a perfectly competitive private-ownership economy leads to, by the end of days (is consistent with), a Pareto-efficient allocation, which is in the public interest if the distribution of income is. Also, A_s includes an assumption – preference strong convexity – that is sharply at odds with introspection and that cannot be weakened to the more plausible preference convexity without collapsing the strong formalization of Smith's argument. Ponder, then, the restric-

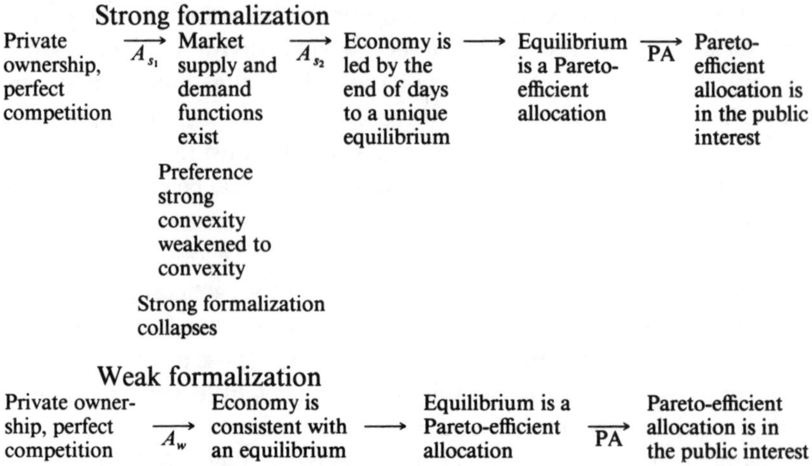

Strong formalization

| Private ownership, perfect competition | $\xrightarrow{A_{s_1}}$ | Market supply and demand functions exist | $\xrightarrow{A_{s_2}}$ | Economy is led by the end of days to a unique equilibrium | $\longrightarrow$ | Equilibrium is a Pareto-efficient allocation | $\xrightarrow{PA}$ | Pareto-efficient allocation is in the public interest |

Preference strong convexity weakened to convexity

Strong formalization collapses

Weak formalization

| Private owner-ship, perfect competition | $\xrightarrow{A_w}$ | Economy is consistent with an equilibrium | $\longrightarrow$ | Equilibrium is a Pareto-efficient allocation | $\xrightarrow{PA}$ | Pareto-efficient allocation is in the public interest |

Figure 8.5. Logical structure of formalizations of Adam Smith's defense of an economy of interest-seeking individuals. A_{s_1}, A_{s_2}, and A_w are sets of assumptions spelled out in the course of the chapter, and PA is a philosophical argument, such as the one by John Rawls (see Section 8.3.3), that income distributions of a certain kind are in the public interest.

tive assumptions and income distribution qualifiers on both formalizations, the "is consistent with" rather than "leads to" form of the weak formalization, and the end-of-days qualifier and introspectively implausible preference strong-convexity assumption of the strong formalization. It is hard not to conclude that GET's ability (at least in its present state of development) to make a case for Adam Smith's defense of an economy where individuals pursue only their own interests is rather overrated.

Summary and conclusion of Part II

We began Chapter 5 by acknowledging the serious difficulties Thomas Kuhn and others have raised for Karl Popper's rationalist reconstruction of science as testing the deduced consequences of theories against observations. Nevertheless, we claimed that an account of classical mechanics along Popper's lines, *together with a catalogue of the account's difficulties,* would be useful in exploring the question, Is microeconomics science? We argued that since classical mechanics *is* science, reasons to believe that microeconomics is *not* science might be found among differences in the catalogues of difficulties of Popper-like reconstructions of the two theories. Our reconstruction of classical mechanics generated a catalogue of seven difficulties: Laws are not logically falsifiable; approximation conventions are needed in prediction statements; circular-like reasoning is entailed; complicated (or even intractable) equations of motion must sometimes be solved; the values of constants are not implied by the theory; the theory's realm of applicability is not always fully observable; scientists' reporting may be imperfect.

We began Chapter 6 by challenging the view that recent work in the philosophy of science offers an affirmative answer to the question, Is microeconomics science? The implicit logic of that view is incoherent: Popper's model of science fails to account for both physics and economics; physics is science, and therefore economics is science too. In contrast to such a view, we extended the argument begun in Chapter 5, developing a reconstruction of microeconomics with a catalogue of difficulties. Each of the seven classical mechanical difficulties had a microeconomic counterpart difficulty. For example, circular-like reasoning was involved in measuring masses in classical mechanics and in identifying supply and demand equations in microeconomics. But there were several microeconomic difficulties that had no classical mechanical analogues: conscious objects of inquiry, observing market equilibria and the distribution of (not just aggregate) income, and variation in consumer tastes (production technologies) across individuals (firms) and time. If

283

microeconomics is not science, we suppose the reasons are in part among these last difficulties, *not* among those difficulties it shares with classical mechanics.

Chapter 6's rational reconstruction of microeconomics involved hypothetical experiments (i.e., those that could in principle be done, but were not actually done). Chapter 7 examined four historical microeconomic experiments, both to see how already identified difficulties arise in practice and in search of possible difficulties not as yet identified. In studying the income maintenance experiments, we discovered several sorts of difficulties not previously catalogued: breaches of deductive and inductive logic, choice of labor-supply functional form that presupposes utility minimization, and failure to seek one particular utility functional form and to model complexities that could have been modeled. Supposing the income maintenance experiments' methodological shortcomings are typical of empirical microeconomic studies, we concluded, on the basis of the newly discovered difficulties, and the form in practice of those catalogued in Chapter 6 (but not shared with classical mechanics), that microeconomics is best understood as an undeveloped science.

While Chapters 5 through 7 sought to account for microeconomics as positive science, Chapter 8 took up the question of microeconomics as normative science (i.e., as an argument for a certain set of social arrangements). Adam Smith was the first political economist to claim that an economy in which individuals pursued their own interests would be led by an invisible hand to an outcome in the public interest. Chapter 8 developed weak and strong versions of Smith's argument that read: Under a set of restrictive assumptions A_s (A_w), a perfectly competitive, private-ownership economy leads to, by the end of days (is consistent with), a Pareto-efficient allocation, which is in the public interest if the distribution of income is. The set of restrictive assumptions for the strong formalization, A_s, includes one (preference strong convexity) that is sharply at odds with introspection. Notice the restrictive assumptions and income distribution qualifiers on both formalizations, the "is consistent with" rather than "leads to" form of the weak formalization, and the end-of-days qualifier and introspectively implausible preference strong-convexity assumption of the strong formalization. These all suggest that microeconomic theory as yet provides paltry advice on how the economy should be arranged, despite the frequent loud assertions to the contrary by Adam Smith's political economic camp followers.

Except for the last step – from Pareto efficiency to the public interest – our account of microeconomics as a normative science is also an abstract account (i.e., without particular utility and production functions) of microeconomics as a positive science. So Chapter 8 leads us back once again to some of the difficulties with microeconomics as positive science that we catalogued in earlier chapters. For example, that the existence of demand functions presupposes preference strong convexity means that our Chapter 6 account of microeconomics as positive science was based on an introspectively implausible assumption, and so raises again the difficulty of a conscious object of inquiry, or introduction of the "law" of supply and demand (differential-equation placeholders involving prices as functions of time) in the strong formalization raises again the difficulty of the role and meaning of time in microeconomics. In our concluding chapter, we turn once again to the conceptual problems that we found, in Chapters 1 through 8, to hamper social experiments.

Conclusion: Some possible barriers
to controlled social experiments as science

C.1   Introduction: The difficulties that may be barriers

C.1.1   *From Parts I and II to the Conclusion*

We considered recent social "experiments," such as the four income maintenance studies, as experiments in two senses. In Part I we viewed them as the sort of random-assignment experiments that R. A. Fisher introduced into agronomy in the 1920s. In Part II we considered them as the kind of tests of hypotheses-deduced-from-theory that Karl Popper argued, beginning in the 1930s, were the hallmark of physics and science more generally. The first sense of experiment is that of what one could call an *empiricist* vision of science, whereas the second is that of what one might label a *rationalist* vision of science. We argued that these visions of science were problematic accounts of agronomy and physics experiments, that the problems could be catalogued, and that reconstructions of recent social experiments in accordance with the two visions of science possessed the same catalogue of problems *and then some*. If the social experiments were not science, we argued, the reasons they were not might be found in the "and then some" portion of their reconstructions' catalogue of problems.

At the base of the empiricist vision of science, we argued, was the drive to know the object of inquiry on the basis of observation alone. We showed that a controlled experiment fails at this task, for randomization-based statistical analysis of cause and effect involves a set of assumptions which are never checked against anything observable (and in some cases can't be). These assumptions always include deterministic causality involving counterfactual conditionals and atomized individuals. They may also include no technical error, perfect (or a particular form of imperfect) reporting, or hypothetical infinite sequences of populations with convergent parameter sequences. Further, we argued that when one shifts from randomization-based inference to inference based on infinite populations, the need for assumptions that go beyond the observations does not diminish. The classical variant of such assumptions (e.g., supposing independent normal distributions in a test of hypothesis that two population means are equal) may be weakened, but then the inferential conclusions that may be drawn must also be weakened.

The problem of empiricism can also be understood as the problem of how the words of the assumptions needed for causal inference "hook onto" the world of our observations and perception. The problem with the rationalist vision of science is its supposition that it has solved this version of the problem of empiricism. In a rationalist reconstruction of classical mechanics, we argued that the problem of empiricism reappears in the form of seven "attachment difficulties": Laws are not logically falsifiable; approximation conventions are needed in prediction statements; circular-like reasoning is entailed; complicated (or even intractable) equation(s) of motion must sometimes be solved; the values of constants are not implied by the theory; the theory's realm of applicability is not always fully observable; scientists' reporting may be imperfect. In reconstructing the income maintenance experiments as Fisher-like randomization-based inference, we noted they shared the problem of empiricism with agronomy experiments. And in reconstructing them as Popper-like tests of hypotheses-deduced-from-theory, we noted they shared the seven attachment difficulties with classical mechanics experiments. So if randomization-based agronomy and hypothetico-deductive classical mechanics experiments are science despite the problem of empiricism and its rationalist form as seven attachment difficulties, social experiments could be science too.

But in reconstructing social experiments as randomization-based inference, and social experiments and microeconomics as tests of hypotheses-deduced-from-theory, we catalogued several difficulties that either were not present at all or seemed to be less problematic in agronomy and physics experiments. We classify these difficulties as (sometimes overlapping) problems of complexity, behavior, and time. We close this book with a deeper examination of these problems, because they are the ones that *might* block social experiments from ever becoming even the imperfect versions of empiricist and rationalist visions of science that agronomy and classical mechanics experiments are.

Under the problem of complexity, we present an overview of many of the difficulties that seem to block microeconomics from becoming a classical-mechanics-like science. Under the problem of behavior, we examine a set of difficulties that seem especially related to the fact that social experiments seek to be part of a science of human behavior, concluding with an illustration of how Chomsky's linguistics avoids similar difficulties by reconceptualizing its object of inquiry away from the language-use behavior of individuals. Under the problem of time, we examine the various ways in which time presents difficulties for social experiments and microeconomics, concluding by modifying an argument of Popper's into an argument that microeconomics in its present form may never become science because it cannot rid itself of history. We conclude our conclusion with an overall conclusion and a consideration of social experiments as a basis for public policy.

C.1.2 *The problem of complexity*

In a number of fields (e.g., computer science) the term "complexity" has been given a formal meaning. We, however, employ the term more or less in its ordinary-language sense of "complicated." For example, in Chapter 7 we argued that the income maintenance researchers failed (unnecessarily) to model their experiments' complex, or complicated, budget constraints. Now we wish to make four points on the complexity of microeconomics in a broad sense. First, microeconomics is a considerably more complex system of propositions/research than is classical mechanics. Second, economists have yet to get the parts of the complex microeconomic proposition/research system coherently assembled and in touch with the world, whereas physicists have accomplished this task for the classical mechanical system. Third, economists often construct research rationales and explanations that somehow manage to obscure the incoherency of microeconomics relative to a discipline like physics. Finally, microeconomics resembles the physics of the historical period between the collapse of classical mechanics and the rise of quantum mechanics.

Whereas physicists seek only to understand nature, economists seek both to understand *and to prescribe for* society. In classical mechanics, a single entity (i.e., God or nature) minimizes one thing (i.e., the action) subject to a constraint, whereas in microeconomics, many different entities (i.e., consumers and producers) each maximize a different thing (i.e., their preferences or profits) subject to a different constraint, and then the individuals' choices must be coordinated and aggregated. Also, in classical mechanics there is only observation against which to check the theory, because no one presumes to know the mind of God or nature directly, whereas in microeconomics there is observation *and introspection* against which to check the theory, because it is not so presumptuous to suppose one can know one's own mind directly. Thus, microeconomics is a considerably more complex system of propositions/research than is classical mechanics.

The classical results of general equilibrium theory require only demand *correspondences,* which are compatible with introspection. Yet stability theorists require stronger, counter-introspective demand *functions,* so as to introduce the "law" of supply and demand and get "leads to" rather than the classical "is consistent with" results. Empirical consumer (labor) econometricians also estimate counter-introspective consumer demand (labor-supply) *functions.* But their functions may not follow deductively from any maximized individual preferences and aggregation. Nevertheless, these econometricians suggest that they are testing consumer (labor–leisure) choice theory against the observations. Putting on their welfare economic hats, economists rarely hesitate to prescribe private-ownership/competition as a cure for inefficiency, even though their stability theory implies that such medicine may not work until the end of days, and if it does work could have the nasty side effect of leaving many individuals starving. Thus, economists have yet to get the parts of the complex microeconomic proposition/research system coherently assembled and in touch with the world, as physicists have for the classical mechanical system.

Milton Friedman suggests that economists should ignore their theory's unrealistic assumptions and proceed to test its deduced consequences against the observations, obscuring the fact that such tests (at least as they are usually done) are no less problematic than direct tests of the realism of some assumptions against introspection. E. R. Weintraub argues that general equilibrium theorists test the consistency of the core propositions of microeconomic theory, whose "derived" theories (e.g., demand theory) are tested against the observations by other, empirical economists. Even if Weintraub is correct, his formulation obscures the fact that the theory versions tested against the observations are sharply at odds with introspection and often are not deductive consequences of any core propositions. Arrow and Hahn develop stability theory and D. W. Hands reconstructs general equilibrium theory in ways that obscure their accounts' counter-introspective foundations.

Quirk and Saposnik (1968) add the placeholder "law" of supply and demand to stability theory and remark: "the economist's ignorance of the specific values of the parameters of the economic system forces him to construct theories that hold under rather general conditions" (p. 171). The remark obscures the fact that keeping stability analysis at the placeholder level blocks the development of a classical-mechanics-like microeconomic dynamics that could be tested against the observations. Finally, D. N. McCloskey weighs in with the word from recent philosophy of science that a long-dominant model of science is no longer in favor, so that economists who worry about their discipline's failure to measure up to that model are "neurotic," as if our failure to give a precise meaning to "science" implies that all disciplines choosing to call themselves "science" offer equally successful accounts of their subject matters!

Extending McCloskey's psychoanalytic metaphor, our view is that each of the formulations alluded to in the previous two paragraphs is an example of avoidance or denial. What these formulations avoid or deny is the fact that microeconomics at the current historical juncture exhibits certain anomalies – beyond those shared with classical mechanics – that prevent it from being a coherent science. Of course, just as Rutherford's avoidance or denial of the fact that his classical model of the atom implied its collapse allowed him to account for scattering, so the avoidance or denial of economists allows them to extend and refine various pieces of microeconomics. But if we are correct in arguing that those pieces together represent intellectual chaos and incoherence (i.e., science in crisis) rather than (or at least in addition to) healthy pluralist veins of research or a division of labor, then McCloskey's complacent view that all is well with microeconomics is a prescription for the discipline's permanent crisis.

Suppose, however, that the passage of time were to prove us wrong, that is, that rather than the historically temporary anomalies of a science in crisis, the difficulties of microeconomics beyond those of classical mechanics represent permanent features of an inexact science. Still, facing those difficulties squarely makes more sense than avoiding or denying them, for the economist who faces them squarely will approach public policy analysis and recommendation knowing full well that he or she knows much less about the workings of the economy than the physicist knows about the workings of nature. The avoider or denier, however, will be a socially dangerous public policy analyst, mistakenly supposing, as he or she does, that his or her knowledge of the workings of the economy is on a par with the physicist's knowledge of the workings of nature.

C.2   The problem of behavior

C.2.1   *Atomization and individuality*

In Chapter 2, when constructing the formalism of the causal inference associated with a random-assignment experiment, we assumed atomized individuals (i.e., that the response of an individual to a treatment didn't depend on the response of any other individual). In Chapter 8, when developing the formalization of Adam Smith's argument for an economy of unfettered individual gain-seekers, we assumed no consumer externalities (i.e., that no consumer's preferences or utility depended on the consumption of any other consumer or the production of any producer). These assumptions were two versions of the same behavioral postulate of atomization (i.e., that the action of an individual doesn't depend on that of others).

When the objects of inquiry in a random-assignment experiment are fields of crops, their responses are crop-yield sizes, and treatments are various types and amounts of fertilizer – as in R. A. Fisher's original agricultural experiments – atomization seems plausible enough. But when the objects of inquiry are human subjects, their responses are behavioral ones, and treatments are various types of social arrangements – as in the income maintenance experiments – atomization seems far less plausible. That is, whereas a field may not respond to the responses of other fields, a human's behavioral responses (or preferences) may depend on the behavioral responses (or preferences) of other humans (e.g., of friends).

In Chapter 8 we saw that atomized preferences could be relaxed to a utility that depends on total consumption as well as the individual's consumption, and the existence of equilibrium would still follow, though that equilibrium need not be Pareto-efficient. Such a result is small comfort for two reasons: the loss of Pareto efficiency, and an individual's utility depending on *total* (in addition to the individual's) consumption seems little more plausible than no externalities. Allowing one individual's responses to depend on those of others seems to collapse Chapter 2's theory of causal inference, for the implicit assumption that an individual has a *single* treatment-group (or control-group) response, on which the chapter's concept of population depended, is lost. Without an implausible atomization assumption, then, it remains unclear how to construct satisfactory sciences of randomization-based causal inference and welfare economics.

In Chapters 6 and 7 we saw that the problem of individuality impedes the development of a microeconomics that is testable in the same sense that classical mechanics is. Certainly the assumption that the tastes and preferences of all individuals are the same is implausible. But relaxing that assumption in even the simplest way meant that the distribution of income, rather than aggregate income, had to be observed. Or it led the income maintenance researchers to ad hoc demographic models, tacked onto a demand functional form, which severed the deductive link between that demand and a utility functional form and so prevented any test of theory in the classical mechanical sense.

Consider again the Chapter 5 test of the law of conservation of momentum (which follows from Newton's laws). We repeatedly launch the *same* two carts at each other on the *same* track and assume an additive normal measurement error model with mean 0. So we suppose that before-collision momentum minus after-collision momentum are observations from a normal distribution and test the hypothesis that the distribution's mean is 0. We could instead repeat the collisions/measurements on tracks of *different* materials with carts of *different* materials and masses and again assume an additive normal measurement-error model with mean 0, but now also an ad hoc additive model involving cart materials and masses. That is, we could suppose that before-collision minus after-collision momentum, minus some function of cart masses and materials and track material, is normally distributed, with mean 0. If we did, though, most physicists probably would judge us to have drifted from the rationalist test of classical mechanical laws in Chapter 5 to an ad hoc empiricist groping for the effects of different materials and masses on before-minus-after-collision momentum.

Yet the income maintenance researchers engaged in precisely that sort of empiricist groping for the effects of different individual characteristics on change in hours worked with an NIT. Their research, then, was not a rationalist test of the microeconomic theory of labor–leisure choice and so was not science in the classical mechanical sense. Rather, labor–leisure choice theory provided two variables (i.e., wages and nonwage income) for an essentially ad hoc model of change in hours worked, but there was no test of a hypothesis deduced from theory as in Chapter 5. The researchers were instead interested in a question generated in the social conflict surrounding a public policy proposal: How would different individuals' hours worked respond to the introduction of an NIT? They sought to overcome the problem of individuality through use of ad hoc statistical models of human behavior.

Ad hoc statistical models may circumvent the problem of individuality, but as a basis for understanding human behavior they leave much to be desired. For example, the models and variables of the income maintenance studies were not always fully spelled out, and/or they were changed, in seemingly haphazard fashion, with each study. Also, the researchers attributed differences in results from experiment to experiment to poorly understood differences in the various sites' labor market structures, the samples' family values, and so forth. All this seems a far cry from a historical sequence of easily replicated, ultimately successful physics experiments ending with an agreed-upon understanding of nature expressed as a set of physical law statements. Rather, the income maintenance articles ended with little collective conviction that we were much closer to understanding the individual human behavior studied than after the first article (or perhaps than even before the experiments began). Thus, to dignify such studies with the label "social *science*" appears premature and serves mainly to debase the term "science."

C.2.2 *Consciousness and will*

The problem of consciousness first arose in our account of social experiments in Chapter 3, with the Hawthorne effect, subject misreporting, and so forth. About such anomalies, the philosopher Ernest Nagel (1961) has argued:

The circumstance that a respondent is aware of being an object of some interest to the interviewer, the consequences he believes his replies may have for matters of concern to himself . . . may bring into operation influences that can radically affect the responses he gives. . . . This difficulty is undeniably a serious one, and there is no general formula for outflanking it; but it is not a difficulty that is unique to the social sciences, and it is not insuperable in principle. Thus, students of natural science have long been familiar with the fact that the instruments used for making measurements may produce alterations in the very magnitude being measured; this fact has received much attention particularly in recent years, in connection with the interpretation of the Heisenberg uncertainty relation in quantum mechanics. (pp. 466–7)

Thus, Nagel likens our Chapter 3 problem of consciousness anomalies to interpretation of the Heisenberg uncertainty principle of quantum mechanics and to the more general natural science problem of measurement instruments altering the objects of inquiry being measured. We shall argue, however, that Nagel's analogies here are not quite on the mark in suggesting that our Chapter 3 problem of consciousness anomalies is parallel to the Heisenberg uncertainty principle in quantum mechanics.

Consider physics first. Around the turn of the century, physicists began recording a series of anomalous "observations" on the subatomic level that called into question the whole theoretical framework of classical mechanics. For example, the initial positions and momenta of "particles," thought to be under no net external force, were measured. Then

Newton's laws, together with the initial conditions, implied later positions for the "particles," but they were not "observed" at the predicted locations at the predicted times. A classical mechanical interpretation of this anomaly was that the act of "observing," or the measurement instrument, altered the locations of the "particles."

Eventually, however, the whole classical mechanical way of thinking about the anomaly was abandoned. The subatomic object of inquiry was viewed not as a "particle" but as a "probability wave," Schrödinger's equation replaced Newton's Second Law, probabilistic causality replaced deterministic causality, and so forth. That is, a new theory-physics – quantum or wave mechanics – replaced the old theory-physics – classical mechanics. The Heisenberg uncertainty principle is part of this new physics. "The principle can be said to have its basis in the wave properties of matter" (Richtmyer, Kennard, and Lauritsen 1955, p. 193). On the new physics, the classical mechanical view's anomalies dissolve. For example, on the uncertainty principle, momentum and position cannot be precisely measured simultaneously; so a subatomic object of inquiry not "observed" in its classically predicted position is no longer an anomaly. Or a "particle," on a classical view, "observed" to penetrate a potential-energy barrier greater than its (apparent) kinetic energy is not an anomaly for quantum mechanical matter or a momentum-position "probability wave." [1]

So random-assignment techniques in *agriculture* are parallel to a classical mechanical framework in, say, astronomy. But when those techniques are applied in *human social* studies, problems of consciousness anomalies arise just as anomalies arise when a classical mechanical framework is applied in the subatomic realm. The technique of double-blindness then is grafted onto random assignment to cope with problem of consciousness anomalies, just as the Heisenberg uncertainty principle is part of the quantum view which dissolves the classical subatomic anomalies. But notice that while quantum mechanics and the Heisenberg uncertainty principle compose an entirely distinct theory from classical mechanics, double-blindness is merely grafted onto the same random-assignment technique from agricultural experiments. A question thus arises: Does double-blindness dissolve the problem of consciousness anomalies in social experiments?

---

[1] For detailed accounts of energy-barrier penetration phenomena that (from a classical, but not quantum mechanical, perspective) are anomalies, see Dicke and Wittke (1960, pp. 40–9).

Social experiments could in principle be researcher-blind, though the income maintenance experiments were not. For example, the modelers were not blind to what the variables represented. Without such blindness, researchers were free to continue trying models, variable redefinitions, and so on until they had "results" consistent with (perhaps unconscious) political worldviews. Because social experimental treatments cannot be concealed with placebos, social experiments seem in principle not subject-blind. Certainly income maintenance subjects knew which treatment group they were in and had an idea of how misreporting, behaving differently than had the program not been experimental, and so forth, would be in their interest. So double-blindness did not "outflank" the Chapter 3 problem of consciousness anomalies in the income maintenance experiments and perhaps *cannot* outflank such anomalies in social experiments more generally.

Nagel (1961) also points out a difference in how anomalies of consciousness arise in the natural and social sciences, but argues that despite the difference the logical nature of the anomalies is the same in both sorts of disciplines:

In both groups of disciplines, the difficulty arises because changes are produced in a subject matter by the means used to investigate that subject matter. However, although in the social sciences (but not in the natural sciences) such changes can in part be attributed to the knowledge men possess of the fact that they are the subjects of an inquiry, this difference bears on the particular mechanism involved in bringing about the changes in one domain; and this difference in mechanism by which the changes are produced does not affect the nature of the logical problem created by the changes. (pp. 467–8)

But Nagel's "changes . . . produced in a subject matter by the means used to investigate that subject matter" is classical mechanical lexicon for subatomic phenomena that quantum mechanics understands differently. Hence, that the economist's object of inquiry is conscious, whereas the physicist's is not, may be a more important difference than is acknowledged in Nagel's metaphor of measuring instruments altering the objects measured. The economist's theory seems obliged to incorporate the consciousness of its objects of inquiry, whereas the physicist's does not.

The second variant of the problem of consciousness arose in Chapter 8's account of general equilibrium stability theory. There the existence of demand functions required an assumption sharply at odds with introspection (i.e., consciousness). So an anomaly arose in microeconomics: Stability theorists and consumer theory econometricians employ demand functions, which are counter to consciousness. Of course, the anomaly may eventually be resolved. For example, economists may succeed in building demand functions on a foundation not at odds with consciousness, stability theory may be reconstituted on a demand correspondence (not function) basis, consumer theory econometricians may begin to specify particular nonstrongly convex preferences and develop statistical techniques for fitting observations to the resulting demand correspondences. But for the moment, microeconomic theory exhibits a problem of consciousness in the consumer realm not unlike that which classical mechanics experienced in the subatomic realm at the turn of the century, and which quantum mechanics eventually "outflanked."

Nagel (1961) also argues that human will is not a problem for social inquiry:

> . . . that beliefs about human affairs can lead to crucial changes in habits of human behavior that are the very subjects of those beliefs – is sometimes presented as if the difficulty it raises for inquiry were unique to the social sciences because of the alleged "freedom of the human will." However, this ancient question is quite irrelevant to the methodological problems of social inquiry. (p. 469)

Nagel then offers an example of an automated physical apparatus: an antiaircraft gun aimed and fired by an adjusting mechanism receiving signaled computer calculations based on a target-sensing radar device. He argues that if the signaled calculations disturb either the adjusting mechanism or target, we have a situation akin to a "suicidal prediction" in social science (i.e., a prediction that would have been true had the consequences of making it not falsified it). And if the radar or computer is defective, still the gun could be accurate because of the signaling disturbance, a situation similar to a "self-fulfilling prophecy" in social science.[2] Nagel's example, then, *at best* demonstrates that human will is not involved in "suicidal predictions" and "self-fulfilling prophecies." (In fact, it is not even convincing on this point.)[3] It hardly shows that the "ancient question" "of the alleged 'freedom of the human will' " "is quite irrelevant to the methodological problems of social inquiry."

---

[2] See Nagel (1961, pp. 469–70) for further details of his automated antiaircraft gun example.
[3] That at a very abstract level a suicidal prediction or self-fulfilling prophecy *in some respects* resembles a strictly physical situation hardly demonstrates that human will is not involved in the social situation because it is not involved in the physical situation. What is required to demonstrate that human will is not involved in the social situation is a successfully tested social theory that excludes the notion of human will. For example, if economists eventually succeed in formulating and testing a "rational expectations" version of microeconomics, then they will have explained suicidal predictions and self-fulfilling economic prophesies without recourse to the notion of human will.

We encountered the problem of human will in the Chapter 3 difficulties of selection and attrition bias. Human beings, unlike particles or plants, may choose not to participate in or to drop out of experiments, and experimenters in a nontotalitarian society must respect such voluntary exercises of free will. Of course, certain remedies may be developed for the resulting attrition-selection bias difficulties. For example, in an income maintenance experiment, one might deal with attrition bias through a two-simultaneous-equations model in which the dependent variables are change in hours worked and did/did not drop out. Such a model, though, moves from randomization-based inference to inference based on ad hoc statistical models, and so is not science in a Part I random-assignment experiment sense.

Nagel, then, is convincing that the problems of human consciousness and will do not constitute valid arguments that a science of human behavior is impossible. The question of whether or not such a science is possible is an empirical rather than a priori question. We shall just have to wait and see whether or not such a science, free of the anomalies of consciousness and will exhibited by present-day social experiments like the recent ones on income maintenance, emerges historically. For the moment, though, such social experiments seem to be poorly developed science in either a random-assignment-based-inference or test-of-hypothesis-deduced-from-theory sense.

C.2.3   Chomsky's linguistics: a nonbehaviorist social science

Now suppose one wished to develop a theory of individuals' language use or behavior. One might begin by seeking the "causes" of the frequency of certain expressions in people's conversations. Suppose one asked college students to volunteer for an experiment consisting of the recording of conversations. Some would refuse, and others would agree, but drop out later. To model the frequency, one would soon find oneself with an ad hoc list of independent variables, such as student's major, ethnic origin, social class, childhood city, and an ad hoc set of (say multiple-regression) models. Supposing a student's expressions might be influenced by his or her roommates, one might toy with the idea of the same list of variables for a student's roommate as part of the student's independent-variables list. And the students, flattered by one's attention, might choose to show off with different speech patterns than they ordinarily employ. In other words, one would be confronted with precisely the same problems of behavior which plague social experiments of the income maintenance sort: atomization, individuality, consciousness, and will.

In fact, since Chomsky's revolution of the mid-1950s, at least one school of linguists has not gone in the direction of the previous paragraph's search for the "causes" of individuals' language use or behavior. Chomsky (1975) shifts the object of linguistic inquiry from the "causes" of individuals' language use or behavior to the rules of the natural languages of the human species for transforming syntactically well-formed sentences. "Individuals of the species . . . under investigation are essentially identical with respect to" (p. 14–15) such an object of inquiry, and so are not influenced by each other. And the rules are "unconscious for the most part and even beyond the reach of conscious introspection" (p. 35). Further:

. . . we should abandon the . . . idea that rules are "chosen" and that we have reasons for these choices. . . . The rules of language are not accepted for certain reasons, but rather are developed by the mind when it is placed under certain objective conditions, much as the organs of the body develop in their predetermined way under appropriate conditions. (p. 76)

Thus, employing the rules is no more an act of human will than is having a heart with four chambers. Hence, with Chomsky's object of inquiry, the problems of behavior – individuality, atomization, consciousness, and will – simply do not arise.

Chomsky (1975) offers an example of how linguistic research proceeds with his object of inquiry. Consider someone making a simple declarative–interrogative transformation: "the man is tall – is the man tall?" (p. 30). One might conjecture the person making the transformation "processes the declarative sentence from its first word . . . continuing until he reaches the first occurrence of the word 'is'. . . he then preposes this occurrence of 'is', producing the corresponding question" (p. 31). Though the conjectured rule works for the given example, it is false, as the following transformation it implies demonstrates: "the man who is tall is in the room – is the man who tall is in the room?" (p. 31). So the correct transformation rule is this: The person "analyzes the declarative sentence into abstract phrases; he then locates the first occurrence of 'is'. . . that follows the first noun phrase; he then preposes this occurrence of 'is', forming the corresponding question" (pp. 31–2).

Chomskyan linguists call the incorrect rule structure-independent and the correct one structure-dependent. They have discovered no human natural languages yet where the declarative–interrogative transformation rule is not structure-dependent, and so they believe that they have learned something about the language faculty of the human species. Notice that "the principle of structure-dependence, can easily be falsified if false" (Chomsky 1975, p. 37). Thus, linguistic research with Chomsky's object of inquiry seems to conform to Kuhn's definition of science (see Section 5.1.2), though the science's structure is not as close to that of classical mechanics as microeconomics' is. The theory, or set of rules, implies individual sentence transformations or predictions. If the theory implies such a transformation that begins with a syntactically well-formed sentence, but ends with one that is not, then the theory has been falsified. So the rules must be altered, and new predicted individual sentence transformations must be tested for syntactically well-formed end sentences.

Of course, Chomsky's linguistic science, like physics and microeconomics, is not without philosophical difficulties. For example, we have left the phrase "syntactically well-formed sentence" undefined because linguists and philosophers have debated and not agreed on how to attach the phrase to sentences of a human natural language.[4] Our guess is that this difficulty and others of Chomsky's linguistics are akin to the attachment difficulties shared by classical mechanics and microeconomics. Our reason, however, for considering Chomsky's linguistics is that it appears to be a social science that manages to avoid the problem of behavior – which microeconomics exhibits, but physics does not – by shifting its object of inquiry away from the language use or behavior of individuals.

Though Chomsky's linguistics demonstrates that a nonbehaviorist social science is possible, its contents make no obvious suggestions as to how microeconomics or recent social experiments could be put on such a nonbehaviorist footing. Microeconomics and recent social experiments are in the unenviable position of having followed physics in conceptualizing their objects of inquiry as the behavior of something. But whereas in physics the something is inanimate, without consciousness and will, and so forth, in microeconomics and recent social experiments it is not. Chomsky (1975) has remarked on attempts to understand language use or behavior:

. . . consider . . . our very limited progress in developing a scientific theory of any depth to account for the normal use of language (or other aspects of behavior).

---

[4] Searle (1974, pp. 24–8) also argues that extending Chomsky's linguistic theory from syntax to semantics entails a circular-like reasoning difficulty not unlike those of classical mechanics and microeconomics that we identified in Chapters 5 and 6. On our account of science, such a circular-like reasoning difficulty would not prevent the semantic extension of the theory from being science.

Even the relevant concepts seem lacking; certainly, no intellectually satisfying principles have been proposed that have explanatory force, though the questions are very old. It is not excluded that human science-forming capacities simply do not extend to this domain, or any domain involving the exercise of will, so that for humans, these questions will always be shrouded in mystery. (p. 25)

Our argument suggests that because of the problem of behavior, Chomsky's remarks here, when applied to microeconomics or recent social experiments, are perhaps a trifle too harsh, but are nevertheless uncomfortably close to the mark.

C.3 The problem of time

C.3.1 *Dynamics and equilibrium*

Our Chapter 6 account of microeconomics involved notions of consumer, producer, and market equilibrium, but no dynamics. We supposed that what we observed in a single market was generated by consumer–producer equilibria and a market equilibrium for that market, and that tastes-preferences and technologies remained fixed with time. We tested these assumptions, in conjunction with others, against the observations. In Chapter 8 we expanded consideration to all markets simultaneously, retained the assumption that tastes-preferences and technologies remained fixed with time, but began to inject a dynamics into microeconomics with the system of differential-equation placeholders we called the law of supply and demand.

Unlike Newton's Second Law in classical mechanics, the law of supply and demand doesn't follow from the theory's optimization procedures. Rather, one can justify the set of differential-equation placeholders as a formalization of the plausible historical musings of political economists that price goes up (down) when demand is greater (less) than supply: It "is what many economists must have had in mind when they proposed 'the law of supply and demand' " (Arrow and Hahn 1971, p. 265). Thus, when one takes the existence of an excess demand function, which *does* follow from the theory's optimization procedures (with a preference strong-convexity assumption), and adds the law of supply and demand, one gets a placeholder relation as the basis of a dynamic theory that differs in structure from classical mechanics but is in no way problematic as science because of that difference.

We saw in Chapter 8, however, that from the *placeholder* law of supply and demand, only the fact that the price vector approaches an equilibrium vector, by the end of days, follows. Such a result cannot be made to confront the observations, just as all other results which followed from placeholders alone could not be. To get a testable dynamic microeconomics here requires specification of a *particular* excess demand function that, in conjunction with the law of supply and demand, will imply a particular relation between the price vector and time that can be tested against the observations. But here a serious difficulty arises: A particular excess demand function that follows from particular utility-production functions under optimization and gives simultaneous differential equations that can be solved is hard to specify.

Consider, for example, the differential equation that follows from the law of supply and demand, $G =$ the identity transformation, and the simple single-market excess demand function of our Chapter 6 example: $(dp/dt) = K_D p(t)^{a_D} - K_S p(t)^{a_S} r_1(t)^{b_S} r_2(t)^{c_S}$. Without the $r_i(t)$ terms, the differential equation would have a simple solution, $p(t) = [(1 - a_D) K_D t]^{1/(1-a_D)} - [(1 - a_S) K_S t]^{1/(1-a_S)}$, though how to fit (p, t) observations to such a functional form is not clear. With the $r_i(t)$ terms, though, one must first specify excess demand functions for the other two commodities and then somehow uncouple and solve the three differential equations. Such a task would appear to be extremely complex, even supposing that the excess demand functions for the other two commodities are as simple as the one specified in our Chapter 6 example.

The problem of uncoupling and solving the three simultaneous differential equations here seems in some ways akin to solving the three-body problem in classical mechanics. That problem is, in general, too complex to solve, but can in certain cases be solved through use of approximations. In deducing the moon's motion around the earth, for example, the $1/r^2$ form of the gravity force allows one to neglect the sun–moon force. Conceivably, then, the other two excess demand functions might be specified in a way that allows the three differential equations to be uncoupled and solved, using approximations. At present, however, empirical microeconomic researchers do not specify particular excess demand functions and a G and then solve the system of differential equations resulting from the conjunction of the particular specifications and the law of supply and demand. That is, for the historical moment, complexity prevents deduction of the relation between price vector and time, and so blocks the development of an empirically testable microeconomic dynamics.

C.3.2 *Time-dependent utility-production*

In the previous section we focused on how microeconomics could be made into a dynamic theory with the law of supply and demand, and how a particular excess demand function was necessary to make the dynamic theory testable against ($\mathbf{p}$, t) observations. We also argued that developing a dynamic microeconomics based on the law of supply and demand, though more complex, seemed akin to solving with approximations the many-body problem in astronomy. But time appears to enter into microeconomic observations in a way not captured with the law of supply and demand and with no analogue yet considered in physics.

The philosopher Karl Popper (1964) has pointed up a difficulty which seems to be characteristic of economics, but not physics:

> . . . it cannot be doubted that there are some fundamental difficulties here. In physics, for example, the parameters of our equations can, in principle, be reduced to a small number of natural constants – a reduction which has been successfully carried out in many important cases. This is not so in economics; here the parameters are themselves in the most important cases quickly changing variables. This clearly reduces the significance, interpretability, and testability of our measurements. (pp. 142–3)

Thus, in astronomy, once the universal gravity constant G^* has been measured, it stays fixed (i.e., we measure it again five years later and we get the same result). In microeconomics, in contrast, the situation is even more difficult than Popper suggests. If we fit an aggregate demand function for a commodity to current-year cross-section observations, we can be virtually certain that two years hence fitting the same function to the new cross-section observations will yield different parameter values. Even worse, a demand functional form different from the one that best fit the first set may best fit the second set of observations. Hence, not only "the parameters are themselves . . . quickly changing variables," but the demand functional forms are likely quickly changing as well.

The difficulty Popper cites is essentially the fact that utility and production, and hence excess demand, functions seem to change with time in microeconomics. Though it introduces a dynamics into microeconomics that was missing in our Chapter 6 account, the previous section's formulation fails to cope with the difficulty because it supposes utility-production functions fixed with time. If, instead, we could write utility and production functions as explicit functions of time, then we could do the same for the excess demand function. So the law of supply and demand placeholder relations would be $(dp_i/dt) = G_i(z_i(\mathbf{p}, t))$ rather than $(dp_i/dt) = G_i(z_i(\mathbf{p}))$. As before, without particular G_i's and z_i's, no $\mathbf{p}(t)$ relation that can be made to confront the observations can be deduced. But $\mathbf{z}$'s explicit dependence on t, though it adds to the complexity, seems to introduce no new difficulties in principle into solving the system of differential equations resulting from the law of supply and demand and particular G_i's and z_i's.

The coupled-differential-equation placeholders of the law of supply and demand with an explicitly time-dependent excess demand function even have a classical physics analogue (electrodynamics, not mechanics, though). For a time-varying electromagnetic field (and hence, implicitly, force), "Maxwell's equations consist of a set of coupled first-order partial differential equations relating the various components of electric and magnetic fields. They can be solved as they stand in simple situations" (Jackson 1962, p. 79). Maxwell's equations are only four, whereas the law of supply and demand has as many coupled equations as there are commodities, but this merely makes the system of the law of supply and demand more complex than that of Maxwell's equations. There appear to be no differences in principle, though, between solving the law of supply and demand and Maxwell equations systems.

Finally, recall from the previous section that empirical econometric research does not yet specify a particular $\mathbf{z}(\mathbf{p})$ and $\mathbf{G}$ from which, with the law of supply and demand, a $\mathbf{p}(t)$ relation that can be tested against the observations follows. Hence, such research has hardly gotten to specifying a particular *explicitly time-dependent* $\mathbf{z}(\mathbf{p}, t)$ and $\mathbf{G}$ from which, with the law of supply and demand, a $\mathbf{p}(t)$ relation that can be tested against the observations follows. Whether or not the "fundamental difficulties" Popper raises can be dealt with by specifying particular explicitly time-dependent utility-production and hence excess demand functions, then, remains an open empirical question. We shall just have to wait and see if empirical econometric work takes such a direction *and*, if so, whether or not the deduced $\mathbf{p}(t)$ relations are corroborated by the observed $(\mathbf{p}, t)$'s.

C.3.3   *History*

Paul Samuelson (1965) distinguishes two kinds of dynamic systems, "dynamic and causal (nonhistorical)" and "dynamic and historical": "a system which is causal from a very broad viewpoint may be regarded as historical if certain movements are taken as unexplained data for purposes of the argument. (In fact, every historical system is to be regarded as an incomplete causal system . . .)" (p. 315). Recall from the previous section that as yet there are no particular time-dependent utility-production functions that, with the law of supply and demand, imply $\mathbf{p}(t)$ relations that have been successfully tested against the observations. Hence, on Samuelson's distinction, the observed phenomenon of utility-production functions that change from year to year remains dynamic and historical rather than causal. And the previous section's argument becomes this: Whether or not such utility-production function changes become dynamic-causal is *ultimately* an *empirical* question. Now, however, we wish to see if there might be any *conceptual* reasons to *doubt* that utility-production function changes can become, in Samuelson's lexicon, dynamic-causal.

Consider first a situation in classical mechanics that *may* be parallel to the one of interest in microeconomics. Suppose a point mass m, constrained to move in only one direction, is attached to a spring which is itself attached to a rigid wall, and let $\mathbf{L}$ ($\mathbf{L_0}$) be the directed (rest) length of the spring. Now suppose $\mathbf{F} = -K(\mathbf{L} - \mathbf{L_0})$ (Hooke's Law) is the particular force exerted by the spring on the point mass, where K is a constant characterizing the spring. Then Hooke's Law conjoined with Newton's Second Law implies $\mathbf{L}(t) \approx A(\mathbf{L_0}/|\mathbf{L_0}|) \cos((K/m)^{1/2}t + \delta) + \mathbf{L_0}$, where A and δ are constants determined by initial conditions. It turns out that as long as the spring is not stretched beyond a limited range, the predicted $\mathbf{L}(t)$ is indeed confirmed by the observations (i.e., the spring performs what physicists call simple harmonic motion).

Like G^*'s in Newton's Law of Universal Gravitation, K's value in Hooke's Law doesn't follow deductively from the theory, but must be measured from the observations once the broad sinusoidal motion of the mass has been confirmed. But unlike G^*, K is not a *universal* constant. That is, G^* characterizes the gravity force between any two masses, whereas K characterizes the force between any mass and only a particular spring. The basic *form* of Hooke's Law, however, remains the same from spring to spring. K for a given spring roughly characterizes its "stiffness" and is determined by its construction, material, and so forth. The physicist takes K's determination by these factors, however, as outside the scope of classical mechanics. Hence, whereas classical mechanics predicts the gravity force that will prevail 20 years hence between two masses, it predicts only the form of the force between a mass and spring (constructed 20 years hence) because it doesn't predict the nature of spring construction, materials, and so forth, 20 years hence.

What remains outside the scope of classical mechanics, then, is the historical development of spring production technology. And here is the possible parallel with our microeconomics situation, for our observed changes in utility-production functions are changes in production technologies (and consumer tastes and preferences). Just as the historical development of spring production technology remains outside the scope of physics, the historical development of production technology (and consumer tastes and preferences) more generally may remain outside the scope of microeconomics. Also, however, whereas the historical development of technology with respect to Hooke's Law means a changing constant, but constant form, with respect to any microeconomic production or utility function, it may mean changing constants *and* form. If so, the word "law" to describe utility-production function forms would appear out of place.

Karl Popper (1964) has argued that *"for strictly logical reasons, it is impossible for us to predict the future course of history"* (p. vi) as follows:

The course of human history is strongly influenced by the growth of human knowledge. . . . We cannot predict, by rational or scientific methods, the future growth of our scientific knowledge. . . . We cannot, therefore, predict the future course of human history. . . . This means that we must reject the possibility of a *theoretical history;* that is to say, of a historical social science that would correspond to *theoretical physics.* There can be no scientific theory of historical development serving as a basis for historical prediction. (pp. vi–vii)

Now production technology is simply production know-how and so certainly seems to be a form of "human knowledge." Hence, on Popper's argument, though Popper doesn't draw this conclusion himself, microeconomics (based on its present set of concepts) seems barred from ever encompassing the historical development of production technology so blocked from scientific prediction over a time space when production technology changes. Thus, if Popper is correct, the development task for microeconomics that we sketched in the previous section *may* be an impossible one.

Of course, mathematical economists have long been aware that history may pose some difficulties for their discipline. Samuelson (1965) suggests one way they seek to circumvent such problems: "Thus we distinguish long run tendencies from shorter run tendencies, and so forth in infinite regression" (p. 320). Such distinctions in part seek to isolate periods over which production technologies *do* remain fixed and hence during which scientific prediction *is* possible. Ernest Nagel (1961) expands the concept of initial conditions from differential equations: "More generally, initial conditions constitute the special circumstances to which the laws included in the explanatory premises are applied" (pp. 31–2). So with "fixed production technology" defined as an initial condition, one can retain the sense of a particular production function as a law. Yet it is difficult to see how either Samuelson's or Nagel's way around the problem of history amounts to much more than a semantic feint. Certainly one can abstractly distinguish a short-run period during which production technologies are fixed, or equivalently add the phrase "for fixed production technology" to a law statement in which time fails to appear. But unless one can a priori conjecture the duration of the short run, or equivalently the period of "fixed production technology," and concretely incorporate that duration in a conjectured particular law statement, there seems no way of formally deducing testable predictions.

Anyhow, repeated annual cross sections suggest that the short run in most cases is sufficiently short to make predictions based on a two- or three-year period next to useless in assessing the consequences over several decades of the sort of major social policy changes on which the income maintenance and other social experiments focused. And the political economist Joseph A. Schumpeter (1950) has characterized our economic system as follows: "Capitalism, then, is by nature a form or method of economic change and not only never is but never can be stationary. . . . The fundamental impulse that sets and keeps the capitalist engine in motion comes from the new consumers' goods, the new methods of production or transportation, the new markets, the new forms of industrial organization that capitalist enterprise creates" (pp. 82–3). If Schumpeter is correct, then, the words "prediction over a short run or for fixed production technology" may fail to attach to our economic world (i.e., apply nowhere or be empirically empty).

In summary, then, microeconomics has a problem in incorporating time into its theory. One route to overcoming the problem would be to specify a particular excess demand function that, in conjunction with the law of supply and demand, would imply a particular $p(t)$ relation that could be tested against the observations. In addition, one might make the excess demand function explicitly time-dependent, seeking to cope with the fact that utility-production functions appear to change over time. Empirical econometric researchers have yet to pursue either of these solutions, perhaps because solving the resultant simultaneous differential equations would be a highly complex matter. Also, however, success on either of these solution paths may be blocked *in principle* by the problem of history, for production function changes with time are actually the historical development of production technology, which seems unpredictable by any form of scientific theory.

C.4   Summary and conclusion

C.4.1   *Overall conclusion*

Parts I and II distinguished two distinct modes of social inquiry. First was the random-assignment controlled experiment pioneered by the statistician R. A. Fisher in the 1920s, which we labeled an empiricist mode. Second was the theory-testing experiment explicated by the philosopher Karl Popper beginning in the 1930s, which we labeled a rationalist mode. Exploring each mode of inquiry, using the income maintenance experiments as examples, we discovered a series of problems, most of which fall into three mutually exclusive categories that we shall now label. Philosophical problems are those that inhere in the mode of inquiry no matter where applied, methodological problems are those that inhere in the mode as presently applied to social phenomena only, and mistakes are problems of particular experiments that could have been avoided.

The empiricist mode of inquiry seeks inference based solely on certain physical acts (e.g., stratification and/or random assignment and/or sampling) and the observations. Yet we demonstrated that the mode's inference always employs, in addition, certain assumptions that go beyond the observations. Further, while a randomization-based test of the hypothesis of no causal effects is possible, the test relies on an arbitrary choice of statistic. And though a randomization-based test of zero mean causal effect is possible, the dependency of the samples allows only an upper bound for the test's p-value. When the empiricist mode shifts the basis of its inference from physical acts to infinite populations, then the weaker the assumptions about the populations, the weaker the inferential conclusions – again frustrating inference based solely on the observations. These, then, are the philosophical problems of the empiricist mode of inquiry (i.e., the problem of empiricism).

The methodological problems of the empiricist mode of inquiry – broadly classed as problems of complexity, behavior, and time – include problems of consciousness and will. Thus, subjects are not blind, and they may choose to participate or not, drop out or not, self-report responses under incentives to lie, and experience the Hawthorne effect. But history is also a methodological problem for the empiricist mode of inquiry. For example, the availability of federal income transfers should have different impacts on divorce before, during, and after a women's social movement has aroused many women to seek equality of social arrangements with men. Finally, certain mistakes were made in the income maintenance experiments that prevented inference in the empiricist mode. For example, a randomization-based unbiased point estimate of the mean of all causal effects could not be calculated, because elaborate stratification occurred, but cell means couldn't be weighted because they weren't reported. And quickly reconciled separations were still counted as separations.

Popper supposed that he had solved the problem of empiricism, but we demonstrated that the problem reappears in his rationalist mode of inquiry as a series of "attachment difficulties" (i.e., problems with hooking the words of theorems, definitions, etc., onto the world of our observations). The seven attachment difficulties are as follows: nonfalsifiable laws; circular-like reasoning; approximation convention required; imperfect reporting; constant values not implied by theory; mathematically intractable equations; the nonvisibility of certain phenomena. These seven attachment difficulties, then, are the philosophical problems of the rationalist mode of social inquiry. Methodological problems of the rationalist mode – again, broadly classed as problems of complexity, behavior, and time – include problems of consciousness and individuality. For example, accounting for tastes and preferences in a way that is consistent with introspection and allows for variation between individuals with time has so far proved too complex a task for this mode. Finally, mistakes that were made in the income maintenance experiments' rationalist-mode analysis include choosing a labor-supply function that implicitly supposes utility minimization and failing to properly model budget constraint nonlinearities.

Frank Hahn (1986) recently noted that "some economists refer to each other as scientists . . . scientists in turn are pictured as Newtonian physicists" (p. 12). The economist Philip Mirowski (1984) not so long ago called attention to the historical origin of much of the conceptual structure of present-day microeconomics in classical physics. Finally, the economist Robert M. Solow (1985) recently observed: "My impression is that the best and brightest in the profession proceed as if economics is the physics of society" (p. 330). In the spirit of these suggestions, we sought in Part II to explore the parallels in the logical structures of classical mechanics and microeconomics. We discovered that microeconomic logic was a partially worked-out version of classical mechanical logic. Hence, we concluded that the income maintenance experiments, as tests of the microeconomic theory of labor–leisure choice, were undeveloped or immature science.

We learned that classical mechanics has precisely the same set of philosophical problems as microeconomics, so that because classical mechanics is evidently mature science despite these problems, microeconomics cannot be immature science because of them. Rather, the *methodological* problems of microeconomics – *none* of which are shared by classical mechanics – lead to logical inconsistencies, gaps, and incoherences that keep microeconomics an undeveloped science. For example, in modeling consciousness, convex (but not strongly convex) preferences meet the test of introspection, but empirical research and stability theory presuppose strongly convex preferences. Or to cope with individuality, the use of ad hoc statistical models leads to a gap between statements that follow deductively from the theory and those that are actually tested. Or complexity leads to the use of estimation procedures for which statistical optimality properties have not been deduced. Or complexity and the problem of time prevent deduction from theory of particular $\mathbf{p}(t)$ relations which can be tested against the observations. Finally, with microeconomics in its present form, history appears to make the "laws" change unpredictably with time.

Some philosophers have claimed that the difficulties we have labeled methodological problems imply that microeconomics can in principle never become "science" in the sense that the natural sciences are "science."[5] We have instead argued that though these difficulties keep microeconomics in its present form an undeveloped science, whether or not economics can ever become a mature science is ultimately an empirical question. The example of Chomsky's linguistics demonstrates that by conceptualizing its object of inquiry to exclude the historically evolving willed behavior of individuals, a field can mature into a science. (That is, if Chomsky's linguistics was based on meaningful words rather than syntactically correct sentences, then it would still be undeveloped, because the dictionary of words meaningful in speech behavior is constantly changing.) Unfortunately, Chomsky's linguistics offers no clues as to how microeconomics might shift its object of inquiry so as to exclude human behavior and history.

Some economists concede that economics has no realm of reliable predictions, but deny that such a realm is a necessary condition for a field to be a science. Their favorite counterexample to the necessity of a realm

---

[5] For an argument along these lines, see Winch (1958).

of reliable predictions is evolutionary theory. Frank Hahn, for example, claims: "If it [i.e., science] means the power to predict accurately then many conventionally taken as scientists would be demoted. Evolutionary theory would not qualify as scientific." [6] But Hahn's claim is based on the misconception that a prediction for evolutionary theory must be of the following form: A given species will evolve in the future in a given fashion. But evolutionary theory instead predicts what sort of fossils of past species' development will be discovered, the rules of Chomsky's linguistics predict which sentence transformations are syntactically correct, classical mechanics predicts the behavior of masses, and so forth. A realm of accurate predictions – which economics does still lack – is indeed a necessary condition for a mature science (see Section 5.1.2).

Many social experimenters continue to be only dimly aware that with our present rather meager conceptual knowledge, methodological problems block their work from being the sort of reliable randomization-based inferences that Fisher's agricultural experiments were. For example, Jerry Hausman and David Wise (1985) sum up their views on social experiments:

Thus analysts generally resorted to estimation . . . that did not distinguish various treatments, or they assumed a structural model that allowed interpolation across individuals assigned to different treatments . . . it subverts the major goal of using random selection and treatment assignment to circumvent the inherent limitations of hypothesized structural models. . . . This is not to say that experimental data should not be used to estimate structural econometric models. These data can of course be used like other survey data for this purpose. But the experiment should be thought of in the first instance as a way to obtain accurate estimates of the effects of particular programs. . . . The major advantage of experiments should not be lost sight of. . . . A lack of confidence in such [structural] estimates is the motivation for the experiments. To use experimental data only to provide more such estimates, or to set up the experiments in such a way that only such estimates are possible, is to travel to Rome to buy canned peas. (pp. 194, 196)

---

[6] Hahn (1986, p. 12). Popper, maintaining that a realm of falsifiable predictions is necessary for a successful science, originally argued that evolutionary theory, because of its historical character, could never be such a science (Popper 1964, pp. 105–19). Economists like McCloskey and Hahn argue that evolutionary theory is indeed a science, though (as Popper says) it lacks a realm of falsifiable predictions. Hence, they imply that Popper is wrong in supposing that a realm of falsifiable predictions is necessary for a successful science. But Popper and McCloskey–Hahn are both muddled on evolutionary theory. A realm of falsifiable predictions *is* necessary for a successful science – Popper is right and McCloskey–Hahn are wrong on this point. And evolutionary theory has such a realm, though it is not the future historical evolution of species that both Popper and McCloskey–Hahn seem to suppose it must be.

As far as they go, these conclusions seem reasonable. Certainly it makes little sense to randomize and then do *no* randomization-based inference, and it makes even less sense, after randomizing, to block (with the way the data are gathered and/or reported) others from doing such inference. But the conclusions of Hausman and Wise seem *most* noteworthy for what they *fail to say.*

Social experiments, including the four income maintenance experiments, are of necessity plagued with problems of consciousness, will, and individuality. For example, in a nontotalitarian society, all such experiments will exhibit a selection bias of unknown size (a difficulty that observational studies may, though need not, suffer from) because of the use of volunteers. If there were researchers at the outset of the income-maintenance-led round of social experiments who hoped to use random selection to circumvent the selection bias difficulty, then they were confused. None of the income maintenance experiments made anything but gestures at dealing with various sources of bias (selection, self-reporting, absence of double-blindness, and Hawthorne effect), not because the researchers were incompetent but because in our present state of methodological knowledge we don't know how to deal with these difficulties.

Hausman and Wise themselves sought to deal with the will-generated attrition bias problem in the Gary experiment. Because there is no way to deal with this problem within a randomization-based inferential framework, they were driven to a clever two-equation structural model with the usual problems of ad hoc functional forms and lists of individual characteristics. In attempting to deal with the experimenter-generated truncation-bias problem, Hausman and Wise were again of necessity driven outside a randomization-based inferential framework (to inference based on infinite normal populations). Similarly, though less of a necessity, they sought to deal with stratification bias with inference based on infinite normal populations.[7] Though all four experiments suffered from attrition, truncation, and stratification bias simultaneously, because of the daunting complexity of the task, neither Hausman and Wise nor any other researchers conducted analysis aimed at dealing with more than one of these three difficulties at a time.

Finally, the problem of time looms over the four income maintenance and other social experiments. For example, as we argued earlier, it is difficult to believe that the effect of a newly introduced NIT on the poor's

---

[7] See Hausman and Wise (1979) on attrition bias, as well as on truncation bias (1976, 1977) and stratification bias (1981).

divorce rates would be the same before, during, and after the emergence in society of a movement for women's equality. And it seems doubtful that the effect of a newly introduced NIT on the poor's work efforts would be the same during a period when the Protestant ethic is often scorned as during a subsequent period when traditional values such as hard work reassert themselves. The divorce rate and work-effort effects of an NIT most plausibly evolve historically and unpredictably. Given social experiments' problems of behavior, complexity, and time, there are few good reasons to believe they can provide "accurate estimates of the effects of particular programs."

C.4.2 *Social experiments as a basis for public policy*

We have argued that those who continue to regard social experiments as providing "accurate estimates of the effects of particular programs" thereby engage in denial and avoidance of problems of behavior, complexity, and time. In an income maintenance experiment done during a period of social upheaval, for example, a reported average work-effort reduction of 8% by male household heads might overestimate an actual 4% average because of non-blind researchers' unconscious political opposition to income redistribution, or it might have been 4% had Protestant-ethic types not selected themselves out to start (attritioned themselves out later), or with the subsequent return of a period of traditional values. Or the 8% figure might underestimate an actual 12% average because of false self-reporting in their interest by non-blind participants, or it might have been 12% in the absence of a Hawthorne effect from an experiment.

Similarly, a reported 40% increase in "marital dissolutions" with an NIT might overestimate an actual 10% average because of the unconscious "important-finding" tenure needs, or political opposition to income redistribution, of non-blind researchers. Or the 40% figure might have been 10% had legal definitions of marriage and divorce been employed, strong families not selected and attritioned themselves out, a "dissolution bonus" not encouraged self-reporting participant families to lie, or in a subsequent historical period when divorce rates had ended a 20-year-long gradual climb. Or the 40% figure might have been 70% had self-reporting non-blind participants not lied in their long-term interest of getting an NIT adopted, or in the absence of a Hawthorne experimental effect. The belief, then, that either social experimental or observational statistical studies provide "accurate estimates of particular programs" seems a result of the defensiveness of researchers who mistake our collective methodological inability to predict historically evolving human behavior for implied accusations that they err in their work.

Another difficulty with social experiments as a basis for social policy is that the possibility of implementing any particular program also evolves historically (and thus unpredictably) with the ebb and flow of political conflict. Thus, Daniel Patrick Moynihan, the Nixon administration NIT bill (FAP) spokesman and advocate, pointed out: "The events leading to and from the proposal of FAP have a conceptual unity that admits treatment as a long range development in social policy. The proposal was made, however, as part of an over-riding short term strategy to bring down the level of internal violence" (Moynihan 1973, p. 12). Thus, FAP was proposed and the income maintenance experiments were begun in the midst of an effort to deal with a revolt of the urban poor. But 10 years later, when the experiments' results were finally ready, the revolt of the urban poor was spent, and in fact had engendered a tax revolt in reaction. Under such conditions the possibility of enacting an NIT bill were nil, *no matter what the experiments' results were.* Thus, the term of an experiment may outlast the period during which the particular program it studies could possibly be implemented.

Also, social experiments for social policy seem sometimes to embody a social class bias. For example, if divorce and sloth are social problems, they are problems for other income classes besides the poor. A choice of income tax system based on the system's impact on divorce and sloth, to be fair, requires knowledge of the effects of various such systems on the divorce rates and hours worked for other income classes besides the poor. For example, besides the questions studied by the income maintenance researchers, one probably should ask, Does the present tax rate on income from capital investments lead recipients of such income to divorce more or work less than would higher rates on such income? As a basis for deciding on an income tax system (including a possible NIT), the income maintenance experiments were biased in the social class sense.

So social experiments for social policy sometimes seem to transform discussion of a redistributional issue into a question of an instrumental effect of a particular program. For example, though *some* initially hoped that an NIT would solidify marriages of the poor, *most* (including Milton Friedman, the program's original proposer) understood an NIT as a redistributional device. But when the (dubious, as we have seen) income maintenance experimental evidence showed that an NIT would have the opposite effect on marriages from the one some had hoped for, the redistributional issue receded from consideration. Certainly the recent historical growth of divorce rates is disturbing, but is divorce *never* a good idea? What if any excess of divorces from an NIT occurs because women are given the courage to leave husbands who beat and abuse their children? A social experiment that transforms the discussion of a redistributional issue into one of an instrumental question runs the risk of having the issue decided on the basis of dubious and incomplete evidence.

Because the problems of behavior, complexity, and time prevent them from providing reliable knowledge of historically evolving human behavior, and because they are frequently plagued with the difficulties of implementation timing, social class bias, and redistributional issues, we regard social experiments as somewhat overrated aids for socially just public policy decision-making. Sometimes, however, there will be general agreement that the goal of a particular program is a given instrumental goal rather than a redistributional goal, but there may be disagreement that the program will achieve the goal – perhaps because of conflicting evidence from observational studies. If the instrumental question has been formulated in a way relatively free of social class bias and there is general agreement to act in accordance with its result, then a social experiment may well be in order. If so, certain experimental steps seem wise (in part taught us by the mistakes of the income maintenance experiments).

Assignment to the various treatment groups should be at random, and data should be gathered and reported so that observed group means, weighted for any stratification, can be calculated. Some, at least, of the data analysis should be randomization-based, and statisticians as well as social scientists should be involved in the experiment's design. If analysis based on microeconomic theory rather than randomization is conducted, then demand and/or supply functional forms that follow from particular utility and/or production functions under utility and/or profit maximization should be chosen. Also, budget constraints should be properly modeled, and estimation procedures with statistical optimality properties should be employed, wherever possible.

Within the limits of our methodological knowledge, steps should be taken to deal with problems of individuality and researcher consciousness. For example, all variables and models – including functional forms – to be employed could be specified in advance of any analysis and then held to. Or extensive analysis of the sensitivity of results to functional forms and included variables could be conducted and the detailed results reported.[8] Researchers conducting randomization-based analysis could

---

[8] See Leamer (1985) for a more extensive discussion of a role for sensitivity analysis.

be blind to which group is which and to what each response and independent variable is. Finally, two groups of researchers, perhaps each with different political views on the issue, might conduct data analysis in isolation from each other and then be compelled to reconcile differences in their results before reporting any findings.[9]

The steps we have suggested would make the evidence from social experiments a more reliable basis for social policy decision-making than was the income maintenance experimental evidence. But our proposed steps cannot overcome the problems of behavior, complexity, and time. Social experimenters must begin to comprehend that these problems are largely outside the scope of our present methodological knowledge and that their work cannot yet provide the sort of understanding of society that physics provides of nature. Once social experimenters begin to learn that lesson – which is taught by the income maintenance experiments – they are likely to become a bit more constrained in claiming that their work can prescribe for social policy. And the easing of the scientific pretensions of those who make a living studying them can only ultimately benefit the poor.

---

[9] One of the reviewers of this book's manuscript suggested the idea of two groups of analysts, though we added the wrinkle of "each with different political views on the issue."

Appendix: Proofs of theorems, lemmas, and propositions

Chapter 2

Referring to Section 2.2.1, notice that randomization induces an equiprobable distribution on the $\binom{N}{n}$ $(\binom{N}{N-n})$ subsets of size n $(N-n)$ of $\{T_1, \ldots, T_N\}$ $(\{C_1, \ldots, C_N\})$. So we have a lemma.

Lemma 2.1. *PEI implies that if* $p(t_1, \ldots, t_n; c_1, \ldots, c_{N-n}) \neq 0$, *then*

$$p(t_1, \ldots, t_n; c_1, \ldots, c_{N-n}) = p(t_1, \ldots, t_n)$$
$$= p(c_1, \ldots, c_{N-n}) = 1/\binom{N}{n}.$$

This lemma will enable us to draw on some standard results from the theory of sampling (without replacement) from a finite population. Now we prove our theory's principal theorem.

Theorem 2.1. *PEI implies that* $\bar{t} - \bar{c}$ *is an unbiased estimate of* $\bar{T} - \bar{C}$.

Proof: By the definition of expectation (defn. expectation), $E(\bar{t} - \bar{c}) = \Sigma (\bar{t} - \bar{c}) p(t_1, \ldots, t_n; c_1, \ldots, c_{N-n})$, where Σ here is over the finite class of possible sets of observations. By Lemma 2.1 and hypothesis, $\Sigma (\bar{t} - \bar{c}) p(t_1, \ldots, t_n; c_1, \ldots, c_{N-n}) = \Sigma \bar{t} p(t_1, \ldots, t_n) - \Sigma \bar{c} p(c_1, \ldots, c_{N-n})$, where the second (third) Σ is over the $\binom{N}{n}$ $[\binom{N}{N-n}]$ subsets of size n $(N-n)$ of $\{T_1, \ldots, T_N\}$ $(\{C_1, \ldots, C_N\})$. But $\Sigma \bar{t} p(t_1, \ldots, t_n) - \Sigma \bar{c} p(c_1, \ldots, c_{N-n}) = \bar{T} - \bar{C}$ (Cochran 1977, pp. 11, 22). Conclusion then follows from defn. unbiased estimate.   Q.E.D.

Next we prove two other simple lemmas.

Lemma 2.2. *For constants* c_j *and estimates* h_j, $E(\Sigma_{j=1}^m c_j h_j) = \Sigma_{j=1}^m c_j E(h_j)$.

Proof: By defn. expectation,

$$E\left(\sum_{j=1}^{m} c_j h_j\right) = \sum \left(\sum_{j=1}^{m} c_j h_j\right) p(t_1, \ldots, t_n; c_1, \ldots, c_{N-n})$$

$$= \sum_{j=1}^{m} c_j \left(\sum h_j p(t_1, \ldots, t_n; c_1, \ldots, c_{N-n})\right)$$

$$= \sum_{j=1}^{m} c_j E(h_j). \qquad \text{Q.E.D.}$$

Lemma 2.3. $\text{Var}(\bar{t} - \bar{c}) = \text{Var}(\bar{t}) + \text{Var}(\bar{c}) - 2\text{Cov}(\bar{t}, \bar{c})$.

Proof: Lemma 2.2 and defns. expectation, Var, and Cov imply that

$$\text{Var}(\bar{t} - \bar{c}) = E[(\bar{t} - \bar{c}) - E(\bar{t} - \bar{c})]^2$$
$$= E[\bar{t} - E(\bar{t})]^2 + E[\bar{c} - E(\bar{c})]^2 - 2\,E[(\bar{t} - E(\bar{t}))(\bar{c} - E(\bar{c}))]$$
$$= \text{Var}(\bar{t}) + \text{Var}(\bar{c}) - 2\text{Cov}(\bar{t}, \bar{c}). \qquad \text{Q.E.D.}$$

Now we can prove another theorem.

Theorem 2.2. *PEI implies*

(i) $\text{Cov}(\bar{t}, \bar{c}) = -S_{TC}/N$,

(ii) $\text{Var}(\bar{t} - \bar{c}) = [(N - n)/nN]S_T^2 + [n/(N - n)N]S_C^2 + (2/N)S_{TC}$.

Proof: By Lemma 2.3, it suffices to prove (i), $\text{Var}(\bar{t}) = [(N - n)/nN]S_T^2$, and $\text{Var}(\bar{c}) = [n/(N - n)N]S_C^2$. But the latter two equalities follow from defns. $\bar{t}$, $\bar{c}$, S_T^2, and S_C^2, Lemma 2.1, and Cochran (1977, pp. 23–4). Lemma 2.2, Theorem 2.1, and defns. unbiased estimate and Cov imply that $\text{Cov}(\bar{t}, \bar{c}) = E(\bar{t}\bar{c}) - \overline{T}\overline{C}$. PEI, hypothesis, and defn. expectation imply that

$$E(\bar{t}\bar{c}) = \sum(t_1 + \ldots + t_n)(c_1 + \ldots + c_{N-n})/n(N - n)\binom{N}{n}.$$

But T_j is observed iff C_j is not observed for all $i = 1, \ldots, N$; $T_i C_i$ appears in *no* assignment-allowed sets of observations for $i = 1, \ldots, N$. Now consider $T_i C_j$ for $i \neq j$. If T_i and C_j are observed, T_j is *not* observed. Hence, $T_i C_j$ appears in $\binom{N-2}{n-1}$ assignment-allowed sets of observations.[1]

---

[1] Notice the role the atomized-individuals assumption plays here. Suppose "atomized individuals" was dropped, so that the value of $T_i C_j$ for any i, j might differ for different possible sets of observations. In the worst case, for example, T_i for any i might have as many as $\binom{N-1}{n-1}$ distinct possible values – one for each possible set of observations in which

Hence,

$$E(\bar{tc}) = [(\binom{N-2}{n-1})/n(N-n)(\binom{N}{n})] \sum_{j=1}^{N} \sum_{i \neq j} T_i C_j$$

$$= [1/n(N-1)][N^2 \overline{TC} - \sum_{i=1}^{N} T_i C_i].$$

So

$$\text{Cov}(\bar{t}, \bar{c}) = [N^2/(N-1)]\overline{TC} - \sum_{i=1}^{N} T_i C_i/N(N-1) - \overline{TC}$$

$$= -(1/N)[1/(N-1)][(\sum_{i=1}^{N} T_i C_i) - N\overline{TC}].$$

But the definition of S_{TC} implies that

$$S_{TC} = \sum_{i=1}^{N} (T_i - \overline{T})(C_i - \overline{C})/(N-1) = [1/(N-1)][(\sum_{i=1}^{N} T_i C_i) - N\overline{TC}].$$

Hence, $\text{Cov}(\bar{t}, \bar{c}) = -S_{TC}/N$.   Q.E.D.

Next we prove the two theorems of Section 2.2.3.

Theorem 2.3.   *PEI implies $E[s_p^2] = \text{Var}(\bar{t} - \bar{c}) - (2/N)S_{TC}$.*

Proof: Lemma 2.1, hypothesis, defns. s_t^2 and s_c^2, and Cochran (1977, pp. 23, 26) imply that $E[s_t^2] = S_T^2$ and $E[s_c^2] = S_C^2$. The conclusion then follows immediately from Lemma 2.2, defn. s_p^2, and Theorem 2.2(ii).   Q.E.D.

Theorem 2.4.   *Under PEI, $B = 1 - 2R_{TC}/[MQ + (1/MQ)]$.*

Proof: Theorems 2.2 and 2.3, hypothesis, and defns. B, M, Q, and R_{TC} imply that

$$B = E(s_p^2)/\text{Var}(\bar{t} - \bar{c})$$
$$= -(2/N)S_{TC}/([(N-n)/nN]S_T^2 + [n/(N-n)N]S_C^2 + (2/N)S_{TC})$$
$$= -2R_{TC}/[MQ + (1/MQ) + 2R_{TC}].$$ Q.E.D.

Then we prove the theorem of Section 2.2.4.

it can appear. Then we could not take the next step in the proof. Of course, assumptions *weaker* than strict atomized individuals would be sufficient for taking the next step in the proof. Chapter 8 considers weakening the atomized-individuals assumption in the context of general equilibrium theory.

Theorem 2.5. *Suppose PEI holds.*

(i) *If* $[N(h_o)/\binom{N}{n}] \leq \alpha$ *for* H_0, *then* H_0 *is false, or* h_0 *is an unusual occurrence of* h,

(ii) *If* $[N(h_0)/\binom{N}{n}] > \alpha$ *for* H_0, *then* H_0 *is false, or* h_0 *is not an unusual occurrence of* h.

Proof: (i) PEI, H_0 is true, and h_0 is not an unusual occurrence of h imply $\alpha < p(|h - E(h)| \geq |h_0 - E(h)|) = N(h_0)/\binom{N}{n}$. Hence if PEI holds, H_0 is false or h_0 is not an unusual occurrence of h. The proof of (ii) is similar.   Q.E.D.

The first two theorems of Section 2.3.1 are easily deduced.

Theorem 2.7. *PBE implies that* $\bar{t}_j - \bar{c}_j$ *is an unbiased estimate of* $\overline{T}_j - \overline{C}_j$ *for* $j = 1, 2$.

Proof: Follows immediately from Theorem 2.1, hypothesis, PBE, PEI, and defns. expectation and unbiased estimate.   Q.E.D.

Theorem 2.8. *PBE and* $(n_j/N_j) = (n/N)$ *for* $j = 1, 2$ *imply that* $\bar{t} - \bar{c}$ *is an unbiased estimate of* $\overline{T} - \overline{C}$.

Proof: Lemma 2.2, hypothesis, Theorem 2.6, and defns. $\bar{t}, \bar{c}, \overline{T}, \overline{C}, \bar{t}_j, \bar{c}_j$, $\overline{T}_j$, and $\overline{C}_j$ for $j = 1, 2$ imply that

$$
\begin{aligned}
E(\bar{t} - \bar{c}) &= E[((n_1\bar{t}_1 + n_2\bar{t}_2)/n) - ((n_1\bar{c}_1 + n_2\bar{c}_2)/n)] \\
&= (n_1/n)E(\bar{t}_1 - \bar{c}_1) + (n_2/n)E(\bar{t}_2 - \bar{c}_2) \\
&= (n_1/n)(\overline{T}_1 - \overline{C}_1) + (n_2/n)(\overline{T}_2 - \overline{C}_2) = \overline{T} - \overline{C}. \qquad \text{Q.E.D.}
\end{aligned}
$$

Now consider a definition.

Definition

$$S_{T_j}^2 = \Sigma_{i=1}^{N} (T_{ji} - \overline{T}_j)^2/(N_j - 1),$$

$$S_{C_j}^2 = \Sigma_{i=1}^{N} (C_{ji} - \overline{C}_j)^2/(N_j - 1), \text{ and}$$

$$S_{T_jC_j} = \Sigma_{i=1}^{N} (T_{ji} - \overline{T}_j)(C_{ji} - \overline{C}_j)/(N_j - 1) \text{ for } j = 1, 2.$$

We can then prove the third theorem of Section 2.3.1.

Theorem 2.9. *If* $(n_j/N_j) = (n/N)$ *and* $[(N_j - 1)/(N - 1)] \approx (N_j/N)$ *for* $j = 1, 2$, *then*

$$\text{Var}_{\text{PEI}}(\bar{t} - \bar{c}) \approx \text{Var}_{\text{PBE}}(\bar{t} - \bar{c})$$
$$+ [1/(N-1)N] \sum_{j=1}^{2} N_j[((N-n)/n)^{1/2}(\bar{T}_j - \bar{T})$$
$$+ (n/(N-n))^{1/2}(\bar{C}_j - \bar{C})]^2.$$

Proof: $(T_{ji} - \bar{T}) = (\bar{T}_j - \bar{T}) + (T_{ji} - \bar{T}_j)$; so $(T_{ji} - \bar{T})^2 = (\bar{T}_j - \bar{T})^2 + (T_{ji} - \bar{T}_j)^2 + 2(\bar{T}_j - \bar{T})(T_{ji} - \bar{T}_j)$; so

$$\sum_{j=1}^{2} \sum_{i=1}^{N_j} (T_{ji} - \bar{T})^2 = \sum_{j=1}^{2} \sum_{i=1}^{N_j} (\bar{T}_j - \bar{T})^2 + \sum_{j=1}^{2} \sum_{i=1}^{N_j} (T_{ji} - \bar{T}_j)^2$$
$$+ 2 \sum_{j=1}^{2} (\bar{T}_j - \bar{T}) \sum_{i=1}^{N_j} (T_{ji} - \bar{T}_j)$$
$$= N_1(\bar{T}_1 - \bar{T})^2 + N_2(\bar{T}_2 - \bar{T})^2 + \sum_{i=1}^{N_1} (T_{1i} - \bar{T}_1)^2$$
$$+ \sum_{i=1}^{N_2} (T_{2i} - \bar{T}_2)^2.$$

So defns. $\bar{T}_j$ and $S_{\bar{T}_j}^2$ imply that

$$(N-1)S_{\bar{T}}^2 = N_1(\bar{T}_1 - \bar{T})^2 + N_2(\bar{T}_2 - \bar{T})^2 + (N_1 - 1)S_{T_1}^2 + (N_2 - 1)S_{T_2}^2,$$

or

$$S_{\bar{T}}^2 = (N_1/(N-1))(\bar{T}_1 - \bar{T})^2 + (N_2/(N-1))(\bar{T}_2 - \bar{T})^2$$
$$+ [(N_1 - 1)/(N-1)]S_{T_1}^2 + [(N_2 - 1)/(N-1)]S_{T_2}^2,$$

or

$$[(N-n)/nN]S_{\bar{T}}^2 = [(N-n)/nN][(N_1/(N-1))(\bar{T}_1 - \bar{T})^2$$
$$+ (N_2/(N-1))(\bar{T}_2 - \bar{T})^2] + [(N-n)/nN]$$
$$[((N_1 - 1)/(N-1))S_{T_1}^2 + ((N_2 - 1)/(N-1))S_{T_2}^2].$$

So Lemma 2.1, hypothesis, defn. Var, and Cochran (1977, pp. 23, 90, 93) imply that

$$\text{Var}_{\text{PEI}}(\bar{t}) \approx [(N-n)/nN][(N_1/(N-1))(\bar{T}_1 - \bar{T})^2$$
$$+ (N_2/(N-1))(\bar{T}_2 - \bar{T})^2] + \text{Var}_{\text{PBE}}(\bar{t}).$$

By similar arguments,

$$\text{Var}_{\text{PEI}}(\bar{c}) \approx [n/(N-n)N][(N_1/(N-1))(\bar{C}_1 - \bar{C})^2$$
$$+ (N_2/(N-1))(\bar{C}_2 - \bar{C})^2] + \text{Var}_{\text{PBE}}(\bar{c})$$

and

$$\text{Cov}_{\text{PEI}}(\bar{t}, \bar{c}) \approx -(1/N)[(N_1/(N-1))(\bar{T}_1 - \bar{T})(\bar{C}_1 - \bar{C}) \\ + (N_2/(N-1))(\bar{T}_2 - \bar{T})(\bar{C}_2 - \bar{C})] + \text{Cov}_{\text{PBE}}(\bar{t}, \bar{c}).$$

So, by Lemma 2.3,

$$\text{Var}_{\text{PEI}}(\bar{t} - \bar{c}) \approx [(N-n)/nN][(N_1/(N-1))(\bar{T}_1 - \bar{T})^2 \\ + (N_2/(N-1))(\bar{T}_2 - \bar{T})^2] \\ + [n/(N-n)N][(N_1/(N-1))(\bar{C}_1 - \bar{C})^2 \\ + (N_2/(N-1))(\bar{C}_2 - \bar{C})^2] \\ + (2/N)[(N_1/(N-1))(\bar{T}_1 - \bar{T})(\bar{C}_1 - \bar{C})] \\ + (N_2/(N-1))(\bar{T}_2 - \bar{T})(\bar{C}_2 - \bar{C})] \\ + \text{Var}_{\text{PBE}}(\bar{t} - \bar{c})$$

$$= \text{Var}_{\text{PBE}}(\bar{t} - \bar{c}) + [1/(N-1)N] \sum_{j=1}^{2} N_j[((N-n)/n)^{1/2} \\ (\bar{T}_j - \bar{T}) + (n/(N-n))^{1/2}(\bar{C}_j - \bar{C})]^2. \qquad \text{Q.E.D.}$$

Referring to Section 2.3.2, notice that random selection induces an equiprobable distribution on the $\binom{N}{N}$ subsets of size N of $\{\mathcal{T}_1, \ldots, \mathcal{T}_{\mathcal{N}}\}$ ($\{\mathcal{C}_1, \ldots, \mathcal{C}_{\mathcal{N}}\}$). So we have a lemma.

Lemma 2.4. *PEII implies that* $p((T_1, C_1), \ldots, (T_N, C_N)) = p(T_1, \ldots, T_N) = p(C_1, \ldots, C_N) = 1/\binom{\mathcal{N}}{N}$.

From the lemma, we can prove the theorem of Section 2.3.2.

Theorem 2.10. *PEI and PEII imply that* $\bar{t} - \bar{c}$ *is an unbiased estimate of* $\bar{\mathcal{T}} - \bar{\mathcal{C}}$.

Proof: Hypothesis, PEI, PEII, and defn. expectation imply that

$$E(\bar{t} - \bar{c}) = \sum (\bar{t} - \bar{c})p(t_1, \ldots, t_n; c_1, \ldots, c_{N-n}) \\ = \sum \left(\sum (\bar{t} - \bar{c})p(t_1, \ldots, t_n; c_1, \ldots, c_{N-n}) \right) \\ p((T_1, C_1), \ldots, (T_N, C_N)),$$

where the inner (outer) Σ is over assignment-allowed $(t_1, \ldots, t_n; c_1, \ldots, c_{N-n})$'s [selection-allowed $((T_1, C_1), \ldots, (T_N, C_N))$'s]. But then Theorem 2.1, Lemma 2.4, defn. expectation, and Cochran (1977, pp. 11, 22) imply that

$$\sum (\sum(\bar{t} - \bar{c})p(t_1, \ldots, t_n; c_1, \ldots, c_{N-n}))p((T_1, C_1), \ldots, (T_N, C_N)) \\ = \sum (\bar{T} - \bar{C})p((T_1, C_1), \ldots, (T_N, C_N)) \\ = \sum \bar{T}p(T_1, \ldots, T_N) - \sum \bar{C}p(C_1, \ldots, C_N) = \bar{\mathcal{T}} - \bar{\mathcal{C}}.$$

Conclusion then follows directly from defn. unbiased estimate (Section 2.3.2).   Q.E.D.

The two theorems of Section 2.3.3 are both simple consequences of Theorem 2.1.

Theorem 2.11. *PEI and $E(\epsilon - \bar{\eta}) = 0$ imply that $\bar{r}(t) - \bar{r}(c)$ is an unbiased estimate of $\bar{T} - \bar{C}$.*

Proof: $r(t_i) = t_i + \epsilon_i$ and $r(c_j) = c_j + \eta_j$ for all $i = 1, \ldots, n$ and $j = 1, \ldots, N - n$. So defns. $\bar{t}$, $\bar{c}$, $\bar{r}(t)$, $\bar{r}(c)$, $\bar{\epsilon}$, and $\bar{\eta}$ imply that $\bar{r}(t) - \bar{r}(c) = (\bar{t} - \bar{c}) + (\bar{\epsilon} - \bar{\eta})$. But defn. unbiased estimate and Theorem 2.1 imply that $E(\bar{t} - \bar{c}) = \bar{T} - \bar{C}$. Hence, $E(\bar{\epsilon} - \bar{\eta}) = 0$ and Lemma 2.2 imply that $E(\bar{r}(t) - \bar{r}(c)) = \bar{T} - \bar{C}$, and the conclusion follows from defn. unbiased estimate.   Q.E.D.

Theorem 2.12. *PEI and $E(\bar{\zeta} - \bar{\xi}) = 0$ imply that $\bar{t}^* - \bar{c}^*$ is an unbiased estimate of $\bar{T} - \bar{C}$*

Proof: Similar to that of Theorem 2.10.   Q.E.D.

The final theorem is also an extension of Theorem 2.1.

Theorem 2.13. *PEI implies*

    (i) $\hat{\alpha}_T - \hat{\alpha}_C$ *is an unbiased estimate of* $\alpha_T - \alpha_C$,

    (ii) *if the X_{Ti}'s (or X_{Ci}'s) are a trait, then $\hat{\beta}_T - \hat{\beta}_C$ is an unbiased estimate of $\beta_T - \beta_C$.*

Proof: (i) is merely a restatement of Theorem 2.1. (ii) Let $D = \sum_{i=1}^{N} (X_{Ti} - \bar{X}_T)^2$ [which by hypothesis $= \sum_{i=1}^{N} (X_{Ci} - \bar{X}_C)^2$], $T_i^- = N(X_{Ti} - \bar{X}_T)T_i/D$, and $C_i^- = N(X_{Ci} - \bar{X}_C)/D$ for $i = 1, \ldots, N$. Then defns. β_T and β_C imply that $\beta_T = \sum_{i=1}^{N} T_i^-/N$ and $\beta_C = \sum_{i=1}^{N} C_i^-/N$. Also, let $t_i^- = (x_{ti} - \bar{x}_t)t_i/D$ for $i = 1, \ldots, n$ and $c_j^- = (x_{cj} - \bar{x}_c)c_j/D$ for $j = 1, \ldots, N - n$. Then defns. β_T and β_C imply that $\beta_T = \sum_{i=1}^{n} t_i^-/n$ and $\beta_C = \sum_{j=1}^{N-n} c_j^-/(N - n)$. So with (T_i^-, t_i^-) and (C_i^-, c_i^-) playing the roles respectively of (T_i, t_i) and (C_i, c_i), the result follows from defn. unbiased estimate and Theorem 2.1.   Q.E.D.

Chapter 3

First, there are two items to prove from Section 3.2.1.

Lemma 3.1.   *PRES implies $E(\bar{\epsilon} - \bar{\eta}) = 0$.*

Proof: Defns. $\bar{\epsilon}$ and $\bar{\eta}$, and PRES, imply that

$$E(\bar{\epsilon} - \bar{\eta}) = E[\sum_{i=1}^{n} (\epsilon_i/n) - \sum_{j=1}^{N-n} (\eta_j/(N-n))]$$

$$= (1/n) \sum_{i=1}^{N} E(\epsilon_i) - (1/(N-n)) \sum_{j=1}^{N-n} E(\eta_j) = 0. \qquad \text{Q.E.D.}$$

Theorem 3.1.   *PRES and PEI imply that $\bar{r}(t) - \bar{r}(c)$ is an unbiased estimate of $\overline{T} - \overline{C}$.*

Proof: Follows directly from Lemma 3.1 and Theorem 2.10.   Q.E.D.

Then we must prove the following theorem.

Theorem 3.2.   *Let PBE hold, and suppose $(n_j/N_j) \neq (n/N)$ for $j = 1, 2$. Then*

 (i)   *$\bar{t} - \bar{c}$ is not an unbiased estimate of $\overline{T} - \overline{C}$,*

 (ii) *$E(\bar{t} - \bar{c}) - (\overline{T} - \overline{C}) = ([(n_1 N - nN_1)\overline{T}_1 + (n_2 N - nN_2)\overline{T}_2]/ nN) - ([(nN_1 - n_1 N)\overline{C}_1 + (nN_2 - Nn_2)\overline{C}_2]/N(N-n)),$*

 (iii) *$\bar{t}^+ - \bar{c}^+$ is an unbiased estimate of $\overline{T} - \overline{C}$.*

Proof: Since (ii) implies (i), it suffices to prove (ii) and (iii). Theorem 2.1 applied to each block implies $E(\bar{t}_i) = \overline{T}_i$, $E(\bar{c}_i) = \overline{C}_i$ for $i = 1, 2$. But

$$\bar{t} = (n_1 \bar{t}_1 + n_2 \bar{t}_2)/n \quad \text{and} \quad \bar{c} = [(N_1 - n_1)\bar{c}_1 + (N_2 - n_2)\bar{c}_2]/(N-n),$$

so

$$E(\bar{t} - \bar{c}) = (n_1/n)E(\bar{t}_1) + (n_2/n)E(\bar{t}_2)$$
$$- [(N_1 - n_1)/(N-n)]E(\bar{c}_1) - [(N_2 - n_2)/(N-n)]E(\bar{c}_2)$$
$$= (n_1/n)\overline{T}_1 + (n_2/n)\overline{T}_2 - [(N_1 - n_1)/(N-n)]\overline{C}_1$$
$$- [(N_2 - n_2)/(N-n)]\overline{C}_2.$$

But

$$\overline{T} - \overline{C} = [N_1\overline{T}_1 + N_2\overline{T}_2]/N - [N_1\overline{C}_1 + N_2\overline{C}_2]/N,$$

so

$$E(\bar{t} - \bar{c}) - (\overline{T} - \overline{C}) = [(n_1/n) - (N_1/N)]\overline{T}_1 + [(n_2/n)$$
$$- (N_2/N)]\overline{T}_2 - [((N_1 - n_1)/(N - n))$$
$$- (N_1/N)]\overline{C}_1$$
$$- [((N_2 - n_2)/(N - n)) - (N_2/N)]\overline{C}_2$$
$$= ([(n_1 N - nN_1)\overline{T}_1 + (n_2 N - nN_2)\overline{T}_2]/nN)$$
$$- ([(nN_1 - n_1 N)\overline{C}_1$$
$$+ (nN_2 - Nn_2)\overline{C}_2]/n(N - n)). \quad \text{Q.E.D. (ii).}$$

Defns. $\bar{t}^+$ and $\bar{c}^+$ imply that

$$E(\bar{t}^+ - \bar{c}^+) = (N_1/N)E(\bar{t}_1) + (N_2/N)E(\bar{t}_2) - (N_1/N)E(\bar{c}_1)$$
$$- (N_2/N)E(\bar{c}_2)$$
$$= (N_1/N)\overline{T}_1 + (N_2/N)\overline{T}_2 - (N_1/N)\overline{C}_1 - (N_2/N)\overline{C}_2$$
$$= \overline{T} - \overline{C}. \quad \text{Q.E.D. (iii).}$$

Chapter 6

There is some value in seeing how some of our particular supply–demand and budget-share models are deduced.

Theorem 6.8. *Let S_4 be: UM and $p_1 q_{1i} + p_2 q_{2i} = Y^i$ for each consumer, PM for each producer, $\mathscr{C}$ and $\mathscr{P}$ hold. Let S_3 be: MUC for each consumer, MPC for each producer, $\mathscr{C}$ and $\mathscr{P}$ hold. Then S_4 implies S_3 implies S_2 implies S_1.*

Proof: $\mathscr{C}$ implies $U^i = \alpha_i q_{1i}^{\beta_{1i}} q_{2i}^{\beta_{2i}}$ for constants $\alpha_i, \beta_{1i}, \beta_{2i} > 0$ for $i = 1, \ldots, N$, which implies

$$(\partial U^i/\partial q_{1i}) = \beta_{1i}\alpha_i q_{1i}^{(\beta_{1i}-1)} q_{2i}^{\beta_{2i}} > 0.$$

Similarly, $(\partial U^i/\partial q_{2i}) > 0$. Also,

$$(\partial^2 U^i/\partial q_{1i}^2)(\partial U^i/\partial q_{2i})^2 - 2(\partial^2 U^i/\partial q_{1i} q_{2i})(\partial U^i/\partial q_{1i})(\partial U^i/\partial q_{2i})$$
$$+ (\partial^2 U^i/\partial q_{2i}^2)(\partial U^i/\partial q_{1i})^2 = -\alpha_i^3 \beta_{1i}\beta_{2i}(\beta_{1i} + \beta_{2i})q_{1i}^{(3\beta_{1i}-2)} q_{2i}^{(3\beta_{2i}-2)} < 0$$

everywhere on the interior of $\mathscr{Q}$ for $i = 1, \ldots, N$. Hence, Axiom 6.1 is satisfied for $i = 1, \ldots, N$. So, that UM and $p_1 q_{1i} + p_2 q_{2i} = Y^i$ hold for each consumer implies that MUC holds for each consumer, by Theorem 6.2(i).

$\mathscr{P}$ implies $q^j = \gamma_j x_{1j}^{\delta_{1j}} x_{2j}^{\delta_{2j}}$ for constants $\gamma_j, \delta_{1j}, \delta_{2j} > 0, \delta_{1j} + \delta_{2j} < 1$, for $j = 1, \ldots, M$, which implies

$$(\partial q^j/\partial x_{1j}) = \delta_{1j}\gamma_j x_{1j}^{(\delta_{1j}-1)} x_{2j}^{\delta_{2j}} > 0.$$

Similarly, $(\partial q^j/\partial x_{2j}) > 0$. Also,

$$(\partial^2 q^j/\partial x_{1j}^2) = (\delta_{1j} - 1)\delta_{1j}\gamma_j x_{1j}^{(\delta_{1j}-2)} x_{2j}^{\delta_{2j}} < 0,$$

and similarly, $(\partial^2 q^j/\partial x_{2j}^2) < 0$. Finally,

$$(\partial^2 q^j/\partial x_{1j}^2)(\partial^2 q^j/\partial x_{2j}^2) - (\partial^2 q^j/\partial x_{1j}\partial x_{2j})$$
$$= (1 - \delta_{1j} - \delta_{2j})\delta_{1j}\delta_{2j}\gamma_j^2 x_{1j}^{(2\delta_{1j}-2)} x_{2j}^{(2\delta_{2j}-2)} < 0$$

everywhere on the interior of $\mathscr{X}$ for $j = 1, \ldots, M$. Hence, Axiom 6.2 is satisfied for $j = 1, \ldots, M$. So, that PM holds for each producer implies that MPC holds for each producer, by Theorem 6.6(i).   Q.E.D. S_4 implies S_3.

$\mathscr{C}$ and MUC for any given i imply $(p_1/p_2) = (\beta_{1i}/\beta_{2i})(q_{2i}/q_{1i})$, which, with $p_1 q_{1i} + p_2 q_{2i} = Y^i$, implies $q_{1i} = (Y_i/p_1)[\beta_{1i}/(\beta_{1i} + \beta_{2i})]$. But by $\mathscr{C}$, $[\beta_{1i}/(\beta_{1i} + \beta_{2i})] = \beta$, say, for $i = 1, \ldots, N$. So

$$q_D = \sum_{i=1}^{N} q_{1i} = (N\beta/p_1) \sum_{i=1}^{N} Y^i \quad \text{or} \quad Q_D = a_{1D}P + a_{2D}Y + a_{3D}R_1$$
$$+ a_{4D}R_2 + a_{5D},$$

where $a_{1D} = -1$, $a_{2D} = 1$, $a_{3D} = a_{4D} = 0$, and $a_{5D} = \ln(\beta)$.

$\mathscr{P}$ and MPC for any j imply $(r_1 x_{1j}/\delta_{1j}) = (r_2 x_{2j}/\delta_{2j})$, which, with $\mathscr{P}$, implies

$$q^j = \gamma_j x_{1j}^{\delta_{1j}}[(\delta_{2j}/\delta_{1j})(r_1/r_2)x_{1j}]^{\delta_{2j}},$$

which implies $q^j = k_j r_1^\delta r_2^\nu p^{-(\delta+\nu)}$, where

$$k_j = 1/[\gamma_j(\delta_{2j}/\delta_{1j})^{\delta_{2j}}\delta_{1j}]^{1/(\delta_{1j}+\delta_{2j}-1)}, \qquad \delta = \delta_{1j}/(\delta_{1j} + \delta_{2j} - 1)$$

for any j (by $\mathscr{P}$), and $\nu = \delta_{2j}/(\delta_{1j} + \delta_{2j} - 1)$ for any j (by $\mathscr{P}$). Thus, $q_S = \Sigma_{j=1}^{M} q^j = kr_1^\delta r_2^\nu p^{-(\delta+\nu)}$, where $k = \Sigma_{j=1}^{M} k_j$, or $Q_S = a_{1S}P + a_{2S}Y + a_{3S}R_1 + a_{4S}R_2 + a_{5S}$, where $a_{1S} = -(\delta + \nu) = -(\delta_{1j} + \delta_{2j})/(\delta_{1j} + \delta_{2j} - 1) \neq -1 = a_{1D}$, $a_{2S} = 0$, $(a_{3S} + a_{4S}) = \delta + \nu = -a_{1S}$, $a_{5S} = \ln(k)$.   Q.E.D. S_3 implies S_2.

For given r_1, r_2, Y, set $K_{2D} = Y + a_{5D}$ and $K_{2S} = a_{3S}R_1 + a_{4S}R_2 + a_{5S}$. Then S_2 implies $Q_D = a_{1D}P + K_{2D}$ and $Q_S = a_{1S}P + K_{2S}$. Q.E.D. S_2 implies S_1.

The proof of Theorem 6.14 is similar to that of Theorem 6.8. Theorem 6.15 follows from Theorem 6.14 and the approximation $\ln(x) \approx 1 + x$ for $0 < x \ll 1$.

Theorem 6.18. *MUC holding for each consumer, and*

$$-\ln U(q_1^k, q_2^k, q_3^k) \approx \eta_0 + \sum_{i=1}^{3} \ln(q_i^k) + \sum_{i=1}^{3} \sum_{j=1}^{3} p_{ij} \ln(q_i^k) \ln(q_j^k)$$

for $k = 1, \ldots, N$, *imply*

$$(p_j Q_j^*/y) \approx \sum_{k=1}^{N} (Y^k/y)[(a_{j0} + \sum_{i=1}^{3} a_{ji} \ln(q_i^k))/(b_0 + \sum_{i=1}^{3} b_i \ln(q_i^k))]$$

for $j = 1, 2, 3$ *for constant* a_{ji}'s *and* b_i's, *where*

$$a_{ij} \approx a_{ji} \quad \text{for } i \neq j \quad \text{and } b_i \approx \sum_{j=1}^{3} a_{ji} \quad \text{for } i = 1, 2, 3.$$

Proof: The theorem's hypothesis and Theorem 6.17 imply

$$(p_j q_j^k/Y^k) \approx [(a_{j0} + \sum_{i=1}^{3} a_{ji} \ln(q_i^k))/(b_0 + \sum_{i=1}^{3} b_{ji} \ln(q_i^k))]$$

for $j = 1, 2, 3$ and $k = 1, \ldots, N$, as well as the relations between the a_{ji}'s and b_i's of the theorem's conclusion. So

$$p_j q_j^k \approx Y^k[(a_{j0} + \sum_{i=1}^{3} a_{ji} \ln(q_i^k))/(b_0 + \sum_{i=1}^{3} b_i \ln(q_i^k))]$$

for $j = 1, 2, 3$ and $k = 1, \ldots, N$, which implies

$$(p_j \sum_{k=1}^{N} q_j^k/y) \approx \sum_{k=1}^{N} (Y^k/y)[(a_{j0} + \sum_{i=1}^{3} a_{ji} \ln(q_i^k))/(b_0 + \sum_{i=1}^{3} b_i \ln(q_i^k))]$$

$$\text{for } j = 1, 2, 3$$

or

$$(p_j Q_j^*/y) \approx \sum_{k=1}^{N} (Y^k/y)[(a_{j0} + \sum_{i=1}^{3} a_{ji} \ln(q_i^k))/(b_0 + \sum_{i=1}^{3} b_i \ln(q_i^k))]$$

$$\text{for } j = 1, 2, 3. \quad \text{Q.E.D.}$$

Chapter 7

Theorem 7.2. *Let* $u(L, y_d)$ *be a worker's utility function which satisfies Axiom 7.1.*

(i) *If Budget Constraint 7.1 is the worker's budget constraint, then* $u(L, y_d)$ *has a UCAM, say* (L_0, y_{d0}), *and*

$$h[T, \quad w_g(1 - t_F), \quad (w_g T + y_g)(1 - t_F) + t_F E, \quad w_g, \quad (w_g T + y_g)]$$
$$= T - L_0$$

is the individual-worker labor-supply function, while

$$v[w_g(1 - t_F), (w_g T + y_g)(1 - t_F) + t_F E, w_g, (w_g T + y_g)]$$
$$= u(L_0, y_{d0})$$

is the indirect utility function.

(ii)   *If Budget Constraint 7.2 is the worker's budget constraint, then*
$u(L, y_d)$ *has one, say* (L_0, y_{d0}), *or two, say* (L_0, y_{d0}) *and* (L_*, y_{d*}),
CAMs, and

$$H[T, w_g(1 - t_F), (w_g T + y_g)(1 - t_F)$$
$$+ t_F E, w_g(1 - t_n), (w_g T + y_g)(1 - t_n) + G]$$
$$= \{T - L_0\} \quad or \quad \{T - L_0, T - L_*\}$$

is the individual-worker labor-supply correspondence, while

$$v[w_g(1 - t_F), (w_g T + y_g)(1 - t_F)$$
$$+ t_F E, w_g(1 - t_n), (w_g T + y_g)(1 - t_n) + G] = u(L_0, y_{d0})$$
$$(= u(L_*, y_{d*}))$$

is the individual-worker indirect utility function.

Proof: (i) Consider the two pieces of Budget Constraint 7.1 as if they were distinct budget constraints. By Theorem 7.1(ii), each piece has a UCAM. Let (L', y'_d) and (L'', y''_d) be these two UCAMs. If $u(L', y'_d) \neq u(L'', y''_d)$, the one of (L', y'_d), (L'', y''_d) whose u value is larger is a UCAM for Budget Constraint 7.2. If $u(L', y'_d) = u(L'', y''_d)$, then Axiom 7.1 is not satisfied, which is a contradiction. The proof for (ii) is the same as for (i), except that if $u(L', y'_d) = u(L'', y''_d)$, we get two CAMs instead of a contradiction.   Q.E.D.

Theorem 7.3's proof first requires a *family* budget constraint version of Theorem 7.1(ii) as a lemma in which Axiom 7.2 replaces Axiom 7.1. The lemma's proof is a straightforward four-dimensional version of Theorem 7.1(ii)'s two-dimensional proof. With the lemma and Budget Constraints 7.3 and 7.4 replacing Theorem 7.1(ii) and Budget Constraints 7.1 and 7.2, respectively, the foregoing proof becomes a proof of Theorem 7.3.

Proof of Theorem 7.4 depends on the following lemma.

Lemma 7.1.   *UM,* $\mathcal{P}_1$, $\Sigma_{i=1}^3 w_{gi}(T_i - L_i)(1 - t) + y_n = y_d$ *for* $0 \leq t < 1$,

and y_d, L_i, $w_{gi} > 0$ for $i = 1, 2, 3$ *imply that* $(L_{10}, L_{20}, L_{30}, y_{d0})$, *where*
$L_{i0} = [\beta_i y_n / w_{gi}(1 - t)] + \beta_i T^*/w_{gi}$, $y_{d0} = (1 - \beta)[y_n + T^*(1 - t)]$ *for*
$i = 1, 2, 3$, *and* $\beta = \Sigma_{i=1}^3 \beta_i$, *is a relative maximum of u.*

Proof: Form

$$V^* = kL_1^{\beta_1}L_2^{\beta_2}L_3^{\beta_3}y_d^{1-\beta} + \lambda[\Sigma w_{gi}(T_j - L_j)(1 - t) + y_n - y_d],$$

so that

$$k\beta_{i_1} L_{i_1}^{\beta_{i_1}-1} L_{i_2}^{\beta_{i_2}} L_{i_3}^{\beta_{i_3}} - \lambda w_{gi_1}(1 - t) = 0 = k(1 - \beta)L_1^{\beta_1}L_2^{\beta_2}L_3^{\beta_3}y_d^{-\beta} - \lambda$$
$$\text{for } i_j = 1, 2, 3 \quad \text{and} \quad i_j \neq i_k \text{ for } j \neq k,$$

which are the first-order conditions for a constrained relative maximum.
Also,

$$\begin{vmatrix} \partial^2 V^*/\partial L_1^2 & \partial^2 V^*/\partial L_1 \partial L_2 & -w_{g1}(1 - t) \\ \partial^2 V^*\partial L_2 \partial L_1 & \partial^2 V^*/\partial L_2^2 & -w_{g2}(1 - t) \\ -w_{g1}(1 - t) & -w_{g2}(1 - t) & 0 \end{vmatrix} = kL_1^{\beta_1}L_2^{\beta_2}L_3^{\beta_3}y_d^{(1-\beta)}(1 - t)^2 \times [-w_{g2}L_1^{-2}\beta_1(\beta_1 - 1) \\ + 2w_{g1}w_{g2}L_1^{-1}L_2^{-1}\beta_1\beta_2 - w_{g1}^2L_2^{-2}\beta_2(\beta_2 - 1)] > 0.$$

Similarly,

$$\begin{vmatrix} \partial^2 V^*/\partial L_1^2 & \partial^2 V^*/\partial L_1 \partial L_2 & \partial^2 V^*/\partial L_1 \partial L_3 & -w_{g1}(1 - t) \\ \partial^2 V^*/\partial L_2 \partial L_1 & \partial^2 V^*/\partial L_2^2 & \partial^2 V^*/\partial L_2 \partial L_3 & -w_{g2}(1 - t) \\ \partial^2 V^*/\partial L_3 \partial L_1 & \partial^2 V^*/\partial L_3 \partial L_2 & \partial^2 V^*/\partial L_3^2 & -w_{g3}(1 - t) \\ -w_{g1}(1 - t) & -w_{g2}(1 - t) & -w_{g3}(1 - t) & 0 \end{vmatrix} < 0,$$

and

$$\begin{vmatrix} \partial^2 V^*/\partial L_1^2 & \partial^2 V^*/\partial L_1 \partial L_2 & \partial^2 V^*/\partial L_1 \partial L_3 & \partial^2 V^*/\partial L_1 \partial y_d & -w_{g1}(1 - t) \\ \partial^2 V^*/\partial L_2 \partial L_1 & \partial^2 V^*/\partial L_2^2 & \partial^2 V^*/\partial L_2 \partial L_3 & \partial^2 V^*/\partial L_2 \partial y_d & -w_{g2}(1 - t) \\ \partial^2 V^*/\partial L_3 \partial L_1 & \partial^2 V^*/\partial L_3 \partial L_2 & \partial^2 V^*/\partial L_3^2 & \partial^2 V^*/\partial L_3 \partial y_d & -w_{g3}(1 - t) \\ \partial^2 V^*/\partial y_d \partial L_1 & \partial^2 V^*/\partial y_d \partial L_2 & \partial^2 V^*/\partial y_d \partial L_3 & \partial^2 V^*/\partial y_d^2 & -1 \\ -w_{g1}(1 - t) & -w_{g2}(1 - t) & -w_{g3}(1 - t) & -1 & 0 \end{vmatrix} > ($$

so the second-order conditions for a constrained relative maximum
are satisfied.[2] Thus, the solution of the first-order conditions, which
with a bit of algebra is $L_i = [\beta_i y_n / w_{gi}(1 - t)] + \beta_i T^*/w_{gi}$, $y_d = (1 - \beta)[y_n + T^*(1 - t)]$ for $i = 1, 2, 3$, is indeed a constrained relative
maximum. Q.E.D.

Theorem 7.4. *Suppose UM, S, $\mathscr{P}_1$, and MEC hold; $G > t_n E$, $t_n > t_F$, and*
$(G - t_F E)/(t_n - t_F) < T^*$; *in the period before (after) an NIT is intro-*

---

[2] See Henderson and Quandt (1971, pp. 403–7) for the first- and second-order conditions
for a constrained relative maximum.

duced, Budget Constraint 7.3 (7.4) is the family budget constraint; y_d, w_{gi}, $L_i > 0$ for $i = 1, 2, 3$; and $w_{g1}L_1 + w_{g2}L_2 + w_{g3}L_3 \neq T^* - E + y_g$, T^* before and $\neq T^* - [(G - t_F E)/(t_n - t_F)] + y_g$, T^* after an NIT is introduced. Then $\Delta h_{ik} = -\beta_i Q_{ik} + c_i$, where $0 < \beta_i < 1$ and $c_i = 0$ for $i = 1, 2, 3$.

Proof: $\mathcal{P}_1$ implies $\beta_i > 0$ and $\beta_1 + \beta_2 + \beta_3 < 1$, which implies $0 < \beta_i < 1$ for $i = 1, 2, 3$.

Case (i). If ik pays no federal tax before and receives a payment during the NIT, S implies (in the lexicon of Lemma 7.1) $t = y_n = 0$ before and $t = t_n$, $y_n = G$ during the NIT. So the theorem's hypotheses and Lemma 7.1 imply $L_{i0} = \beta_i T^*/w_{gi}$ before and $= \beta_i G/w_{gi}(1 - t_n)$ during the NIT. Hence,

$$\Delta h_{ik} = [\beta_i G_k/w_{gik}(1 - t_n)] + (\beta_i T^*/w_{gik}) - (\beta_i T^*/w_{gik})$$
$$= \beta_i G_k/w_{gik}(1 - t_n).$$

Case (ii). If ik pays federal taxes and receives a payment during the NIT, then S implies (in the lexicon of Lemma 7.1) $t = t_F$, $y_n = t_F E$ before and $t = t_n = G$ during the NIT. So the theorem's hypotheses and Lemma 7.1 imply $L_{i0} = [\beta_i t_F E/w_{gi}(1 - t_F)] + \beta_i T^*/w_{gi}$ before and $= [\beta_i G/w_{gi}(1 - t_n)] + \beta_i T^*/w_{gi}$ during the NIT. Hence,

$$\Delta h_{ik} = [[\beta_i G_k/w_{gik}(1 - t_n)] + (\beta_i T^*/w_{gik})] - [[\beta_i t_F E_k/w_{gik}(1 - t_F)]$$
$$+ (\beta_i T^*/w_{gik})]$$
$$= \beta_i[[G_k/w_{gik}(1 - t_n)] - [t_F E_k/w_{gik}(1 - t_F)]].$$

Case (iii). If ik pays federal taxes before and during (and receives no payment during) the NIT, then S implies (in the lexicon of Lemma 7.1) $t = t_F$, $y_n = t_F E$ before and during the NIT, and so the theorem's hypotheses and Lemma 7.1 imply $L_{i0} = [\beta_i t_F E/w_{gi}(1 - t_F)] + \beta_i T^*/w_{gi}$ before and during the NIT. Hence, $\Delta h_{ik} = 0$. Therefore, by definition of Q_{ik}, $\Delta h_{ik} = -\beta_i Q_{ik} + c_i$, where $0 < \beta_i < 1$ and $c_i = 0$ for $i = 1, 2, 3$.

Q.E.D.

Theorem 7.5 follows from Case (i) of Theorem 7.4's proof.

Theorem 7.6. *Let $u(L, y_d)$ be an individual utility function of a worker facing a budget constraint $wh + y = y_d$, $h + L = T$. Suppose that $u(L, y_d)$ satisfies Axiom 7.1, that UM holds, $L \neq 0$, T, and that $\partial h(w, y)/\partial y = \beta$ and $[\partial h(w, y)/\partial w]|_{u=constant} = \alpha$ for constants α and β for $w, y > 0$. Then*

(i) $u(L, y_d) = y_d - [(L^2/2) - TL]/\alpha,$

(ii) $h(w, y) = \alpha w$,

(iii) $\beta = 0, \alpha > 0$.

Proof: $\partial h/\partial w = [\partial h/\partial w]|_{u=\text{constant}} + h(\partial h/\partial y)$, by Theorem 7.1(iv), with the theorem's hypothesis, implies $\partial h/\partial w = \alpha + \beta h$. But $\partial h/\partial y = \beta$ implies $h = \beta y + f(w)$; so $\partial h/\partial w = \alpha + \beta h$ implies $\partial h/\partial w = \alpha + \beta[\beta y + f(w)] = \alpha + \beta^2 y + \beta f(w)$ implies $h = \alpha w + \beta^2 yw + \beta w f(w) + g(y)$; so $\beta y + f(w) = \alpha w + \beta^2 yw + \beta w f(w) + g(y)$ implies $\beta = g(y) = 0$; so $h(w, y) = \alpha w$, which are (ii) and the first part of (iii).

Form $V^* = u(L, y_d) + \lambda[w(T - L) + y - y_d]$, so that MUC is $(\partial u/\partial y)w - (\partial u/\partial L) = 0$. But $h = T - L$, so that $\alpha w = T - L$ or $w = (T - L)/\alpha$ if $\alpha \neq 0$. So $(\partial u/\partial y_d)[(T - L)/\alpha] - (\partial u/\partial L) = 0$ implies $[dL/-1] = [dy_d/(T - L)/\alpha]$ or $dy_d = [(L - T)/\alpha] dL$ or $y_d = [[(L^2/2) - TL]/\alpha] + K$, which implies $u(L, y_d) = y_d - [(L^2/2) - TL]/\alpha$ (Martin and Reissner 1956, pp. 213–16), which is (i).

Then Axiom 7.1 implies $\partial u/\partial y_d = 1$, $\partial u/\partial L = (T - L)/\alpha$ both > 0 and

$$(\partial^2 u/\partial L^2)(\partial u/\partial y_d)^2 - 2(\partial^2 u/\partial L \partial y_d)(\partial u/\partial L)(\partial u/\partial y_d) + (\partial^2 u/\partial y_d^2)(\partial u/\partial L)^2 = -1/\alpha < 0$$

implies $\alpha > 0$, which is the second part of (iii).   Q.E.D.

Theorem 7.7's proof is similar to that of Theorem 7.4. Theorem 7.9's proof resembles that of Theorem 7.8, except that it involves three distinct cases (rather than one), like the proofs of Theorems 7.4 and 7.5, to account for the budget constraints' *piecewise* linearity.

Lemma 7.2. *Suppose UM, $\mathscr{P}_2$, and y_d, L, $w_g > 0$ hold. Then*

(1) $w_g(T - L)(1 - t) + y_n = y_d$ for $0 \leq t < 1$ implies (L_0, y_{d0}), where $L_0 = T - \alpha w_g(1 - t)$, $y_{d0} = y_n + \alpha w_g^2(1 - t)$, is a relative maximum of u,

(2) $a'L^2 + b'L + c' = y_d$ for a', $c' > 0$, $b' < 0$, implies (L_0, y_{d0}), where $L_0 = (b'\alpha + T)/(1 - 2a'\alpha)$, $y_{d0} = a'[(T + b'\alpha)/(1 - 2a'\alpha)]^2 + b'[(T + b'\alpha)/(1 - 2a\alpha)] + c'$ is a relative maximum of u.

Proof: (1) Form $V^* = y_d - [(L^2/2) - TL]/\alpha + \lambda[w_g(T - L)(1 - t) + y_n - y_d]$, so that $1 - \lambda = 0 = [(T - L)/\alpha] - \lambda w_g(1 - t)$, which implies that $L = T - \alpha w_g(1 - t)$ and $y_d = y_n + \alpha w_g^2(1 - t)$ are the first-order conditions for a relative maximum. Also,

$$
\begin{vmatrix}
\partial^2 V^*/\partial L^2 & \partial^2 V^*/\partial L \partial y_d & -w_g(1-t) \\
\partial^2 V^*/\partial y_d \partial L & \partial^2 V^*/\partial y_d^2 & -1 \\
-w_g(1-t) & -1 & 0
\end{vmatrix} = 1/\alpha > 0,
$$

so that the second-order condition for a constrained relative maximum is satisfied.

(2) Form

$$
V^* = y_d - [(L^2/2) - TL]/\alpha + \lambda[a'L^2 + b'L + c' - y_d],
$$

so that $1 - \lambda = 0 = [(T - L)/\alpha] + \lambda a'L + \lambda b'$, which implies that $L = (b'\alpha + T)/(1 - 2a'\alpha)$ and $y_d = a'[(T + b'\alpha)/(1 - 2a'\alpha)]^2 + b'[(T + b'\alpha)/(1 - 2a'\alpha)] + c'$, are the first-order conditions for a relative maximum. Also,

$$
\begin{vmatrix}
\partial^2 V^*/\partial L^2 & \partial^2 V^*/\partial L \partial y_d & 2a'L_0 + b' \\
\partial^2 V^*/\partial y_d \partial L & \partial^2 V^*/\partial y_d^2 & -1 \\
2a'L_0 + b' & -1 & 0
\end{vmatrix} = 1/\alpha > 0,
$$

so that the second-order condition for a constrained relative maximum is satisfied.[3]   Q.E.D.

Theorem 7.10.   *Suppose UM, S^*, $\mathscr{P}_2$, and MEC hold, $(t_e/2) > t_F$, $G > (t_e/2)E$, and $2r(G - t_F E) > t_e[(t_e/2) - t_F]$, w_g, $y_d > 0$, and the worker faces Budget Constraint 7.1 and $L \neq 0$, $T - (E/w_g) + (y_g/w_g)$, T in the period before the NIT is introduced.*

(1) *That $w_g T[(t_e/2) - t_F] > (G - t_F E)$ and the worker faces Budget Constraint 7.5 and $L \neq 0$, $T + (y_g/w_g) - (G - t_F E)/[(t_e/2) - t_F]w_g$, $T + (y_g/w_g) - (t_e/2rw_g)$, T in the period after the NIT is introduced imply $(h_b/h_{NIT})_k = -\alpha Q'_k + c^* R'_k + c$, where $\alpha > 0$, $c^* = 1$, $c = 0$.*

(2) *That $w_g T[(t_e/2) - t_F] \leq (G - t_F E)$ and the worker faces Budget Constraint 7.6 and $L \neq 0$, $T + (y_g/w_g) - (t_e/2rw_g)$, T in the period after the NIT is introduced imply $(h_b/h_{NIT})_k = \alpha Q''_k + c^* R''_k + c$, where $\alpha > 0$, $c^* = 1$, and $c = 0$.*

Proof: (1) Before the NIT, k either pays no tax or pays federal tax; so S^* implies (in the lexicon of Lemma 7.2(1)) $t = 0$ or t_F. So the theorem's

---

[3] Ibid.

hypotheses and Lemma 7.2(1) imply $L_0 = T - \alpha w_g$ or $T - \alpha w_g(1 - t_F)$; so $h_b = \alpha w_g$ or $\alpha w_g(1 - t_F)$. During the NIT, k's gross income is in $(0, t_e/2r)$ or $(t_e/2r, (G - t_F E)/[(t_e/2) - t_F])$, or k pays federal tax (and receives no payment). So S^* implies (in the lexicon of Lemma 7.2) $a' = w_g^2 r$, $b' = -w_g[(1 - t_e) + 2r(w_g T + y_g)]$; $t = (t_e/2)$ or t_F, and so the theorem's hypotheses and Lemma 7.2 imply

$$L_0 = (-\alpha w_g[(1 - t_e) + 2r(w_g T + y_g)] + T)/(1 - 2\alpha w_g^2 r)$$
$$\text{or} \quad T - \alpha w_g[(1 - (t_e/2)] \quad \text{or} \quad T - \alpha w_g(1 - t_F).$$

So

$$h_{\text{NIT}} = [\alpha w_g[(1 - t_e) + 2r y_g]]/(1 - 2\alpha w_g^2 r) \quad \text{or} \quad \alpha w_g[1 - (t_e/2)]$$
$$\text{or} \quad \alpha w_g(1 - t_F).$$

So there are 6 (=2 × 3) cases.

(i) If k pays no federal tax before the NIT and gross income is in $(0, t_e/2r)$ during the NIT, $(h_b/h_{\text{NIT}})_k = (1 - 2\alpha w_{gk}^2 r)/[(1 - t_e) + 2r y_{gk}]$.

(ii) If k pays no federal tax before the NIT and gross income is in $(t_e/2r, (G_k - t_F E_k)/[(t_e/2) - t_F])$ during the NIT, $(h_b/h_{\text{NIT}})_k = 1/[1 - (t_e/2)]$.

(iii) If k pays no federal tax before the NIT and pays federal tax during the NIT, $(h_b/h_{\text{NIT}})_k = 1/(1 - t_F)$.

(iv) If k pays federal tax before the NIT and gross income is in $(0, t_e/2r)$ during the NIT, $(h_b/h_{\text{NIT}})_k = (1 - t_F)[1 - 2\alpha w_{gk} r]/[(1 - t_e) + 2r y_{gk}]$.

(v) If k pays federal tax before the NIT and gross income is in $(t_e/2r, (G_k - t_F E_k)/[(t_e/2) - t_F])$ during the NIT, $(h_b/h_{\text{NIT}})_k = (1 - t_F)/[1 - (t_e/2)]$.

(vi) If k pays federal tax before and during the NIT, $(h_b/h_{\text{NIT}})_k = 1$.

Hence, by definition of Q_k' and R_k', $(h_b/h_{\text{NIT}})_k = -\alpha Q_k' + c^* R_k' + c$, where $\alpha > 0$, $c^* = 1$, and $c = 0$.   Q.E.D. (1).
    The proof of (2) is similar to that of (1).   Q.E.D.

Theorem 7.11. *Suppose a worker faces a budget constraint $wh + y = y_d$, $h + L = T$, and $L \neq 0, T$. Then*

(i) *there is no twice differentiable $u(L, y_d)$ for which UM implies*
$h = \gamma + \delta w + \eta y$,

(ii) $h = \gamma + \delta w + \eta y$, *with $\delta > 0$, $\eta < 0$, and utility minimization imply* $\mathscr{P}_3$: $u(L, \quad y_d) = [(\eta y_d - (\delta/\eta) + \gamma)/(T - L - (\delta/\eta))] + \ln[T - L - (\delta/\eta)]$.

Proof: Suppose there is such a $u(L, y_d)$. Let $u(h, y_d) = u(T - h, y_d)$. Then UM and the budget constraint imply $[\partial u/\partial h/\partial u/\partial y_d] = -w$. And $[\partial u/\partial h/\partial u/\partial y_d] = -w$, with the budget constraint and the labor-supply relation, imply $w = [y_d - (h/\eta) + (\gamma/\eta)]/[h - (\delta/\eta)]$. So $(\partial u/\partial h)(\eta h - \delta) + (\partial u/\partial y_d)(-h + \eta y_d + \gamma) = 0$, which implies $dy_d/(\eta y_d - h + \gamma) = dh/(\eta h - \delta)$, or $(\eta h - \delta) \, dy_d + (-\eta y_d + h - \gamma) \, dh = 0$. Let $A' = \eta h - \delta$ and $B' = -\eta y_d + h - \gamma$; so $\partial A'/\partial h = \eta$, $\partial B'/\partial y_d = -\eta$, and $[(\partial B'/\partial y_d) - (\partial A'/\partial h)]/A' = -2\eta/(\eta h - \delta)$. Set $\mu(h) = \exp[\int - 2\eta/(\eta h - \delta) \, dh] = [h - (\delta/\eta)]^{-2}$. So $\mu(h)[(\eta h - \delta) \, dy_d + (-\eta y_d + h - \gamma) \, dh] = 0$, which implies $\eta[h - (\delta/\eta)]^{-1} \, dy_d + [h - (\delta/\eta)]^{-2}[-\eta y_d + h - \gamma] \, dh = 0$. Then let $M' = \eta[h - (\delta/\eta)]^{-1}$ and $N' = [h - (\delta/\eta)]^{-2}[-\eta y_d + h - \gamma]$; so $\partial M'/\partial h = -\eta[h - (\delta/\eta)]^{-2}$ and $\partial N'/\partial y_d = -\eta[h - (\delta/\eta)]^{-2}$. So if $\partial F/\partial y_d = M' = \eta[h - (\delta/\eta)]^{-1}$, then $F = \eta[h - (\delta/\eta)]^{-1}y_d + k'(h)$, which implies $\partial F/\partial h = -\eta[h - (\delta/\eta)]^{-2}y_d + dk(h)/dh = N' = -\eta[h - (\delta/\eta)]^{-2}y_d + (h - \gamma)[h - (\delta/\eta)]^{-2}$, which implies $dk'(h)/dh = h[h - (\delta/\eta)]^{-2} - \gamma[h - (\delta/\eta)]^2$. So $k'(h) = \ln[h - (\delta/\eta)] - (\delta/\eta)[h - (\delta/\eta)]^{-1} + \gamma[h - (\delta/\eta)]^{-1} + $ constant, which implies $u(h, y_d) = \eta[h - (\delta/\eta)]^{-1}y_d + \ln[h - (\delta/\eta)] - (\delta/\eta)[h - (\delta/\eta)]^{-1} + \gamma[h - (\delta/\eta)]^{-1}$, which implies $u(L, \quad y_d) = [(\eta y_d - (\delta/\eta) + \gamma)/(T - L - (\delta/\eta))] + \ln[T - L - (\delta/\eta)]$. But then

$$\partial u/\partial L = [\eta y_d + \gamma - (T - L)]/[T - L - (\delta/\eta)]^2,$$
$$\partial u/\partial y_d = \eta/[T - L - (\delta/\eta)], \quad \partial^2 u/\partial y_d^2 = 0,$$
$$\partial^2 u/\partial L \partial y_d = \eta/[T - L - (\delta/\eta)]^2,$$
$$\partial^2 u/\partial L^2 = [L - T - (\delta/\eta) + 2\eta y_d + 2\gamma]/[T - L - (\delta/\eta)],$$

which implies

$$\begin{vmatrix} \partial^2 u/\partial L^2 & \partial^2 u/\partial L \partial y_d & w \\ \partial^2 u/\partial y_d \partial L & \partial^2 u/\partial y_d^2 & 1 \\ w & 1 & 0 \end{vmatrix} = -\eta^2/[\eta T - \eta L - \delta]^2 < 0,$$

which implies that $(L, y_d) = (T - \delta w - \eta y - \gamma, \delta w^2 + \eta w y + \eta w + y)$ *minimizes $u(L, y_d)$.* Q.E.D.[4]

---

[4] The proof's implications are from Martin and Reissner (1956, pp. 213–16, 37–42) and Henderson and Quandt (1971, pp. 403–7).

The proofs of Theorems 7.12 and 7.13 are similar to those of Theorems 7.4 and 7.5, respectively.

Theorem 7.14. *Let $u(L, y_d)$ be an individual utility function of a worker facing budget constraint $wh + y = y_d$, $T = h + L$. Suppose that $u(L, y_d)$ satisfies Axiom 7.1, UM holds, $L \neq 0$, T, and $h = kw^\alpha y^\beta$, with $k > 0$, $\alpha \geq 0$, and $\beta < 0$. Then*

(i)   we can't specify a closed-form $u(L, y_d)$ for each α, β,

(ii) $v(w, \ y) = k(w^{1+\alpha}/(1 + \alpha)) + y^{1-\beta}/(1 - \beta)$, where v is the worker's indirect utility function,

(iii) if $\alpha = 0$, then $u(L, y_d) = [y_d/(T - L)] + (T - L)^{(1/\beta)-1}/[(1/\beta) - 1]k^{1/\beta}$.

Proof: See Burtless and Hausman (1978, pp. 1111–13) for proofs of (i) and (ii).

(iii)   Let $u(h, y_d) = u(T - h, y_d)$. Then UM and the budget constraint imply $[\partial u/\partial h/\partial u/\partial y_d] = -w$, and the budget constraint and labor-supply relation imply $w = [y_d - (h/k)^{1/\beta}]/h$; so $h[\partial u/\partial h] + [y_d - (h/k)^{1/\beta}](\partial u/\partial y_d) = 0$, which implies $(dh/h) = dy_d/[y_d - (h/k)^{1/\beta}]$ or $dh[y_d - (h/k)^{1/\beta}] - h dy_d = 0$. Let $A' = y_d - (h/k)^{1/\beta}$ and $B' = -h$; so $\partial A'/\partial y_d = 1$, $\partial B'/\partial h = -1$, and $[\partial A'/\partial y_d - \partial B'/\partial h]/B' = 2/(-h)$. Set $\mu(h) = \exp[\int - (2/h) dh] = h^{-2}$; so $\mu(h)[[y_d - (h/k)^{1/\beta}] dh - h dy_d] = 0$, which implies $[y_d h^{-2} - (h^{(1/\beta)-2}/k^{1/\beta})] dh - (h^{-1}) dy_d$. Then let $M' = y_d h^{-2} - (h^{(1/\beta)-2}/k^{1/\beta})$ and $N' = h^{-1}$; so $\partial M'/\partial y_d = h^{-2} = \partial N'/\partial h$. So if $\partial F/\partial h = M' = y_d h^{-2} - (h^{(1/\beta)-2}/k^{1/\beta})$, then $F = -y_d h^{-1} - (h^{(1/\beta)-1}/[(1/\beta) - 1]k^{1/\beta}) + k'(y_d)$, which implies $\partial F/\partial y_d = -h^{-1} + dk'(y_d)/dy_d + N' = -h^{-1}$, which implies $k'(y_d) = $ constant, which implies $u(h, y_d) = (y_d/h) + (h^{(1/\beta)-1}/[(1/\beta) - 1]k^{1/\beta})$, which implies $u(L, \ y_d) = [y_d/(T - L)] + [(T - L)^{(1/\beta)-1}/[(1/\beta) - 1]k^{1/\beta}]$. But then $\partial u/\partial L = [y_d/(T - L)^2] - (T - L)^{(1/\beta)-2}/k^{1/\beta}$, $\partial u/\partial y_d = 1/(T - L)$, $\partial^2 u/\partial y_d^2 = 0$, $\partial^2 u/\partial y_d \partial L = 1/(T - L)^2$, and $\partial^2 u/\partial L^2 = 2[y_d/(T - L)^3] + [(1/\beta) - 2](T - L)^{(1/\beta)-3}/k^{1/\beta}$, implying

$$
\begin{vmatrix}
\partial^2 u/\partial L^2 & \partial^2 u/\partial L \partial y_d & w \\
\partial^2 u/\partial y_d \partial L & \partial^2 u/\partial y_d^2 & 1 \\
w & 1 & 0
\end{vmatrix} = -(T - L)^{(1/\beta)-3}/\beta k^{1/\beta} > 0,
$$

which implies that $(L, \ y_d) = (T - ky^\beta, \ wky^\beta + y)$ maximizes $u(L, y_d)$.[5]   Q.E.D.

[5] Ibid.

The proof of Theorem 7.15 is similar to that of Theorem 7.6.

Proposition 7.3. *If* $a_{ji}^{x_i} \approx 1 + x_i \ln(a_{ji}) + (x_i \ln(a_{ji}))^2/2!$ *for j = 1, 2,*[6] *then*

$$A^*: \beta_i^*(\overline{\alpha}, \delta) \approx 1 + [2/\ln(y_{1i}y_{2i})]$$
$$- [2e^{z\delta}(w_{2i}^{1+\overline{\alpha}} - w_{1i}^{1+\overline{\alpha}})]/(1 + \overline{\alpha})[(\ln(y_{1i}))^2 - (\ln(y_{2i}))^2].$$

Proof: With $a_{2i}^{x_i} - a_{1i}^{x_i} = C_i(\delta, \overline{\alpha})x_i$ and $a_{ji}^{x_i} \approx 1 + x_i\ln(a_{ji}) + (x_i\ln(a_{ji}))^2/2!$ for $j = 1, 2$, then

$$C_i(\delta, \overline{\alpha})x_i = a_{1i}^{x_i} - a_{2i}^{x_i} \approx (1 + x_i\ln(a_{1i}) + [(x_i\ln(a_{1i}))^2/2!])$$
$$- (1 + x_i\ln(a_{2i}) + [(x_i\ln(a_{2i}))^2/2!]),$$

which implies

$$C_i(\delta, \overline{\alpha})x_i \approx x_i(\ln(a_{1i}) - \ln(a_{2i})) + (x_i^2/2)[(\ln(a_{1i}))^2 - (\ln(a_{2i}))^2],$$

which implies

$$C_i(\delta, \overline{\alpha}) \approx (\ln(a_{1i}) - \ln(a_{2i})) + (x_i/2)[(\ln(a_{1i}))^2 - (\ln(a_{2i}))^2],$$

which implies

$$x_i \approx [2C_i(\delta, \overline{\alpha}) - 2(\ln(a_{1i}) - \ln(a_{2i}))]/[(\ln(a_{1i}))^2$$
$$- (\ln(a_{2i}))^2] = [-2/\ln(a_{1i}a_{2i})] + 2C_i(\delta, \overline{\alpha})/[(\ln(a_{1i}))^2 - (\ln(a_{2i}))^2],$$

which implies

$$\beta_i^* = 1 - x_i \approx 1 + [2/\ln(y_{1i}y_{2i})] - 2e^{z_i\delta}(w_{2i}^{1+\overline{\alpha}} - w_{1i}^{1+\overline{\alpha}})/(1 + \overline{\alpha})[(\ln(y_{1i}))^2$$
$$- (\ln(y_{2i}))^2].   \text{Q.E.D.}$$

---

[6] See Chemical Rubber Company (1959, p. 372) for the series on which this approximation is based.

References

Arrow, K. 1975. "A Difficulty in the Concept of Social Welfare," in *Microeconomics Selected Readings,* E. Mansfield (ed.), Norton, New York, pp. 466–82.

Arrow, K. J., and F. H. Hahn. 1971. *General Competitive Analysis,* Holden-Day, San Francisco.

Avery, R. 1977. "Effects of Welfare 'Bias' on Family Earnings Response," in *The New Jersey Income-Maintenance Experiment,* Vol. III, H. Watts and A. Rees (eds.), Academic Press, New York, pp. 303–21.

Basu, D. 1980. "Randomization Analysis of Experimental Data: The Fisher Randomization Test," *Journal of the American Statistical Association* 75(371):575–82.

Berkeley, G. 1957. "A Treatise Concerning the Principles of Human Knowledge," in *Berkeley Selections,* M. W. Calkins (ed.), Scribner, New York, pp. 99–216.

Bickel, P. J., and K. A. Doksum. 1977. *Mathematical Statistics,* Holden-Day, San Francisco.

Bishop, J. H. 1980. "Jobs, Cash Transfers and Marital Instability: A Review and Synthesis of the Evidence," *Journal of Human Resources* 15(3):301–34.

Bohm, D. 1957. *Causality and Chance in Modern Physics,* Routledge & Kegan Paul, London.

Bohr, N. 1913. "On the Constitution of Atoms and Molecules," *Philosophical Magazine and Journal of Science* 26(151):1–25.

Burtless, G., and J. A. Hausman. 1978. "The Effect of Taxation on Labor-Supply: Evaluating the Gary Negative Income Tax Experiment," *Journal of Political Economy* 86(6):1103–30.

Chemical Rubber Company. 1959. *Standard Mathematical Tables,* CRC, Cleveland.

Chomsky, N. 1975. *Reflections on Language,* Pantheon, New York.

Christensen, L. R., D. W. Jorgenson, and L. J. Lau. 1975. "Transcendental Logarithmic Utility Functions," *American Economic Review* 65:367–83.

Cobb, C. W., and P. H. Douglas. 1928. "A Theory of Production," *American Economic Review* 18(1)(Suppl.):139–65.

Cochran, W. G. 1976. "Early Development of Techniques in Comparative Experimentation," in *On the History of Statistics and Probability,* D. B. Owen (ed.), Dekker, New York, pp. 3–25.

1977. *Sampling Techniques,* Wiley, New York.

Commerce Clearing House. 1969. *1970 U.S. Master Tax Guide,* CCH, Chicago. 1971. *1972 U.S. Master Tax Guide,* CCH, Chicago.

Conlisk, J., and M. Kurz. 1972. "The Assignment Model of the Seattle and Denver Income Maintenance Experiments," Research Memorandum 15, Center for the Study of Welfare Policy, SRI International, Menlo Park, Calif.

Copas, J. B. 1973. "Randomization Models for the Matched and Unmatched 2 × 2 Tables," *Biometrika* 60(3):457-76.

Crowther, E. M. 1936. "The Technique of Modern Field Experiments," *Journal of the Royal Agricultural Society of England* 97:54-81.

Daniel. 1970. "The Book of Daniel," *The New English Bible,* Oxford and Cambridge University Presses, pp. 1251-73.

Darwin, C. 1958. *The Origin of Species,* New American Library, New York.

Debreu, G. 1959. *Theory of Value,* Wiley, New York.

1962. "New Concepts and Techniques for Equilibrium Analysis," *International Economic Review* 3(3):257-73.

Descartes, R. 1955. "Of the Principles of Human Knowledge," in *Philosophical Works of Descartes,* Vol. I, E. S. Haldane and G. R. T. Ross (trans.), Dover, New York, pp. 219-53.

Dicke, R. H., and J. P. Wittke, 1960. *Introduction of Quantum Mechanics,* Addison-Wesley, Reading, Mass.

Feller, W. 1966. *An Introduction to Probability Theory and Its Applications,* Vol. II, Wiley, New York.

Findley, J. 1942. "Goedelian Sentences: A Non-Numerical Approach," *Mind* 51:259-65.

Fisher, F. M. 1966. *The Identification Problem in Econometrics,* McGraw-Hill, New York.

Fisher, R. A. 1966. *The Design of Experiments,* Oliver & Boyd, London.

1970. *Statistical Methods for Research Workers,* Hafner, Darien, Conn.

1972. "The Arrangement of Field Experiments," in *Collected Papers of R. A. Fisher,* Vol. II, J. H. Bennett (ed.), University of Adelaide, pp. 84-94.

Friedman, M. 1953. "The Methodology of Positive Economics," in *Essays in Positive Economics,* University of Chicago Press, pp. 3-43.

1957. "The Permanent Income Hypothesis," in *A Theory of the Consumption Function,* Princeton University Press, pp. 20-37.

1962. *Capitalism and Freedom,* University of Chicago Press.

Fussell, G. E. 1935. "The Technique of Early Field Experiments," *Journal of the Royal Agricultural Society of England* 96:78-88.

Garfinkel, I. 1977. "Effects of Welfare Programs on Experimental Responses," in *The New Jersey Income-Maintenance Experiment,* Vol. III, H. Watts and A. Rees (eds.), Academic Press, New York, pp. 279-301.

Gilbert, J. P., and F. Mosteller. 1972. "The Urgent Need for Experimentation," in *On Equality of Educational Opportunity,* F. Mosteller and D. P. Moynihan (eds.), Vintage, New York, pp. 371-83.

Gödel, K. 1970. "On Formally Undecidable Propositions of *Principia Mathematica* and Related Systems I [with related abstract and note]," in *Frege and*

Gödel, J. van Heijenoort (ed.), Harvard University Press, Boston, pp. 86–108.

Goldstein, H. 1959. *Classical Mechanics.* Addison-Wesley, Reading, Mass.

Goodman, N. 1979. *Fact, Fiction, and Forecast,* Hackett, Cambridge, Mass.

Haas, H., H. Fink, and G. Härtfelder. 1963. "The Placebo Problem," *Psychopharmacology Service Center Bulletin* 2(8):1–65.

Hacking, I. 1965. *Logic of Statistical Inference,* Cambridge University Press.

Hahn, F. 1986. "On Some Common Mistakes in Economic Theorizing," in *Conference Papers Book I, Conference on the Rhetoric of Economics,* Wellesley College, pp. 1–16.

Halsey, H. I. 1980. "Data Validation," in *A Guaranteed Income, Evidence from a Social Experiment,* P. K. Robins, R. G. Spiegelman, S. Weiner, and J. G. Bell (eds.), Academic Press, New York, pp. 33–55.

Händler, E. W. 1982. "The Evolution of Economic Theories: A Formal Approach," *Erkenntnis* 18:65–96.

Hands, D. W. 1985. "The Logical Reconstruction of Pure Exchange Economies: Another Alternative," *Theory and Decision* 19:259–78.

Hannan, M., N. Tuma, and L. P. Groenveld. 1976. "The Impact of Income Maintenance on the Making and Breaking of Marital Unions: Interim Report," Research Memorandum 28, Center for the Study of Welfare Policy, SRI International, Menlo Park, Calif.

1977. "Income and Marital Events: Evidence from an Income Maintenance Experiment," *American Journal of Sociology* 82(6):1186–211.

Hausman, J. A., and D. Wise. 1976. "The Evaluation of Results from Truncated Samples: The New Jersey Income Maintenance Experiment," *Annals of Economic and Social Measurement* 5(4):421–45.

1977. "Social Experimentation, Truncated Distributions, and Efficient Estimation," *Econometrica* 45(4):919–38.

1979. "Attrition Bias in Experimental and Panel Data: The Gary Income Maintenance Experiment," *Econometrica* 47(2):455–73.

1981. "Stratification on Endogenous Variables and Estimation: The Gary Income Maintenance Experiment," in *Structural Analysis of Discrete Data with Econometric Applications,* C. F. Manski and D. McFadden (eds.), M.I.T. Press, pp. 365–91.

1985. "Technical Problems in Social Experimentation: Cost versus Ease of Analysis," in *Social Experimentation,* J. A. Hausman and D. Wise (eds.), University of Chicago Press, pp. 187–208.

Hayek, F. A. 1979. *The Counter-Revolution of Science,* Liberty Press, Indianapolis.

Henderson, J. M., and R. E. Quandt. 1971. *Microeconomic Theory,* McGraw-Hill, New York.

Hildebrand, F. B. 1962. *Advanced Calculus for Applications,* Prentice-Hall, Englewood Cliffs, N.J.

Huang, K. 1963. *Statistical Mechanics,* Wiley, New York.

Huber, P. J. 1972. "Robust Statistics: A Review," *Annals of Mathematical Statistics* 43(4):1041–67.

Hume, D. 1955. *An Inquiry Concerning Human Understanding*, Liberal Arts Press, New York.

Jackson, J. D. 1962. *Classical Electrodynamics*, Wiley, New York.

Keeley, M. C., and P. K. Robins. 1980. "Experimental Design, the Conlisk-Watts Assignment Model, and the Proper Estimation of Behavioral Response," *Journal of Human Resources* 15(4):480–98.

Keeley, M. C., P. K. Robins, R. G. Spiegelman, and R. W. West. 1977. "The Labor Supply Effects and Costs of Alternative Negative Income Tax Programs: Evidence from the Seattle and Denver Income Maintenance Experiments, Part I, The Labor Supply Response Function," Research Memorandum 38, Center for the Study of Welfare Policy, SRI International, Menlo Park, Calif.

1978. "The Labor-Supply Effects and Costs of Alternative Negative Income Tax Programs," *Journal of Human Resources* 13(1):3–36.

Keeley, M. C., R. G. Spiegelman, and R. W. West. 1980. "Design of the Seattle/ Denver Income-Maintenance Experiments and an Overview of the Results," in *A Guaranteed Annual Income, Evidence from a Social Experiment*, P. K. Robins, R. G. Spiegelman, S. Weiner, and J. G. Bell (eds.), Academic Press, New York, pp. 3–31.

Kehrer, K. C. 1979. "The Gary Income Maintenance Experiment: Introduction," *Journal of Human Resources* 14(4):431–3.

Kehrer, K. C., E. K. Bruml, G. T. Burtless, and D. N. Richardson. 1975. "The Gary Income Maintenance Experiment: Design, Administration, and Data Files," Mathematica Policy Research, Princeton, New Jersey.

Kehrer, K. C., J. F. McDonald, and R. A. Moffitt. 1979. "Final Report of the Gary Income Maintenance Experiment: Labor Supply," Project Report 80-2S, Mathematica Policy Research, Princeton, New Jersey.

Kershaw, D., and J. Fair. 1976. *The New Jersey Income-Maintenance Experiment*, Vol. I, Academic Press, New York.

Keynes, J. N. 1891. *The Scope and Method of Political Economy*, Macmillan, London.

Knudson, J., R. A. Scott, and A. R. Shore. 1977. "Household Composition," in *The New Jersey Income-Maintenance Experiment*, Vol. III, H. Watts and A. Rees (eds.), Academic Press, New York, pp. 353–74.

Kuhn, T. S. 1970. *The Structure of Scientific Revolutions*, University of Chicago Press.

1972a. "Logic of Discovery or Psychology of Research?" in *Criticism and the Growth of Knowledge*, I. Lakatos and A. Musgrave (eds.), Cambridge University Press, pp. 1–23.

1972b. "Reflections on My Critics," in *Criticism and the Growth of Knowledge*, I. Lakatos and A. Musgrave (eds.), Cambridge University Press, pp. 231–78.

Kurz, M., R. G. Spiegelman, and J. A. Brewster. 1973. "The Payment System for the Seattle and Denver Income Maintenance Experiments," Research Memorandum 19, Center for the Study of Welfare Policy, SRI International, Menlo Park, Calif.

Lakatos, I. 1972. "Falsification and the Methodology of Scientific Research Programmes," in *Criticism and the Growth of Knowledge,* I. Lakatos and A. Musgrave (eds.), Cambridge University Press, pp. 91–196.

Landau, L. D., and E. M. Lifshitz. 1976. *Mechanics.* Pergamon Press, Oxford.

Laplace, P. S., M. de. 1917. *A Philosophical Essay on Probabilities,* F. W. Truscott (trans.), Wiley, New York.

Leamer, E. A. 1985. "Sensitivity Analysis Would Help," *American Economic Review* 75(3):308–13.

Lehmann, E. L. 1950. "Some Principles of the Theory of Testing Hypotheses," *Annals of Mathematical Statistics* 21(1):1–26.

1959. *Testing Statistical Hypotheses,* Wiley, New York.

1975. *Nonparametrics: Statistical Methods Based on Ranks,* Holden-Day, San Francisco.

Levi, I. 1983. "Direct Inference and Randomization," in *Proceedings of the 1982 Meeting of the Philosophy of Science Association,* Vol. 2, P. D. Asquith and T. Nickles (eds.), Philosophy of Science Association, East Lansing, Mich., pp. 447–63.

McCloskey, D. N. 1983. "The Rhetoric of Economics," *Journal of Economic Literature* 21:481–517.

McCord, W., and J. McCord. 1959. *Origins of Crime: A New Evaluation of the Cambridge-Somerville Youth Study.* Columbia University Press, New York.

Mackie, J. L. 1973. *Truth, Probability, and Paradox.* Clarendon Press, Oxford.

1974. *The Cement of the Universe,* Clarendon Press, Oxford.

Martin, W. T., and E. Reissner. 1956. *Elementary Differential Equations,* Addison-Wesley, Reading, Mass.

Mayo, E. 1933. *The Human Problems of Industrial Civilization,* Macmillan, New York.

Mill, J. S. 1973–4. *A System of Logic Ratiocinative and Inductive,* Vols. VII and VIII of *Collected Works of John Stuart Mill,* University of Toronto Press.

Mirowski, P. 1984. "Physics and the 'Marginalist Revolution,' " *Cambridge Journal of Economics* 8:361–79.

Moffitt, R. A. 1979. "The Labor Supply Response in the Gary Experiment," *Journal of Human Resources* 14(4):477–87.

Morganstern, O. 1976. "Pareto Optimum and Economic Organization," in *Selected Economic Writings of Oskar Morganstern,* A. Schotter (ed.), New York University Press, pp. 253–66.

Moynihan, D. P. 1973. *The Politics of a Guaranteed Income,* Vintage, New York.

Muraka, B. A., and R. G. Spiegelman. 1978. "Sample Selection in the Seattle and Denver Income Maintenance Experiments," Technical Memorandum 1, Center for the Study of Welfare Policy, SRI International, Menlo Park, Calif.

Nagel, E. 1961. *The Structure of Science,* Hartcourt Brace & World, New York.

Neuberg, L. G. 1977a. "The Limits of Statistics in Planning Analysis," *Quality and Quantity* 11(1):1–26.

1977b. "Two Issues in the Municipal Ownership of Electric Power Distribution Systems," *Bell Journal of Economics* 8(1):303–23.

1978. "Municipal Ownership of Electric Utilities and the Fiscal Problems of the Cities," *Review of Radical Political Economics* 10(4):35–47.

1986. "What Can Social Policy Analysts and Planners Learn from Social Experiments?" *American Planning Association Journal* 52(1):60–74.

1987. "Randomization-Based Causal Inference and the Two-Sample t-Test," presented at the Harvard Statistics Department Colloquium, February 25, 1987, and at the M.I.T.–Harvard Econometrics Colloquium, March 19, 1987.

In press "Distorted Transmission: A Case Study in the Diffusion of Social 'Scientific' Research," *Theory and Society.*

Neyman, J. 1935. "Statistical Problems in Agricultural Experimentation," with comments by R. A. Fisher and others, *Supplement to the Journal of the Royal Statistical Society* 2(2):107–80.

Nicholson, W. 1977. "Differences Among the Three Sources of Income Data," in *The New Jersey Income Maintenance Experiment,* Vol. III, H. W. Watts and A. Rees (eds.), Academic Press, New York, pp. 353–74.

Nozick, R. 1974. *Anarchy, State, and Utopia,* Basic Books, New York.

Orcutt, G. H., and A. G. Orcutt. 1968. "Incentive and Disincentive Experimentation for Income Maintenance Policy Purposes," *American Economic Review* 58:754–72.

Polanyi, K. 1968. "The Self-Regulating Market and the Fictitious Commodities: Land, Labor, and Money," in *Primitive, Archaic, and Modern Economics,* G. Dalton (ed.), Beacon Press, Boston, pp. 26–27.

Popper, K. R. 1964. *The Poverty of Historicism,* Harper, New York.

1968. *The Logic of Scientific Discovery,* Harper, New York.

Putnam, H. 1984. "After Ayer, After Empiricism," *Partisan Review* 51(2): 255–75.

Quirk, J., and R. Saposnik. 1968. *Introduction to General Equilibrium Theory and Welfare Economics,* McGraw-Hill, New York.

Rawls, J. 1971. *A Theory of Justice,* Harvard University Press, Cambridge, Mass.

Richtmyer, F. K., E. H. Kennard, and T. Lauritsen. 1955. *Introduction to Modern Physics,* McGraw-Hill, New York.

Robins, P. K., and R. W. West. 1980. "Program Participation and Labor Supply Response," *Journal of Human Resources* 15(4):499–523.

Roemer, J. E. 1982. "Property Relations vs. Surplus Value in Marxian Exploitation," *Philosophy and Public Affairs* 11(4):281–313.

Roethlisberger, F. J., W. J. Dickson, and H. A. Wright. 1939. *Management and the Worker,* Harvard University Press, Cambridge, Mass.

Ross, H. 1966. "A Proposal for a Demonstration of New Techniques in Income Maintenance," Data Center Archive, Institute for Research on Poverty, University of Wisconsin, Madison.

Rubin, D. B. 1978. "Bayesian Inference for Causal Effects: The Role of Randomization," *Annals of Statistics* 6(1):34–58.

1980. "Comment [on Basu (1980)]," *Journal of the American Statistical Association* 75(371):591–3.

Rutherford, E. 1911. "The Scattering of α and β Particles by Matter and the

Structure of the Atom," *Philosophical Magazine and Journal of Science* 21(125):669–88.

Samuelson, P. A. 1965. *Foundations of Economic Analysis*, Atheneum, New York.

Savage, L. J. 1976. "On Rereading R. A. Fisher," *Annals of Statistics* 4(3):441–83.

Sawhill, I. V., G. E. Peabody, C. A. Jones, and S. B. Caldwell. 1975. "Analysis Using Data from the New Jersey Income Maintenance Experiment," in *Income Transfers and Family Structure*, The Urban Institute, Washington, D.C., pp. 45–72.

Scheffé, H. 1956. "Alternative Models for the Analysis of Variance," *Annals of Mathematical Statistics* 27(2):251–71.

Schrödinger, E. 1935. "Indeterminism in Physics," in *Science and the Human Temperament*, J. Murphy and H. W. Johnston (trans.), Norton, New York, pp. 52–80.

Schumpeter, J. A. 1950. *Capitalism, Socialism and Democracy*, Harper, New York.

Searle, J. 1974. "Chomsky's Revolution in Linguistics," in *On Noam Chomsky: Critical Essays*, G. Harman (ed.), Anchor Books, New York, pp. 2–33.

Seidenfeld, T. 1981. "Levi on the Dogma of Randomization in Experiments," in *Henry E. Kyburg, Jr. and Isaac Levi*, R. J. Bogdan (ed.), Reidel, Dordrecht, pp. 263–91.

Sen, A. K. 1970. *Collective Choice and Social Welfare*, Holden-Day, San Francisco.

Sills, D. L. (ed.) 1968. *International Encyclopedia of the Social Sciences*, Vol. 7, Macmillan and Free Press, New York.

Smith, A. 1961. *The Wealth of Nations*, Bruce Mazlish (ed.), Bobbs-Merrill, Indianapolis.

Solomon, R. L. 1949. "An Extension of Control Group Design," *Psychological Bulletin* 46:137–50.

Solow, R. M. 1985. "Economic History and Economics," *American Economic Review* 75(2):328–31.

Spiegelman, R. G., and K. E. Yaeger. 1980. "Overview [of the Seattle and Denver Income Maintenance Experiments]," *Journal of Human Resources* 15(4):463–79.

Stegmüller, W. 1979. *The Structuralist View of Theories*, Springer-Verlag, Berlin.

Stigler, S. M. 1974. "Gergonne's 1815 Paper on the Design and Analysis of Polynomial Regression Experiments," *Historia Mathematica* 1:431–47.

Stone, C. J. 1977. "Consistent Nonparametric Regression," *Annals of Statistics* 5(4):595–645.

"Student." 1937. "Comparison Between Balanced and Random Arrangements of Field Plots," *Biometrika* 39:363–79.

Suppes, P. 1970. *A Probabilistic Theory of Causality*, North Holland, Amsterdam.

Thomas, G. B., Jr. 1960. *Calculus and Analytic Geometry*, Addison-Wesley, Reading, Mass.

U.S. Bureau of the Census. 1984. *Statistical Abstract of the United States: 1985,* USBC, Washington, D.C.

U.S. Department of Health, Education, and Welfare. 1976. "The Rural Income Maintenance Experiment," SR10, USDHEW, Washington, D.C.

Vance, E. P. 1955. *Unified Algebra and Trigonometry,* Addison-Wesley, Reading, Mass.

Watts, H. W., R. Avery, D. Elesh, D. Horner, M. J. Lefcowitz, J. Mamer, D. Poirier, S. Spilerman, and S. Wright. 1974. "The Labor-Supply Response of Husbands," *Journal of Human Resources* 9(2):181–200.

Watts, H., and D. Horner. 1977. "Labor-Supply Response of Husbands," in *The New Jersey Income-Maintenance Experiment,* Vol. II, H. Watts and A. Rees (eds.), Academic Press, New York, pp. 57–114.

Watts, H., and J. Mamer. 1977. "Analysis of Wage-Rate Differentials," in *The New Jersey Income-Maintenance Experiment,* Vol. III, H. Watts and A. Rees (eds.), Academic Press, New York, pp. 341–51.

Weintraub, E. R. 1985. "Appraising General Equilibrium Analysis," *Economics and Philosophy* 1:23–37.

Whitehead, T. N. 1938. *The Industrial Worker,* Vols. I and II, Harvard University Press, Cambridge, Mass.

Winch, P. 1958. *The Idea of a Social Science,* Routledge & Kegan Paul, London.

Wonnacott, T. H., and R. J. Wonnacott. 1970. *Econometrics,* Wiley, New York.

Wood, T. B., and F. J. M. Stratton. 1910. "The Interpretation of Experimental Results," *Journal of Agricultural Science* 3(4):417–40.

Zelen, M. 1979. "A New Design for Randomized Clinical Trials," *New England Journal of Medicine* 300(22):1242–5.

Symbols and abbreviations

Entries are arranged in order of first appearance in the text.

Chapter 1

1.2.1

$P(X_t)$	probability of event X occurring at time t
$P(A_t|B_{t'})$	probability of event A occurring at time t given that event B occurs at time t'
$c(B_{t'})$	event B not occurring at time t'

1.2.2

IPES	*inductio per enumerationem simplicem* (form of induction explained by Mill)
A, B, C, D	causes
a, b, c	effects

1.3.1

h	placeholder for height from which mass is dropped
g	gravitational constant
h_0	value of h
DLW	deterministic law of working

1.3.3

GDL	gambling-device law
$\mathbf{v}_1, \mathbf{v}_0$	initial velocities
$\mathbf{p}_0$	initial position

Chapter 2

A bar (hat) above a symbol indicates a mean (estimate), S^2 (or s^2) a variance, and R a correlation. For example, $\overline{T}$ is the mean of the T_i's, $\hat{\alpha}_T$ is an estimate of α_T, s_c^2 is the variance of the c_j's, and R_{TC} is the correlation of the (T_i, C_i)'s.

345

2.2.1

$\binom{N}{n}$	number of ways of choosing n objects from N
S	vector of subjects
A	vector of subject treatments
T	vector of subject responses
r	vector of reported subject responses
N	number of subjects
n (or $N - n$)	number of observed treatment (or control) responses
T_i (or C_i)	subject i's response to treatment (or control)
t_i (c_i)	the ith observed treatment (control) response
PEI	first principle of equiprobability (distribution created by simple random assignment)
p	probability of

2.2.2

h	a function of the observations
H^*	a function of the T_i's and C_i's
E	expectation of
Var (Cov)	variance (covariance) of

2.2.3

s_p^2	pooled sample variance
B	$E(s_p^2)/\mathrm{Var}(\bar{t} - \bar{c})$
M	ratio of control- to treatment-group size
Q	$S_T S_C$
$\sim H$	not H
H	hypothesis of no mean causal effect

2.2.4

H_0	hypothesis of no causal effects
α	significance level
$N(h_0)$	number of possible observation sets with values of h at least as extreme as h_0
$\tilde{t}$ (or $\tilde{c}$)	median of $t_1, \ldots, t_n$ (or $c_1, \ldots, c_{N-n}$)
U	observed value is an unusual occurrence of
$\vee, \wedge$	or, and

2.2.5

In Section 2.2.5 only, the subscript N indicates for or from the Nth population of a hypothetical infinite sequence of populations of increasing size.

Π_i	ith of hypothetical infinite sequence of populations of increasing size
z_N	statistic for comparing treatment–control means

$U_{NT}(U_{NC})$	mean of T_{Ni}'s $(C_{Ni}$'s)
$F^2_{NT}(F^2_{NC})$	variance of T_{Ni}'s $(C_{Ni}$'s)
$T(C)$	limit of $\overline{T}_N(\overline{C}_N)$ as N approaches ∞
$V^2_T(V^2_C)$	limit of $S^2_{NT}(S^2_{NC})$ as N approaches ∞
$Y^2_T(Y^2_C)$	limit of $F^2_{NT}(F^2_{NC})$ as N approaches ∞
R	limit of R_{NTC} as N approaches ∞
γ	limit of (n/N) as n and N approach ∞
$N(0, \sigma^2)$	normal distribution with (mean, variance) $(0, \sigma^2)$
p_0	test result's p-value

2.3.1

N_j	number of subjects in block j
n_j (or $N_j - n_j$)	number of observed treatment (or control) responses in block j
T_{ji} (or C_{ji})	treatment (or control) response of ith subject of block j
t_{ji} (or c_{ji})	the ith observed treatment (or control) response of block j
PBE	principle of blocked equiprobability (distribution created by stratified random assignment)
Var_{PBE} (or Var_{PEI})	variance under PBE (or PEI)

2.3.2

$\mathcal{N}$	number of volunteers
$\mathcal{T}_i$ (or $\mathcal{C}_i$)	volunteer i's response to treatment (or control)
$\mathcal{H}^*$	a function of the $\mathcal{T}_i$'s and $\mathcal{C}_i$'s
PEII	second principle of equiprobability (distribution created by simple random selection from volunteers followed by simple random assignment)

2.3.3

r	reported value of
ϵ_i (or η_j)	reporting error $r(t_i) - t_i$ (or $r(c_j) - c_j$)
$\bar{r}(t)$ (or $\bar{r}(c)$)	mean of the $r(t_i)$'s (or $r(c_j)$'s)
t^*_i (or c^*_i)	the ith observed treatment (control) response (when there are technical errors)
ζ_i (or ξ_j)	technical error $t^*_i - t_i$ (or $c^*_j - c_j$)

2.3.4

X_{Ti} (or X_{Ci})	subject i's second response to treatment (or control)
x_{ti} (or x_{ci})	the ith observed second treatment (control) response
α_T (or β_T)	intercept (or slope) from ordinary-least-squares regression of T_i on X_{Ti}
α_C (or β_C)	intercept (or slope) from ordinary-least-squares regression of C_i on X_{Ci}

Chapter 3

A bar above a symbol indicates a mean; for example, $\overline{C}$ is the mean of the C_i's.

3.2.1

N	number of subjects
n (or $N - n$)	number of observed treatment (or control) responses
T_i (or C_i)	subject i's response to treatment (or control)
t_i (c_i)	the ith observed treatment (control) response
r	reported value of
ϵ_i (or η_j)	reporting error $r(t_i) - t_i$ (or $r(c_j) - c_j$)
$\bar{r}(t)$ (or $\bar{r}(c)$)	mean of the $r(t_i)$'s (or $r(c_j)$'s)
PEI	first principle of equiprobability (distribution created by simple random assignment)
F	probability distribution of reporting errors
PRES	principle of reporting-error symmetry

3.2.2

G_1, G_2	groups into which all patients are randomized under Zelen's design
A, B	treatments under Zelen's and conventional designs

3.2.3

N_j	number of subjects in block j
n_j (or $N_j - n_j$)	number of observed treatment (or control) responses in block j
T_{ji} (or C_{ji})	treatment (or control) response of ith subject of block j
t_{ji} (or c_{ji})	the ith observed treatment (or control) response of block j
PBE	principle of blocked equiprobability (distribution created by stratified random assignment)
$\bar{t}^+$ (or $\bar{c}^+$)	average of observed treatment (or control) responses weighted by block sizes

Chapter 4

The pair (μ, σ^2) designates the (mean, variance) of a probability distribution; a bar over a symbol (s^2) indicates a sample mean (variance), and a hat over a parameter indicates an estimate. For example, (μ_t, σ_t^2) is the (mean, variance) of the probability distribution generating the t_i's, $\bar{t}$ (s_t^2) is the sample mean (variance) of the t_i's, and $\hat{\theta}$ is an estimate of the parameter θ.

4.2.1

T_i (or C_i)	random variables with value t_i (or c_i)
n, m	sample sizes

p		probability of
Π		product

4.2.2

DLW		deterministic law of working
π_r		proportion of r imagined repetitions for which
LRFA		limiting relative-frequency account

4.2.3

S_r		sum of r Bernoulli trial results

4.2.4

$\mathscr{P}$		class of probability distributions
Θ		parameter space of indices θ for elements of $\mathscr{P}$
H (or Θ_H)		subclass of $\mathscr{P}$ (or Θ) for which hypothesis H is true
ϕ, ϕ^*		test procedures for H
R_ϕ (or A_ϕ)		acceptance (or rejection) region for ϕ
α		significance level
β		probability of accepting H when it is false
Φ_α^H		class of level-α tests for H
UMP		uniformly most powerful
Φ_α^{HU}		class of unbiased level-α tests for H
UMPU		uniformly most powerful unbiased

4.2.5

c_0		a known constant
t^*		statistic for comparing two population means
s_p^2		pooled sample variance
ϕ^+		Student's two-sample t test of H: $\mu_t = \mu_c$
$t_{(i)}$		density function of Student's t distribution with i degrees of freedom
C_α		critical value for ϕ^+
δ		$(\mu_t - \mu_c)/\sigma[(1/n) + (1/m)]^{1/2}$
$t_{(d,\,(i))}$		density function of noncentral t distribution with (noncentrality parameter, degrees of freedom) (d, i)

4.3.1

F, G		probability distribution functions
Δ		shift parameter
R_i (or S_i)		rank of t_i (or c_i) among the $t_1, \ldots, t_n; c_1, \ldots, c_m$
Φ^{HR} (or Φ_α^{HR})		class of all (or level $-\alpha$) rank tests for H
[ ]		greatest integer $\leq$

Φ_α^{HRU-} class of unbiased level-α rank tests for H for which β is a
decreasing function of Δ for any continuous F

4.3.2

$N(0, \sigma^2)$ normal distribution with (mean, variance) $(0, \sigma^2)$
ϕ_i test of H: $\mu_t = 0$
C_α^i critical value for ϕ_i
g standard normal density function
g_δ normal density function with (mean, variance) $(\delta, 1)$

4.3.3

ρ third absolute moment of F

4.3.4

$D(Y, X)$ bivariate probability distribution function
E expectation of
$E(Y|X = x), f(x)$ conditional expectation of Y given $X = x$
a (or b)   intercept (or slope) of $f(x)$ when it is linear in x
$\mathcal{L}(\theta)$ likelihood function of parameter θ
MLE   maximum-likelihood estimate
W_i^*, W_i ith weight

Chapter 5

A dot above a symbol means differentiation with respect to time. $X = |\mathbf{X}| =$ length of $\mathbf{X}$ for any listed vector $\mathbf{X}$. The subscript i on a symbol indicates particle or time, depending on context.

5.1.3

$\mathbf{F}, \mathbf{F}^*$ forces
NSL   Newton's Second Law
m (or t)   mass (or time)
$\mathbf{r}$ position, or position of particle 2 with respect to 1
(x, y, z) Cartesian coordinates of $\mathbf{r}$
$\nabla \cdot$ $[\partial/\partial x + \partial/\partial y + \partial/\partial z]$
V potential energy
$\mathcal{L}$ (or $\mathcal{S}$)   Lagrangian (or action)
HP   Hamilton's Principle
$\mathbf{g}$ gravity constant
EFH   empirically falsifiable hypothesis
M^* (or F^*)   the particular mass (force) function is so and so
M all masses are constant

5.2.1

v	average speed in x direction
r^*	reported value of
ϵ, η	reporting errors
E	expectation of

5.2.2

$\mathbf{F}_{iext}$	external force acting on particle i
$\mathbf{F}_{ji}$	internal force on particle i due to particle j
(r, θ)	polar coordinates of $\mathbf{r}$
$'$	indicates post-collision value

5.2.3

G	gravity constant
NLUG	Newton's Law of Universal Gravitation
μ	reduced mass
λ	angular momentum with respect to center of mass
T^+	kinetic energy with respect to center of mass
E^*	total energy
dA/dt	areal velocity
(a, c)	(semi-major axis, distance from center to focus) of ellipse
e	c/a
λ_0, E_0^*	constants over time
τ	period
τ_{1j}	planet j's period (or semi-major axis) for an orbit around the sun

5.2.4

R	radius of earth
x^*-y^*	plane perpendicular to $\mathbf{r}$ and tangent to earth
z^*	height of m_2 above earth's surface
h	height from which m_2 is dropped
x_0^*, y_0^*	constants over time

5.2.5

$\mathbf{T}$	tension force
$\mathbf{L}$	directed pendulum length
β	angle of pendulum with vertical
β_{max}	largest value β takes
$\mathbf{u}$	unit vector perpendicular to $\mathbf{L}$

5.3.1

$\sim$	is distributed as
N	number of molecules or particles

(R^*, V^*, T^*, P)	(moles, volume, temperature, pressure)
$N(K_i, \sigma^2)$	normal distribution with (mean, variance) (K_i, σ^2)
K	Boltzmann's constant $\times$ Avogadro's number

5.3.2

q_i	charge
ME	Maxwell's Equations
$\dot{r}_0$	m_1's initial velocity
s	perpendicular distance from m_2 to $\dot{r}_0$
E^*_{Rad}	radiated energy
φ	scatter angle
N_φ	number of particles hitting solid angle $\Delta\varphi$
d	foil thickness

5.3.3

e^*	electron charge
c	speed of light
v	frequency of emitted radiation

5.3.4

CL	Coulomb's Law
n	positive integer
h^*	Planck's constant
W_i	energy level i
EFC	Einstein Frequency Condition
R^+	$2\pi^2\mu e^{*4}ch^{*3}$
U_i	ith line in series of spectral emission lines
$d(U_i, U_j)$	distance between lines U_i and U_j

Chapter 6

Subscript 0 on listed symbol designates value of variable. Subscript $j (\geq 1)$ on R_i, Y, P, Q, R_i^*, or Y_i^* designates jth observed value. For example, x_{i0} and Q_0 are values of x_i and Q, respectively, and R_{13} is the third observed value of R_1.

6.2.1

q_i (or p_i)	quantity (or price) of ith consumption commodity
Y^+	consumer's money income
u (or U)	consumer's utility
$\mathcal{Q}$	consumption possibility set
R^1	the set of real numbers
UM	principle of utility maximization

MUC	marginal utility condition
NSL (or HP)	Newton's Second Law (or Hamilton's Principle)

6.2.2

UCAM	unique constrained absolute maximum on $\mathscr{D}$
$q_i(p_1, p_2, Y)$	individual consumer demand for the ith commodity
SE	Slutsky's Equation
F	a monotonic differentiable function
Y	the distribution of income $(Y^1, \ldots, Y^N)$
$q_j^i(p_1, p_2, Y^i)$	ith individual consumer's demand for the jth commodity
q_{Dj}	market consumer demand for jth commodity

6.2.3

x_i (or r_i)	quantity (or price) of ith production input
$\mathscr{X}$	production input possibility set
q (or p)	quantity (or price) of production output
Π (or f)	firm's profits (or production function)
PM	principle of profit maximization
MPC	marginal product condition
UAM	unique absolute maximum on $\mathscr{X}$
$q(p, r_1, r_2)$	firm's supply function
q_S	firms' market supply function

6.2.4

q_{ki}	ith consumer's consumption of kth commodity
q^j	jth firm's production
q_D	market consumer demand for production output
$\mathscr{C}$ (or $\mathscr{P}$)	condition that specifies N (or M) log-linear utility (or production) functions with equal marginal utility (product) ratios
U^i	ith consumer's utility
α_i, β_{ki}'s (or γ_i, δ_{ki}'s)	parameters of ith consumer's (firm's) log-linear utility (production) function
x_{ki}	ith firm's quantity of kth production input
(R_i, Y, P, Q_D, Q_S)	$(\ln(r_i), \ln(\Sigma_{i=1}^N Y^i), \ln(p), \ln(q_D), \ln(q_S))$
S_2 (or S_1)	a market demand and supply function (or curve)
a_{iD}'s (or a_{is}'s)	parameters of market demand (or supply) function or curve of S_2 or S_1
K_{2D} (or K_{2S})	intercept of market demand (or supply) curve of S_1
S_4 (or S_3)	UM and budget constraint (or MUC) for each consumer, PM (or MPC) for each producer, and $\mathscr{C}$ and $\mathscr{P}$ all hold

6.3.1

MEC	market equilibrium condition
q^*	observed amount of commodity
Q	$\ln(q^*)$
D	determinant with rows $(Y_i, R_{1i}, R_{2i}, 1)$ for $i = 1, \ldots, 4$
D_j (or C_j)	D with the jth column replaced by P_i's (or Q_i's)

6.3.3

ϵ, v	measurement errors
E	expectation of
r^*	reported value of
Q^+ (or R_i^*)	$Q + P - Y$ (or $R_i - P$)
Δ	determinant with rows $(R_{1i}^*, R_{2i}^*, 1)$ for $i = 1, 2, 3$
Δ_j	Δ with the jth column replaced by Q_i's
OLSQ	ordinary least squares

6.3.4

$\mathscr{C}'$	$\mathscr{C}$ for two consumers without the restriction of equal marginal utility ratios
Y_1^* (or Y_2^*)	$\ln(Y^1)$ (or Y^2/Y^1)
β_i	$\beta_{1i}/(\beta_{1i} + \beta_{2i})$
S_2'	a second market demand and supply function
a_{iD}''s (or a_{iS}''s)	parameters of market demand (or supply) function of S_2' or S_2''
K_{6D}	constant term in market demand function of S_2'
S_4' (or S_3')	S_4 (or S_3) with $\mathscr{C}$ replaced by $\mathscr{C}'$
S_i''	S_i' with approximate market demand function
θ (or x)	small angle (or positive number less than but close to 1)
D'	determinant with rows $(Y_{1i}^*, Y_{2i}^*, R_{1i}^*, R_{2i}^*, 1)$ for $i = 1, \ldots, 5$
D_j' (or C_j')	D' with jth column replaced by P_i's (or Q_i's)
$\theta_{\max}$	maximum value of θ

6.3.5

η_i's, ρ_{ij}'s	parameters of translog consumer utility function
a_{ji}'s, b_{ji}'s	parameters of budget-share equation
Q_i^*	total amount of ith commodity consumed

Chapter 7

A bar (hat) above a symbol indicates a mean (estimate). Subscripts 0 and $*$ designate the value of a variable, p and b (e and NIT) indicate before (during) NIT.

A dot " $\cdot$ " indicates a dot product, and Δ indicates a change in. Pieces of all listed piecewise functions depend on before- and during-NIT tax brackets of subscript-indicated party. In some cases, a symbol has a meaning for certain sections only (as indicated).

7.2.1

MUC (or MPC)	marginal utility (or product) condition
(T, L, y_d, h, w, y)	(available hours, leisure hours, disposable income, work hours, wage rate, nonwage income)
u, u^+, u_2, u^*	utility from leisure and disposable income
$\mathscr{L}$	leisure–disposable-income possibility set
R^n	direct product of n real lines
UM	principle of utility maximization
q_i (or p_i)	quantity consumed (or price) of ith commodity
u_1	consumer's utility
CAM (or UCAM)	(unique) constrained absolute maximum
v	indirect utility
$\vert_{u=\text{constant}}$	value for constant utility
SE	Slutsky's Equations
R^+, r_i	calculus and analytic geometry variables
f	function of
$\mathscr{R}$	a subset or region of R^2
c	constant

7.2.2

(E, t_F)	federal income tax (exemption level, fixed rate)
(w_g, y_g)	pre-tax (tax rate, taxable nonwage income)
(t_n, G)	negative income tax (rate, guarantee level)
$h[\ \ ]$ (or $H[\ \ ]$)	individual labor-supply function (or correspondence)

7.2.3

$(L_i, h_i, T_i, w_{gi}, c_i)$	(L, h, T, w_g, c) for ith family member, where $i = 1, 2, 3$ are wife, husband, and child, respectively
D_i	determinant obtained by bordering $(i + 2)$th principal minor of Hessian determinant of u's second-order partials by row and column containing u's first-order partials, and putting 0 in southeast corner
$\mathscr{C}$ (or $\varnothing$)	set of CAMs of u (or empty set)

7.3.1

T^*	$w_{g1}T_1 + w_{g2}T_2 + w_{g3}T_3$
Δh_i	family member i's change in hours worked with NIT

MEC	market equilibrium condition
S	no (w_{gi}, T_i) change with NIT, $y_g = 0$, perfectly elastic market demand for low-income labor
$\mathcal{P}_1$	specifies log-linear family utility function $u(L_1, L_2, L_3, y_d)$
k, β_i	parameters of $\mathcal{P}_1$'s family utility function
subscript k	indexes families or individual workers
Q_{ik}	piecewise function of (G, t_n, w_g, t_F, E) for ith member of kth family
β_i (or $\mathbf{x}_i$)	vector of parameters (or demographic variables) for ith family member
ϵ_i	measurement-error term for ith family member
$\sim N(a, b)$	distributed normally with (mean, variance) (a, b)
(L_f, L_m, L_o, Y)	(L_1, L_2, L_3, y_d) (Section 7.3.1 only)
$(t, \Delta H_m, \beta_m)$	$(t_n, \Delta h_2, \beta_2)$ (Section 7.3.1 only)
W_m	w_{g2}
Q	$G/(1 - t_n)w_{g2}$
$(\Delta H, \mathbf{B}'_1, \overline{\mathbf{X}}_1)$	$(\Delta h_2, -\boldsymbol{\beta}_2, \mathbf{x}_2)$
$(\mathbf{B}'_1\overline{\mathbf{X}}_1, c_m, e)$	$(-\boldsymbol{\beta}_2 \cdot \mathbf{x}_2, c_2, \epsilon_2)$
$\mathbf{B}'_0$ (or $\mathbf{X}_0$)	vector of parameters (or demographic variables)
$\mathbf{B}'_0\overline{\mathbf{X}}_0$	$\mathbf{B}'_0 \cdot \overline{\mathbf{X}}_0$

7.3.2

NSL (or NLUG)	Newton's Second Law (or Universal Law of Gravitation)
W	observed wage rate
A^* (or X_{0j})	age
(E, L) (or (E_m, L_m))	dummy variables for husband's (education, health) (or missing (education, health) status) (Section 7.3.2 only)
P	weeks worked by husband in year prior to experiment (Section 7.3.2 only)
H_0	hypothesis of 0 distributional mean (Section 7.3.2 only)
C_w (or Y)	vector of parameters (or demographic variables)
ζ	measurement-error term
Var	Variance of
OLSQ	ordinary least squares
AFDC-UP	aid to families with dependent children and unemployed father

7.3.3

Δh	change in worker's hours worked with NIT
S^*	no (w_g, T) change with NIT, perfectly elastic market demand for low-income labor

$\mathscr{P}_2$	specifies individual-worker utility function linear (quadratic) in $y_d\,(L)$
α, β	parameters of $\mathscr{P}_2$'s individual-worker utility function
Q_k^+	piecewise function of (w_g, t_n, t_F) for kth individual worker
$\boldsymbol{\alpha}$ (or $\mathbf{x}$)	a vector of parameters (or demographic variables)
ϵ'	measurement-error term
(U, H, Y_d, Y_n)	(u, h, y_d, y) (Section 7.3.3 only)
k_1, k_2	constants
η_i	a function of $(\Delta r_1, \Delta r_2)$
R_k	piecewise function of $(H_p, w_g, t_n, G, t_F, E)$ for kth individual worker
FABOVE	dummy variable for family payment eligibility and above breakeven level
BREAK	breakeven level of family earnings
EARNABV	amount of family earnings beyond breakeven level
r, t_e	constants (Section 7.3.3 only)
h_{NIT} (or h_b)	hours worked during (or before) the NIT
Q_k', Q_k'' (or R_k', R_k'')	piecewise functions of (w_g, r, t_e, y_g, t_F) (or (r, t_e, y_g, t_F)) for kth individual worker
c^* (or $\mathbf{c^*}$)	parameter (or vector of parameters)
DECLINE	dummy variable for declining tax rate family
C (or M)	set of assignment-control (or manpower) variables
A_2	conditions on utility and labor-supply functions; UM, S^*, and MEC hold
B_2, B_2', B_2''	conditions specifying different budget constraints
b_i's	coefficients in Seattle-Denver ΔH relation
ϵ	error term

7.3.4

W	gross wage rate
t_p (or t_e)	pre-NIT (or during NIT) tax rate
$\mathbf{b_1^*}$ (or $\mathbf{C}$)	vector of parameters (or control variables)

7.3.5

r (or t)	average non-NIT (or NIT) tax rate on earnings (Section 7.3.5 only)
N	non-NIT nonwage income
B_0	NIT benefit per month at zero hours
$\mathscr{P}_3$	specifies an individual-worker utility function
w_{nbk} (or $w_{n\mathrm{NIT}k}$)	piecewise function of (w_g, t_F) (or (w_g, t_n, t_F)) for kth individual worker

y_{nbk} (or $y_{n\text{NIT}k}$)	piecewise function of (y_{gb}, t_F, E) (or $(y_{g\text{NIT}}, t_n, G, t_F, E)$) for kth individual worker
γ	vector of parameters
θ (or x)	small angle (or positive number below but close to 1)
$\delta', \delta'', \eta', \eta''$	parameters of once-modified H relation
D	dummy variable for preenrollment observation
a through h	parameters of twice-modified H relation (Sections 7.3.5 and 7.3.6 only)
$\mathbf{a}$	vector of parameters
A	$L \neq 0, T$
H^*	condition on hours-worked function
B, B^*	conditions specifying budget constraints before and during NIT

7.3.6

SMSA	standard metropolitan statistical area
TOBIT	estimating procedure for regression coefficients

7.3.7

k, α, β	parameters of log-linear individual worker's hours-worked function (Sections 7.3.7 and 7.3.8 only)
V	condition specifying individual-worker indirect utility
z	vector of demographic variables
$\alpha^+, \beta^+, \delta$	vectors of parameters
subscript i	indexes individual workers (Sections 7.3.7 and 7.3.8 only)
δ^*	parameter
$\epsilon_\alpha, \epsilon_\beta, \epsilon_\delta, \epsilon^*, \epsilon_1, \epsilon_2$	stochastic error terms
K^*	condition specifying k as varying with individual-worker tastes and preferences
exp	exponent of
$(z_i, \delta, z_i\delta)$	$(\mathbf{z}_i, \boldsymbol{\delta}, \mathbf{z}_i \cdot \boldsymbol{\delta})$
σ_2^2	variance of ϵ_2
$\bar{\alpha}$	constant value of α_i for every i
B^+	condition specifying β_i stochastically
$\mathcal{TN}(0, \sigma_1^2)$	truncated normal distribution with mean 0 and variance σ_1^2
$(\mu_\beta, \sigma_\beta^2)$	(−truncation point, variance) of truncated distribution
(w_1, y_1, w_2, y_2)	$(w_g(1 - t_n), y_g(1 - t_n) + G, w_g(1 - t_F), w_g(1 - t_n) + t_F E)$
β_i^*	value of random variable β_i
$(-\infty, \beta_i^*), (\beta_i^*, \infty)$	open intervals
$C_i(\delta, \bar{\alpha})$	$e^{z_i\delta}(w_{1i}^{1+\bar{\alpha}} - w_{2i}^{1+\bar{\alpha}})$
$1 - \beta_i^*$	x_i (Section 7.3.7 only)
a_{1i} (or a_{2i})	y_{1i} (or y_{2i})

x_i^*	closed-form solution of equation (7.1)
A^*	condition specifying approximation for $\beta_i^*(\delta, \bar{\alpha})$
$\varphi(\ )$ (or $\Phi(\ )$)	standard normal density (or distribution) function
H	hours worked (Section 7.3.7 only)
PC (or PNC)	probability of observing H before (during) NIT
(w_3, y_3) (or $\tilde{\tilde{H}}$)	(w_g, y_g) (or $(E - y_g)/w_g$)
BL (or BU)	function of (w_3, y_3) (or (w_2, y_2))
w_j (or y_j)	$\ln(w_j)$ (or $\ln(y_j)$) for $j = 1, 2, 3$
$R(a, b, c)$	function of algebraic unknowns a, b, c (Section 7.3.7 only)
$X(BU, BL)$	function of (BU, BL) (Section 7.3.7 only)
H^+	condition specifying log-linear hours-worked function
J	UM, S^*, V, MEC, A^*, conditions on (G, T, E, t_n, t_F, w_g)
A_1^* (or B_1^*)	$\alpha = \alpha^+ \cdot \mathbf{z}$ (or $\beta = \beta^+ \cdot \mathbf{z}$)

7.3.8

δ_j	jth component of δ
IRS	Internal Revenue Service
$w_s(\alpha')$	spouse's wage rate (parameter) in log-linear hours worked function

Chapter 8

The term "$a \le b$" means that a is less than or equal to b if a and b are real numbers, a is a subset of b if a and b are sets, and every a's ith component is less than or equal to b's if a and b are vectors. A dot "$\cdot$" indicates dot product. A colon means "for which."

8.2.1

R^n	Cartesian product of n real lines
X, Y	subsets of R^n
$\mathbf{x}, \mathbf{y}$	points in R^n
$\in$	is an element of
Ω	set of elements of R^p whose components are all nonnegative
$\{\mathbf{x}_n\}_{n=1}^{\infty}$	an infinite sequence of elements of R^n
λ	a real number between 0 and 1, inclusive
$\varnothing$	the empty set
$\mathscr{P}(Y)$	the set of subsets of Y
r	an ordered binary relationship on a subset of R^n
$\mathbf{x}_g$	greatest element of X for r
Q_i	consumption possibility set of consumer i
$\mathbf{x} \le_i \mathbf{y}$	$\mathbf{x}$ is not preferred to $\mathbf{y}$ by consumer i

$\mathbf{x} \sim_i \mathbf{y}$	$\mathbf{x} \leq_i \mathbf{y}$ and $\mathbf{y} \leq_i \mathbf{x}$
$\mathbf{x} <_i \mathbf{y}$	$\mathbf{x} \leq_i \mathbf{y}$, and it is not true that $\mathbf{y} \leq_i \mathbf{x}$
$(\mathbf{p}, \mathbf{Y})$	consumers' (prices, income distribution)
$\mathbf{q}_i, \mathbf{q}_i', \mathbf{q}_i^1, \mathbf{q}_i^2$	elements of Q_i
$\mathbf{q}_i \geq_i$ (or $>_i$) $\mathbf{q}_i'$	$\mathbf{q}_i' \leq_i$ (or $<_i$) $\mathbf{q}_i$
t	a real number between 0 and 1, exclusive
PS (or CPP)	principle of preference satisfaction (or "parametric" consumer prices)
Y_i	ith component of $\mathbf{Y}$ or consumer i's income
$\mathbf{A, B, C}$	elements of a three-dimensional consumption possibility set
(q_{hi}, q_{jk})	element of a two-dimensional consumption possibility set
u_i	ith consumer's utility
S_i	set of all $(\mathbf{p}, \mathbf{Y})$ for which there is at least one $\mathbf{q}_i \in Q_i$ satisfying consumer i's budget constraint
T_i	nonempty set
ϵ_i	consumer i's individual consumer demand correspondence
ϵ	market demand correspondence

8.2.2

$\mathbf{0}$	0 element of R^n
$\mathbf{x} > 0$	each component of $\mathbf{x} \geq 0$, and at least one component of $\mathbf{x} > 0$
Q_i^*	production possibility set of producer i
$\mathbf{q}_i^*, \mathbf{q}_{i0}^*$	elements of Q_i^*
Q^*	entire production sector's possibilities
S^p	set of possible price vectors
PM (or PPP)	principle of profit maximization (or "parametric" producer prices)
t_0	real number > 1
max	maximum of
η_i	producer i's firm supply correspondence
η	market supply correspondence

8.2.3

ω_i	consumer i's initial endowment
ω	economy's initial endowment
θ_{ij}	consumer i's share of profits in producer j
E (or POE)	private-ownership economy
$\mathbf{q}_i^+, \mathbf{q}_i^0$	elements of Q_i
$\mathbf{q}_j^{*+}$ (or $\mathbf{p}^+$)	elements of Q_j^* (or S^p)
$a \cap b$	all elements in both sets a and b

| (P1, P2) | two producers |
| (C1, C2) | two consumers |

8.3.1

A	set of attainable states of a private-ownership economy
$\mathbf{a} \leq \mathbf{b}$	$\mathbf{b}$ is Pareto-superior to $\mathbf{a}$, where $\mathbf{a}$ and $\mathbf{b}$ are in A
I_0	an individual

8.3.2

$(\omega_{i1}, \omega_{i2})$	ω_i for a two-commodity, two-consumer economy
P	the set of Pareto-efficient allocations
$O_1 O_2$	contract curve
BC	subset of contract curve
M_1, M_2, M_1^*, M_2^*	attainable states of two-commodity, two-consumer economy
$\mathbf{p}^*$	a price vector

8.3.4

$\mathscr{S}$	action
$(\mathbf{r}_i, \dot{\mathbf{r}}_i, m_i)$	(position, velocity, mass) of particle i
t^*	time
S_*^p	set of possible price vectors whose components add to 1
$\mathbf{z}(\ )$	excess demand function
k	real constant > 0
E^*	set of equilibria
GS	gross substitutes
z_i	ith component of $\mathbf{z}$
G_i	sign-preserving function of z_i
$\mathbf{p}(0)$	initial value of $\mathbf{p}$

8.3.5

D, D'	assumptions
A_w (or A_s)	set of weak (or strong) assumptions
$A_{w'}$	weaker set of assumptions than A_w
WGS	weak gross substitutes

8.4.1

| GET | general equilibrium theory |

Conclusion

C.3.1

| G | identity transformation |
| K_D, K_S | constants |

a_D, a_S, b_S, c_S	constants
$p(t)$	price of consumption commodity at time t
$r_i(t)$	price of production commodity i at time t
G	vector whose ith component G_i transforms the ith component of excess demand, but preserves its sign

C.3.2

$(\mathbf{p}, t)$	(prices, time)
G^*	universal gravity constant
$z_i(\mathbf{p}, t)$	ith component of excess demand
G_i	ith component of **G**

C.3.3

m	point mass
L (or $\mathbf{L_0}$)	spring directed length (or rest length)
K	spring constant
F	force exerted by spring on m
$\cos(\ )$	cosine of
A, δ	constants
$|\mathbf{L_0}|$	magnitude of $\mathbf{L_0}$

Index

approximation, 105–7, 133–4, 172, 174,
175, 302
Arrow, K., 247, 257, 260, 272, 273, 278
assumptions: in causal inference, 39–42,
54–6, 59t, 288; in modern statistical
inference, 10, 112–13; in Neyman–Pearson theory, 96, 98, 100–7, 110; in
regression point estimation, 107–10;
statistical, in classical mechanics, 126,
128, 147; in welfare microeconomics,
10, 274–80
attachment difficulties: in classical
mechanics, 144–6; in income maintenance experiments, 200–5, 214–18,
223–6, 232–4, 237–41, 244–5; in
microeconomics, 163–77, 179–81,
244–5; rationalist mode of inquiry and,
309

behavior, problem of: atomization and
individuality and, 291–4, 312;
Chomsky's linguistics and, 298–301;
consciousness and will and, 294–8, 312

causal inference: assumptions for, 39–42,
54–6, 59t, 288; difficulties with, in
controlled social experiments, 61–4; and
hypothesis testing, 47–51; stratified
assignment in, 50–2, 71–3; summary of
theory of, 59–60t
causality: counterfactual conditionals and,
25–6, 115; inconsistency and circularity
and, 19–22; observations and, 22–3; two
concepts of, 17–19
causal point estimation: with disproportionate stratified assignment, 71–3;
extended, 50–8; impediments to, in controlled social experiments, 64–73;
impediments to, in income maintenance
experiments, 73–82; theory core of, 43–5
Chomsky, N., 288, 299–301, 310
classical mechanics: approximation in,
133–4, 302; attachment difficulties in,
144–6; Bohr's hydrogen model and,
140–1; classical hydrogen model in,

139–40; falling bodies in, 132–3;
Hooke's Law in, 305–6; ideal-gas law in,
135–6; law of conservation of linear
momentum in, 127–9; Newton's laws
in, 122–7, 129–32; parallels between
microeconomics and, 10, 179t, 237t,
301–8; pendulum motion in, 133–4;
placeholder relations in, 124, 182–3;
Rutherford scattering in, 136–8;
statistical inference in, 148f; summary
of formalization of, 141–4
collocation, 26–7
complexity, 288–91, 302
controlled social experiments: as basis for
public policy, 313–16, difficulties in,
61–4, 287–91; good and true in conflict
in, 67–71; impediments to causal point
estimation in, 64–73; income maintenance experiments as, 5; problem of
behavior and, 291–301, 312; problem of
time and, 288, 289, 301–8, 312–13; two
senses of "experiment" and, 3–4; *see
also* income maintenance experiments

Debreu, Gerard, 247, 249–53, 257
deductive logic: breaches of, in income
maintenance experiments, 196–9,
205–14, 218–23, 234–7; in classical
mechanics, 141–4; comparison of, in
classical mechanics, microeconomics,
and income maintenance experiments,
237t; in Gary experiment second study,
226–32; in microeconomics, 178–9; in
science generally, 119–21, 141–2
deterministic law of working (DLW), 27, 42
die roll, 30–2, 93
distributions, *see* probability distributions
DLW, *see* determinstic law of working
(DLW)
double-blind experiments, 61–3, 65–7

empiricism, 10, 21–3, 116, 287, 308–9;
attachment difficulties and, 288;
beginning immanent critique of, 34f;
Hume and, 89

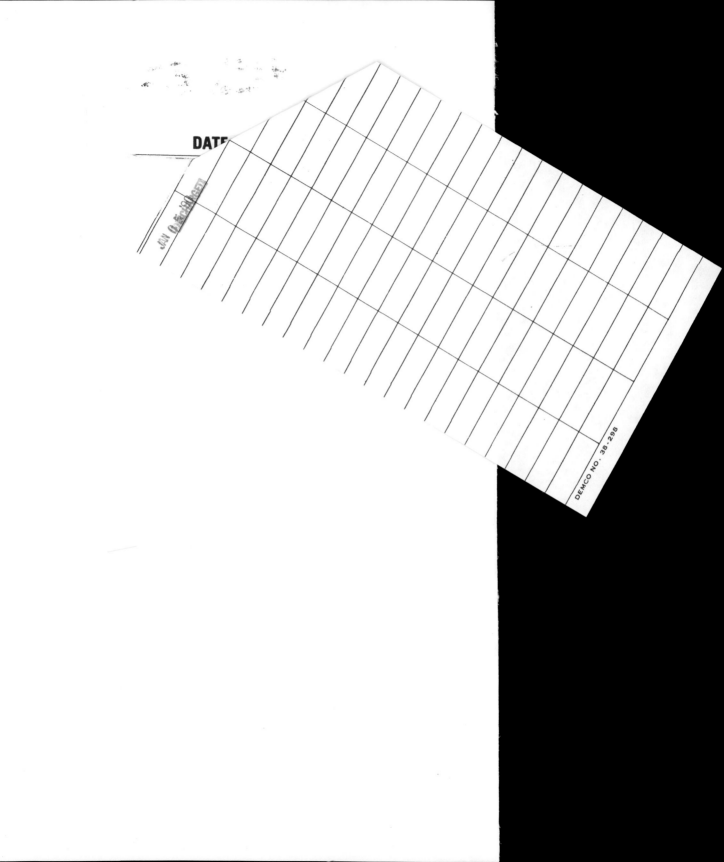

DATE

DEMCO NO. 38-298

The
Destructive
Achiever

The Destructive Achiever

POWER AND ETHICS IN THE AMERICAN CORPORATION

Charles M. Kelly

ADDISON-WESLEY PUBLISHING COMPANY, INC.
Reading, Massachusetts   Menlo Park, California   New York
Don Mills, Ontario   Wokingham, England   Amsterdam   Bonn
Sydney   Singapore   Tokyo   Madrid   San Juan

For permission to reprint previously copyrighted material, grateful acknowledgment is made of the following:

The American Management Association, for material adapted from Charles M. Kelly's article "The Interrelationship of Ethics and Power in Today's Organization," *Organizational Dynamics,* Summer 1987. © 1987, American Management Association, New York. All rights reserved.

The American Society for Training and Development, for material reprinted from *Training and Development Journal.* Copyright 1984, American Society for Training and Development. Reprinted with permission. All rights reserved.

The Regents of the University of California, for material reprinted/condensed from David A. Whitsett and Lyle Yorks's article "Looking back at Topeka: General Foods and the quality-of-work-life experiment." © 1983 by the Regents of the University of California. Reprinted/condensed from the *California Management Review,* Vol. 25, No. 4. By permission of the Regents.

Dow Jones & Company, Inc., from excerpts from *The Wall Street Journal.* Reprinted by permission of *The Wall Street Journal.* © 1985, Dow Jones & Company, Inc. All rights reserved.

Executive Enterprises, Inc., for the excerpt from Charles M. Kelly's article "The fusion process for productivity improvement," *National Productivity Review,* Volume 2, Spring 1983. Copyright 1983, Executive Enterprises, Inc., New York, New York.

The Foundation for the School of Business at Indiana University, for excerpts from Richard Chewning's article "Can Free Enterprise Survive Ethical Schizophrenia?" in *Business Horizons.* Copyright © 1984 by the Foundation for the School of Business at Indiana University. Reprinted by permission.

Continued on page vi

Library of Congress Cataloging-in-Publication Data

Kelly, Charles M.
   The destructive achiever : power and ethics in the American
corporation / Charles M. Kelly.
     p. cm.
   Includes bibliographical references and index.
   ISBN 0-201-09039-2
   1. Business ethics. 2. Success in business. I. Title.
  HF5387.K46 1988
  658.4'094—dc19   87-31910
                                        CIP

Sponsoring editor, Scott Shershow
Production supervisor, Perry McIntosh
Cover design by Marge Anderson
Text design by Diana Eames Esterly
Set in 10 point Galliard by Compset Inc., Beverly, Massachusetts

ISBN 0-201-9039-2

ABCDEFGHIJ-DO-898
First printing, March 1988

Contents

Introduction

An Important Definition:

Culturally we are suffering from the ravages of a *metaphysical* can-cer—a psychological rejection mechanism that questions the pos-sibility of anyone's being able to know right from wrong. . . . The new ethic is at its very heart an *anti-ethic*.

—*Richard C. Chewning*
Professor of Business Ethics
University of Richmond

Despite our recently renewed interest in business ethics, and several much-pub-licized scandals in a variety of industries, we are usually assured that ethics is no more of a problem for today's organizations than it has ever been. Robber bar-ons, it is argued, have always been with us. Indeed, they may even be a healthy stimulus to our free enterprise system. The fact that so many organizational misdeeds make the headlines today, according to many, is merely the result of better surveillance and reporting.

But I disagree with these assumptions. Today, the business climate is dif-ferent. The robber barons of the past were individuals on the make, and they functioned outside of the accepted norms of business and society. Today, the new generation of robber barons—the Destructive Achievers—are being devel-oped on the groupthink level. They are the product of our educational and organizational systems. Values and behaviors that were once condemned, or considered unseemly, are, in today's organization, considered essential.

I recently conducted a quality improvement session in a major corporation with mid- and lower-level personnel who, a month before, had attended a company workshop on "image projection." There they were told by the consultants that, for advancement, performance doesn't matter. All that counts is "image." In effect, the corporation told its employees through its consultants that any improved technical performance on their part would not be rewarded!

It was a difficult group. Some members were clearly hostile toward any outside attempts to make their organization more effective by addressing its quality problems. Others simply seemed resigned to what they regarded as just one more exercise in futility—one that would have no real impact on their politically-dominated environment. Earlier, the same group had gone through a "survey-feedback" program and a series of nonproductive problem-solving meetings. Some of the employees had expressed their anger, to the point of shouting, about the destructive politics and the way the wrong persons were getting rewarded in their company. And what was the result? The company hired consultants to tell the employees that they were just naive about corporate life, and to teach them how to better project an "image" of power and leadership. The consultants succeeded. But what did their new insights about the apparent reality of corporate life do to the employees' values? In other words, we're becoming "realistic" about our failure to reward the right people for the right reasons. We're teaching people how to exercise power and influence and trying to make everyone equal in so-called political "sophistication." But the problem with this egalitarian effort is that those who benefit most are the "naturals"—the egocentric individuals who are already devoted to their own success, regardless of the consequences for the groups and organizations of which they are a part.

Apparently, our society today has almost reached William Scott and David Hart's ominous 1979 projection of our "probable future":

> It has been the residue of commitment to the individual values of our past that has prevented the complete domination by the modern organization. This compulsion to eradicate what remains of individualism in values is bringing America to the edge of a modern revolution.[1]

Richard Chewning called this erosion of individual values an *anti-ethic,* a psychological rejection mechanism that questions the possibility of anyone's being able to know right from wrong. (Throughout this book, the term "anti-ethical" does not mean *un*ethical, or morally wrong. Instead, it refers to the assumption that moral issues are not a significant concern.)

Before our culture contracted metaphysical cancer we operated with an ethical system that enjoyed a deep and wide base of acceptance. This old ethic has been called by many names: the Work Ethic, the Protestant Ethic, the American Ethic, the Capitalistic Ethic, the Yankee Ethic, and others. . . . This old, long-standing ethic was dislodged from its central and consensus-molding position following World War II. A new ethic—new for the masses—emerged, but it did not become the basis for a new consensus because the new ethic is at its very heart an *anti-ethic*.[2]

The problem with an anti-ethic, according to Chewning, is that it denies our ability to productively discuss those assumptions that profoundly shape our self-identity, our sense of purpose, and, subsequently, our behaviors. These assumptions involve timeless issues, such as the nature of existence itself; how we "know" anything—empirically, rationally, existentially, or revelationally; how we order and deal with all of the information we encounter; and what constitutes ethical behavior.

Business schools have finally acknowledged the importance of such assumptions—certainly a welcome, although belated, development. Although a formal study of ethics is an excellent and necessary first step, it should not be of highest priority.

Many schools have already announced courses that will give students the tools and experiences necessary to help them deal with moral dilemmas: for example, in a personnel cutback, do you dismiss recently hired blacks, or whites with tenure? At what point do safety improvements in an automobile become uneconomic and morally unnecessary? What represents an appropriate level of profit for the shareholders of a public utility? These issues are important, but they pale in significance when compared to the daily violations of standards that we already *know* are unethical: promoting a political supporter or a friend instead of the most qualified; awarding a contract to the person who gives the biggest kickback; taking unreasonable risks (to the organization) in order to realize personal advancement; and on and on.

Business schools, and the behavioral sciences as well, need to analyze their own roles in creating management systems and procedures that naturally lead to the anti-ethic. Robert Denhardt got to the heart of the matter in 1981:

We originally sought to construct social institutions that would reflect our beliefs and our values; now there is a danger that our values may reflect our institutions. Here we encounter a most serious problem: as we permit organizations to structure our lives, rather than vice versa, we may become locked in their grasp. We may begin innocently enough, engaging in organizational activities which we hope will promote useful social goals, yet wind up doing certain things not because we choose to do them, but because "that's how things are done" in the world or organization.[3]

That is exactly what has happened to management education, and to the philosophies of senior corporate managements. Elaborate theories which we have never really understood now prevail. Traditional beliefs that do not lend themselves well to statistical analysis are overturned in favor of isolated short-term studies. The machinations of management and behavioral science experts are given precedence over basic human values. This turnabout in values has to be a major cause of our organizational problems. Peters and Waterman cited 43 "excellent" companies in their book, *In Search of Excellence*. A subsequent cover story in *Business Week* reported that 14 of them had "lost their luster." In a remarkably candid response to the analysis, Peters and Waterman concluded, "If you're big, you've got the seeds of your own destruction in there." The excellent companies seemed to be just big corporations that were "losing less fast."[4]

How can this pessimistic view be possible when we have never before been so knowledgeable in the managerial and behavioral sciences? Why is stagnation so often the fate of corporations and institutions as they get larger, older, or more complex? Even with the best consultants, and the latest theories, the same degenerative problems emerge and increase. More specifically, what is it about commonly accepted "professional" management systems and practices that deaden ethical sensibilities? How have we managed to con ourselves into adopting them? What do we need to do to correct them? That's what this book is about.

First, unethical behaviors exact tremendous long-term penalties not only on individual organizations but on society in general. Although we instinctively know this, it has not been well documented by research, or at least many managers are reluctant to acknowledge it. Scott and Hart commented, "It is difficult to decide which condition is worse: the sense of helplessness among some about their inability to change organizations, or the insensitive belief among others that all is well within the organizational firmament."[5]

Part One of this book is directed at Scott and Hart's second option. Indeed, all is not well in the organizational firmament. Those who have convinced themselves that egocentric self-interest isn't all that bad need to hear what their subordinates and associates are saying. There are some disgusting things being done by certified members of the club, all in the name of teamwork. We need to take a much closer look at the persons and the behaviors we are promoting. Our moral standards affect more than the quality of the leaders we select. They also determine the kinds of leaders that we are developing for the future. At this point, the Destructive Achiever (DA), whose impact on the organization is disastrous, has a disproportionate competitive edge. This leadership type is also the least understood candidate for promotion in the modern organization. Therefore, I attempt in Part One to paint a picture of the behaviors of DAs for their easier and earlier recognition by top management.

Part Two describes why sophisticated research and development efforts can paradoxically lead to systems and procedures that are detrimental to organizational health. In trying to abstract from reality models of good leadership and good organizations, we have managed to present the models themselves as reality. As a result, our attention has been directed toward issues that are only superficial, and we have depreciated those that are far more important. We automatically contribute to Chewning's anti-ethic when we focus on the *forms* of leadership and organizations, not their *substances*. Systems, procedures, and controls take precedence over values, integrity, and commitment.

Part Three is speculation about the organizational factors that can encourage ethical behaviors. The primary requirement is a climate of equality, with a commitment to fair and objective problem solving. This, in turn, depends on a constructive kind of confrontation that brings all significant issues to the surface. Although we know there are many systemic obstacles that keep these ideals from being realized, we also know of at least a few hopeful approaches that may reduce them.

A TIME FOR PHILOSOPHY

This book is more philosophy than science—a result of personal observation, not rigorous testing. In my view, it is the inadequacies of the latter approach that are half the problem. Our progress in the physical sciences has outstripped our progress in moral behavior. We know more about how to kill each other than we do about how to live together harmoniously. These axioms are generally recognized, but what is not is that our progress in management and the behavioral sciences has outstripped our progress in developing ethical behaviors for utilizing it. We have created theories with inadequate philosophical bases or, what is worse, with bases that are anti-ethical.

In addition, we continue to turn to systems and procedures in our efforts to solve problems that are essentially ethical and ideological. We do this from the vantage point of what we think people and organizations *are*. Although this must obviously be the first step, we must also see organizations and people according to what they are capable of *becoming*. This requires trust, commitment, and experimentation—to discover new ways of solving "unsolvable" problems.

Footnotes

1. William G. Scott and David K. Hart, *Organizational America.* Houghton Mifflin, Boston, 1979, 209.
2. Richard Chewning, "Can free enterprise survive ethical schizophrenia?" *Business Horizons,* March-April 1984, 5–6.
3. Robert B. Denhardt, *In the Shadow of Organization.* Regents Press of Kansas, Lawrence, Kans., 1981, 32.
4. Cover story: "Who's excellent now?" *Business Week,* Nov. 5, 1984, 77.
5. Scott and Hart, 9.

PART ONE

Power, Ethics, and Leadership Type

The eternal question of democracy: How can a free people guarantee that only worthy persons move into the positions of leadership in both the public and private sectors? The answer is, they cannot. There are no guarantees. But that is the essence of democracy, that a free people must constantly search for virtue, in themselves and their leadership.

—*William G. Scott and David K. Hart*
Organizational America

1

Five Types of Manager

Probably the best known comment on power is Lord Acton's: "Power tends to corrupt and absolute power corrupts absolutely." Will and Ariel Durant qualified his observation: "Power dements even more than it corrupts, lowering the guard of foresight and raising the haste of action." Statesmen, philosophers, and historians have treated the use of power with awe and distrust, recognizing its necessity, yet warning of its almost inevitable malignancy. Although modern behavioral scientists also warn of the danger involved in the process of acquiring and using power, they have a far kinder attitude toward it. In fact, a great deal of training and self-help materials today are devoted to assisting the modern executive in the quest for greater influence, control, and prestige in the organization.

Encouraging a more utilitarian view of power, David McClelland observed that "America's concern about the possible misuse of power verges at times on a neurotic obsession."[1] John Kotter echoed the point when he told executives, "I have seen many extremely able people lose a key promotion or lose favor with their superiors, making a turning point in their careers, because of their unrealistic ideas about power."[2] Opinions like this are the result of behavioral and organizational research as well as direct observations of the realities of human institutions. Both research and observation result in essentially the same valid conclusions: Power must be sought to be acquired, there are predictably successful ways of getting it, and its use is necessary and inevitable. Unfortunately, however, in acquiring and using power, ethics and values suffer, not because their importance is denied, but because they are assigned a role that is distinctly subservient to the realities of human behavior. Although writers obligingly refer to ethics, and even to the power advantages of ethical behavior, they usually emphasize the more appealing aspects of getting and using power.

I suggest that our recently acquired knowledge about power has failed to improve our ability to use it wisely. Moreover, we use our knowledge of power to justify our more cynical exploitation of it. In fact, our increasing sophistication in the practical use of power could well be a major cause of the steady decline we now see in all kinds of maturing institutions—corporations, the family business, and even society as a whole.

The problems we face today in our large corporations can be best understood in light of the interaction between each organization's ethical foundations and the specific individuals to whom the organization gives power. I have documented that interaction in well over a thousand interviews conducted over the past 25 years. These interviews were with the superiors, peers, and staff of managers in a wide range of organizational settings. Sometimes, in writing up the interviews, I disguised specific situations to protect confidentiality. But the quotations in the following pages are accurate, because I feel that it is important for you to "hear" the kinds of things being said today by people in organizations. These observations of people's values and working behaviors, along with many other discussions and one-on-one conversations, form the basis of the viewpoints I expressed below.

Five Profiles

My work with people in organizations suggested to me that there are certain basic values and orientations that people take toward work and their working relationships. These orientations appear to cluster around the following five types: *Leader, Builder, Destructive Achiever, Innovator,* and *Mechanic.* The major difference between these categories of leadership and the ones in the past is that the present categories do not represent personality traits, or "styles," inherent in the individual. Nor are they necessarily stable qualities fixed early on in one's life or career. Instead, they represent the image managers present to their superiors, peers, and subordinates in terms of their ethical behavior and net impact on the organization.

The "Leader" is seen as an ethical manager with charisma who effectively leads a group for the long-term benefit of the organization. The Leader receives "high" ratings for performance from superiors, peers, and subordinates and is perceived as having unlimited career potential. Usually found at a high level in the organization relative to tenure, the Leader is noted for a genuine sense of responsibility to all constituencies. Because the Leader is a team player, superiors, peers, and subordinates give genuine support.

The "Builder" is also an ethical and effective manager, but without charisma. He or she has most of the same traits as the Leader, but not in the same quantities or combinations. Perhaps the Builder simply has not yet been acknowledged as a Leader. Superiors usually rate Builders "adequate" to "high," but see them as dead-ended in their careers, or as having only limited leadership potential. On the other hand, peers and subordinates believe that the Builder's experience, talents, and past contributions warrant a higher level position. Like the Leader, the Builder is seen as a genuine team player, and is supported by other segments of the organization.

The "Destructive Achiever" has the Leader's charisma but not the latter's operational values. Hence, the net effect on the long-term welfare of the organization is negative. This type receives adequate to high ratings from superiors, and may be seen as having unlimited potential. Superiors credit the Destructive Achiever with "getting the job done"; but they may or may not perceive the individual as having high ethical standards or a net positive long-term impact on the organization. Most subordinates and some peers rate the Destructive Achiever as "poor" to "very poor." They see this type as being committed primarily to his or her own short-term personal goals and career, even at the expense of the total organization. The DA's support comes from those who respect power, as well as from those who conscientiously make the compromises they see as necessary for the group's benefit. Peers and subordinates may openly speculate about how such an individual could ever succeed in a sophisticated organization with an excellent reputation for good management.

The "Innovator" is at the vanguard of his or her profession and contributes to its creative breakthroughs. Superiors give this person "poor" to "high" ratings, depending on the group's recent contributions. The Innovator may or may not be seen as being a good manager, as having high ethical standards, or as having further leadership potential. Subordinates and most peers also see the Innovator as being highly creative, making many personal contributions, and being tops in his or her field. They may view the net impact on the organization as disruptive or highly positive, frequently both. But they do not see the Innovator as being a leader of others, at least not in a developmental sense. This type is, in effect, a nonmanaging manager. (I would define the Innovator as a Leader or a Builder if he or she were an effective manager of others. If the net influence were negative, I would label him a "Destructive Achiever".) The Innovator's loyalty is to his or her own technology and chosen followers, sometimes to the deliberate exclusion of other technologies or individuals. The Innovator may or may not be considered trustworthy, depending upon the circumstances. This individual's technical confidence may reach the point of arrogance, which can arouse the competitive instincts in others. Often the support is grudging. Innovators can be found at any level of an organization.

"Mechanics," who make up the majority of the organization, are those persons who are considered to be professionally competent. A Mechanic could be a chemist, a lawyer, even a janitor. No matter what level they occupy in the organization, they maintain the state of the art but don't advance it significantly. Unfortunately, as managers, they never make the transition from being a specialist to being a Leader or a Builder. Superiors give the Mechanic "poor" to "good" ratings, depending on the kind of maintenance and support he or she provides the system. Peers and subordinates describe the Mechanic as being a nonmanaging "pass-through" manager, trying to do what others expect, especially those with the most power. This individual exerts little personal influence on intergroup or interlevel issues. The Mechanic may be regarded as fundamentally trustworthy, but usually is seen as having a weak ethical backbone. The Mechanic won't fight for what is right, and the impact he or she has on the organization is perceived to be slightly negative to slightly positive, depending upon what pressures he or she reacts to. Usually found at low levels of management, under the right circumstances (i.e., a stagnant, bureaucratic environment) this individual can rise quite high in the organization.

In the following paragraphs I analyze the ethics of the Destructive Achiever as they contrast with those of the Leader and Builder. The Destructive Achiever is easily confused with the Leader, and his or her impact on lower-level Leaders, Builders, Mechanics, and especially Innovators is disastrous. On the other hand, the Builder is bureaucracy's most unrecognized, certainly its most underdeveloped, resource. I include detailed descriptions of Innovators and Mechanics here as a basis for my later discussion of the impact Destructive Achievers, Leaders, and Builders have on other types of organizational members.

The Destructive Achiever (DA)

The Destructive Achiever could be any one of a number of types: Harry Levinson's "Abrasive Personality," Michael Maccoby's "Gamesman," or David McClelland's "Personalized Power" individual, among others. They all have one thing in common: some combination of strengths and weaknesses that causes them to seek and use power in ways that achieve short-term results at the expense of long-range effectiveness. In a few cases, this combination appears to be a basic personality or character flaw; but usually it is the result of the person's developmental environment. For a variety of reasons, the person uses power in a way that destroys trust and commitment, suppresses innovation, and gradually changes a vibrant organization into the classic systems-bound, survival-oriented bureaucracy.

THE ABSOLUTIST DA

The DA is not necessarily intentionally unethical. In fact, a common type of DA is the self-righteous manager who strongly believes in his or her own absolute value system. One staffer noted:

> You can disagree with him about *almost* any technical subject. In fact, he seems to enjoy it; he brags about how his staff will fight him all the time. But *never* disagree on *some* issues, especially about people. If a mistake is made, it's never an honest mistake. The guy was either careless, lazy, or incompetent. And if you defend him, you open yourself to attack—of even being disloyal. Your own personal standards are suspect.

Or, as another subordinate in a different situation put it:

> He genuinely can't stand yes men. But don't disagree with him.

Self-righteous managers have integrity in the moral sense, in that they consciously live by the strict standards they have set for themselves and everyone else. But their values—respect for others, honesty, and the other commonly accepted tenets of good management—are not integrated into their actions. The absolutism of the DA prevents management from objectively analyzing problems or discovering the real causes of success or failure. A peer comments:

> With him things are never confusing—they are either all good or all bad. He does not see things in shades of gray. His main vulnerability is that he can become emotionally involved in a project and not be able to cut it off when it turns bad. He will commit himself to a course of action and it is almost impossible to turn him off. His reaction to bad news is so threatening that he doesn't get it. What he describes as a problem and what is actually going on are different. What he hears is filtered through so many levels, and he is not in touch with what is being done.

THE DISHONEST DA

Of course, a small number of DAs have a self-serving orientation well beyond the normal—to the point of dishonesty. One subordinate commented:

Last week he took a break from a meeting and gave me a call. He said that he told them that we had already done the tests, and that it would be my neck if I didn't get them back by 11:30. I can't believe that he hasn't been caught doing things like that.

A peer cited one of the most often mentioned behaviors of the DA:

He will give as his own the results that were achieved by others. He takes credit for ideas that others originated. When he writes a monthly report, he gives the impression that his group did the work that was done by a support staff.

This kind of DA may be blind to his or her own dishonesty, just as absolutist DAs are blind to their ethical violations. Associates report cases in which acknowledged DAs actually took credit publicly in situations in which they *had* to know that others were aware of their misrepresentations. Why did they do it? The best explanation seems to be that their own egocentrism made them believe they were telling the truth. The same is true of other kinds of situations. A DA once recommended that a new plant be located in a specific city, because the work force there had a high basic skill level. Another executive pointed out that, on the contrary, the city was known for a *low* skill level. Later, in the same meeting, the DA observed that the city's low skill level would be a benefit, "because we will be able to train them according to our own standards." Apparently he was completely unaware of the contradiction between his two statements.

THE DEVELOPED DA

The majority of DAs were originally potential Builders and Leaders who learned the wrong lessons in anti-ethical organizational environments. As a result of formal training and job experiences there, they lost their idealism and so became realistic, bottom-line achievers. A marketing associate relates:

He told me that he had instructed his production people to cooperate with us in making the necessary product changes. Yet one of his subordinates told one of mine that the message was, "cooperate if you feel it is needed, but you're still going to be judged on the basis of present production schedules."

A subordinate describes another common tactic of the DA:

(DA) is a real politician. He is undercutting [another support staff] in doing work for the plants. He wants to become a plant manager, and is doing everything for his own needs—which means what the plant needs. Once [the other support staff] was requested to do something, and they responded that it would take three months. He said that we could do it in one month. We almost made the deadline, but while we were achieving this objective our support for other groups [that were felt to have less influence in the performance appraisal system] went to hell.

These DAs learned to become independent entrepreneurs within their organizations. The focus of each is in the success of his or her own unit—a plant, a department, a marketing district, and so on. The welfare of the corporation as a whole comes second. This happens whenever a conflict of interest arises between a DA's own objectives and those of the total organization.

The Leader (L) and the Builder (B)

Leaders and Builders are more similar than they are different. Indeed, except for the different kinds of recognition they receive, they are essentially the same. In describing Leaders and Builders, a good place for me to begin is to modify the "trait" and "situation" theories developed by others: Neither the traits of, nor the situations for, leadership are as predictable or as clear-cut as we would like to believe. First, leaders and followers have the same traits, but the unique combination of traits that each has varies. And it is this variation that determines the person's leadership appeal. Second, although the situation may be key in determining who rises to leadership, it is always controlled by those who hold the power to confer that leadership. As an illustration, compare the following two supervisors' descriptions of two different staff members:

> (Perceived Leader) does an excellent job for us in public relations. People are always after him to give talks to various organizations. By and large he does an excellent job, but he also invariably makes some people mad. He just can't seem to go through a talk without swearing or telling an off-color joke. No matter how much we talk to him about the negative effects, he goes ahead and does it. He sees people enjoying his style and just decides, "to hell with it."

> (Perceived Builder) has a very outgoing personality and everybody likes him. But I'm not sure how much they respect him. Whenever someone gets transferred and we have a roasting, (B) can tell some really raunchy jokes—the kind where people groan. He's just never been able to grow out of his "good-ole-boy" ways.

Both individuals were rated high by their subordinates and peers. Subordinates of the Builder, however, couldn't understand why he had apparently peaked in his career. In the Leader's case, his indiscretions were acceptable to senior management, within the context of his total performance. In the Builder's case, however, weaknesses of his performance were considered reason enough not to promote him to the next level up. Were the perceived differences due to traits in the subordinates or in the superiors? Or were they dependent upon the situation? Were the differences simply a matter of the Leader's style appealing to a particular superior—or did he truly have more leadership qualities, which allowed him to benefit in a specific situation? Were the evaluations correct? There is no way to know, but, in each case, the determining factor was the decision by senior management.

There is another dimension to this same issue. Dave, a training coordinator in a Fortune 500 company, had consistently received good or excellent performance ratings by his immediate supervisors and was held in high regard by virtually everyone associated with him. However, senior management found him to be too "low key," and to lack the necessary interpersonal skills required to work with higher levels of the organization, in spite of the fact that he was a major in the Army Reserve at the time. Another corporation in the same city somehow heard of this jewel-in-the-rough and offered him a significant salary increase and a higher management position. Dave had been "programmed" by his company to believe that he was not high-caliber management material. Despite this, or maybe because of this, Dave risked leaving his secure job at the Fortune 500 company and joined their competitor. The last I heard, Dave was made vice-president.

This is common. But why? Is it due to the traits of the subordinate, or the traits of the superiors? Or were the situations truly different? And who made the mistake, the first company, or the second? Or was a mistake even made? All we know for sure is that those who have the net balance of power—whether subordinates (who, in some cases, can withhold support from their bosses), peers, or superiors—are the ones who always determine who will lead, or who will *not* lead, rightly or wrongly. And anyone who wishes to challenge the legitimacy of promotional decisions must confront the predominant forces that made them.

I consider this to be an important point, because today we tend to distinguish not only between leaders and followers, but between leaders and mere "managers." There is a certain mystique about leadership that is quite separate from the nitty-gritty details of its exercise. No wonder opinion polls find that the general public truly respects so few of our political or business leaders. The kind of charismatic, powerful leader they have been led to respect exists only in make-believe or in the products of Hollywood and Madison Avenue.

With this manufactured mystique about leadership, we simply pander to our leaders' desire for power, which is intoxicating enough in itself. This is exactly the wrong approach. We must learn to see our leaders, not as superior beings, but simply as fallible individuals in positions of power and responsibility. The false distinction between "leader" and "manager" contributes to false or lowered expectations about what either can do for the group.

Clearly, there are some people who do possess the kind of personality traits seen as representing leadership, people who attain positions of power and acclaim. But it is important to make the following points:

First, there is no assurance that an acclaimed leader is positively contributing to the long-term welfare of his or her group or society.

Second, given the right developmental opportunities, other people may have been able to achieve as much, or even more, for the group, although possibly in a different and less grandiose manner.

Third, presently acknowledged leaders are not necessarily desirable models for the development of future leaders.

Actually, managers at all levels ought to be able to lead, or they cannot *manage* people—they can only *administer* the bureaucratic system. Effective managers exemplify the values of the group, yet they are willing to challenge the system when it is wrong. They must command attention, serve as role models, monitor results, and do all those things that leaders do. If they are not doing these things, at least to some degree, they should not be in management positions. This may frighten some, those who feel that these requirements are unrealistically demanding. Where would we find enough people to meet our managerial needs? I agree that we don't have enough true Builders in most organizations today, but that's because we haven't been willing to recognize and develop them early in their careers. The potential is there, and it is abundant.

The only real difference between a traditionally defined leader and a good manager is this: for whatever reasons, the leader was recognized by those in power. Beyond this, the leader may *seem* to have charisma, *possibly* based on real traits lacking in other executives. To maintain this distinction, I refer to these individuals as *Leaders* and *Builders* (I use "leaders" and "managers" only in the traditional sense). Builders, then, are essentially the same as Leaders, but, *in their own specific group,* they are either unrecognized or don't in fact have the same quantity or types of strengths that the recognized Leaders do. But they *are* ethical, and they do have the leadership qualities that are necessary to maintain *their* areas of responsibility. Associates perceive both Leader's and Builder's sum total of performance to be "good" to "very good." The net impact they have on the organization is positive. Yet any one specific individual will do many things well, and some things not so well, or even poorly. When compared to those of the Destructive Achiever, Innovator, or Mechanic, three strong qualities of Leaders and Builders seem to stand out:

- A genuine interest in subordinate development, based on concern for the future of the individual and the organization.
- Integrity, including an openness and honesty in communications.
- A problem-solving orientation, a sincere effort to meet the legitimate needs of the organization.

Other types may have some of these same qualities, but not as strongly, as consistently, or as objectively. A Destructive Achiever *may* take an interest in subordinate development, but others see the DA as doing it for his or her own record, not for the subordinate. A Mechanic *may* have high ethical standards, but not be able to adhere to them in threatening situations. The Innovator *may* have a strong problem-solving orientation, but once he or she takes a stand, objectivity may be lost.

When a member of an organization tries to describe the strengths and weaknesses of an associate, it is frequently difficult to label the associate as a Leader or a Builder until a comprehensive profile evolves. Therefore, unless stated otherwise, the following traits and abilities could apply to either a Leader or a Builder.

DEVELOPMENT OF SUBORDINATES

It may be possible to fool unsophisticated subordinates for a time, but, in most organizations, no one understands a manager better than they do. They are the ones who usually see managers at their worst, as well as at their best. Builders and Leaders get results because of the positive impact they have on others. Subordinate development is an important part of that impact. Two subordinates comment:

(B) is not only very conscious about the importance of us learning to do his job, but he also lets us know that our subordinates need to know how to do our jobs. Once when I came back from leave of absence, he talked with me about how my backup replacement did while I was gone. We discussed in detail what the subordinate did well and what he needed to do to improve. He seems to be well aware of what people are doing, even two levels down, without seeming to spy or pry.

(L) is a very good listener and always gives the impression that he wants to help the person he's talking with. When we go to talk to him, we know we'll get an honest response from him. He considers subordinates' welfare as one of his managerial duties and gives us an opportunity to shine. He is people oriented and is very concerned about how decisions will affect people in his unit.

Contrast the two observations above with the following one, made by a subordinate of a Destructive Achiever:

(DA) is looking upward too much, too hard—too concerned about promotions and his own career. He's kind of an odd mixture in that he probably has excellent leadership characteristics. Has a manner and bearing about him that would allow him to be a good leader. But he doesn't listen, and he doesn't seek out time to spend with us. He really doesn't hurt others, doesn't try to—just a matter of ignoring them.

In general, subordinates are usually dissatisfied with the appraisal process. But subordinates of Builders and Leaders react the most favorably toward it. These subordinates respect the conscientiousness and genuine concern of their Builders and Leaders, whether or not the appraisal itself is done well.

When (B) gave me my first appraisal, he really shot me out of the saddle. He didn't do it at all like they tell you you're supposed to. I was depressed and couldn't do anything constructive for a whole week. Then he noticed changes that I was making and gave me some encouragement. Now I realize that, although he was too blunt and direct, he did an excellent job with me. He's very aboveboard and tried to give a good picture of what's right and what's wrong. It was the most profitable discussion I've had since I've been with the company.

(B) puts more thought into the appraisal than anyone I've ever worked for. He knows he's dealing with subordinates' futures and handles this responsibility quite well. He takes into account not only what he thinks, but what others in top management think, and he goes deeply into the subordinate's background. He doesn't have a fixed format, his procedures will vary, but he is always thorough, organized, and conscientious.

A subordinate had this to say of a Destructive Achiever:

(DA) has a good feel for the strengths and weaknesses of people, but he engages in very little discussion about them. He's not really interested in people; his personnel decisions are either dictatorial or he avoids them. The appraisal is just a form to be filled out.

INTEGRITY

No management text is complete without a discussion of the importance of integrity. Therefore, every manager must create at least a minimally acceptable image—but Leaders and Builders deserve it. A peer notes:

> (L) in meetings is highly cooperative, direct, honest, not a wheeler-dealer. Even in competition with his peers for promotion, he is very friendly and doesn't let this affect his cooperation. He also doesn't pass the buck in his area—isn't uptight about accepting criticism.

And a subordinate notes:

> (B) is very honest and aboveboard. People feel at ease when getting in their status reports. Whether to the person's favor or disfavor they trust him to handle it right. He understands his people very well and you can always count on him to do what he says he'll do. Whatever he tells you, you know he's telling others the same thing.

Of the Destructive Achiever, a peer had this to say:

> (DA) looks out primarily for himself. Wants me to do things for him before others—a very selfish individual. He even tried to get me to distort a report to make his position on an issue appear to be the best. Puts himself first, takes credit when he gets the opportunity, and does it more by omission than by denial, in that he doesn't give credit to others when due.

PROBLEM-SOLVING ORIENTATION

Objective problem solving is the essence of teamwork, and Builders and Leaders do it best. A manager needs to ensure that all relevant elements of a situation are considered, including the needs of the individuals and groups involved. A subordinate comments:

> (L) never lets a manager go off on his own, without considering the others. He makes sure that people listen to each other, and if a subject is controversial, he makes sure everyone gets his input. Too often, in such a large plant like this, a department manager will take off on his own without checking, and if the plant manager isn't on top of this it could lead to a lack of teamwork. This is probably the prime reason that (L) has been able to build such a cohesive group here.

And of the Destructive Achiever, a peer comments:

> (DA) is aggressively defensive in our meetings, no matter what we try to do. For example, if three customers are having a problem with a product, and seven are not, he will say, "What is wrong with the three customers?" In other words, instead of working to explore the problem in an open-minded way, he immediately wants to blame the customers as having done something wrong, which may or may not be the case.

Objective problem solving is highly important for organizational effectiveness. It will be developed more fully in Chapter 2.

Of course, Builders and Leaders can be vulnerable, too. They have the same weaknesses others have, and in practically every aspect of management. However, their weaknesses are usually a matter of personality or skill, and are almost never associated with a lack of integrity. For example, one Leader was labeled by his subordinates as a "cold fish," but his interest in them—their performance and their future careers—was considered genuine and effective, although purely professional. (They suspected that he actually had real feelings about them, but he never showed it in any obvious ways.) In another case, a Builder was said by peers to do a poor job of leading problem-solving meetings; they lasted far too long, and they rambled. Yet they were honestly led, and when they were over, everyone felt they had had a chance to contribute, and everyone knew what to expect from everyone else.

The Innovator (I)

Leaders and Builders both manage; but the Innovator is a nonmanaging manager. That is, the Innovator is in a position of management only because of his or her technical expertise. Technicians have little power in typical organizations. Whatever influence they may have in the decision-making process, it is out of proportion to their creative talents. A senior research associate may be at an organizational level equivalent to the manager of a staff service group. However, it is unlikely that the research associate would have the same amount of power as the manager, who controls more resources. As a result, most Innovators choose management (if they *have* a choice) when they want to influence the technical direction of the company. Because their technical value is so well acknowledged, they often do become quite successful, not only for themselves, but for their organization as well. Their success, however, has little to do with their human resource leadership skills.

To make the most of the talents of the Innovator manager, supervisors, subordinates, and peers alike need a lot of understanding and flexibility. These groups soon discover that it's well worth the effort, because Innovators contribute much to the organization and its members. But adjustment is a one-way street, because many Innovators also have a remarkable immunity to coaching, personal development workshops—and even threats. (If contributions are not significant enough to make adjustments worthwhile, then of course a personnel decision must be made.)

Note: Highly innovative managers who behave like Leaders, Builders, or Destructive Achievers, are referred to as such. I refer to those who do not as "Innovators" simply to indicate another kind of *manager*. Therefore, most of the following descriptions suggest innovation but no leadership skills.

To begin with, it is helpful to remind ourselves of the Innovator's talents. Associates comment:

> (I) is the tower of technical strength in [company] and not much is going to get by him. He can get to the heart of a subject in a meeting and can bring out things that weren't there before.

> He's probably one of the two best authorities in the entire world in his area, and I think he is the very best in [specific technical area]. He is one of those rare birds that not only understands the theory, but he also knows how to apply it and make it commercial.

> He's simply a damned smart, intelligent problem solver. He amazes others with his ideas, and is able to give different options for a solution to a problem. Not a one-solution guy.

Because of their high level of expertise, Innovators either short-circuit the management process because they want to get things done in a hurry or they create a bottleneck because they always want to approve others' actions first. Innovators use their organizations as a personal laboratory or shop for the purpose of implementing their own ideas, or perhaps the ideas of others that they approve of. As a subordinate put it:

> Sometimes he interacts with people three levels down; even group leaders have received instructions from him which were counter to the original plan. And if we have to get back in touch with him and he is hard to reach, we have to wait on his approval.

In disagreements, when things go wrong, or when they are having a hard time selling an idea, Innovators tend either to become overly aggressive or to withdraw, depending upon their personality. A full range of response is possible:

silence, retreat into theory or philosophy, or attack. Four different peers describe four different Innovators:

> (I) tends to be opinionated and arrogant, unswayed by others' arguments. He communicates that, "Hell, I've looked at it and that's the way it is." He has a low volatility point. Sparks will fly if there is another strong personality in the meeting. He even got to the point once where he got mad, jumped up, and threw his chair into the corner of the room.

> He doesn't try to sell his ideas, doesn't want to listen, seems defensive. He just presents his case and that's it. He frequently makes a public washing of hands—"Well, I've given my position, if it blows up in our faces, it's not my fault."

> (I) doesn't keep others informed nearly as much as he should. He'll let a unit go astray. He'll see a peer making a mistake and he'll warn them once and then watch them flounder if they don't heed his advice immediately. He expects people to react immediately to everything he says.

> He's actually a very likable guy once you get to know him and know what to expect. He once was called onto a project after it was well underway— it was the prototype for a new double-line conveyor belt system. With several persons gathered around, he walked around the equipment a few times without saying a single word. He thought a minute and then said, "Just what I expected, all ——ed up." I could hardly contain myself; I wanted to laugh so bad I had tears in my eyes. But, of course, the engineers were absolutely devastated.

Innovators' first loyalty is to their own technology and its appropriate application to the organization's needs, as they determine them. This can be extended to personnel, in that Innovators tend to favor those persons who share their viewpoint. Even Innovators who are not particularly assertive don't seem to have much flexibility in this regard. For example, it is difficult to get their cooperation if it requires a compromise, or if they must make significant adjustments in their game plan. A subordinate remarks:

> (I) has a very broad grasp of the technology, but sometimes has technical prejudices. When he is in the formulation stage of his ideas, he sucks up information like a vacuum cleaner. He gets all the data he possibly can and is very receptive to others' ideas. But once his mind is made up, he doesn't want to change it—closes himself off to other considerations.

A peer rates some of the Innovator's subordinates higher than the Innovator does himself:

(I) is a perfectionist and his staff is never good enough for him. But his people are really good. He does have a good staff, but he judges them harshly. He also gives higher ratings to the subordinates he selects himself. If he is *given* a subordinate, he tends to rate him lower.

The Innovator's development needs were best summed up by an associate:

He has such very strong opinions, he makes it very hard on himself to change his mind and come up with a different opinion. He has a fantastic batting average; he's right, say, 90 percent of the time. But if he had a more genuine participative approach he would have a better chance of discovering the times when he is wrong, and have maybe a 95 percent average.

And, for some persons, the Innovator remains an enigma. As a supervisor explained:

(Company president) sees him as a complete mystery. Doesn't have any kind of rapport with him at all.

The Mechanic (M)

Like the Innovator, the Mechanic is also a nonmanager in a managerial position, but the Mechanic's failure to manage is due to personality and experience. Mechanics retain their managerial position only because they reliably follow technical or administrative procedures (in a sort of holding pattern). Or they may give senior management some appreciated nonmanagerial service, for which they wish to reward the Mechanic beyond standard policy guidelines. Of course, there is also the possibility that senior management is simply unwilling to make a difficult personnel decision that would involve replacing the individual in question.

Although Mechanics are usually very knowledgeable and skilled in the technology of their profession, their expertise is not at the level of the Innovator. And Mechanics who are truly innovative usually go unrecognized. In either case, Mechanics fail to make the "vital shift" from professional to manager, or have any significant impact on the work group. One superior commented:

(M) is willing to work nights or weekends—whatever it takes to give good staff service to the rest of the plant. He's a very good administrator, and, any time you need to know something from the records, he can get to it immediately. His major problem is that he seems to lack the ability or the skills to solve the problems that are unique in nature. The upper level managers have quit going to him because he doesn't go beyond giving them the standard answers.

Subordinates are demoralized by their supervisors' unwillingness to use the power of their position:

> (M) is very talented technically, but a conservative decision maker. He tries to avoid conflict in all directions, both one-on-one and in meetings. In a one-on-one situation, he prefers to simply let the problem go by. In meetings, he prepares for a disagreement in case it comes up, but won't initiate it if it doesn't.

There is one type of Mechanic who gives the impression that the group should manage itself. This type prefers to simply serve as a model, instead of trying to exert any direct influence on the group. Three subordinates express their views:

> (M) should do a better job of informing me in advance of things I need to know. This would include just about everything. I've frequently found out about upcoming meetings from other people instead of getting it from him. He delegates too much; he lets people have the reins almost without any kind of guidance at all.

> Sometimes (M) may speak over the head of a subordinate. We talk for extended periods, and I realize at the end that we're not even talking about the same thing. Counseling is usually after-the-fact, or after the panic begins. It should occur before projects are worked on.

> (M) is more of a doer than a manager. He is still the super engineer. Basically, his problems are planning, directing, prioritizing, and communicating. He would be better off in a high-level, one-man, staff position.

Peers of the Mechanic sympathize with the Mechanic's subordinates:

> (M) needs to get more involved in his group's projects. He doesn't seem to know where they stand. There have been times when it was obvious to me that he learned where his group stood in certain activities at the same time I did.

> In the last appraisal process (M) over-rated all his subordinates, and then the ratings were changed in later discussions. Therefore, he got out of kilter with both his peers and his subordinates. We felt that he over-rated his people, and they felt that he didn't stand up for them. Actually, he couldn't defend them because he was unfamiliar with what they were doing.

It is clear from these remarks that the Mechanic exerts little personal influence on the system. Unhappily, the result is that a vital part of the organizational renewal process becomes inoperative. Subordinates are not developed; moreover, an important power position is missing in the problem-solving process. There is one less viewpoint to consider when competing forces collide in the decision-making process.

The Problem with Labels

"Leader," "Builder," "Destructive Achiever," "Innovator," and "Mechanic" are merely labels I use to make the task of understanding managerial behaviors easier. If they can just do that much they will have been useful. There are no pure types, and each and every manager is some mixture of all five. Moreover, the proportionate mix of the five types can change within each individual over time, in striking contrast to previously identified leadership categories based on personality or managerial "style"—more or less fixed characteristics.

I would classify most managers as Builders; and some, I'm sure, have the potential to achieve that more lofty image of Leader. They simply need to have the opportunity—or the developmental experiences—early enough in their careers. But, once caught in their organization's degenerative political climate, even these Builders can become Destructive Achievers, a type which is increasing in proportion to the other five. Most DAs, at one time or another in their careers, had the potential to become either a Builder or a Leader. Some appear to be inherently destructive, unable to change; but they are relatively few in number.

Next are the Mechanics, some of whom could function as Builders if the work environment they are in were more amenable to participation than to conflict. With proper coaching, and with developmental opportunities early enough in their careers, a few are able to do so. But if the environment is an unethical one, there is a strong tendency for Mechanics to become DAs.

According to my definitions, we see few Leaders and Innovators in modern organizations. Moreover, there is a very real danger that these types could become Destructive Achievers as their self-confidence increases, and as their personal contact with their constituencies is lost.

History clearly demonstrates that increased leadership responsibility is all too often accompanied by the arrogance of power. Leaders, above all, must maintain a careful watch over the value and ethical bases of their behaviors. As I indicate in Chapter 4, the temptation is to gradually move away from a problem-solving orientation that is ethically based to a power orientation based on Chewning's *anti-ethic*, the assumption that moral issues are of little concern.

Footnotes

1. David C. McClelland, "The two faces of power." *Journal of International Affairs,* 24, 1, 1970, 44.
2. John P. Kotter, *Power in Management.* AMACOM, New York, 1979, 6.

CHAPTER 2

The Uses and Abuses of Power

Some may attribute the tragedy of the space shuttle *Challenger* to some incredibly bad decision making on the part of NASA management. Actually, however, it is typical of what frequently occurs, even in Peters and Waterman's "excellent" organizations. It's also an example of the natural tendency of aging institutions to become increasingly biased in favor of short-term priorities, even when the risks are enormous.

According to newspaper accounts,[1] Morton Thiokol engineers were adamant when they unanimously recommended that the *Challenger* not be launched because of the cold weather conditions (24 degrees) the night before. They felt so strongly about it that Allan McDonald, director of the solid rocket motor program, said, "I made the direct statement that if anything happened to this launch, I sure wouldn't want to be the person to stand before a board of inquiry and explain why I launched it." In spite of this, "company management recommended the launch based on an assessment that temperature data were inconclusive." Later, a report quoted laboratory test results: "At 100 degrees Fahrenheit the O-ring maintained contact. At 75 degrees Fahrenheit the O-ring lost contact for 2.4 seconds. At 50 degrees Fahrenheit the O-ring did not reestablish contact in 10 minutes, at which time the test was terminated." When asked in a telephone interview to comment on the test data, "Lawrence Mulloy, who headed the booster rocket program at the Marshall Center, said the information from the test 'was inconclusive' because 'it would seem to say that the joint would fail at 75 degrees also, and we knew that wasn't the case' from previous shuttle flights."

Mulloy also was "quoted as responding to the engineers' concerns by saying, 'My God, Thiokol, when do you want me to launch, next April?' " There was no official pressure forcing anyone to go against their own best judgments. Still, this was the sort of bad news that would cause the management of a company whose contract was due to expire at the end of the year some measure of concern. Intended or not, in effect, Mulloy's message was clear: Thiokol would be blamed if the launch were delayed, because the booster system they had designed was not hearty enough to withstand cold weather conditions.

What was originally a problem-solving process had become a power process. Instead of an objective decision being based on a total group assessment of the probabilities, what happened was that a subjective decision was made, based purely on a personal (or subgroup) interpretation of the situation. In defense of NASA management, what happened occurs daily in virtually every large, complex organization in which some persons have more power than others. The difference is that such mistakes usually don't blow up in public view. Also, it must be recognized that more often than not the long-term results are beneficial.

A large, complex organization simply can't always convene a total group for a lengthy problem-solving discussion to reach consensus. There are also times when managers must overcome group indecision, disagreement, or excessive caution and take the steps necessary for productive action. As one company president once said: "R&D is never ready to go on-stream. They always want to do one more trial, make one more improvement. There comes a time when you have to put your foot down and say, 'We can't test forever, now is the time to try it so that we can see what we *do* have at this point. It doesn't have to be perfect; if we can beat the market it'll pay off.' "

Clearly, there are merits to the power side of the power–problem-solving continuum (see Exhibit 2.1, page 27). However, historically, *the natural tendency of leadership is constantly to progress in the direction of power.* The result is that the problem-solving side tends to fall away. To illustrate, it might be instructive if we speculate about what would have happened to the NASA decision-making process if the shuttle had been launched successfully: The engineers who warned of disaster would once again be proved wrong; senior management would become more confident of its rationalized approach to risk taking; and the rest of the engineers, or their supervisors, would be more cautious and less adamant in future disagreements with senior management. Unfortunately, this is the usual course of events. Conservative risk takers will be proved wrong more often than right, say four times out of five, at least as far as potential O-ring failure in marginal weather conditions is concerned. But that fifth time is what proves them right in the end. Too bad it takes a disaster to remind management of its original intent—to err on the side of caution.

As managers become more and more successful in beating the odds, they get increasingly comfortable with risk; hence, they are tempted to yield to even subtle pressures for quick results. They know that organizations that fail to deliver don't get contracts, and that overly cautious persons who lack a "can-do" image don't get promoted. Lower-level personnel watch senior management promote the engineer or manager who delivers what is requested, not the lower-level individual whom management sees as "negative," "naive," "unrealistic," or "too theoretical." The situation is made worse when the latter are punished. NASA senior management's first reaction to the tragedy was to reassign Allan McDonald, the person who had been among the most vocal in opposing the shuttle launch, to a job with lesser responsibilities. It was only after congressional outrage that he was offered the job of directing the redesign work on the shuttle's solid-rocket boosters.[2] People who oppose short-term opportunities rarely get credit even when they later are proved to be right.

The cost of the reckless use of power is rarely evaluated appropriately. In our huge economy, a corporation can go out of business and cause little stir. The lesson to be learned is largely ignored by others engaging in the same kinds of negative behaviors that caused its demise. The implications for world survival are enormous: The proliferation of nuclear projects, acid rain, the greenhouse effect, and the widening discrepancies between the haves and the have-nots are ominous issues today. In any one of them we could all too easily reach the point where a DA, or a collection of DAs, would push us over the edge. It seems already to have happened in some societies, such as Northern Ireland and South Africa, at least for several generations to come.

In *Power and Influence,* John Kotter contrasted the leader who uses power wisely with the totalitarian leader. He concluded that our challenge is to learn how to distinguish more clearly between the two: "I fear that many of us today are not skilled at making this distinction."[3] The difficulty in making such a distinction is understandable. It is quite possible that a person whose fundamental style appears to be one of a "Gamesman," or even a "Jungle Fighter," according to much of Michael Maccoby's description, may also be a Leader, by our description. On the other hand, a person whose style appears to be close to Robert Blake and Jane Mouton's "9.9," as measured by a questionnaire or ratings by management, may be seen as a Destructive Achiever by his associates and subordinates. It is therefore becoming vitally important to quickly and accurately recognize the uses and abuses of power, which are epitomized by the Builder and the Destructive Achiever. The advantages and disadvantages of their contrasting behaviors can be more fully appreciated if we

- understand the practical implications of their relative positions on the power–problem-solving continuum.

- understand the appropriate and inappropriate use of position, expertise, and charismatic power.

- acknowledge key behavioral differences that help us to distinguish between the two managerial types.

Power–Problem-Solving Continuum

To understand the difference between the use and abuse of power, the first step is to understand the relationship of power to problem solving. This relationship is represented as a continuum (see Exhibit 2.1). The use of power always includes the presumption of superiority, typically by virtue of position, expertise, or charisma. The individual assumes the right to persuade, manipulate, or coerce others in order to achieve the objectives associated with his or her superior official status, profession, or role in the informal network. Even though the person may be pursuing organizational objectives, the way he or she uses power personalizes those objectives. Personal behaviors, or the role the person plays in determining organizational behaviors, reflect the power of his or her own position, expertise, or social influence. In exercising it, personal objectives may be entirely consistent, or very inconsistent, with the legitimate needs of the group.

This exercise of power automatically suggests a selectivity in interacting with individuals and groups and in picking issues. It is focused on the specific resistances to be overcome. Information, including assumptions and premises, is judiciously either shared or withheld. Specific individuals or groups are either included or excluded, either as concerns of or as participants in the process. On the other hand, a cooperative problem-solving attitude presumes equality. Each individual in the problem-solving process assumes the obligation to fully inform and to be fully informed; to be operationally open and obvious. Each has an equal part in determining and achieving those objectives that are most consistent with the total welfare of the group.

If all members are to share group objectives equally, then relevant issues and interests must all somehow be represented in the process. The point of problem solving is not to attack and overcome resistances to group objectives by force, but to remove them from consideration by means of mutual, informed, agreement. Of course, power and problem solving in their pure states never exist in reality. Power cannot be exercised without an element of problem solving, and problem solving never occurs in the absence of power.

Destructive Achievers, however, make inordinate use of the power at their disposal. Their emphasis is on control—its systems, methods, and techniques. Conversely, Builders are likely to rely too much on the problem-solving approach and fail to develop, or take realistic advantage of, their power potential. Leaders bridge the gap very well as they pursue the full range of their responsibilities. They recognize that the key to group success is objective problem

Exhibit 2.1   POWER–PROBLEM-SOLVING CONTINUUM

Power	Problem Solving
Superiority of individual membership, based on organizational position, expertise, or personal characteristics	Equality of individual membership, regardless of organizational position, expertise, or personal characteristics
Direction Persuasion Manipulation Coercion	Participation Data sharing Openness Objective analysis
Personal objectives	Group objectives
Selective individuals, groups, issues (assumptions and premises strategically communicated)	All relevant individuals, groups, issues (including openly communicated assumptions and premises)

←—————————————————————————————→

| Destructive Achiever | Leader | Builder |

Innovators' and Mechanics' average performance can be at any point on the continuum.

solving, but they also exercise good judgment in using power when the problem-solving process is ineffective or inappropriate.

Clearly, there are some subtle differences between those who abuse power and those who use it in the best interest of the group. These differences will become more apparent in the next section, where I apply the power–problem-solving continuum to the most commonly defined sources of power: position, expertise, and charisma.

Position, Expertise, and Charisma

In itself, the power–problem-solving continuum does not have an ethical dimension. It is ethically neutral. There are times when the participants in a problem-solving situation should be fully involved—say, in listing criteria for selecting a site for the city dump. But there are also times when someone must

be forced into doing or not doing something—say, when an executive is required to report a violation of securities laws.

Either power or group problem solving, therefore, can be used or abused.
It's not always easy to tell what is actually happening, unless one understands
the net impact of an individual's behaviors. In the case of a DA, the use or abuse
is usually on the power side of the continuum, especially when it comes to the
DA's position in the company, which requires little explanation or justification.
Two subordinates describe the DA's attitude about *position power*:

> Sometimes (DA) gives the impression that he doesn't want to know things
> I bring up. If I bring up something a second time, he indicates that he is
> tired of hearing it, "Here we go again." He only wants to discuss issues
> that are important to him. He'll say, "We'll do this" and we say, "We can't
> because . . ." and he'll say, "We'll do it" and not say why.

> Everything keys on (DA), and if he doesn't support it it doesn't get done.

Whereas a heavy hand may be used with subordinates, with peers and superiors,
the DA must maintain control of his or her position in a more subtle way—by
capitalizing on the formal organization's abhorrence of "bypassing." An example
of bypassing is when a person decides to deal directly with someone two or
more levels away, without going through the intermediate manager. Usually this
kind of bypassing does not include casual, informal contacts, but rather those
contacts that relate to significant changes in plans or operational procedures.

In the DA's unit, however, bypassing can include almost any kind of contact, and is considered by the DA to be a violation of his or her position integrity. For example, market research executives may suddenly discover that no
longer are they able to discuss sales projections with sales managers the way they
did when they had a Builder for their district director. Now they must interact
solely with the new director—all in the name of professional management. A
senior market research executive explained it this way:

> (DA) basically seems to be insecure in the job and is overprotective of his
> area. He's very much afraid of being bypassed—very sensitive to other peo-
> ple talking to his subordinates without going through him. He apparently
> fails to appreciate the distinction between lines of information and lines of
> authority. In meetings, he should let his subordinates have their say, and
> not take it away from them. He tends to cover subjects his subordinates
> could cover. This deprives them of the opportunity to look good.

DAs typically view spontaneity as a threat, and so use their position in the company to control information flow. Whereas reports to management used to be
made by lower-level personnel, now they are made by the DA. If a lower-level

person does give a report, either it is a formal, polished report or the DA makes sure that he or she is present at the time.

Builders also use position power, but, in the context of total performance, they create a much different impression. A subordinate states:

> People will exert effort to go beyond the ordinary when they work for (B). He can get a whole group of people behind him—even though 70 percent of the people are for a project and 30 percent are against it.

The characteristic that most distinguishes Builders from Destructive Achievers is that Builders make use of their position power but only after they have taken adequate measures to ensure a fair and objective decision—even though some persons may disagree with it. DAs, on the other hand, use their position power to advance personal interests first—then to solve the problem within the context of that personal interest.

Expertise power can also be used or abused, and it's not always easy to tell which, unless the total impact on the company is considered. Associates of the Innovator believe that his or her use of power is legitimate. They trust the individual's ethics and feel that his or her expertise is genuine. A subordinate credits his supervisor with having the right to use his expertise to get the best option accepted:

> (I) comes off very strong with tremendous self-confidence. He presents data clearly and thoroughly. However, he is not compelled to hedge his bets all the time—does not feel a pressure to be completely honest. That is, he will do whatever he has to to get the right way accepted, even if this means that he is not completely honest in doing it. He may give incomplete information or a one-sided presentation, to make sure the technical solution he sees as being desirable is accepted. And I approve of this; sometimes that's the only way to get things done around here.

A peer explains the power of the expertise of the Innovator, especially when the Innovator's integrity and competence have already been demonstrated:

> (I)'s technical background is too strong for others to fight. In an argument, everyone else just gives up, shrugging their shoulders, saying, "He must be right." Others just can't match up to him technically.

I would not like to have to defend the ethics of these Innovators in a court of law. However, if expertise *power* means anything at all, it must mean something other than consensus problem solving. Innovators who try to educate others in order to enable them to make a decision objectively, are engaging in problem solving, not power. Often, there's no time for such adequate education, so Innovators make the decision themselves and use their power to get it accepted. This is why it is so important for senior management to develop a rapport with the expert's subordinates (they may be dealing with a Destructive Achiever). They realize that any abuse of expertise power will first be observed by the expert's own staff. As one subordinate explained:

> (DA) has created a credibility gap with his own people. He tends to make things more clear than they really are. For example, he'll say, "We thoroughly analyzed this situation and have concluded that . . ." and we know damn well that we haven't thoroughly analyzed it and don't have sufficient reasons for concluding something. He tends to bluff in all directions. Pretends that there are mounds of data to support his remarks, but we know that there aren't mounds of data.

Charisma, or the appropriate use of *charismatic power,* is especially difficult to assess. As a matter of fact, senior management, and sometimes peers, are often completely wrong in the case of the consummate Destructive Achiever or the unsophisticated Builder. A description a peer gave of a DA shows how easy it is to mistake sophistication for charisma:

> (DA) is a very good communicator—extremely good at discussing controversial or sensitive issues in a calm, logical manner. He has a mature manner and is very much in control of his emotions. And able to avoid unnecessary and disruptive conflicts for the same reasons. But he doesn't act like a team member, and works too hard to become successful; it seems to dominate his thinking. He's reluctant to go to others for help because it is a sign of weakness to ask for assistance. He also won't volunteer to help others, even when he is given the opportunity. If he thinks one of his peers is doing something wrong, he'll let him figure it out for himself. He's not really interested in his subordinates; very much a politician.

And the description a subordinate gives of a Builder shows how a *lack* of sophistication can be mistaken for a lack of downward and lateral charisma:

> (B) could do a better job in speaking in front of a group, as he comes off as being nervous, although this has improved with time. However, he is quite strong in motivating subordinates. He has a particular enthusiasm for his job that is contagious. He has put new life into this side of the house; it had been lethargic before, and he has come in and really fired people up to do a better job. He's been a great help in getting new programs started and he makes you want to get the job done.

Unfortunately, senior management frequently mistakes an outgoing personality, or demonstrated interpersonal skills, for charisma, giving no particular consideration to the ethical use that a manager makes of these qualities. Many currently popular training programs do the same. In the case of subordinates and peers, however, ethics is the main consideration.

The DA versus the Builder

The first priority of the Destructive Achiever is to impress senior management with up-to-the-minute knowledge of the status of projects, not only in his or her own area, but other areas as well. DAs can do this without genuine involvement with other organizational subgroups, simply by designing efficient information and reporting systems. As a result, the DA *appears* well informed, organized, capable of getting results, and effective in leading a unit that adapts quickly to senior management's requests. A supervisor sees only the image, and so gets only a hint of the DA's problem:

> (DA) does an excellent job of knowing what his subordinates are doing, what projects they are on top of. Follows up very well with them. He runs a very effective organization in terms of achievement—quality, cost control, and production. His major strengths are his ability to organize, set goals, and achieve objectives. Runs a tight ship, but is somewhat autocratic.

Conversely, the Builder's first priority is to manage a unit effectively. He or she devotes less time to appearances and risks revealing personal weaknesses. Subordinates appreciate the hands-on attention the Builder gives to issues that are important, as well as his or her willingness to delegate. However, if the Builder is not politically sensitive, the result may be a comment like the following one made by a details-oriented supervisor:

> Occasionally (B) demonstrates that he has not done his homework. A couple of times he has been caught in a rather embarrassing situation where I felt he didn't know as much as he should have. This involved possibly relying too much on work of subordinates who didn't come through with as much information as they should have.

The supervisor acknowledged this Builder's general competence. However, the flaw he perceived was, from the viewpoint of the Builder's subordinates, more of a strength than a weakness. They were free to act without constantly getting involved in meetings, phone calls, and so on, as were the subordinates of other managers.

COMMUNICATION: FORM OR SUBSTANCE

The DA enjoys an image of being well informed, but the performances of others suffer as a result. By the DA's insistence on extensive information flow, communication is disrupted at lateral and lower levels. This comes from the DA's being specifically briefed on, instead of conscientiously involved in, a particular issue. Two different subordinates discuss the manager who wants information for its own sake:

> (DA) is a nit-picker, but he sees it as a necessity because of the management we have. He's sometimes questioned about things, and he feels that he has to anticipate questions and be able to answer them. Sometimes he dominates meetings because he doesn't understand what's going on and we have to update him. We should be spending more time working on the problems at hand.

> (DA) is not a good listener and sometimes doesn't understand what he's being told, primarily because he tries to keep track of too many balls in the air at the same time. He's never around when you need him. His boss will ask him something, he then asks me, he hears what I say and reports back to his boss, who asks more questions. (DA) realizes that he still doesn't quite know the answers and comes back to me, and the cycle continues.

A peer describes how a DA retains a competitive advantage by his unilateral control of communication:

> (DA) likes to be sure upper management knows what he is doing. Some peers help you in your job and don't advertise the fact. You wonder what he's talking about when he talks to your boss. Communication with him is all one way. You give him information and feedback and none comes back.

On the other hand, the Builder's information is the result of involvement in issues that genuinely require his or her attention. Because subordinates and peers don't have to spend time and effort on political communications and strategies, they enjoy greater flexibility and freedom to act. But sometimes this puts the Builder in an embarrassing situation. A peer offered this advice for a Builder associate:

> (B) should require his subordinates to do a better job of keeping him informed. He is content to let them run their jobs, and occasionally he's caught off-guard in a meeting. He needs to key them in to what he needs to know to protect himself. Senior management expects you to know these things.

Subordinates, however, saw him differently:

> (B) stays well informed about what his subordinates are doing. Always available to talk to, no matter what his own time demands are. If he is reading something when I go into his office, he'll put his own paper down and give me full attention. Stimulates a creative environment with his group.

> (B) communicates the impression that he really wants to understand the other guy's problems—really wants to get the details from them. By virtue of this trait, he knows all the people who work for him very well. In a meeting, he is a good problem solver. He's not the least bit timid about admitting that he doesn't understand something, and even if it takes all day, he's going to make sure he understands it before he takes another step.

> One thing (B) does regularly is to tour the area first thing in the morning to see his subordinates, to rake up things we're concerned about. Then we don't have to run him down.

Senior management often complains that the upward communication of the Builder is inadequate. Their complaints may be valid; however, the total communication impact is positive. The DA's communication performance, although more impressive, has a very negative impact on peers and subordinates—the ones who are getting the job done.

DEFENSIVENESS

DAs guard against their vulnerabilities. By projecting organization and adaptability upward, DAs create disorganization and rigidity downward, as subordinates are autocratically whipsawed from project to project. No matter what the contingency, there is always an action plan in place to meet management's perceived concerns: a committee has been formed, research is being conducted, guidelines have been developed and published, recommendations have been made, and other persons or groups have been alerted. A subordinate describes a supervisor who defends himself by covering all bases:

> (DA) must have a program for everything, whether he really does have one or not. He's afraid to say that he doesn't have a solution to a problem. He finds out what upper management wants and then goes that way. Never wants to be caught unprepared. If there are five possible ways to do something, he'll quickly try all five so he can have a definite opinion.

A peer explains how an opposite kind of DA defends his programs, even when changes are called for:

(DA)'s greatest problem is his dedication to his objectives and his commitments. Either he has assumed commitments, or they were given to him, and he is determined to look good and stand by them no matter what. He always tries to give the impression that "my work is more important than your work." Talking with him is like pushing a button—you get a response, so why push the button. He is too predictable and follows the party line.

Although the above DAs followed opposite approaches, the objective was the same—self-defense rather than genuine problem solving.

All managers take care to exemplify sound management practices, and to present a good case for the objectives they pursue. However, DAs will exemplify only those practices they know management is monitoring. Moreover, they will only take a stand and disagree with management when they feel it is safe to do so. No DA wants to win an argument and later be proved wrong. DAs take no chances. But Builders will. Builders will sometimes allow themselves to be vulnerable to senior management's displeasure, in order to maintain work group effectiveness. A Builder may project an image of being uncooperative, or even disloyal. By sometimes appearing disorganized, or rigid, the Builder provides his or her subordinates with the stability they need for longer-term achievements. Here is one supervisor's perception:

> Probably one of (B)'s biggest problems is that he wants to do things independently. Of all the people who report to me, he puts the lowest priority on reacting to what I require. When he follows through on projects that I give him, he follows his own time schedule and ignores my own. [Yet] he wants meetings to start on time with no messing around; he becomes very impatient if others appear to be wasting his time.

A subordinate, however, views these same qualities favorably—as a necessary stabilizing influence:

> Occasionally (B) decides to do what is right instead of what upper management tells us to do. He's been caught and chewed out for it a few times. He may be too open and push truth too far in disagreements with higher management. However, he also has taken a few gambles that have really paid off, such as working on anticipated problems or ordering equipment prior to their approval. He's even decided not to start work on a few projects that he felt sure management would abandon later, and they did.

In presenting these examples, I am not endorsing insubordination. My point here is, when a manager risks bucking the system in order to benefit the organization, associates consider that manager to be a Leader or Builder. They see the person as a necessary buffer between themselves and the rigid political environment. But when a manager, in order to benefit himself, chooses to support a system when the system is making a serious mistake, associates consider that manager to be a Destructive Achiever.

CRISIS MANAGEMENT

The DA has an exaggerated sense of urgency. And to a DA, any high-visibility, short-term objective is a crisis. As a result, lower-level organizational needs are ignored. Subordinates may find it difficult, even impossible, to get a commitment on significant objectives they have identified. If they do succeed in getting the DA's attention, and if projects are actually started, the projects may be abandoned as soon as the DA perceives that another crisis is about to occur. This exaggerated sense of urgency can sometimes be seen in the way a DA will resort to intimidation, or will overcontrol a situation. In either case, the benefits of such behavior come at a cost of gamesmanship later, and inefficiency at lower levels. Three subordinates describe it this way:

> (DA) is very good at handling crises and pushes effectively to get things done. But he functions on a crisis-to-crisis basis. Today's and tomorrow's projects are more important than the long-term projects. All his attention is on the projects in the spotlight.

> Everyone knows that you have to have a priority list—you have to rank-order things to be done. (DA) gives the impression that everything has to be done at once; everything is important and has to be done tomorrow. Obviously, some things have to be done tomorrow, and I would like to know which things really do. The second-priority items should be put smoothly into the system and you shouldn't shake up your organization to get something done today just for the sake of getting it done faster. I don't want to transmit a sense of urgency down the line when it's not needed.

> (DA) doesn't motivate subordinates, he threatens them. The project you are on at the time is the most important one, and if you fail at it, it could mean the end of your career. But he says this on almost every project. He's running scared from Atlanta [corporate office] and fears making mistakes; if he didn't have self-motivated people he would be in trouble.

Three peers observe different dimensions of the DA's crisis mentality:

> When (DA)'s group installs a prototype, they overmonitor the operation and get artificially good results. They see an operator make a mistake and they overreact. They simply couldn't get that good a result if they didn't

constantly watch the process. Then they leave and we get the blame when the process fails in normal conditions.

(DA) will go to the [staff group] people and say, "This needs to be done right now," even though it is 4:30 on a Friday afternoon. They may want to wait until Monday, but they know that if they don't work on it, he will blame them if things go wrong and will talk to others about it. He must get his own way and must solve his own problems, regardless of how it affects others.

(DA) can't take the pressure of having a problem on his hands. If he has a problem, he wants others to solve it for him very quickly.

As they have in other ways, some Builders increase their vulnerability to criticism by serving as a buffer between a short-term-oriented senior management and their subordinates. While others swirl madly about them, Builders provide a sense of calm. Two subordinates express their view:

(B) is the best boss I've ever had. He makes for a good working environment. My own subordinates are more committed, now that we have less of a crisis atmosphere. (B) has been able to create a cool, calm environment because of his general manner. I feel very grateful working under someone with (B)'s management style.

(B) is strong at motivating his group—a soft-sell kind of guy. He is direct in that he doesn't try to pull the wool over anyone's eyes, and he attacks things directly. He says "Will you do something?" rather than "Do it." Yet, I still feel like I know how he will react in a given situation.

A supervisor recognizes his subordinate Builder's strengths, but wishes the Builder could make those strengths more obvious to senior management:

(B) always has his case, but he could be more emotional—an outward sign of urgency. People wonder, "Will he throw himself into it?" He *does*, as you find out later.

A DA peer, oriented to a crisis environment, expresses impatience with an acknowledged Builder:

(B) needs to be more dynamic both in action and manner. He tends to set a pace for himself regardless of what the situation is. No matter how touchy things get, (B) is pretty much the same in speed, reaction time, and everything else. A very good influence on the kind of situation where you are discussing controversial and sensitive issues, however.

Another peer summarizes the Builder's stabilizing influence:

(B) is open, honest, doesn't play politics. Can handle anything that comes his way. Good relationships with his peers. He does a good job when he gives bad news to people—doesn't attack them, and he informs the people who should be informed. Assesses a situation quite well. Is interested in my problem and is willing to discuss it. When he agrees to cooperate, he does it with vigor, and there is no problem with who is going to get the credit.

INNOVATION

The most negative impact a DA has is on innovation and creativity, no matter what the area may be—technical research, human resource training and development, or some other department or unit. Mastery of the system is the DA's major accomplishment, not meaningfully improving its personnel, product, or service. And the focus is on visible results, not the most effective solution to a problem. Whereas a Builder appreciates, protects, and helps the Innovator, the DA views the individual as an impediment to his or her short-term goals and disruptive to the system.

If the DA and the Innovator happen to have the same area of expertise, the DA may see the Innovator as a constant personal threat, especially if they had been associates before. The DA, in reports to top management, presents the Innovator as a nonentity, omitting accomplishments as if they didn't exist. The DA may even go so far as to remove the Innovator from direct, influential communication networks. An Innovator subordinate talks about a DA:

The first thing he did when he took over was to stop the two most successful projects I was working on—that my previous boss and I had agreed to. Without even asking about them or trying to understand them, he dismissed them as being too theoretical and unlikely to get any practical results. Then he gave me several picayune projects of his own that any new college-hire could do.

On the other hand, the DA and the Innovator may have different areas of expertise. DAs whose technical competence isn't threatened simply put greater pressure on Innovators for quick results on present projects. Some DAs shift Innovators to projects with faster pay-offs. One subordinate states:

We had a great group when (L) was here. We were making solid progress in several areas and were achieving a lot. Every member was involved. (DA) now totally ignores the projects that don't have high visibility and doesn't have the slightest idea what a couple of his people are doing. He's pushing hell out of the rest of us. All our results are overblown. He reports favorable outcomes before we've actually analyzed the data. He's literally raping the group of its accomplishments over the last three years; I feel sorry for the next guy that inherits it.

The DA is more of an extractor from than a contributor to the organization. DAs use up the human resources of the group they inherit, then leave an inflated set of expectations for the manager who follows.

By definition, innovation is unexpected, to the point of sometimes appearing illogical. Innovation is usually difficult for the unsophisticated person to appreciate upon first exposure to it. Therefore, under the DA, "innovation" takes on a new meaning. Creativity means to adopt the latest fad or technological development that already has proved workable. The patently obvious becomes "innovative" and "creative," because it gives the DAs something new, but something they already understand and are able to implement with predictable results. A subordinate describes a DA:

(DA) is a conservative risk-taker. If a person is working on an innovation, it must show up as a plant benefit within six months if it is to be considered successful. He also doesn't want to deviate from the recommendations of others, especially from the higher Hay Points. And he has too much faith in emulating [another prestigious organization]'s results.

A peer comments:

(DA) needs to pursue more innovative ideas. He needs to realize when his group has hit the 80 percent level of the Pareto Curve. They get hung up with pet projects and carry them too far, as opposed to seeking more creative approaches.

When Innovators' efforts are unappreciated, or their creativity frustrated, they take their talents elsewhere—to another company, or to their non-work-related interests. Once an Innovator is rendered harmless, or checks out of the organization, either actually or in spirit, one of the better Mechanics becomes the new Innovator. This individual is seen as being more cooperative in supervising innovation.

Bureaucratic Stagnation

Everyone who reports to a Destructive Achiever must, to some degree, behave like a Mechanic. Mechanics tend to acquiesce, no matter what the DA demands. The DA may demand longer and harder working hours, or a changed performance evaluation of an innovative subordinate. Two subordinates report some problems created by a Mechanic, common in a DA-dominated environment:

> (M) is too easily snowed by people who want to present a case to him. He should ask tougher questions; and pursue an issue to find out what is really true or realistic. He doesn't seem to be shooting for higher goals; he needs to have a greater hunger to be right.

> (M) needs to get our inputs more about objectives—he should ask for more staff and facilities for us. He assumes too many projects and responsibilities, and he should work to narrow down the range of activities he has us doing. Over time, he has become clearly less assertive than he used to be. Now he appears to be gun-shy.

A peer had participated in some appraisal discussions with a Mechanic. He had this to say about the Mechanic's contribution to bureaucratic stagnation:

> (M) will try to read what higher management's opinions are about his subordinates, and when he finds out that they don't rate his subordinates as highly, he will tend to back down. In our discussions, he should be more willing to take a stronger stand—defend them if they stand out.

Leaders and Builders react to DA supervisors in various ways. It depends on the strength and flexibility of their ethical principles. It also depends on how significant are the benefits of retaining tenure—pension plans, vacation time, and so forth—and on how they perceive their chances for influencing the system favorably. Under the supervision of a DA, a Leader or a Builder may decide to leave or to stay and adapt to the DA-dominated environment.

And adapt they must. For the DA is a natural inbreeder, and *any* lower-level person is vulnerable to the DA's disfavor. Once perceived by the DA to be unsupportive or disloyal, he or she soon discovers that any characteristic weakness, or even an unappreciated strength, can be fatal—a sufficient reason for dismissal, demotion, or delegation to obscurity. If a Leader or a Builder criticizes the actions of a DA, no matter how legitimately, the criticism will be presented to upper management as evidence of a negative attitude. If subordinates oppose the DA's attempted action, the DA will see such opposition as resistance to change; but if subordinates themselves want to attempt something new, the DA will consider them too theoretical. And so on.

The more DAs and Mechanics enter the power network, the less influential the Leaders, Builders, and Innovators become. Organizational decisions are based on their implications for systems and less on real technical merit, and what may once have been a problem-solving climate is now a political one. Thus it becomes crucial, if the organization is to survive, that those who have good ideas develop the power and skills to get them acknowledged and accepted. This, of course, is no easy task. Many meetings soon degenerate into verbal sparring matches in which the intrinsic merits of different solutions may have little to do with the decisions made. Problem solving becomes secondary to control for its own sake. People take sides, and either you win or you lose. A marketing manager complains:

> We spend endless hours on the phone and in meetings to develop better sales forecasts. Our record for accuracy has been terrible. When we've done the best we know how, we submit the forecasts to the directors. They immediately say that they're unacceptable, and that we know we can do better than that. Now we have two additional problems: How can we inflate which figures and do the least harm to manufacturing; and how can we explain later why we were wrong?

A maintenance manager:

> My people are working 10 to 12 hours a day and getting tired of it. It's never going to end. We haven't done any preventive maintenance in two years. The time and money lost when a machine breaks down in the middle of a run is tremendous, and it'll get worse. But to (DA), every priority is geared to today's production.

An industrial relations manager:

> Right now, this very minute, we're in the process of firing engineers who have a superior rating (top 35 percent), because of the head-count cutback. Now (DA) tells us that *every* facility *must* recruit at least two new college-hires this year. There is a principle somewhere that you can't have a year without bringing in new blood! Explain that one to me.

A production manager:

> We have been told that we have no choice—it's a matter of survival. We either have to meet our production and cost reduction goals or admit that

we can't compete and get out of the business. The trouble is, we've been told that three times in the last ten years. It's boom, expansion, bust, and cutback. And at each stage of the cycle we have to prove that we are loyal soldiers who can bite the bullet and go the extra mile.

An R&D group leader:

There is no way we can meet all our objectives. Objectives are only added to the list, never taken off. On each project, (DA) wants us to explore every possibility that upper management has suggested, no matter how illogical. Anything they want, he assures them we can do; and if we don't deliver, it's our fault. The situation has got so impossible, people don't care any more.

The tragic solution to unproductive situations like these is often either a change in management or a total reorganization. Of course, those in charge of the reorganization are the same ones who caused the problems in the beginning. In such a situation, a Builder will likely be given a lesser role, or will be replaced by a DA. Senior management will again be inclined to pick other DAs with "can-do" attitudes, ones who will support *without reservation* whatever they want done. They look for DAs who apparently have what it takes to quickly whip a group back into shape. A DA supervisor expressed a very common criticism of Builders:

(B) is a very conservative manager, very deliberate. Very much his own man and will not join others in petty battles. There are times when he sees [two staff groups] letting him down, and he doesn't take the necessary steps to correct it quickly. He prefers to take it slowly and maybe spend a month correcting a situation, instead of taking a chance of a blow-up and trying to correct it in a week or so. Sometimes I've had to step in and take the risks myself, when he should have done so. Once he decides on a course of action, it's difficult to get him to change his mind. Tends to be noncommittal in group conflicts, as opposed to being supportive of me. Sometimes I have to be assertive and demanding to get him to go along with me.

Another supervisor describes a sad reality of the power politics environment:

(B) is too reserved, not outgoing enough, not energetic enough. There are situations where you have to sell yourself. He needs to get people caught up with him. He'll always be a second fiddle to the high flyers.

Once the cycle is well entrenched, it tends to continue.

Perspectives

The descriptions above are for illustration only. No single behavior is in itself a sign of a specific managerial type. For example, although the net impact of a Leader is positive, he or she may exhibit a few destructive behaviors as well. In a similar way, a DA may have only a few of the characteristics attributed to DAs in general. Yet associates may still rate that individual's impact on the organization negatively.

In other cases, of course, the negative characteristics of a DA may be clearly discernable to associates. I once helped conduct a one-week, in-house management seminar for newly promoted executives. On the last day, we had a senior executive of the corporation as a guest lecturer. He talked to the group about the personal characteristics and actions of the "ideal" executive. When he finished and asked for questions, a participant asked: "What kinds of things do you look for when you promote a person to higher management?" The speaker essentially repeated some of the things he had just said. Another participant then asked, "What do you look for in promoting a person to Branch Manager?" Obviously unsettled, the speaker raised his voice and said, "I just *told* you," and said that there was no difference. The discussion mercifully ended at that point.

By the time the speaker left, he may have figured out that what he actually had heard was an indirect criticism of the organization's last three promotions to Branch Manager. In later discussions it became clear that there was a consensus among the workshop participants that all three persons selected had had strong reputations as DAs *at all locations where each had been.* This senior management was either out of touch with its own organization, or it was unwilling to integrate its promotional decisions with its stated values. The consensus was too general and too strong.

On the other hand, I'm sure that even the best management or consultant cannot detect the consummate DA with any degree of assurance. Indeed, there is likely to be strong disagreement throughout the lower levels as to whether or not a DA *is* a DA. Two concerned subordinates explained:

> (Possible DA) is a very genuine, likable person, friendly, does not knowingly deceive others, but he talks too much and is a dominating person. Very enthusiastic, but he gets caught up in his own enthusiasm and he doesn't realize how much he is asking others to do.

> (Possible DA) does a good job communicating with people. Always seems to be out among people, talking with them. He's a very good listener. He will also volunteer for projects to the extent that sometimes he'll take the ball away from you without even thinking about it. May just go into someone else's area and take over a project.

And a peer:

> He confronts conflict very well. He is very positive and enthusiastic, and it's difficult to be at odds with him. You never feel you are in conflict with him, and you never feel threatened. He never tries to put people down, but his dominating attitude puts people down indirectly.

This type of person may take over a "breakthrough" technical project and, from development to installation, turn it into a profitable operation for the company. In the process, many persuasive and charismatic leadership characteristics, and few DA characteristics, were seen. So how could this person possibly be a DA?

But in-depth discussions may reveal problems, which had either gone unnoticed or had been ignored or unappreciated by senior management. For example, other research groups may find that, during the development of the breakthrough technology, their projects had lower priorities and longer turnaround times than those of service groups. The other research groups had become, in a sense, second-class citizens, but not because of any malicious intent. It was a matter of sponsors monitoring just the DA's high-visibility project, not theirs.

When the DA's project was first put into production, it was a three-month disaster. Personnel from groups such as engineering, quality improvement, and maintenance were diverted from other projects in a crash effort to make it succeed. Many were associated with the breakthrough project, but many more were not; and those who were not felt that too much credit was being given to those who were, and at their expense. After all, the project *had* to succeed, by virtue of all the resources thrown at it. As for those who were associated with the project, some saw the possible DA as a Leader, who got more out of them than they thought they had in themselves. But others felt that they had been manipulated, and that those who had actually been responsible for the project's success were not the ones who were getting the credit for it.

These perceptions are not likely to disturb the DA or the DA's supporters. The DA's only responsibility is to achieve his or her own objectives, which have been endorsed by senior management. Because competition in the corporate marketplace is acknowledged as a fact of life, it is up to others to advance their own objectives in whatever ways *they* can. If others allow the DA to consciously or subconsciously command a bigger share of resources, it is due to *their* weaknesses, not the DA's abuse of power (whether of position, expertise, or charisma). In other words, the DA need not exercise self-restraint. It is up to others to restrain the DA from his or her own egocentrism.

The problem with a situation like this is that it is virtually impossible to assess the total cost/benefit ratio later on. The project itself may have been very profitable, but at what cost? What was the effect on other research projects, on the standard product lines, on future employee commitment to maintain a stable, quality business? Did the project represent the *best* use of resources? These are issues that must be considered, because the consummate DA is going to succeed no matter what. And he or she is likely to *accept,* and champion vigorously, any goal presented, regardless of its long-range implications or its impact on other organizational units. Therefore, senior management had better be sure that it is the right goal to begin with, and that they are willing to live with the results, especially in nuclear, chemical, and other industries that might affect large segments of future populations.

Some DAs are visionaries who select and pursue their own idealized goals. These DAs may appear to some to be examples of Abraham Maslow's "self-actualized person" or Harold Leavitt's "pathfinder." However, their devotion to their own cause blinds them to the long-term costs to society at large.

For example, who knows for sure whether those who favored the early manned space flights over more extensive prior development of space robots were right or wrong? The real question is: Were they doing what they *thought* was right, what they *wanted to think* was right, or what was most *personally* self-actualizing? And, what is most important, to what extent did they make a conscientious attempt to consider other viewpoints? Only through an extensive analysis of their behavior could we determine whether they were functioning as Leaders or consummate Destructive Achievers. The same could be asked of those involved in the Strategic Defense Initiative, any number of costly corporate acquisitions, or any decision in which large amounts of resources, power, and self-interest are involved.

To conclude, DAs range from the obviously destructive, at least as perceived by peers and subordinates, to the DA who borders on being a Leader. The latter is by far the most dangerous, because there is such a fine line between this type's use and abuse of power. Blessed with an unsophisticated senior management—or a naive public—there is no limit to how much power the consummate Destructive Achiever can acquire, even in a fully open system.

Footnotes

1. Quotations in this sequence are from articles in the *Charlotte Observer,* Feb. 2, 1986, 1A, 10–11A; and May 13, 1986, 2A.
2. *Charlotte Observer,* June 1, 1986, 9A.
3. John Kotter, *Power and Influence.* Free Press, New York, 1985, 87.

CHAPTER 3

The Image of Leadership

Many readers may have found the descriptions of Builders in Chapter 2 unsettling. They don't fit the conventional model of the successful executive. But that is why they were selected—to point out a disturbing fact: in our society, management favors the Destructive Achiever over the Builder. It is the DA who is considered to be the "successful executive," not the Builder. It is a cultural bias and therefore extremely difficult to change.

In comparing Builders to Destructive Achievers, traditional management has the risk/reward ratio all wrong. When the DA fails, both the project and the ethical standards of management suffer. When the Builder fails, the project suffers but standards are left intact. When the DA succeeds, the ethical culture may be permanently damaged, especially if the DA is wrongly rewarded by a promotion. If the Builder succeeds, the ethical culture is strengthened. Management instinctively knows this, yet the traditional bias persists.

A senior executive of a Fortune 500 company suddenly had to leave a meeting in which management was discussing plans to dismiss a large number of employees. The executive, highly aware of his obligation to stockholders, and genuinely concerned about the welfare of his employees as well as that of the total organization, became physically ill. Yet, this same executive later promoted to plant manager an individual who did not meet his ethical standards. Known to be bottom-line oriented in the negative sense, this new plant manager viewed employees as expendable and considered only short-term demands.

Consciously or subconsciously, even such a well-intentioned Leader as this may be inclined to favor a Destructive Achiever over a Builder (and, occasionally, over a subordinate Leader) for promotion when the two are approximately equal. Comfortable with their own values, they feel safe delegating power to those who will most likely further their own goals, regardless of the candidate's questionable ethics. On the other hand, those whose ethics are unquestioned, but whose ability to deliver appears less certain, usually come out second-best. This accounts partially for the barrage of publicity on the misdeeds of corporate personnel. In the case of a producer of computer chips for sensitive military equipment, over 100 persons had been involved in, or must have known about, the falsification of quality control testing. Yet, senior management claimed no knowledge of the illegal, or at least highly unethical, practices. Other misdeeds include a deliberate approval of defective welds in a nuclear plant, falsification of expense charges by a defense contractor, check kiting in a brokerage firm, and omission of test data in a drug manufacturer's reports.

The DA's Competitive Edge

The net impact of the DA on the long-term welfare of the organization is negative. Despite this, he or she has a significant competitive edge over the Builder, the Innovator, the Mechanic, and, occasionally, even the Leader. By promoting DAs, management can "benefit" from their short-term value orientation yet remain insulated from blame when the longer-range ill effects inevitably show up. Those that reach the newspapers are the more blatant ones. Far more costly is the unpublicized day-to-day deterioration of employee morale and commitment, and the resulting loss of quality productivity.

In all fairness we should note that the bias favoring DAs is not limited to management. For example, in a highly competitive environment, subordinates prefer a DA who has the power to advance them and their unit. In fact, if a unit is successful under a DA, morale can be quite high, with many kudos and monetary rewards to be shared. The unit develops the same kind of pride and spirit a winning football team has, no matter how it won the game. In fact, when circumstances are favorable, and the DA is getting results, he or she functions almost as a genuine Leader. It's when times get tough that the differences between the DA and the Leader become painfully evident to subordinates.

THE PROFESSIONAL IMAGE

Leaders are vulnerable to this promotional bias because of the personal rapport they have with DAs. DAs *seem* more like themselves than do Builders, Mechanics, or Innovators. DAs have the charisma of professional image: that is, they dress and speak well; they are persuasive, knowledgeable, energetic, and competitive; and they have a sense of urgency and are willing to work long hours. A supervisor comments on a DA who quickly delivered what he had requested:

> (DA) took over a job that was all messed up and immediately developed the right measurements to organize the unit. He set up a system for feedback and accountability and turned the department around.

But others saw him differently. A peer observed:

> (DA) is outstanding in communications—writes and speaks very clearly. Basically, he does what he says he'll do, in terms of dependability. But he doesn't put on the big hat. Just those things that make him look good. All he's interested in are his own objectives that have high visibility. He actually attends too many meetings, but only if the right people are there. If they're not, he won't attend. He's only interested in promotion—not his peers or subordinates.

Also, DAs *express* the same values one would expect of Leaders: the production of a quality product; service to customers; high ethical standards; and the development, involvement, and participation of lower-level personnel. These leadership behaviors and expressed values are easily observed, even in the brief contacts that managers have with personnel several levels down. A staff member describes how the DA can successfully, but superficially, meet senior managers' expectations:

> (DA) will give credit to "his guys" for coming through for him. But it's all phony. He does it in such a way that he's the one who always comes out looking good. He probably is best in his relationships with people outside the facility. Does a good job of representing us to higher management— very smooth in that regard. He's very personable, meets people easily, a good representative. He's also a good man to have around during management training sessions.

Many DA managers are fundamentally sophisticated. They are well read, and they know the language of the well-adjusted and effective executive. They know how to respond, and so they are able to fool even trained psychologists and management consultants, who occasionally become unwitting accomplices in a DA's image projection. But subordinates and most peers are in a position to distinguish between the *image* and the *substance* of effectiveness.

Despite the suggestions of writers of current training materials, it is difficult for many Builders (or Mechanics and Innovators who are capable of becoming Builders) to develop such charismatic traits to the degree necessary to compete with the DA. It is certainly harder for them than for those who come by it naturally, by virtue of personality or family or cultural background. A direct supervisor describes why it is so difficult for some Builders to meet management's expectations:

> (B) is a very good communicator when he is at ease. However, when he is under stress, when he gets questioned by people like [the president] and [vice-president], he gets very nervous and falls apart. He tends to withdraw, doesn't look others in the eye, gets tense, doesn't smile. His throat gets dry as a ditch. He tends to be self-effacing—he's his own worst PR man.

Poor upward communication and poor professional image are the two most frequently mentioned failings of Builders who are rated as nonpromotable by direct supervisors. The following kinds of comments appear on their appraisal forms with maddening frequency.

> George continues to demonstrate superior technical ability. He can break down problems into basic terms that lead to practical solutions. He has excellent rapport with his peers and subordinates. Communication upward and his upward image continue to be his problems.

> Fred is well respected by all those who work with and for him. Superiors, especially those some distance from him, are not sure of his abilities. He needs to work to improve their confidence.

> His outspoken manner sometimes gives upper management the wrong impression of a negative attitude.

> Good downward and lateral communications. Needs to devote more time and effort to upward communication.

> His projects and results don't get the exposure and publicity that they deserve. He needs to generate and project more self-confidence, sharpen his political skills.

Wendell Johnson wrote that our judgments of others are inevitably descriptions of ourselves. They represent "a projection of our own tastes, standards, feelings and knowledge, or lack of understanding."[1] The above assessments identify genuine individual weaknesses; but when the weaknesses appear in excessive number within an organization, they are an indictment of management. Management has created a culture in which: (1) many solid performers cannot be fully effective in their day-to-day interactions with others, especially upper management; and (2) a personality profile, and its accompanying values, continues to be underrepresented in management's decision making.

Organizational attempts to help Builders confront such an environment (through one-on-one coaching, assertiveness training, effective presentation, persuasive interviewing) may actually add to the cynicism the Builders feel. They believe the problem lies with higher management's values, not their own skills.

POLITICAL ORIENTATION

The number of employees who have an old-fashioned dislike of organizational politics and distrust of self-serving power is greater than generally recognized. Many are literally turned off by organizational pressure to conform. They consider it demeaning to have to devote an inordinate amount of their time to political communications designed primarily for their own or their superior's personal advancement.

The following is a reaction I got to a coaching session with an individual who was an effective technical group leader. Versions of it are not uncommon. He and I had just finished discussing a book that had a chapter in it on political communications, when he said:

> I know what you're saying is good advice. But I'm just not going to be that way. I'm respected in this plant. People know they can come to me and I won't take advantage of them or take undue credit. I'd just as soon not do all that "communicating" if that's what it takes to get promoted.

The surprising thing about his reaction was that we were discussing communications of the "squeaking wheel" variety, which were ethically acceptable. I had made what I thought was an innocuous suggestion, that he send in more status reports, more often, and to more persons. He said too many were doing this already and he didn't wish to add to the overburdened propaganda mill.

Subordinates and peers alike respect the Builder's stabilizing influence and commitment to nonpolitical behaviors. Yet, at the same time, peers are almost apologetic in warning Builders of the personal dangers involved in being politically naive:

> (B) is a consistent, predictable communicator, and you know he will always tell you what you need to know. You never hear it from your own boss before you hear it from him. He also tells you before he tells his own boss. One of his strengths is that he can ask some damned perceptive questions. I hate to recommend this, but sometimes he's too willing to risk showing his ignorance. He can come off bad by sticking his neck out and asking a question that demonstrates that he doesn't understand the technical intricacies of a problem.

> (B) is generally informal and does well at the peer level, but excessive openness may hurt him. Over time he'll learn that the upper managers aren't

that smart. He has to learn how to play the game, and if necessary, be more aggressive.

Subordinates have the same respect, and concerns, for the Builder's lack of political orientation:

> (B) might let out too much information too often, but I prefer it to the closed approach we used to have under my old boss. You get the feeling that nothing is sacred with him, and you worry about what he may be discussing with others. However, he really opened up with [another department], and they returned the openness. It could have been a bad situation, but it didn't turn out that way. He discussed potential plans that could really unsettle people, and it did at first, but it turned out well. They reacted positively to being let in on the inside dope.

> I rate (B) as an extremely honest manager—almost too honest. He will admit mistakes in his budget, in his project management, in engineering performance, and so forth. If he's going to have a project overrun, he'll admit it. However, he's very effective and gets things done through others.

Not only do many Builders balk at the idea of political communications, some have such a strong aversion to it that it seems they almost make themselves look bad deliberately. Mark McCormack described why many executives "are often found languishing in middle and lower-middle management":

> . . . their instincts are bad. What they pick up perceptively they always manage to misuse. Deep down inside they know what should or should not be said and when or when not to say it, but they can't help themselves. They blurt out some indiscretion, or can't check their need to "tell it like it is," even when they are aware that it is in their own worst interest to do so.[2]

Unfortunately, some Builders have already done irreparable damage to their careers by the time they learn McCormack's lesson. Even in companies that supposedly value openness, political communications are important and the dangers of indiscretions are real. A supervisor expresses his frustration with a determinedly nonpolitical Builder:

> I warned (B) *not* to bring up the subject of open appraisals again when Fred [a manager two levels up] was present. I know everyone else considers it important, but Fred just doesn't want to hear about it any more. Sure enough, after the actual meeting was over, and [another upper-level manager] asked if there was anything else we needed to discuss, (B) hit on the need for open appraisals.

Builders who are habitually open call into question the quality of their judgment. A fine line separates informed dissent from a "negative attitude," and many Builders cross the line all too often. Although subordinates and peers rate their integrity and net performance high, senior management will often punish them for it. A supervisor wishes his subordinate would learn to be more devious:

> (B)'s problems are with two factors. First, he tends to be negative about new ideas. Second, he tends to be a little rigid after he has decided to do something. I recognize that he may be correct in taking a stand on various things, but he doesn't have to show his rigidity as much as he does. That is, if he could be as rigid as he is but not show it so much, he could be a lot more effective.

The DA, on the other hand, creates a more favorable impression. DAs know that image is an important part of performance, and they cultivate it to meet management's expectations. DAs develop a keen sense of which lines not to cross. If a DA does cross any lines, he or she does it very carefully, always knowing when to back off. A peer describes the advantages of a chronically positive attitude:

> Dallas [location of company headquarters] likes the positive approach, and if they advocate a certain action, (DA) will see only advantages to the new approach. He won't point out the negatives because he knows Dallas doesn't want to hear them. He *will* take a stand, but only if it is a positive stand. If they want something done and it's a little controversial, he'll say, "I will get it done; I will make it happen." If he thinks nothing should be done, he won't say so because that's being negative, and being negative is bad for the image.

DAs keep their faults under cover and publicly display only those "cosmetic" improvements they know their superiors want to see. Conversely, Builders somehow seem to have a kind of perverse motivation to remain true to themselves, even to the point of publicly displaying their faults, regardless of the consequences. Why? Some writers suggest it's because there are just so many naive persons in our society. Others note that many Builders are "counterdependent." Psychologists define "counterdependency" as an appearance of independent behavior as a reaction against innate but unwanted feelings of dependency. This may be true. On the other hand, Builders may be crying out for management's attention, trying to say through their behaviors that they think the system is too political, too power-oriented, and not working the way it is supposed to. By assuming significant and obvious personal risks, they try to get the sophisticated power brokers to realize the damage they are doing to the organization.

MANAGEMENT DENIAL

Senior management prides itself on being a "good judge of horseflesh," yet will often deny the obvious faults of those they promote to managerial positions. It's important that the record show that they evaluated personnel correctly, and so they tend to view subordinates' future behaviors as confirming their evaluations. This is a common fault. Obviously all five managerial types are going to cut across the broad spectrum of personality traits and behavioral styles—and of managerial skills.

A person considered to be a Leader may be noted for poor conference leadership skills. Conversely, an untrustworthy DA may be given credit for initiating a fundamentally sound MBO system. An Innovator may be an aggressive attacker or a withdrawn theoretician. A Mechanic may be a highly caring individual or a "cold fish." There is such a wide variety of possible combinations of strengths and weaknesses that it is easy for senior managers to be misguided by their biased interpretations of performance results. It is usually difficult to determine to what extent a group's success is due to its manager. Did the supervisor provide planning, delegation, and guidance? Or were the subordinates responsible for results, despite poor supervision?

Builders suffer the most from this nagging dilemma. Their reluctance to advertise their good points turns a strength into a weakness. A common reason given for not promoting a Builder is that the Builder was successful because of strong subordinates, not because of strong leadership skills. A direct supervisor explains why he couldn't sell senior management on his subordinate:

> (B) is very low key. His group has made solid, long-range, and fundamental improvements; not just fire-fighting. He's better than he thinks he is, but he needs to communicate more. Call on the phone more often and update others. He has a lot of successes to talk about. Senior management has trouble separating what he does from what his group does—who should get the credit.

A peer commented on a Builder's image weakness, which is also a strength:

> (B) may go overboard in letting his subordinates have the limelight, in giving them too much credit, and in not taking credit himself.

This is a frequent criticism, in terms of the Builder's self-interest. Another is that Builders allow subordinates to give reports they could give themselves.

Associates sometimes refer to the Builder as "always a bridesmaid but never a bride." Time and time again the individual is recommended for promotion by lower-level managers and then vetoed by senior management, so much so that it becomes embarrassing. Usually this Builder is promoted *only* when there are virtually no Leaders or Destructive Achievers with comparable experience available. When this Builder is eventually promoted and becomes successful, management, although somewhat surprised, is pleasantly relieved. They don't quite understand how such an unimpressive person is able to meet the urgencies of the business, especially at the next higher level, no matter what that level may be. Consider these superiors' reactions:

> John's group gets results, and I find it difficult to criticize his performance. I know this sounds crazy, and I apologize for it, but if he would just look more *worried* when we have a crisis, I would feel a lot more comfortable. He doesn't seem to have that sense of urgency that's necessary to get things done.

> I know Chuck is doing a fine job where he is. But he's working in a confined environment. For the life of me, I can't see him taking on that crowd we've got in marketing.

Management is also surprised when a DA fails. The DA seemed to "have it all"—the trappings of a Leader. The DA "looks like" a Leader, even when accepting blame for failure. Acceptance is done with "class" and in such a way as to make it obvious that the circumstances just were not right. The DA knows where he or she made mistakes and now will be an even better manager. A superior describes the DA who has convinced senior management of his good intentions, positive attitude, and desire to improve:

> (DA) needs to improve his overall diplomacy. People need to be more comfortable in expressing ideas to him. Needs to keep an open mind longer before deciding an issue. But it wouldn't take much for him to turn this around. He is very motivated to improve and is willing to try new things. He has the capacity to make the change to become a more participative manager.

On the other hand, when a Builder fails, management's suspicions are confirmed. The Builder's influence in the power structure is now permanently reduced. But the DA who fails is frequently transferred or promoted, much to the astonishment and dismay of subordinates and peers (and sometimes an entire facility). Two superiors tell how senior management favors short-term results over the welfare of their human resources:

> We're transferring (DA) to head up the Southeastern Division. He's got what it takes to get things done but he's going to need help. We know that he created some real morale problems where he was. He's got to learn that you just can't go into a district and replace everyone who's only average.

We don't have that many superior people around. We *all* have to learn how to manage a pretty typical group of people.

We're moving (DA) to a staff job where he'll have no one reporting to him. And now *he'll* be the one who has to get cooperation from the plants. He needs to appreciate the negative effects that his arrogant attitude has on others. Let him see what life is like from the other side for a change.

In spite of his very poor reputation, the DA was later promoted over his peers in the line organization.

There are many reasons why senior management often denies the shortcomings of the people they promote. It's hard to admit to error. And there are pressures to retain, albeit temporarily, those subordinates who deliver rapid improvements at the bottom line, whatever the long-term risks. And, of course, there are always guilt feelings to contend with when a DA plays a significant part in senior management camaraderie. A subordinate describes the social strengths of some DAs:

(DA) is a completely different person in the plant than when he is outside. You couldn't find a better guy for a round of golf, or when he has a cookout for us at the lake. Something happens to him when he's at work—maybe it's the pressure for results. But people have quit trying to do new things and are now just trying to stay out of trouble.

This DA is likely to do very well at traditional senior management gatherings at popular resorts. The mixture of golf, tennis, conferences, discussions, and cocktail parties do not make the same demands on individuals as do missed deliveries, production mishaps, or lost customers. Those who want to know about his or her impact on subordinates and peers at the plant are blinded by his star performance in the clubhouse. Also, superiors fail to understand a basic truth: Personnel two or three levels down can't interact with the DA the way they do. The power relationship is entirely different. As one naive senior manager saw it:

If they [a DA's subordinates] would learn to confront him, they wouldn't have these problems. They need to forget that he is a vice-president and just talk to him. Sure, he'll yell and stomp around, but that's just (DA). He'll forget it the next day. He needs a strong, open staff who can stand up to him and say "Whoa," or he is going to be taking action when he shouldn't. He doesn't want people who will withhold bad news from him— that's a cardinal sin in his eyes.

A different viewpoint was expressed by a staff member:

> You have only two options with (DA) in a disagreement: back down, or face a tough confrontation. If you are strong enough to stand up to him, (DA) may change his mind. If you don't, he thinks you had no conviction to begin with. When a subordinate makes a mistake, he will come down hard as a first reaction. He may try to come back later and clean it up, but it's awfully tough on an individual. Eventually you get the feeling that you've used up your chances—you can go to the well only so often with (DA). It's very demoralizing.

Some Leaders and Builders exhibit similar behaviors. This makes it easier for senior management to deny the severity of the DA's problem. The behaviors can be described *almost* the same way, but subordinates see them very differently. As one put it:

> (L) gets emotionally involved in situations and displays his anger and frustrations too much. Some of the things he does seem inappropriate for a Director. Yet these are also the very things that give him his strengths—people really rally behind him.

In each of the two preceding cases, senior management knew that the manager expressed anger and verbally attacked subordinates. But they also knew through the grapevine that the subordinates in the first case (the DA) were hurting, well beyond the level of simple job dissatisfaction, and that the subordinates in the second case (the Leader) were not. Only their denial, or a rationalized business decision, would cause them to retain the DA.

Changing Management's Bias

The term "management bias," as used here in only an organizational context, can obviously not be limited to today's organizations, to our culture, or even to this century. The problems cited here are built into the human experience, and they have contributed to the decline of all kinds of organizations and societies throughout history. To address them, we need to make some fundamental changes in how we define and select leaders. To begin with, we need to agree that a leader without ethics is no Leader, even for the short term. The more difficult challenge is to reassess the ways in which we have always viewed the only alternative to the Destructive Achiever—the Builder (assuming all obvious

Leaders have been identified). Our bias against Builders is so ingrained that we see their strengths as weaknesses. A direct supervisor describes why a Builder may have to adopt counterproductive behaviors if he is to improve his image with senior managers:

(B) could be an R&D director, or he could handle [the vice-president]'s job. In general, he needs a more aggressive image with [two persons in top management]. However, if he were to be more aggressive, it could be detrimental to his relationships in the plants. He has very good relationships with all of them, as well as with the marketing group.

Even though a Builder may not have what it takes to become a charismatic Leader, he or she is always a better choice than the charismatic DA. Once properly developed, many Builders achieve at higher levels in the same environment. Certainly they deserve the same chance to do so as the DA. Peers cite two different Builders who fit this description:

(B) is not the leader of the pack. He is not assertive or demanding enough. He is politically astute, but he doesn't use his astuteness. Too nice a guy— not going to hurt anybody. He is a person you can depend on to get the job done. He's capable and does his homework. Over time, his managerial style has improved. He participates more in meetings, and expresses his views more. He gets more people involved in his projects and is more forceful in expressing his opinion than he used to be.

(B) did more in 12 months in (DA)'s old job than (DA) did in three years. [(DA) had been promoted from the job; (B) was considered not promotable and eventually went to work for another company.]

Effective managers range from the Mechanic, who lacks leadership skills and actually should not be in management, to the Leader. Between the two fall three levels of Builders. At the lowest level are those whose leadership skills are adequate for their position. They do fine where they are and make positive contributions to the organization. Next are those Builders who could achieve higher leadership positions within a definable environment; they are able to grow and to develop the political skills and power required by future work associates. At the highest level are the Builders who are able to become recognized Leaders by effectively interacting with outside forces in a variety of environments.

These are theoretical distinctions; such clearly definable persons are never obvious, except after the fact. The point, however, is that senior managements need to appreciate four things about Builders:

1. *Any* level of Builder will make a more positive long-term impact on the organization than the most sophisticated and impressive DA. (There may conceivably be exceptions to this statement; for example, by violating environmental or securities laws, a DA may allow an organization to survive a financial or legal crisis. Of course, it was probably a DA who led the organization to the critical juncture in the first place.)

2. It is likely that significant numbers of the middle and highest levels of Builder exist in every large organization right now—unrecognized, undeveloped, and underutilized.

3. Some Builders who are not recognized may lower their standards. If they conclude that the organization requires it, and they have no other choice, they become more like DAs.

4. Whereas natural Leaders and DAs are capable of developing themselves in a variety of environments, most Builders need a proactive developmental environment early in their careers.

Accepting these four premises won't be easy for traditional managers. They undoubtedly have already found out that many Builders are less enjoyable—and more frustrating—to work with than are Leaders and DAs. Builders are not instinctively "members of the club"; nor are they concerned with the refinements of management protocol. However, if senior managers truly wish to value and develop their human resources, they have no choice but to develop Builders. As will be explained in Chapter 5, it isn't much of a problem in young, small, dynamic organizations. There is a sense of community under those conditions, and a variety of persons can be effective and have rapport with each other. Management's batting average for promoting Builders is initially good in these circumstances.

But to promote a DA is always a serious mistake whose effects will be long-term and cumulative. Every time a DA is promoted over someone whose ethical superiority is apparent to others, the value base of the organization is further diminished. No actions of management have greater ethical consequences than its personnel decisions—who will be promoted, who will be terminated, and who will be allowed to stagnate.

In Part One, the focus has been on the Destructive Achiever and the Builder: their ethics, and their long-term impact on the organization. Now we need to consider how these two different kinds of manager affect, and are affected by, our educational programs and our organizational systems. This interaction between people and systems appears to be caught in a downward spiral, and so we had better begin to understand it. That's what the next five chapters in Part Two are all about.

Footnotes

1. Wendell Johnson, *Your Most Enchanted Listener.* Harper & Bros., New York, 1956, 61.
2. Mark H. McCormack, *What They Don't Teach You at Harvard Business School.* Bantam Books, New York, 1984, 60–61.

PART TWO

Systems and Substance

Where Goethe is said to have cried when dying, "Light, light, more light!" and Miguel Unamuno . . . has responded with his "No—warmth, warmth, more warmth! for we die of cold and not of darkness. . . ."; we today seem driven to plead for "Air, air, more breathing space, for we die by the outbreathing of our own poisons."

We have more light, more knowledge than Goethe ever dreamed of possessing. We have warmth and passion enough to spare behind the political and social ideologies of our time. We seem almost to be broiled in a surfeit of this light and warmth as though we had been strapped under a sun lamp and could not pull away. Yet the lungs of our spirit fail us in the self-poisoned air where we breathe and rebreathe the projections of our own fear and greed and provincialism.

—Douglas V. Steere

CHAPTER 4

Research and Education

As stated in Chapter 1, most of the current generation of Destructive Achievers are not, like the robber barons of yesterday, *inherently* destructive. Their behaviors are instead a *consequence* of how they were formally trained to manage and how they learned to manage through experience.

In research and training in management and the behavioral sciences, the mechanics and procedures of human interaction are easy to observe, evaluate, teach, and apply. On the other hand, *operational* ethics and values are difficult to observe or measure and impossible to instill in the classroom. As a result, the verifiable expediencies of today become the standard for tomorrow. In spite of good intentions, our abilities and opportunities to manipulate our environment take precedence over the development of the values we know we are supposed to have. This leads to what philosopher Abraham Kaplan called the " 'ordinal fallacy' . . . first I will achieve power, then use it for the public good."[1]

To understand why this happens, we need to look first at the subjective nature of the research on various management and behavioral science theories. It is easy to document whatever management philosophy one *wants* to believe. Although a Leader or a Builder can "prove" the necessity for one set of values and behaviors, the Destructive Achiever, increasingly it seems, has access to a contradictory set. Research, then, is inadequate as a foundation for behavioral standards. Moreover, there is a tendency in training and development to pursue expedient solutions for *symptoms* of behavioral problems, rather than address causes.

Research Validity

If we could ask him, Attila the Hun would probably give Genghis Khan high marks for leadership style. This would be a "valid" response: that is, it would accurately reflect Attila's view of success and his understanding of leadership, as he would define the term. *Validity* is an aspect of management and behavioral research that is generally misunderstood, creating all sorts of problems.

Validity is *always* subjective. To determine validity, one asks the question: Are you observing what you think you are observing? Not long ago, when a drowned person was taken out from under the ice of a frozen lake, we made what we thought was a valid observation of death. Subsequently, having revived a few such persons, we have discovered the "mammalian diving reflex," which protectively shuts down the human system for as long as 40 minutes. This, as well as many other medical breakthroughs, have taught us that no longer can we conclusively determine "death" on observation. Its outward signs are no longer "valid" as criteria.

In the same way, when, in research, we identify a "successful" executive, what standards are we going by? The standards of senior management (remember Attila)? The standards of subordinates (in what way did they personally benefit from their superiors' behaviors)? Or the bottom-line results (were they short-term, at the expense of long-term)? Each measure is valid to the extent that we agree on the definition of success. We must also understand very clearly exactly *what* we are measuring, which is remarkably hard to do. In implementing the results of management and behavioral research, we tend to forget that our research was based on *unique* organizational and behavioral systems; that is, systems that were

- unique to a particular time period, which could be an entire work force generation. Research only tells us about the past and is therefore limited to the *perceived validity* of the accepted definitions at that time.

- unique to a particular society—Japan, Western Europe, Southeastern United States, the steel industry, an R&D facility, a manufacturing plant.

- unique to a particular ideological environment—traditional, progressive, directive, participative, competitive, cooperative.

To illustrate, consider General Foods' Topeka Project, begun in 1971. With the exception of the Western Electric Hawthorne Studies, the Topeka Project is possibly the most discussed study in modern management. Although it is generally recognized as an important contribution to management theory, it has been criticized (as were the Hawthorne Studies) for having been inadequately and inaccurately reported. After re-examining the Topeka experience and the literature about it, David Whitsett and Lyle Yorks made some pertinent observations about case studies:

> On occasion, these case studies are used not as sources of hypotheses for future testing but as anecdotal arguments for ideological positions.
>
> It politicized the Topeka system to the extent that representatives of opposing schools of thought found it necessary to either protect it or repudiate it. Understanding it became a secondary issue. In the process, myths about Topeka became institutionalized into textbooks as case reports became part of the body of knowledge of management. . . . Under the traditional rules of science, Topeka cannot be offered as a generalizable example of organizational innovation—the conditions there were unusual enough to severely limit the extent to which we can safely generalize. . . . The nature of the case study simply does not permit generalizing to a wider sample of organizations. Walton wrote more confidently than was warranted. Further, the study was lacking in detail. In social situations very small differences are often important.
>
> Field applications of management principles are always situation specific.[2]

The only problem I have with the excellent analysis by Whitsett and York is their implication that, "under the traditional rules of science," there *are* generalizable examples of organizational innovation, even though Topeka isn't one of them. Surely there *are* absolutes in human and organizational behavior, but how do we ever know when we have discovered one? The point is that *all* research, whether case study or controlled experiment, reflects the ideology of the researcher. The researcher is like the walleyed pike put into an aquarium divided by a glass pane into two sections—the pike on one side, smaller fish on the other. Every time he tried to eat another fish, he bumped into the glass. Eventually he quit trying, even when the glass was removed. His previous environment conditioned his behaviors and limited his perceived options.

In the same way researchers are limited by their values and their understanding of reality, from the initial determination of objectives to the definition of variables. Results are important, but it is just as important to understand the researcher's unique definition of terms and ideology. Whenever we try to generalize from anyone's research-derived conclusions about leadership, power, or success, we must do so from a sound ideological base of our own. For scientific research is *anti-ethical*. As explained in the introduction, the term refers to an absence of ethical concern in which subjective value interpretations are discouraged. Scientific research is concerned with what *is,* not what *should be,* from a moral viewpoint. It deals with verifiable, repeatable experiences. Therefore, if we base our actions on what works, and not on what *should* work, or what we can *make* work, then our values will follow whatever expediencies the research dictates.

I'm not saying that we should ignore research, or that we should never change our values. Rather, we should be aware of our values, what we want to change, and how we're going to do it. Otherwise, we become slaves to the past and present, limited to manipulating our way through an environment that is increasingly results-based (in a materialistic way), and not value-based. Therefore, researchers' recommendations should be interpreted in the context of the values that, in the observer's judgment, seem appropriate. The same qualified interpretation should be applied to any proposed explanation of the realities of organizational life, however derived.

Two issues are especially relevant to the anti-ethical nature of research— power versus problem solving and personality versus performance. These issues are heavily influenced by the ideologies of researchers and so serve as perfect examples of how so-called valid conclusions can contribute to the development of either a Destructive Achiever or a Builder, depending upon the predisposition of the research user.

First, the power-versus-problem-solving issue centers on opposite views of how to favorably change organizations. The power orientation appears to be gaining in popularity and is, in fact, becoming the wisdom of the day. Second, the personality-versus-performance issue deals with our ability to identify personality types and to predict their behaviors in different situations. The possibility of reliably selecting and controlling individual behaviors by applying scientifically derived personality profiles is indeed appealing.

POWER VERSUS PROBLEM SOLVING

Compare John Kotter's view of power[3] and Chris Argyris's view of problem solving.[4] It makes a good comparison, because Kotter presents an excellent case for a power-oriented ideology and Argyris an equally excellent case for a problem-solving ideology. Also, both base their theories on extensive research with personnel in modern organizations. Kotter's views of power are valid for today's organizations and societies. Although he warns of the potential danger of power leading to authoritarianism (among other negative behaviors), he still sees the occasional necessity for the leader to use manipulation, or even coercion, to achieve an important objective.[5] He also criticizes those who would rely primarily on the development of problem-solving skills to the exclusion of power skills as the way to arrive at rational decision making.[6]

Argyris also makes some valid observations about the deficiencies of today's organizations. He sees a need to develop better problem-solving values in future organizations and to reduce traditional management's reliance on persuasion, not to mention manipulation and coercion. According to Argyris, today's managers have a strong desire to control their environments. This prevents them from discovering and accurately analyzing the most important problems and leads to faulty decision making.[7]

As strange as it may seem, I have no argument about the validity of either view. Each one is based on an accurate perception of organizational realities, albeit from different perspectives. However, I must agree with Argyris's view, as the more desirable ideology. Manuel Velasquez describes an ideology as expressing a ". . . group's answers to questions about human nature, . . . about the purpose of our social institutions, . . . about how societies actually function, . . . and about the values society should try to protect."[8]

Although Kotter talks of power and influence, he sometimes specifically describes, and his ideology suggests, power *politics*. (*Webster's* defines "power" as "a possession of control, authority, or influence over others," and "power politics" as "politics based primarily on the use of power as a coercive force rather than upon ethical precepts."[9]) In describing how two managers, both effective, used their formal authority, he cites the advantage of personalizing the power inherent in the position:

> Two managers with the same formal authority can have very different amounts of power entirely because of the way they have used that authority. For example:
> - By sitting down with employees who are new or with people who are starting new projects and clearly specifying who has the formal authority to do what, one manager creates a strong sense of obligation in others to defer to his authority later.
> - By selectively withholding or giving the high-quality service his department can provide other departments, one manager makes other managers clearly perceive that they are dependent on him.
>
> On its own, then, formal authority does not guarantee a certain amount of power; it is only a resource that managers can use to generate power in their relationships.[10]

Certainly, Kotter does not condone unethical behaviors; in fact, he specifically condemns them. Yet, in the above example, what ethical right do managers have to withhold service to other departments in the same organization simply to increase their personal power beyond their formal position? Such behavior, if not unethical, is certainly *anti-ethical*; that is, *not* based on ethical precepts. It is, moreover, a clear indication that many persons are now getting more than their official share of power. What counts in today's organizations is *who* will have more power—not *what* is just, or what the long-term utilitarian effects on the total group will be.

In contrast to Kotter, Argyris cites the negative long-term effects that result from the way many executives attempt to solve problems. He suggests that those who wish to gain credibility must do more than learn the new skills; they must also internalize a new set of values.[11] These new values incorporate, among other things, a more genuine respect for valid information, free and informed choice, and internal commitment to the choice.[12] They also require less of a win-lose orientation and less effort invested in attempts to control the purpose of group meetings or personal encounters.[13]

Kotter's recommendations for power and influence appear to be more scientific than those of Argyris. Based on the realities of today's organizations and managers, they are valid and well intentioned. Nevertheless, their appeal is to those in positions of power who are seeking more predictable methods and justifications for exercising personal control. Even when presented within an ethical context, the idealization of power at the expense of problem solving tends to become anti-ethical—that is, essentially value-free, quite naturally appealing to the Destructive Achiever in each of us.

Argyris is more concerned with who we should become and how we should behave—a much less scientifically defensible position, but a better ideology for developing an ethical culture. It appeals to Builders who are frustrated with political conditions that prevent objective problem solving. We need to progress in that direction, but the question is, can we get there from here?

A new generation of organizational development professionals doubt that our power-oriented culture can be changed. In answering the question "Is Organization Development Possible in Power Cultures?" Peter Reason points out that collaboration and consensus are not likely to occur in a power culture:

> If a small group of people hold a position from which they can impose their definitions of reality, any move to explore the reality of the organization and any suggestion of renegotiation must be seen by them as a threat.[14]

Those in power continue to reject problem-solving approaches based on valid data and open communications. Andrew Kakabadse goes so far as to charge:

> The data feedback theorists, such as Bennis (1966) and Nadler (1977), who seem to assume the existence of the one fundamental, rational truth—data are merely the vehicle by which people will jointly recognize the truth—or the sharing, caring, participation philosophies, as epitomized by Argyris (1970), not only were and are unrealistic, but further are irresponsible.[15]

Kakabadse feels that those who want to improve organizations should first recognize the reality of power politics and then identify the holders of power. Further power strategies include isolating uncooperative managers from positions of influence, making deals, and strategically withholding information. History tells us that the power advocates are right—but that they shouldn't be. Unless we can devise ways to convince those with power to ethically and objectively solve problems, we are trapped in a social and historical downward spiral, making the same mistakes we have always made.

PERSONALITY VERSUS PERFORMANCE

A companion to the power ideology is the ideology based on control through personality. It is also anti-ethical and suppresses the development of problem-solving behaviors. As with power, the apparent realities of personality, as indicated by both research and personal observation, take precedence over values.

In summarizing the debate among psychologists about the predictability and changeability of behaviors, Stanton Peele concluded,

> One problem with personality research is that it examines and rates whatever traits the researchers are interested in at the time, conscientiousness, emotional stability and so on. But when we describe someone in real life, we don't consider an array of personality measures; instead we focus on a few distinctive traits that sum up the essence of the person.[16]

David Funder gave the debate additional perspective.

> It is undeniably true that *every* behavior is situationally caused, in the sense that everyone's actions are responses to the world around him or her. . . . Similarly, every behavior is *dispositionally* caused, in the sense that people do things differently from one another . . . and that these differences connect with behaviors at other times.[17]

In another article he aptly concluded that

> There seems to be little need to argue further about whether personality "exists" or is "overwhelmed by situations," . . . such arguments are as between straw men, and more constructive issues remain to be explored.[18]

Here again we find that contradictory opinions can be valid, depending on one's perspective and values. For some organizational positions, personality success profiles can be statistically reliable, yet exceptions occur, because we are all unique. We all sport never-before-measured combinations of traits, skills, and technical abilities that interact successfully with specific situations. Not only that, it may be the exceptions to the profile that turn out to be the most important source of innovation and system correction—simply because they don't conform to the system-defined mold.

Psychologists vary in their orientation. I once attended a meeting in which a consulting university psychologist presented the results of an opinion survey.

He had found a high level of middle management dissatisfaction with the authoritarian style of upper management. Upper management's response was that middle management should appreciate the necessity of quickly reacting to marketing pressures and management directives. The psychologist—clearly control-oriented—then suggested to upper management that they consider some new guidelines for promoting those middle managers who would be more willing to take orders. He himself could develop the guidelines by interviewing and testing the middle managers.

To me such a solution seemed unbelievable. Compare that of other consulting groups whose orientation is toward personal development. They stress the importance of understanding, appreciating, and utilizing individual differences, thereby achieving a natural balance of personal styles in which strengths and weaknesses complement each other. When the use of psychological research data is oriented toward control rather than development, at least two problems are created: (1) an inordinate insistence on upper-level/lower-level interpersonal compatibility; and (2) the development of standardized, although possibly unofficial, personality success profiles.

Once it becomes axiomatic in the upper echelons of management that there be a personality fit between boss and subordinate, abuses increase. To dismiss a "deviant" employee becomes an unquestionable right of the manager, who feels less need to adjust to the deviant, sometimes very effective and innovative, behavior of the subordinate. According to Lee Iacocca, the reason given to him for his dismissal from Ford was "Well, sometimes you just don't like somebody."[19]

The necessity for personality compatibility is a half-truth that has some ominous implications. It means that some managers are in positions of power because they have personalities that are more compatible than others, possibly despite the fact that they are less sophisticated, even incompetent, in the technology, service, or business. In the early seventies the AMA surveyed employee attitudes toward personality versus ethics in organizations. The results were shocking: *88 percent* of all respondents felt that "a dynamic personality and the ability to sell yourself and your ideas" was more important to career advancement than was "a reputation for honesty or firm adherence to principles."[20]

Immediate subordinates conform to the idiosyncratic expectations of their superiors. But so do subordinates several levels below. Pleasing the boss has become a cultural norm, and it has priority over objective problem solving. It is *expected,* successful people *do* it, and the only way to effect change is to work within the *realities* of the organization—which have, by default, become its operational system of ethics. Granted, management must ensure a quality work environment. It cannot allow a few individuals to disrupt it unduly. But, to remain competitive, management must realize that it has a wide range of persons at its disposal and must make use of those who can best contribute.

It is especially damaging to organizational excellence when professional image is based on superficial personal qualities, including, unfortunately, such seminar topics as "image" itself, dress, general style, and sophistication of formal communication skills. These are qualities that Destructive Achievers develop quite readily but that some Builders find difficult or distasteful.

Top management determines the level of substance that eventually permeates the organization. A personnel director complains about a senior manager whose first concern was the image of the job applicant:

> (DA) spent just 15 minutes with the candidate. His only comment was "I'm not going to waste my time interviewing someone for a job at that level who is dumb enough to show up in a polyester suit."

Even the most conscientious managers may be trapped into sacrificing substance in the name of professional decision making. They develop, as guides to personnel evaluation, success profiles that are incomplete and not related to actual performance. As organizations grow, these profiles become more important than performance itself, especially if the philosophy of management is that a key position cannot be held by someone they feel is "nonpromotable."

The problem is exacerbated by the rapid movement of managers upward. It means that many more of them are responsible for sophisticated technologies they don't understand. Because personal demeanor is so much easier to assess than individual contribution to long-range effectiveness, relatively inexperienced persons with "potential" are chosen over those of proven performance. Unfortunately, DAs appear to fit the psychological behavioral profiles of successful executives as well as Leaders do and, in some situations, even better.

When a person's power orientation and personality are correct, but he or she is getting only short-term results, the less observable (from the distant viewpoint of top management) long-range effectiveness of proven performance becomes, unfortunately, unimportant to the system. In business, our attitudes on power versus problem solving and personality versus performance form the core of our values. They determine our orientation to the world we live in (the present) and how we will go about creating the world we *want* to live in (our idealized future)—a goal we cannot reach if we allow our attitudes to keep us fixed in the present.

Training and Development

Jack Grayson, chairman of the American Productivity Center, was the organizer of the White House Conference on Productivity in 1983. He recently criticized business schools for teaching managers that they are the allocators and controllers of resources and that people, like materials, and energy, can be controlled.

"It's a one-way focus that assumes that the resources will behave in the way the manager wants them to."[21] He went on to say:

> I was trained at the Harvard Business School. I've taught at Stanford. I've been the dean of two business schools, and I can say that I was part of the group that helped to lead current managers into some of the ways we now manage. Now I'm trying to say we've all made a mistake. I think the initiative for the shift that's needed must come from management because it still directs the decision-making and control apparatus. It is extremely difficult to change a situation from below. There are too many odds to overcome—too much inertia and too much power. . . . The more power you [management] give up, the more you gain, but it's a different kind of power. You might lose legitimate institutional power derived from the organizational structure, but you will gain the power to lead people to greater productivity and higher quality.[22]

Grayson's analysis is accurate, but it doesn't go far enough. It fails to explain how difficult it is to get managers to initiate change from the top. Their biases regarding power have been built into the training systems of our organizations and the development methods we use. We will have to look beyond conventional management training and development methodologies for effective remedies, if there are any.

POWER-DOMINATED TRAINING

There is a catch-22 in organizational development: When managers use power and persuasion to create a climate of equality and problem solving, they prevent the latter from occurring. In other words, autocratic methods do not generate participative behaviors. Even when progressive managements believe that greater equality, more involvement, participation, and so on are needed, their attempts to bring about change are almost always unrealistic. Regardless of their new intentions, their behaviors reflect their old values. They let change occur only if they can control it in ways they are comfortable with. They want the societal and organizational benefits, but not the personal risks. As a result, they are blind to the fact that lower-level personnel are making all the adjustments and taking all the risks. This leads to some incredible situations in which senior management attempts, paradoxically, to correct organizational problems they themselves caused by improving the performances of lower management.

Many leading universities and consultant groups conduct workshops on management. Their advertisements indirectly describe the organizational climates that some senior managers have created:

Assertiveness (four different seminars):. . . for men and women in any phase of management or staff assignment who want to develop—and be recognized—as truly effective leaders. . . . You'll learn how to handle conflict wih strength and decisiveness. . . . If you're going to be a successful manager, you must be assertive. . . . You'll learn how to cope with criticism, anger, manipulation. . . . effectiveness in dealing with and influencing peers, subordinates, and superiors.

Negotiation (three different seminars): How to get more of what you want. . . . The only way to judge the success of a negotiation is to determine what you got versus what you gave, plus the long-term effect of the agreement. . . . This program can give you the "little edge" that might make a *big* difference . . . for every situation . . . an arsenal of strategies and tactics, tells you how and why they work and how to defend yourself against them.

Communication skills (three different seminars): In any setting, effective delivery is the key to success. . . . Not only what you say, but how you say it is important. . . . Many top executives state, unequivocally, that communication finesse makes the difference between promotion and dead-end career jobs.

One university business school analyzed its audience well in an advertisement for a training program. The ad began with "The bottom line starts at the top." A course for top management? Of course not. It then went on:

As a manager, you know that the responsibility for maintaining and improving the bottom line begins with *you*. How? By making sure that all those employees under you—especially your first-line supervisors—do their jobs. And do them well.

These descriptions are good examples of courses that probably meet the individual needs of many persons in today's organizations. Paradoxically, such courses create, at the same time, an even greater organizational need. It is a training version of a pyramid scheme: As each person improves his or her skills of assertiveness, negotiation, and communication *performance* (which is not the same thing as *communication*), a new level of power-oriented behaviors becomes the norm, and the reservoir of persons committed to rational problem solving is reduced.

The scenario begins when senior management realizes that it has created, probably inadvertently, a power-oriented conflict environment. In trying to change the fundamental culture for more productive cooperation, it fails to address the root cause. Instead, it chooses to remedy the situation by upgrading the skills of the lower-level persons who must bring the improvements about.

If management doesn't listen, if good ideas are being lost, or if poor recommendations are chosen over better recommendations, then improve the presentation skills of lower-level personnel. If management stifles upward communication, then teach assertiveness. If the climate is overly competitive, then teach negotiation. Unfortunately, such remedies only serve to solidify and legitimize the existing problems. Like drug addicts, we learn to manage the symptoms of an undesirable condition, but we lose our incentive to remove its root cause, which continues to grow. Management has accepted the environment it has created as reality. A situation once considered abnormal and in need of change is now considered normal and unchangeable.

Nowhere is this sad fact more evident than in the proliferation of personal and organizational success-through-power formulas. Even in highly respected academic journals we read about how important it is to dress for success, to cultivate the behavioral manners of those in the power structure, to make sure of a personality "fit" between boss and subordinate, to actively seek power (and to make sure that upper management *notices*), *ad nauseam*. It is all good advice in the sense that it works. But when it is anti-ethically incorporated into company-endorsed training programs, we then have a situation in which managers *expect* subordinate conformity to these "proven qualities of effective executives." In assessing subordinates, it is easier to assess their knowledge and their skills in personal demeanor than it is to assess their operational values or their actual contribution to long-range work group effectiveness. This is especially true when managers are inexperienced in their subordinates' areas of expertise.

The following recommendations for, or descriptions of, the use of power suggest the kinds of behaviors that increasingly are being considered "normal" simply because they are prevalent in today's organizations. Unfortunately, when managers are considering candidates for promotion, they may see these behaviors as *desirable qualities* of the sophisticated executive.

> Declare yourself to be an involved hands-on manager. Image is very important in the making of a successful executive.[23]

> Failure to adhere to dress and general appearance codes prevents many well-qualified executives from reaching the top. . . . CEOs and other executives in positions of power tend to surround themselves with men and women whose appearances please them—this usually means individuals who follow the top executives' appearance codes. . . . Individuals who make a point of looking like successful executives have usually increased their chances of reaching those top-level positions.[24]

> The rising executive being groomed for the top-level job is aware of the pressure from peers and superiors to fit into the ranks. Knowledge of cor-

porate custom and respect for the deference given to rank cannot be overemphasized.[25]

The only point to a "good" idea in business is to get favorable visibility. The manager with a "hot" idea really doesn't care if his plan brings in millions for the company, unless he is the owner. The real motivation is that the idea may result in a promotion, recognition, or a salary increase. . . . Quite a few top executives have proposed ideas that sounded great on paper but ultimately failed. The initiators didn't care. They had long moved on to higher positions, frequently on the basis of the visibility they received when they proposed the plans.[26]

Research and observation suggest that the most important source of power is absence of routinization. From the production worker to the executive suite, the task that is routinizable, that is, highly predictable and regularizable, has less power. . . . If you watch any reasonably cohesive work group, you'll see them seeking to make their external (as distinct from internal) relationships unpredictable to increase their power. They don't want to be taken for granted, to be treated like a low-status group that jumps to respond when asked to perform.[27]

Some may think that these are isolated, or extreme, examples of counterproductive attitudes and behaviors. If so, they need to review the promotional literature of leading universities and consultant groups throughout the United States. A further review of popular books for the executive on the make will reveal far more disturbing reality-based advice than has been cited here.

In 1977 Kotter wrote:

Power as a topic for rational study and dialogue has not received much attention, even in mangerial circles. If the reader doubts this, all he or she need do is flip through some textbooks, journals, or advanced management course descriptions. The word *power* rarely appears. This lack of attention to the subject of power merely adds to the already enormous confusion and misunderstanding surrounding the topic of power and management.[28]

If the number of books, articles, and course descriptions is an accurate index of under- or over-attention to a concept, then we have raised the power ideal to too high a level since 1977. Today, in popular business magazines, professional journals, and books on everything from leadership to power lunching, power is pushed as a high priority goal for the individual as well as for the organization. This swing of the pendulum, combined with the fact that it is easier to stimulate power-seeking behaviors than it is to stimulate their opposite—rational problem-solving behaviors—is a disturbing state of affairs. Apparently, our natural inclination is to progress toward more power, not less. And in so doing, too many move away from the ethical toward the anti-ethical, and ultimately to the *un*ethical use of power. We can see this trend in the literature. Writers are more realistic today in their assessments of modern organizational power cultures.

There are two forms of naiveté regarding power. The most publicized form today is the naiveté of those who wish, and act as though, power were not a factor in human affairs. This is certainly a dangerous attitude for the unwary. The other, however, that the abuse of power does *not* lead to success, is equally naive. This attitude, when it is internalized within the system, is even more dangerous to long-term organizational health. It goes something like this, and it is a standard part of today's training programs:

> Persons get power by gaining support from others in ethical, legitimate ways. This allows them to achieve worthy objectives for their own benefit, as well as for the organization. People who abuse that power do not get the support they need, are rarely successful, and are usually discovered early in their careers.

The first part of the statement represents the ideal use of power and is only partially true. But the second part is an unrealistic attempt to legitimize power advocacy: We need not fear the abuse of power because people who abuse power do not get the support they need to be successful. It simply is not true, given the number of top-level executives and governmental officials who have been caught in illegal, let alone unethical, activities.

Unfortunately, the power skills we give managers to attack the system, exacting from them no promise for their ethical use, are the same ones we should be defending the system against. The use of pure power is sometimes necessary to get things done. But it can be extremely costly in terms of rational problem solving and decision making. In addition, it is a major reason for the cynical disenchantment of today's corporate personnel. When conscientious persons lose out in the promotional and reward systems, especially to someone who is less productive, or even counterproductive, it is no wonder that they lose their sense of commitment to the well-being of the organization.

The appropriate use of power, of course, remains valid. Ethical people will benefit personally from training in the use of power, and will be able to use the power skills and strategies they learn to improve their organizations. However, long-term organizational benefits will depend *entirely* on senior management's willingness to strengthen the organization from the top, that is, to make the *need* for such power skills less important. If it fails to do so, the Destructive Achievers, who are already politically oriented and comfortable with the use of power, will predominate.

To sum up, there are *two* power-related qualities to develop in today's corporate personnel. The first is the ability to acquire and use power (reflecting our traditional orientation toward authority and a presumption of inequality). The second, and far more important, is the ability and *commitment* to make the use of power ultimately unnecessary in conducting organizational affairs (reflecting an orientation toward consensus problem solving, and a presumption of equality). It is not that today's power-dominated training is invalid, or that it fails to meet the needs of today's organizations. The problem, rather, is the *extent* to which power is idealized within an organization. If the organization already relies heavily on power to solve problems, then this form of training will simply increase the tendency to make decisions based on position and influence instead of on free and open communication. Accordingly, both the organization's effectiveness and its ethical standards will suffer.

PROBLEM-ORIENTED TRAINING

In one way or another, training always involves an organization's systems, such as communications, appraisal and promotion, and so forth. In practice, these systems practically never work as intended. Obviously, many of the problems are matters of *form*, that is, administration and interpersonal skills. But the problems that matter relate to *substance*—the philosophy and values of the organization. These are made clear by managers' behaviors and actual organizational conditions.

Development of personnel is most effective when individuals participate in addressing important organizational issues of substance. These issues can be addressed effectively only within the organization. Outside workshops do allow for cross-fertilization of ideas, but they usually amount to little more than entertainment for employees, and don't do much for their quality of performance once they are back within the organization. The employees frequently have problems transferring the general concepts they learn outside to their own organization, and find that they have little support there for initiating new behaviors.

Even the value of in-house training will be only cosmetic if it is limited to skills and administration and fails to address organizational philosophy and values. Because it doesn't seem to employees to be addressing issues that matter, they view it as management hype. Such training, because it contributes to the cynicism of employees, can actually make conditions worse. Those most likely to benefit from implementing the mechanics of a program without suffering under the burden of having to attend to the ethics involved are the Destructive Achievers. Builders, on the other hand, are more likely to depreciate the Mechanics. They do this at their peril, while trying to practice the values—which are not monitored and which may even be discouraged by the system.

Another reason training programs can be ineffective is that most of them are designed and administered by professionals who see themselves as salespersons of packaged, state-of-the-art concepts. In an antiseptic classroom, they try to stimulate productive behaviors, but they fail to realize that the behaviors will be transferred to a culture that has counterproductive values. Participants learn currently popular training clichés. They are taught that the opinions of selected experts and researchers are the sources of information and development. They are not taught how to improve their abilities to function within the set of options their own environment provides. The confrontations that occur in such a setting will involve theoretical concepts only, because influential managers, those who can do anything about changing the culture, are absent. This means that all levels of management must attend at least some of their organization's more important training sessions. They need to appreciate the extent to which their major systems are working in the ways they were intended to work. Managers need to hear directly how their philosophies and values are perceived by those who must implement them. All participants, including managers, must be encouraged to positively confront even the most basic assumptions and practices of the organization (see Chapter 10). Only through open discussion can persons understand the inevitable deficiencies of *any* system, and learn how to effectively function within their own. It also helps employees to appreciate others' genuine efforts to correct problems. If this is to occur, provisions must be made to create a learning atmosphere based on problem solving. Power and personality must be left at the door.

The issue of problem-oriented training is complicated, because it involves subtle philosophical issues as well as some fundamental biases of our culture. These issues and biases will be explored further in Chapters 6, 7, and 8, which deal with the three most basic human resource systems: MBO, appraisal, and communications. The ways an organization presents these subjects have much to do with whether it develops Destructive Achievers or Builders.

Perspectives

A supervisor of a small technical group in an R&D environment was transferred to a manufacturing plant. His new assignment was to stabilize a very sensitive production process. It was the toughest unit in the plant to manage, it had the most problems, and he was acknowledged to be the strongest technical person available. He also had excellent rapport with the personnel of the manufacturing unit. In addition, it was generally felt that the unit would benefit from his long-term orientation and his methodical approach to solving problems.

When senior management first offered him the position, he turned it down. He didn't feel that he was ready to handle this kind of pressure-cooker environment, and he wasn't sure that he ever would be. But upper management said they understood his concerns, assured him that he was the best person for the job, and persuaded him to accept the transfer. Within three months the new superintendent was having serious problems in his unit. There were several production disruptions, and he was under great stress trying to cope with them. Several attempts were made to help him salvage his position. They eventually failed, and he was transferred back to a technical position.

This episode demonstrates the kinds of dilemmas that anyone who seeks an easy solution for improving an organization might face. To begin with, the superintendent didn't have the quick-reacting, power-oriented style that the environment required for short-term success. He let problems go too long before he sought help from others. While he focused on one problem, two others got worse. The wide array of resources available to him were a source of power, but he didn't know how to use them. Moreover, any psychologist would agree with him that his personality did not fit the typical success profile for this kind of job. He was a reflective, thorough person, more interested in systematic analysis, not a fast-reacting decision maker. He had always been allowed to focus on one project at a time, more so anyway than was possible in that particular manufacturing environment.

At first glance, this example seems to support the case for a power-oriented development of personnel and the assignment of individuals to jobs on the basis of their personality profile—and, in a sense, it does. If one is to succeed in a power-oriented environment, then one must have the skills and the personality for it. But if we accept that conclusion, are we not also committing ourselves to maintaining the cultural *status quo*? That is, the ones who are most likely to have the personality, values, and ideology needed for fundamental cultural change are also the very ones who are least likely to succeed. Or, put another way, the more a system becomes like itself, the more it loses its ability to accommodate diversity, and to change to something different—even when it wants to.

Of course, there are many combinations of situations and individuals, from rigid situations that few or none can change without forcible, severe disruption to flexible situations that almost anyone can favorably influence. Therefore, to increase its ability to make favorable changes, management must develop power skills in a diversity of persons, and in the types of environments where they will be expected to perform. Each person needs to learn how to retain his or her individuality, yet fit the mold. The time to test and develop persons' leadership desires, abilities, and ethics is early in their careers, not when they are needed most. This issue is treated in detail in Chapter 7. The most important priority for management must be to pursue vigorously a problem-solving ideology. By their own behaviors and training efforts they must make power skills less necessary and model a respect for individual differences and rights.

As George Bernard Shaw reminds us, every profession is a conspiracy against the laity. Today every school of thought—in management, in the behavioral sciences—conspires to sell its ideology at a profit to the general public, while discrediting every other school. (This book is not an exception.) Some would have us adapt, wholesale, Japanese methods of productivity and quality improvement, regardless of our different cultures. Others would have us reject a successful methodology simply because, although we originated it, it was the Japanese who refined it.

There is a basic problem with research and training in our organizations today. We spend an inordinate amount of time trying to figure out which experts are right, rather than exploring and determining our own values. Even apparently contradictory research results and expert opinions are usually valid. But they are useful, not because they give us answers, but because they present us with options we may not have considered before. They increase the number of sophisticated questions we can ask. But the fundamental questions will always remain: Who are we? What do we value? Who do we want to become? And, after we have answered these, how do *we* get to where *we* want to go?

Footnotes

1. Abraham Kaplan, *The New World of Philosophy*. Vintage Books, New York, 1961, 60.
2. David A. Whitsett and Lyle Yorks, "Looking back at Topeka: General Foods and the quality-of-work-life experiment." *California Management Review*, 25, 4, Summer 1983, 93–108.
3. John P. Kotter, *Power and Influence*. Free Press, New York, 1985.
4. Chris Argyris, "The executive mind and double-loop learning." *Organizational Dynamics*, Autumn 1982, 5–22; and "Double-loop learning in organizations." *Harvard Business Review*, September-October 1977, 115–125.
5. Kotter, 73–74.
6. Ibid., 10.
7. Experiences and cases are described throughout both of Argyris's articles.
8. Manuel G. Velasquez, *Business Ethics*. Prentice-Hall, Englewood Cliffs, N.J., 1982, 144.
9. *Webster's Ninth New Collegiate Dictionary*, G. & C. Merriam, Springfield, Mass., 1983, 922.

10. John P. Kotter, "Power, dependence, and effective management." *Harvard Business Review,* July-August 1977, 132.
11. Argyris, "The executive mind. . . ," 22.
12. Ibid., 19.
13. Ibid., 12.
14. Peter Reason, "Is organization development possible in power cultures?" In Andrew Kakabadse and Christopher Parker (Eds.), *Power, Politics, and Organizations: A Behavioural Science View,* Wiley, 1984, 190.
15. Andrew Kakabadse, "Politics of a process consultant." In Andrew Kakabadse and Christopher Parker (Eds.), *Power, Politics, and Organizations: A Behavioral Science View,* Wiley, 1984, 182.
16. Stanton Peele, "The question of personality." *Psychology Today,* December 1984, 55.
17. David C. Funder, "On the accuracy of dispositional versus situational attributions." *Social Cognition,* 1, 3, 1982, 216.
18. David C. Funder, "Three issues in predicting more of the people: A reply to Mischel and Peake." *Psychological Review,* 90, 3, 1983, 288.
19. Lee Iacocca, with William Novak, *Iacocca.* Bantam Books, New York, 1984, 127.
20. Dale Tarnowieski, *The Changing Success Ethic.* An AMA Survey Report, New York, 1973, 33.
21. Patricia Galagan, Interview, "Staying alive, Jack Grayson on the American productivity crisis." *Training and Development Journal,* January 1984, 60.
22. Ibid., 61.
23. Christopher P. Andersen, "Running your own meetings." *Working Woman,* April 1986, 125.
24. Richard Conarroe, "Climbing the corporate success ladder: A self-marketing program for executives." *Management Review,* February 1981, 42.
25. James Gray, Jr., *The Winning Image.* AMACOM, New York, 1982, 99.
26. Charlene Mitchell and Thomas Burdick, *The Right Moves.* Macmillan, New York, 1985, 181.
27. Leonard R. Sayles, *Leadership.* McGraw-Hill, New York, 1979, 95–96.
28. John P. Kotter, "Power, dependence, and effective management." *Harvard Business Review,* July-August 1977, 125.

CHAPTER 5

Technology and Training

Our most counterproductive management assumption is that an organization's administrative systems are more important than its products or its employees. Of course, it's subconscious, and most managers will deny it. Yet it expresses itself in many ways. In time it pervades all other assumptions. It expedites the natural degeneration of any organization that already has Destructive Achievers in key positions. And, even in organizations that are managed by Leaders and Builders, it places a very real, although perhaps unrealized, stress on personnel at all levels.

Leaders and Builders, as they move up the corporate ladder, may, in an attempt to become more "professional," distance themselves from their products (a technology dimension) and the producers of their products (an equality dimension). They become communications experts, information handlers, and financial analysts. If Leaders are not aware of their vulnerabilities, their orientation becomes destructively egocentric. They would rather preserve the system and their place in it than solve problems of substance. Eventually, it makes no difference to them whether they produce plastic toys or paper towels. Human resources are an expense, and expenses should be minimized. As a result, values throughout the organization suffer, and even ethical Leaders tend to create anti-ethical systems (anything legal is acceptable; justice and social welfare naturally evolve via the free market). Anti-ethical systems, in turn, cause the development of Destructive Achievers and unethical, even illegal, behaviors.

Others have identified some of the factors that relate to the systems/substance assumption, but not, as here, in the context of the Destructive Achiever. William Ouchi pointed out the beneficial role that equality plays in productivity.

Robert Hayes and William Abernathy explained how the pseudo-professionalism of managers was detrimental to technical effectiveness. Many others have cited the necessity of training all levels of personnel in their specialized areas.

Now we need to consider how the interactions among these separate factors affect organizational ethics. These are not just productivity issues; these are *ethical* issues. It's not just a matter of avoiding organizational ineffectiveness—it's a matter of preventing an anti-ethical environment.

The Technology-Equality Connection

The gap between the organization's systems and its product or service is related to two interdependent elements: respect for technology and respect for human equality. Neither is possible without the other. To the degree that *either one* is absent, the systems/substance gap will increase, and the ethical climate of the organization will degenerate.

Hayes and Abernathy described the technical half of the gap when they referred to "pseudo-professionalism" in a classic *Harvard Business Review* article. They expressed this view of the direction of modern management:

> Success in most industries today requires an organizational commitment to compete . . . over the long run by offering superior products. Yet, guided by what they took to be the newest and best principles of management, American managers have increasingly directed their attention elsewhere. These new principles, despite their sophistication and widespread usefulness, encourage a preference for (1) analytic detachment rather than the insight that comes from "hands-on" experience and (2) short-term cost reduction rather than long-term development of technological competitiveness.[1]

> In recent years, this idealization of pseudo-professionalism has taken on something of the quality of a corporate religion. Its first doctrine, appropriately enough, is that neither industry experience nor hands-on experience counts for very much. . . . It encourages the faithful to make decisions about technological matters simply as if they were adjuncts to finance or marketing decisions.[2]

Joseph Limprecht and Hayes continued this theme. They noted that Germans do not view management as a separate profession, and one of their primary concerns is the maintenance of technical expertise at all management levels.[3] The basic orientation of the opinion expressed above was toward the importance of technical competence itself. The following, however, highlights the motivational benefits of such competence at all levels:

Technical competence at all organizational levels makes possible a bond between management and the shop floor altogether different from that achieved by periodic, highly publicized efforts to show that management cares about workers or is interested in what they think about process or quality improvements. . . . Several conditions must be present for such a relationship to work: (1) technically competent managers; (2) managers who take the time to initiate serious dialogues with workers about their work; and (3) a skilled, competent work force with a tradition of pride in what it produces.[4]

As strange as it may seem at first glance, William Ouchi's description of the theory Z organization is closely associated with the concept of technical competence at all levels. It clarifies the equality aspect of the gap between the systems and the substance of an organization:

It is a consent culture, a community of equals who cooperate with one another to reach common goals. Rather than relying exclusively upon hierarchy and monitoring to direct behavior, it relies upon commitment and trust.[5]

Type Z companies generally show broad concern for the welfare of subordinates and of co-workers as a natural part of a working relationship. Relationships between people tend to be informal and to emphasize that whole people deal with one another at work, rather than just managers with workers and clerks with machinists. This wholistic orientation . . . inevitably maintains a strong egalitarian atmosphere.[6]

Every organization that produces a product or service has a technology that encompasses it, and there are technical decisions that are appropriate for every level. This is true even when top corporate executives decide which technologies to divest from their organization. This has two implications. First, *equality* relates to competence in fulfilling one's role in the problem-solving process, a "wholeness" that Ouchi describes throughout his book. If one is not competent in the position held, then one cannot be considered equal, either by oneself or by others. Neither self-respect nor respect for others is possible when competence is lacking.

Second, it is condescending to suggest to another that his or her role in the process is less equal because his or her *kind* of competence is less important. It is *especially* galling when the latter individual deals directly with the product or service itself and considers the person with the more valued competence as a hindrance rather than a help.

Unfortunately, this usually occurs when organizations and their systems expand. As managerial skills begin to relate more to the system than to the technology, the system predominates at the expense of the technology. What is more important, employees lose respect for the organization when they see that many systems experts are Destructive Achievers and are better rewarded than the solid contributors—the Builders, Innovators, and Mechanics.

Even when managers honestly use sound leadership practices, the practices may be perceived as ritualistic machinations if the managers are not technically sophisticated. This is the reason popular models of successful management are so misleading. They leave out the 95 percent of hard work and the development of values and mutual respect that is necessary before the 5 percent of slogans, pep rallies, walking-around, skunk-working, and "intrepreneuring" can be effective.

The technology-equality connection most clearly expresses itself in the type of training given to lower-level personnel. In turn, the importance management places on training indicates the extent of its long-term commitment to treat its employees ethically.

The Training-Ethics Connection

The process of growth, maturation, and decline of the American polyester industry illustrates how ethical Leaders can pursue excellence and economic success—and yet create an anti-ethical environment. It also provides an instructive comparison with Japanese industry, which places greater emphasis on employee training at all levels.

In trying to make positive contributions to society, Leaders are seduced by pressures from the competitive environment and the quarterly report. The process is a subtle one, and each degenerative step along the way has its economic rationale. From the very beginning of the cycle, management's attitude toward the development of lower-level employees sets the stage for future organizational health. The following analysis does not relate to a single specific facility.[7] However, it is a fairly accurate composite summary of what has evolved in a number of cases.

EXPLOSIVE GROWTH

In the initial stages of growth, polyester facilities were small. Everyone was on a steep learning curve, and participation was the name of the game. Although many decisions were made at the top, they were preceded by discussions that included everyone directly involved with a project. Such discussions frequently included several levels of the organization, because no one level had all the relevant information. To a large degree, all managers were functioning as technical professionals.

As new equipment was installed and put into operation, new employees were hired and given minimal training (in order to get production as soon as possible), and new projects were initiated. Because of rapid project development, much operator time was devoted to monitoring equipment performance, correcting it when it deviated from specifications, and seeking help if a problem was unusual. In other words, manning requirements were high, because the process itself was not designed to be interruption-free in the first place.

Management development was on-the-job, focusing more on the improvement of technical knowledge than on managerial skills. Because the core start-up team was developing as well as learning new technology, they knew the technical aspects of their jobs as well as anyone else in the industry. New employees, however, were taught only the essentials of what their jobs required at the time of hiring. If they ran into problems beyond their skills, more experienced persons, usually at a higher level, had to intervene.

There was a strong sense of community, and morale was high. Everyone was involved and had a personal feel for the climate of the organization. Top-level managers had working relationships with persons several levels lower. Results were rapid and observable, and operations were profitable. Promotions and the movement of people were frequent. Organizational excitement was at a peak.

MATURATION

As market demand increased, facilities got larger and more complex. Plant managers no longer knew all the employees, even the foremen, by name. Due to the technology explosion, higher levels of management were unable to make knowledgeable decisions as well as they had in the past. Yet they didn't trust the technical sophistication of the younger managers and technical professionals. In turn, the younger managers and technical professionals resented not having the authority to make important decisions without approval. When approval was required (which was frequently), it significantly slowed down the process. It became obvious that a new approach was needed. Enter professional management development and systematic management. Both internal and external consultants were used extensively to help management be actual "managers," rather than "super engineers."

Management was persuaded that it needed to delegate more of its decision making and authority. Planning and the achievement of objectives had to be more systematic. Proper management development also would require broader experience for key personnel (which was just about everyone in a supervisory capacity), and personnel at all levels needed to improve their interpersonal and problem-solving skills. It was generally conceded that these new managerial directions were effective. The theories made sense and received wide support, both from company personnel and the current literature. Supposedly, everyone was "managing by objectives," and there were numerous examples of individuals who became better managers and of natural work groups that quickly became more effective teams. But somehow within this fast-paced and apparently improving environment there was also dissastisfaction. In the opinion of many, especially at lower levels, there wasn't enough participation, promotions were going to the wrong persons, and people were being moved too much. Different groups felt alienated from each other: operators from foremen, foremen from lower management, technical from production, and so on. People were "getting into their foxholes."

Senior management complained about the lack of competence of its middle managers, who were unwilling to make decisions or try anything new. Nagging technical problems seemed to defy solution, and "high flyers" were shifted from project to project in order to turn around deteriorating situations. People seemed to know more about management and interpersonal relationships, but specific improvements seemed short-lived. Morale was average, and, for many persons not in the power networks, organizational excitement was replaced by apathy.

DOWNSIZING

Deteriorating market conditions for polyester products began to accelerate in 1981–1982. This caused cutbacks in capacity and personnel in U.S. companies. For some, however, home-grown deficiencies in quality and productivity helped reduce customer demand for specific lines. Morale was at its lowest, and whole plants were being permanently shut down.

The orientation of some managements was, for the first time, self-critical. Some seriously tried to analyze the effectiveness of how they went about managing and solving problems. This led to an exchange of ideas and information with companies in other countries. To say that these exchange visits were eye-openers to U.S. managers and professionals would be an understatement. The quality and productivity they saw in some Japanese facilities were in themselves enough to stimulate almost immediate improvements in U.S. polyester production.

FIVE ETHICAL QUESTIONS

The insights gained from critical self-analysis and the successful experiences in quality and productivity improvement projects suggest five pertinent questions about the way management theories are put into practice. At first glance these questions seem deceptively simple and commonplace. They are not at all, in spite of the fact that they are among the first questions asked in a productivity improvement effort.

Do people know how to do their jobs?

Do lower levels of the organization know what upper levels expect?

Is the production equipment developed to its full potential?

Are people organized in the most effective way?

Do people want to do their jobs?

The answers to these questions are fundamental to management's philosophy and strategy. They bridge the gap between organizational systems and organizational substance, and their implications range well beyond the effectiveness of technical training itself. Answers relate to the style of management delegation and involvement, the type and amount of employee participation, job rotation practices, management development and promotion policies, the rate of growth that is pursued, and MBO practices. Above all, they relate to management's view of the lower-level employee: as an expendable part of the machinery, or as a valuable human resource with the rights of full membership. And, if that's not enough, the answers to the five questions are among the most important clues we have to the quality of the organizational culture and its likely production of Destructive Achievers or Builders. I dwell on the first question longer, because it is basic and has implications that are applicable to the remaining four.

Do People Know Their Jobs?   There is a significant difference between those who are technically competent and those who know the technology. When new engineers are hired from the top 20 percent of their classes and later demonstrate obvious technical competence in their conversations with others, it is easy to assume they can do the job. A similar assumption can be made about a middle manager who successfully takes over a new assignment, or about an operator who quickly detects and corrects an equipment malfunction.

These can be dangerous assumptions because they allow a management to grow and expand without having to face up to its deficiencies. Only when compared to what is demonstrated as possible are such assumptions seen to be wrong. American employees at all levels lack knowledge and skill for two reasons, both caused by management: (1) the rapid movement of personnel at all levels; and (2) the philosophy (at least in practice) that a supervisor is primarily a manager or administrator, almost to the exclusion of being a teacher or technical problem solver and decision maker.

A common dilemma in large, fast-growing companies is the newly appointed manager with four subordinates averaging less than a year in their jobs who are also managing inexperienced people. This person, if inclined to become a Destructive Achiever, will retreat into the role of professional manager. Because people like this cannot educate or help in technical decisions, they delegate and hold subordinates responsible for their results. Instead of becoming true team members, they elect to criticize subordinates for not being willing to make decisions or accept responsibility. The DA plays the professional manager role exceptionally well. He is superficially involved with subordinates before the fact, when options and outcomes are unclear, and becomes truly involved only after the fact, when results are in.

Assuming that the modern manager is not *expected* to be technically sophisticated, failures are justly blamed on subordinates' incompetence. The DA then fulfills his or her staffing responsibility by replacing the technically smart but inexperienced subordinate with a proven "high flyer," and the same kind of situation is exacerbated elsewhere.

On the issue of the movement of personnel, Limprecht and Hayes noted:

> German firms [have not] adopted the practice, common in America, of moving managers from place to place or job to job every two or three years. . . . In our judgment, the gulf between plant management and the shop floor, which exists to some degree in every factory, tends to be widest in those American companies that *do* move managers rapidly from location to location.[8]

The German practice of longer job tenure in a specific technology is shared by at least one Japanese polyester manufacturer, who compared its management strategies with those of an American company. The average Japanese manager worked in only one functional area, even through a series of three or four promotions, for an average of thirteen years before being broadened by actual job assignment into other areas. At each level, the primary responsibility for technical training was the supervisor's, who also taught persons what they needed to know about other functions. The operator-to-foreman ratio was 5:1. Problems were solved at the lowest possible level. But, if required, several organizational levels could effectively participate in discussions, and all had the technical training and experience commensurate with their particular level. This contrasts with the frequently heard complaint in American industry that upper management participation in problem-solving discussions is disruptive as often as it is beneficial.

The advantages of management tenure in a specialized area are significant. Although much data can be stored in technical libraries, and libraries *are* important, they tend to be neglected. For reasons of political defensiveness, bureaucratic red tape (levels of approval, prescribed formats, etc.), and a lack of personal incentive, many persons simply fail to take the time to input into the official technical records. There are too many examples of unrecorded data and reinventing the wheel to discount the importance of having data in workers' heads.

It is true that managers need to delegate, leave technical decisions to subordinates, manage objectives versus details, and so on, but only if subordinates have the prerequisite experience and competence. A major mistake in any industry is to make the transition from management-by-technicians to management-by-managers without taking the time to go through the education phase.

Actually, neither extreme is acceptable. There are always technical decisions that can be delegated, and yet there are also strategic product or process decisions appropriate for every level of the organization. In the same vein, the education of subordinates shouldn't be a phase, although it almost has to be, given the fast start-up and growth strategies of highly competitive companies.

The old bromide that all development is self-development, that development is the responsibility of the individual, is a concession to the preferences of managers on the move who prefer the sink-or-swim approach. It is true, of course, in the sense that no one can develop anyone else. But it also makes it too easy for supervisors to fail to consider the education of their subordinates as their responsibility. (In some cases it is virtually impossible, with an operator-to-foreman ratio of 20:1.)

Management must strike a balance between training people in depth within a given functional area and cross-training persons for greater flexibility and breadth of understanding. To achieve a balance like this, a number of factors become crucial: the rate of growth, the teaching role of supervisors, promotional policies (short-term–oriented persons currently have a tremendous advantage), and the size and complexity of the organization. The larger and more complex it is, the more lower-level persons must know their jobs and have the authority to do them *and* the more upper management must be able to support and judge the substance of their performances.

Do Lower Levels Know What Upper Levels Expect?   When rapid growth and fast turnover of personnel prevail, continuity is hard to maintain. Moreover, we see a deterioration of effective communication between levels, especially of values and strategy. With complexity, and when personal identification with the product or service is lost, subtle negative behaviors are easily absorbed by the system. Destructive Achievers go unmonitored, unnoticed, and suffer no consequences.

Unfortunately, the DA does have many of the skills that are necessary for organizational expansion. Once senior management exhausts its supply of obvious Leaders, they look to who is next in line—the DAs. They can't pass up an opportunity for growth, or a chance to increase market share, so the DA is promoted, quickly adapting to the new situation and doing whatever it takes to get the job done. The product or service is delivered on time and within budget, but at great future cost. The company's ethical base and employee commitment will never again be the same. Its quality standards, its reputation—and possibly its concerns for employee and public safety—all suffer.

When management later realizes what it did, it may make a genuine commitment to improve quality, engage in comprehensive "awareness" programs, and actually provide funds to support the effort. Yet, in organizational development sesions with lower level employees, one frequently will hear the complaint that management is only interested in production, not quality. Somehow the message of management's good intentions does not get through.

Labor economist Audrey Freedman of the Conference Board noted that managements of Japanese companies in the United States were getting sharply improved quality and productivity out of their American workers. Being a newcomer to the U.S. is an advantage for the Japanese, he noted, because "They're not dragging an anchor of past practice. It's a major problem for American management."[9]

Because management demonstrates a short-term orientation at all levels, employees understandably distrust announcements of a new philosophy. They have seen too many Destructive Achievers promoted and solidly ensconced in the middle ranks on the basis of previously held organizational values. There is a limit to how much and how quickly management can revive the lost values of employees through such statements of change. Employees need to see real change in the behaviors of the managers they are expected to follow before any long-range improvements in meeting senior management's expectations can be realized.

Is the Equipment Developed to Its Full Potential?   Americans who visited Japanese companies were surprised to see that the basic technology and equipment of the Japanese were not much different from ours. They expected them to be fundamentally superior, because of their very high conversion efficiencies and quality levels. However, they found that the Japanese did more with their technology by means of minor, incremental modifications and innovative alternatives, as opposed to our less creative and more expensive solutions to problems. We invest more time and effort monitoring and correcting process variances; they spend more time analyzing and reducing their causes.

For example, a network of supply lines for a given chemical process may create problems because the tube diameters are too small. To replace them would be prohibitive, so our quick solution is to train the operator to take corrective actions at certain times. There are alternatives, however. For example, the lines could have been insulated so that the chemical does not cool down so fast. Or an automatic drain valve could have been inserted at a certain point to remove sediment. The point is that we manage problems or replace equipment at great expense, rather than taking time at the beginning to remove the cause.

During the growth of the polyester industry the strategy of management was to improve productivity mainly through technological breakthroughs. The philosophy was to devote a bigger share of time and money to a few large-scale projects, instead of to many smaller ones. In the opinion of an obviously cynical lower-level Builder, it was "easier to get approval for a $500,000 blue-sky research project than to get approval to replace $50,000 worth of outdated pumps that are causing production disruptions."

Of course, genuine breakthroughs did occur and the strategy was profitable during the growth stage of the industry. To some extent, it still is. Unfortunately, the successes in research were often achieved at the expense of day-to-day technology. Too little technical effort was put into improving the existing equipment. DAs quickly learned to focus on high-visibility projects, because they knew they would get little or no credit for achieving more mundane objectives.

As we now know, and as frustrated Builders have been telling us all along, it is just as important, if not more so, to constantly improve existing equipment—the core of the business. This requires a managerial reorientation. Without it, the creative talents of our self-made experts, and even of many of our solid professionals and managers, go untapped. As things stand now, we train and develop only the winners, and the basic business remains the loser.

Are People Organized Effectively?   Some organizational structures are the result of a perceived need to quickly train persons to do simple jobs. This too is an issue that is closely tied to management values and philosophy. Reducing jobs to their simplest functions results in fast training and a greater flexibility in moving people. But it also communicates to workers that management sees them as extensions of the equipment. They aren't expected to use their brains. In addition, when little has been invested in employees, their utility value, in the sense of an ethical obligation, is reduced. That is, economic considerations then outweigh, in the ethical equation, the factors of justice or individual employee rights. (The major ethical principles of rights, justice, and utility are explained in greater detail in Chapter 10, page 170.)

The syllogism goes something like this:

We haven't invested money or time in you.

Your economic value to the total organization's welfare is small.

Therefore, in a crunch, the organization is economically justified (the ethical principle of utility) when it does not consider your personal interests.

The alternative would be to give workers more complete tasks that require judgment, provide them with the necessary training, and allow them to develop stable work relationships. This management value recognizes the potential of workers, not only to handle the product, but to creatively improve the way the product is produced. This issue is not limited to production workers but is true of layers of managers who have become transient discussion leaders, coordinators, and information handlers. Once the company truly invests in its employees, management cannot so easily justify its disregard of employee welfare on the basis of economic necessity. (Not that it *needs* justification. Judging from the number of employee cutbacks, rehirings, and retrainings that are already going on, most organizations have no appreciation of the actual costs of these practices, even with poorly trained employees.)

Do People Want to Do Their Jobs?   The four questions above have a lot to do with how much employees feel that their environment is an ethical one. This ethical quality, or lack of it, can be seen by employees in every act of management; it therefore has a profound effect on individual motivation—desire to do one's job well.

The Destructive Achiever sees scientific management and behavioral science methodologies merely as a way of getting others to do their work better. This is an anti-ethical view of human resources, and employees easily recognize it. Short-term results and the DA's own self-interest are the bottom line. The DA does not see employees as long-term contributors to the organization. The job at hand is to maximize profits, and ethical obligations are limited to legal requirements. Justice and social welfare will naturally evolve from the marketplace, according to the DA. This is why we see so many short-lived successes—and outright failures—of organizational efforts to improve productivity, the interpersonal climate, and team effectiveness. Workers see themselves as victims of the system, instead of as fully accredited participants in cooperative efforts of substance.

In 1985, Thomas O'Boyle reviewed indicators of employee loyalty. He concluded that

> Studies have found that job satisfaction for all employees, regardless of rank, is lower today than it ever has been. . . . Loyalty—that intangible yet indispensable asset—is waning at a growing number of corporations. The concept is difficult to measure, but surveys, anecdotal evidence and interviews with executives across the country indicate that many managers and workers who once devoted all their energies to their jobs are now concluding that such devotion was misplaced.[10]

Using employee honesty as one indicator, O'Boyle noted that

Thirty-two percent of the employment applicants screened [by a psychological testing firm] last year admitted on signed statements that they had stolen from a previous employer, up from 12% two decades ago. Similarly, a recent survey of 9,000 employees by the Justice Department found that one-third admitted stealing from their current or previous employers.[11]

Many employees are simply "turned off." A Japanese visitor touring several U.S. facilities talked with personnel at all levels. He found that American workers lacked a sense of inquiry—"not being interested enough to ask why." When he asked them about aspects of a process, or why certain things were done, they often could not answer him with the level of sophistication of Japanese workers in comparable positions.

It's no mystery why workers are not curious about their jobs. They lack technical skills, are frequently transferred, and are never asked for their opinions in the first place. This is a management problem. Enough suggestions (job enrichment projects, sociotechnical efforts, quality of worklife programs) are generated under properly designed conditions to indicate that curiosity and desire are there. They always surface in those groups that *are* turned on, even temporarily. Unfortunately, this is not the usual management style. It only happens when time and opportunity for it are designed into the schedule. Management tries to direct motivation and participation into the system through programs for others, rather than allowing these qualities to become an active part of the system itself. The message is not lost on employees. There is a fundamental detachment on the part of management from the substance of the business. As one employee put it in an especially bad situation, "They don't even try to hide it any more."

Countering the Cycle

Whether hand-picked or run-of-the-mill, managers and employees alike seem to be very productive in the initial phases of an innovative organizational development (OD) effort, an *ad hoc* project group, or even a traditional plant start-up. One of the interesting characteristics of turned-on work groups is that, except for their obviously strong identification with their jobs and each other, they are strikingly average personalities. They have the full range of personality quirks; "characters" are more a source of enjoyment than of irritation when persons are productively working together.

As the system ages, however, and personnel are changed, cooperation is replaced by the administration of results. Image becomes more important than personal contribution, and social status and function become more formalized, *even among the original members of the team.*

During an OD session that I conducted, an unusually insightful work group speculated about why it is difficult for effective systems to survive the accommodation of even *good* new managers. They complained that management had lost touch with employees; they were promoting the wrong persons. I knew that upper management was trying to improve conditions, so I asked what management could do that it wasn't already doing. They had no answer—management was already going through all the motions; there was some program or plan of action to deal with each significant issue that came up. Since they had been considered a high-morale group a few years earlier, I then asked when the problem began.

After some soul-searching discussion, some of the old-timers ventured the following scenario: During the plant start-up, all members of the team knew each other and morale was high. They ate together and worked together. Whole departments or natural work groups met periodically for various purposes. Occasionally there were parties. When the organization began to grow, and the larger departments became more distinct, people still socialized in mixed groups. Then some senior managers were replaced. At first, the new replacement managers made an effort to become acclimated to the organization and attended large group activities. But eventually they either stopped coming or, if they came, left early. They had other pressing commitments.

A negative form of managerial leverage then came into play. When senior management quit participating in the general meetings, some of the middle managers quit also. Those who did come formed cliques, people associated primarily with those with whom they felt most comfortable. Eventually, "who you knew became more important than what you did."

This scenario rings true. Even under the best of circumstances it is difficult for a new manager to become a part of the substance of any established system. The new manager can never have the experience of developing the relationships that yielded the original values. Moreover, the manager does not appreciate the group's technological history, including who had done what. Training and indoctrination can help, but they won't produce the *same* values and understanding.

ESTABLISHMENT PRESSURES

It is especially difficult for an innovative and successful group to absorb a new manager who has traditional experiences and biases. Unfortunately, this usually happens *by design* in large corporations. Corporate pressure for innovation to return to the norms of the dominant culture is the rule, not the exception. Systems are designed to perpetuate their own strengths and, inevitably, the weaknesses that accompany those strengths. The more powerful the system, the harder it is to change. To operate outside the system is an affront to it, and to be exceptionally *successful* outside the system borders on disloyalty.

Even good systems begin to distance persons from each other. When persons are already distant from each other, the situation is aggravated even more. When there is no substance in the relationships corporate management has with its facilities, it tries to overcome the situation by means of standardized controls and by replacing innovative managers with traditional managers.

For example, a manufacturing plant developed a very effective employee participation program for identifying and correcting problems of productivity and quality. Its corporation later initiated an organization-wide program to do the same thing; and exerted pressure on the plant to reorganize its structure and procedures according to the recommended corporate format. The plant resisted. This was embarrassing, because the plant was getting excellent results on its own and was not using the recommended structure or procedures. Then the plant was pressured to change the *name* of its program ("Profit Improvement Program") to the name of the corporate program ("Q1 Program"). That at least would allow the corporation to include the plant's quality improvement data when it documented the effectiveness of its new program.

Successful or not, whenever a subgroup such as this deviates from the norms of the dominant group, corrective action is taken. The rationale for the prevailing corporate solution is simple: Select a properly professional replacement manager who can learn the innovative system and capitalize on its strengths. *In addition,* he or she will be able to correct its weaknesses—that is, make it more successful by restoring disciplined corporate values.

This strategy has two serious drawbacks: (1) The newly appointed manager knows that he or she will get no credit for continuing someone else's innovation, even when the corporation recognizes its value; and (2) he or she will be able to quickly identify opportunities for improvements and, not realizing their incompatibility with the innovative system, will insist on, and get, immediate results. That is exactly what occurred in the above example.

Again, this strategy is the rule and not the exception. David Whitsett and Lyle Yorks's analysis of the General Foods Topeka project presents a mirror image of the above case.[12] Briefly, the Topeka plant utilized many progressive principles of organizational development. In fact, it became one of the "first generation" of innovative plants to serve as a model for others to follow. In spite of widespread acknowledgment of its success with the public at large, however, its influence on its corporate parent was quite different, even reversed.

Two things occurred that are pertinent to this discussion. First, instead of the successful innovations at Topeka diffusing into the corporate environment, the corporation's traditional practices diffused back into Topeka.[13] Second, when it was time to replace a key person responsible for the innovations at Topeka, the corporation selected a manager whose personal style and management philosophy were incompatible with the Topeka system.[14]

The practice of replacing innovative managers with conservative managers is especially common when the offending group is considered too informal. There appear to be many opportunities to gain control of apparently undisciplined behaviors. In a situation similar to those above, a manufacturing facility had successfully begun an impressive sociotechnical effort, when a new manager was appointed to replace the initiating manager. To introduce the replacement manager to the process, he was invited to observe a meeting of one of the more successful work groups. Unfortunately, on this specific occasion, the group spent an entire two hours talking about foreign competition. In spite of its success, the program died in its tracks; he felt there obviously were better uses of time.

NARROWING THE SYSTEMS/SUBSTANCE GAP

The irony of systems, even our innovative ones, is that they become ineffective to the degree that they become important in the eyes of those who control them. They become progressively resistant to challenging new data, to changes in the environment, and to perceived threats to their theoretical purity. The emphasis shifts away from the realities of the product or service, and the individuals who produce it, to the maintenance of the system and the security or advancement of those who control it.

This is not to say that there are intentionally devious or evil forces at work. It is a natural characteristic of the combination of power and its supporting structures, and we need to develop better ways to counter its effects. Some suggestions will be made in Part Three of this book, but there are some obvious things that can be mentioned here that relate to emphasizing substance more and systems less.

First, no organizational development efforts ever "last"; they need to be constantly renewed. But renewal is not the equivalent of disruption, and certainly not of devastation. There may be short-term crisis situations where extreme measures are the only ones that will work, but a successful operation is not one of them. Therefore, senior management should select replacement managers who have a commitment to problem solving, not power politics. In either case, they *will* change the system, but in quite different ways. Traditional power-oriented managers will implement top-down control procedures that are consistent with their biases. Indeed, senior management selects them to do just that. These measures will benefit the organization some, and will appear to be successful, but the benefits of the previous system will be lost. The staff view themselves as serving the manager and his or her system, instead of vice versa. If they also see the net long-term effects as negative, they will perceive the manager as being a Destructive Achiever and power politics will be firmly entrenched.

In contrast, the problem-solving manager will first seek to discover and preserve the strengths of a previous system, then add the strengths of his or her own insights. The evolving new system will then be the *result* of the organization's substance. The staff sees the system as serving the needs of the organization, instead of vice versa, so they will perceive the manager as a Builder or Leader. The environment will maintain its commitment to excellence through objective problem solving.

Second, managers should constantly upgrade the technical competence of personnel at all levels. And they should do it in such a way that the substance of respect for human dignity is enhanced. For example, operator training, traditionally, is limited to handling equipment and practice in building physical endurance. Only those designated as technicians or mechanics receive more sophisticated training or promotional kinds of job changes on a nonsupervisory basis. Yet the typical plant has operators who have been loyal employees since start-up—10, 20, 30 years ago. They retain an extensive technical history of the facility, and, if given half a chance, identify with the local organization far more strongly than do the transient managers and professionals. They often are tied to their communities by virtue of their family connections and a personal loyalty to the area. Moreover, they may feel that their particular skill might not be in demand somewhere else. They know how dependent they really are on the success of their company.

In contrast, except for the foremen, most of whom are hourly workers, the average tenure for a manager in the facility may be only five or six years. In the more dynamic corporations, it may be closer to two to four years at the higher levels. In facilities that close due to business conditions, hourly employees and first-level supervisors are hurt the most. They are also the most angry. To them, the disaster was clearly brought about, or at least aggravated, by the horrendous management practices of the Destructive Achievers. In OD sessions they express their tremendous frustration at the managers' lack of quality standards, lack of discipline (in letting cronies get away with slacking off), and lack of the other values that management supposedly espouses and supposedly wants employees to uphold.

The employee complaint of "favoritism" is usually misinterpreted by management. Managers who do not require sensible discipline are resented far more than those who simply display favoritism. It is the organizational ineffectiveness that *results* from favoritism that is the real cause of poor morale, not just that some employees unfairly receive a few extra benefits.

To more effectively capitalize on the talents of traditionally trained operators, some manufacturing plants have successfully implemented "certification workshops." In such a workshop, training deals with the relevant theoretical aspects of the process and equipment and goes well beyond standard operating procedures. Besides operators, it can include new engineers, technicians, supervisors, and R&D personnel, as well as relevant managers. There is a much greater emphasis on "why" and "how come." Such training is usually a two-way affair. Experienced engineers who are the instructors learn new insights from the questions posed by the experienced operators, and together they explore profitable solutions to problems. Better communications and rapport on the job also result, if for no other reason than that people have discovered their practical values.

Employees report that a single workshop such as this, in which a mix of personnel both participate and instruct, does far more for interlevel and intergroup cooperation than does a typical interpersonal skills workshop for supervisory personnel. Genuine respect for individual competence does far more for self-esteem and a sense of community than a hundred "attaboys" (which are important also). Of course, interpersonal skills workshops are valuable, too. But the lower-level managers who attend the endless series of management and leadership training programs are expected to manage persons who more often than not have had *no* training in basic human interaction. Again, it is a matter of management's values. Managements don't train operator-level personnel in human relations because they don't expect it to pay off, even though an employee may end up working at a single facility for 30 years.

Perspectives

Thomas Friedman and Paul Solman, in their review of the debate over the long-term versus the short-term orientation of American management, concluded:

> They come up against an age-old dilemma of capitalism itself: how to efficiently harness the self-interest on which the system itself runs.[15]

The dilemma will worsen if managers continue to view their profession as an abstraction, something apart from their products and their producers. Any management that has lost touch with its technology cannot effectively supervise, evaluate, or fairly reward the performances of its personnel. The Destructive Achiever's promotional advantages greatly increase, and the organization's systems become more important than its substance. And once growth and complexity are seen primarily as an issue of management control, problems get worse. The situation can be especially destructive when the wrong—technically inexperienced but "professional"—manager is appointed to "turn around" a group that is having serious technical problems.

This is not an indictment of MBO, financial controls, management systems, or the currently popular productivity programs. What concerns me is why they are chosen, and why some of the persons meant to implement them are chosen. Somehow, a greater sense of community, in an ethical climate of equality, must be developed in our organizations. This cannot be accomplished without meaningful personal contact among organization members. The amount of contact does not have to be great, but it must be of substance, enough so that persons at all levels can identify with the larger group, rather than remain limited to their own, narrowly defined special interests.

In this chapter I emphasized the gap we see between an organization's systems and its substance and the impact that has on organizational ethics and effectiveness. In Chapters 6, 7, and 8 I describe how these concepts apply to the three most basic human resource systems: MBO, appraisal, and communication. Management-by-objectives is the epitome of systematic management; in a sense, it *is* management. The appraisal and promotion system is the ultimate determinant of employee behaviors. And the nature of the communication system affects all other systems.

Footnotes

1. Robert H. Hayes and William J. Abernathy, "Managing our way to economic decline." *Harvard Business Review,* July-August 1980, 68.
2. Ibid., 74.
3. Joseph A. Limprecht and Robert H. Hayes, "Germany's world-class manufacturers." *Harvard Business Review,* November-December 1982, 138.
4. Ibid., 142.
5. William Ouchi, *Theory Z.* Addison-Wesley, Reading, Mass., 1981, 83.
6. Ibid., 79.
7. Some of the material in this section was adapted from Charles M. Kelly, "Five questions, and a value, for productivity." *Industrial Management,* September-October 1983, 1–6.
8. Limprecht and Hayes, 137–138.
9. Carey W. English, "How Japanese work out as bosses in U.S." *U.S. News & World Report,* May 6, 1985, 75.
10. Thomas F. O'Boyle, "Loyalty ebbs at many companies as employees grow disillusioned." *The Wall Street Journal,* July 11, 1985, 27.
11. Thomas F. O'Boyle, "More 'honesty' tests used to gauge workers' morale." *The Wall Street Journal,* July 11, 1985, 27.

12. David A. Whitsett and Lyle Yorks, "Looking back at Topeka: General Foods and the quality-of-work-life experiment." *California Management Review,* 25, 4, Summer 1983, 104.
13. Ibid.
14. Ibid., 106.
15. Thomas Friedman and Paul Solman, "Is American management selfish?" *Forbes,* January 17, 1983, 77.

CHAPTER 6

Management-by-Objectives

All effective human action, whether in business organizations or anywhere else, involves the setting and fulfilling of specific goals and objectives. The formalization of this process, in the theory and practice known as "Management-by-Objectives" (MBO), has for some years been a widely accepted model for managing people in organizations. By now, MBO may be seen as the very essence of modern professional business management. Yet no other system of management contributes so directly to an organizational focus on short-term, bottom-line results. MBO is, in fact, a system tailor-made for Destructive Achievers.

To understand this paradox, we have to look at individual values and understand why they must precede and determine systems, not vice versa. As usually practiced, MBO is the classic example of an inappropriate use of management research and training. In MBO, procedures and systems are easily confused with values and philosophy, and the results of good management are confused with its causes. That is, in training managers to use MBO, we inevitably focus on skills that are observable and teachable, instead of on qualities that can only be developed in the work group through discipline, experience, and a commitment to the welfare of the organization.

To begin with, MBO, as an approach to good management, is not unique. It never was, except as a distraction *from* good management.[1] (Everyone manages by objectives in the sense that one always has objectives, whether they are limited to the day's activities or extended to the end of the year, whether they are precise or vague, admitted or denied, communicated or withheld, observable or not.) MBO becomes unique only when a management group decides to term it so, to call attention to it as a "program," and to use the program to cure all of the organizational ills—from poor cooperation and planning between groups to poor appraisal systems. And that is when the trouble begins. For, contrary to what most top managers believe, most formal MBO systems don't work the way they were intended to, even to the point of being detrimental to the organization. Long an outspoken critic, Harry Levinson summarized the fundamental problem: "The simple fact is that most . . . MBO systems generally don't work. The reason is that they deal only with results, ignoring the means by which results are achieved."[2]

Everyone Manages by Objectives

MBO is the *sine qua non* of all management, both good and bad. It is inherent in the word "management" itself, in that everyone manages by objectives. Moreover, everyone's management system gets results, that is, achieves its objectives. It isn't a matter of whether or not you *have* a system; it's a matter of what kind of system you have and the results it generates. For example, consider two hypothetical managers in a company that does not profess to have a formal MBO system.

Mr. Smith is typical of the effective manager who led Peter Drucker to describe what he called "management by objectives and self-control" in his classic *The Practice of Management*.[3] It is part of the managerial style of Mr. Smith to clearly define his role and the roles of his associates; to give priorities to expected results as far into the future as is practical, complete with estimated dates and costs; to communicate his expected results to his boss, peers, and subordinates; and to evaluate his performance, and the performance of his group, on the basis of the results actually achieved.

In other words, Mr. Smith "manages by objectives," and gets the kinds of results he wants. His associates are effective in that they help him to succeed, because they know specifically what he wants, when he wants it, and at what cost. Roles are clearly defined, expectations are mutual, and activities are coordinated. The reason subordinates know their jobs is because their tasks are "appropriate"—they make sense to them and to the organization. Mr. Smith believes that his success is due to his ability to anticipate problems and develop and implement plans to deal with them. Being a member of an organization of people, he has all kinds of support to draw on. Very simply, Mr. Smith is a perceptive planner and a clear communicator.

Mr. Smythe, on the other hand, has a different managerial style. He prides himself on reacting quickly to only the most significant problems of the day, expeditiously communicating to others and evaluating his own performance and that of his group according to how things turned out. He also manages by objectives. His objectives are short-range, in some cases not extending beyond the month, week, or day. They often are not precisely defined, and they are not well understood by others. But he *does* have them. And he *does* get the results *he* wants, consciously or not. But because he is the only one who clearly understands his objectives, subordinates and others are unable to make decisions. He must remain in control. By being in control he retains the right to make last-minute changes. His organization is ready to "turn on a dime" at his direction. In sum, Mr. Smith's style is long-term, clearly communicated and monitored, and encourages systematic teamwork. Mr. Smythe's is short-term, unclear, and arbitrarily evaluated. His effectiveness greatly depends on the quality of his own personal decision making.

To suggest that MBO is not unique, that everyone manages by objectives, does not disparage the many books, articles, and training programs that put forth sound reasons for consciously utilizing its principles. Rather, my purpose is to focus attention on some basic assumptions, concerning MBO, which can make or break efforts to improve the effectiveness of a managerial group by means of systematic approach.

When viewed as a unique approach to management, the term "MBO" implies a "program," in that

- predetermined, regularly scheduled meetings are held.

- standard MBO forms and formats are developed.

- written objectives, action plans, and agreements are to be developed in prescribed ways and during specific time periods.

- predetermined performance standards are established and progress reviews are scheduled in advance.

The elements of the program are standard throughout the organization, regardless of how appropriate they may be in any given situation. (Unfortunately, this is exactly the way most programs have been set up.) So, to lower-level personnel, the new program belongs to top management, or to a consultant, and two-way communication is absent from the very beginning. The message, although subtly conveyed, is clear: Significant changes are about to be made. They will require the application of new principles, tools, and skills that are lacking in the group. Managers wonder why top management is insisting on more formal documentation of plans and expected results. Why this program now? Wasn't the group effective before? Is this just a way to put more pressure on employees? By formalizing an already dynamic system, is top management saying that it no longer allows flexibility?

When MBO is viewed as a process, on the other hand, and as an activity that is always present in any group, it suggests a style of management that *can* be unique to an organizational subgroup *if it is controlled by that subgroup*. Instead of having a new program imposed on it, a management subgroup can decide for itself to analyze its own existing practices and revitalize them along whatever lines appear desirable. It can ask the following questions:

> For optimum coordination and cooperation, are our objectives specific enough to be clear to others?
>
> Do our objectives represent the best use of all our resources?
>
> Does each level and function have appropriate control over its activities in order to reach common objectives?
>
> Are objectives developed and communicated to the right persons in time to be useful?
>
> Is the time span appropriate to our specific line of business?
>
> Are the goals challenging but attainable?

Add to these all the questions that represent the usual considerations found in most standard texts dealing with MBO or with management in general.

But the issue here is not so much one of procedure or content but of philosophy and values. Obviously management practices cannot be improved by means of a process that is slipshod, unsystematic, and casual. It's the *way* the process is introduced that is critical. It sets the tone for the type and quality of organizational involvement to follow. If the organization's philosophy is that MBO is *not* unique, that everyone manages by objectives, then it can develop good managers and a better organization through a fresh analysis of the management process. But if its philosophy is that MBO *is* unique, then any new program merely teaches people how to look like good managers and is actually a step backward.

MBO Program Weaknesses

If an MBO program is introduced to an organization with a series of training sessions that cover only such topics as job descriptions, major areas of responsibility, writing good objectives, monitoring results, and so forth, then it will probably be just another management fad. The mechanics of good management will be implemented, but will lack the necessary underlying philosophy, attitudes, and managerial integrity.

In many cases, the group that buys a mature program from a consultant hasn't had a chance to develop the maturity required to use it well. (By group maturity I mean growth within a specific set of circumstances, not overall management sophistication.) It becomes, not a planning tool, but a whipping tool. It helps the group become, not more systematic, but more political. And, in spite of best intentions, the climate is allowed to gradually deteriorate until it is next to impossible to get something done, unless that something is high on somebody's priority list.

The following are some employee concerns about either already established or just implemented MBO systems. They were generated in organizational development sessions. Although the data are pooled from several sources, there is almost no substantial difference between one group and another. Any group confronting a new, formalized, nonparticipative MBO program is likely to have about the same reactions. They are listed in an approximate order of priority.

1. *"How does MBO affect the way my performance is judged? Is there any room left for judgment, or is it purely results, no matter what the circumstances? Must I achieve 100 percent of my objectives? Suppose I have inadequate power to control results?"*

MBO is purported to increase the level of objectivity in any appraisal process, so these are quite natural concerns. Consider three managers of manufacturing plants: Manager A exceeds his expected production results by 10 percent; Manager B just meets her production targets; and Manager C misses his major production objectives by 5 percent. Which manager should receive the best performance rating?

There is absolutely no way to tell from these data. The production results serve only as a base point. Closer inspection may show that manager A exceeded his objectives by using his training budget for production. He stopped many preventive maintenance procedures and compensated by reacting more frenetically to production upsets. The full negative effects of short-changing maintenance may not be seen for another year. Through intimidation and threats, he required his exempt staff to work long hours. Those who entertained any hopes of promotion, or even of continued employment, transmitted the same negative behaviors downward in like manner.

Manager C, on the other hand, may have inherited a plant from Manager A. After establishing his objectives, he found that some of the most experienced engineers had already left the company for more tolerable working conditions. The middle managers were feeling stress, and the hourly workers under their supervision were taking the brunt of it and ready to unionize. Manager C had little flexibility in adjusting to changing product mix requirements because new employees had not been adequately cross-trained. Product quality suffered because employees had learned to circumvent testing procedures, in order to meet the unrealistic production goals that had been previously set.

There are many reasons for success or failure in judging performance.

Therefore, MBO is unique only insofar as it provides an objective *framework* on which to base judgment. But judgement, by its own definition, is *subjective*. It doesn't, however, have to be *arbitrary,* or based on willful, after-the-fact considerations. In appraising a person's performance, it is important to remember that only involved and ethical managers can make the MBO system work as intended. As plans, objectives, and performance evaluations become more precise, managers must work even harder to improve the quality of their judgments. But despite the objective nature of the criteria by which employees are judged under MBO, trust, conscientiousness, and honesty are perhaps even more important than in any other appraisal system. Objective frameworks lose their credibility when uninformed Leaders, Builders, or deliberately arbitrary Destructive Achievers make bad judgments in looking at the myriad influences that affect final results.

In effect, then, the employee concerns expressed above cannot be answered at the front end of an MBO effort. The answers to the questions are independent of the system. They depend entirely on the ethics, conscientiousness, and abilities of the implementers.

2. *"Will MBO be used as a club? As time progresses, lists of objectives seem only to get longer; it is hard to get objectives taken off one's list. If you don't deliver results for your boss's objectives, will you be punished?"*

These concerns, like those above, cannot be answered outside the context of the values and real intentions of management. To illustrate, let's first consider the natural course of events in an MBO system when objectives are being met. Then we will compare them to those that occur when results are deteriorating. Manufacturer A of, say, industrial adhesives is successful. There are any number of reasons—the actual performance of its plant staff, or the outstanding product, which can be sold at below current market prices. Because of technological breakthroughs and a new generation of production equipment, it may even be difficult for Manufacturer A to fail. Objectives are being met or exceeded, the organization is expanding, morale is high, and the quarterly reviews of results are occasions to be enjoyed.

Theoretically, there are no rewards or punishments in an MBO system, only natural consequences. That is, persons are not rewarded or punished on a personal basis; they receive the natural consequences of their objective performance. The natural consequences in good times, then, are that many rewards are distributed, such as bonuses, promotions, and exciting job assignments. Punishments are limited to the absence of rewards. When times are good and results are improving—for whatever reasons—MBO principles can work about as intended for almost any management group.

The situation changes drastically when results are deteriorating, again, regardless of the causes. A competitor, Manufacturer B, is suddenly the one with the superior technology and equipment. This makes Manufacturer A noncompetitive, through no fault of its members. New efforts are made to experiment with and develop improved adhesive formulations. The experimentation inevitably leads to lowered conversion efficiency and escalating costs. The quality of the main line of industrial adhesives deteriorates because of the diversion of resources to experimental purposes. Market share continues to decline.

Suddenly the quarterly reviews occur twice a month and no one is enjoying them. On specific projects, or in crisis situations, the reviews may occur daily over the phone. Persons are constantly asked to explain their actions, especially their failures. Lower-level personnel begin to complain that "if they would just leave us alone we could get something done." Pressures, which easily become threats and intimidation, are exerted on all personnel—"It's a matter of survival." The natural consequence is punishment, through no one's fault at the plant site. The plant eventually closes and most personnel lose their jobs. In such a situation, it is a rare senior management that can recognize a specific manager's superior performance and arrange for a positive consequence for that individual.

Of course, essentially the same scenario will occur when an organization does *not* have a formal MBO system. The point is, however, that fundamentally sound management practices can become additional punitive weapons in the hands of managers who don't have the courage, wisdom, or integrity to implement them properly. MBO meetings and documents are used in order to more effectively intimidate, blame, self-defend, and punish. Short-term expediencies replace long-term values.

3. *"Will MBO introduce conflict? Will there be group-to-group disputes because of different functional priorities, even though supposedly the groups have the same objectives? Will there be boss-subordinate conflicts because of pressures from above."*

MBO principles should help to reduce conflict, simply because front-end efforts are made to plan and coordinate activities. There should be fewer reasons for conflict between groups, because the groups have anticipated problems and have already designed steps to prevent them.

The Destructive Achiever, however, uses these front-end planning efforts to enhance his or her personal career, not to reduce conflict. For example, a plant manager, his technical manager, and his production manager may meet to update objectives for the coming six months. They reach agreements based on what is best for the plant. But the technical manager wants a promotion to the corporate R&D facility. He realizes that his career ambitions depend more on his cooperation with the R&D staff than on his cooperation with the production manager. In the same way, the production manager wants to become a plant manager. His future depends more on his support of the corporate vice-president of manufacturing than on his support of the technical manager. If either manager is inclined to be a Destructive Achiever, under-the-table conflicts begin.

4. *"The MBO process takes a lot of time; will it hurt our flexibility? Once objectives are written, can they be changed? Will it create more inertia in trying to get things done? Will there be more meeting time, administration, reporting, and documentation? There seem to be many steps in the system, level-to-level, function-to-function."*

The very nature of an MBO program suggests more meeting time. For example, there are those who would argue that the manufacturing vice-president and the R&D director above should participate with the plant manager, production manager, and technical manager in setting objectives. But where does the network end? In consensus, no matter where one stops, a DA will be able to search out and capitalize on documented (legalistic) requirements or omissions. As a matter of fact, the more complex the system, and the more it requires political skills, the greater the DA's advantages.

In addition, the more elaborate the MBO network, the more difficult it becomes to get it to adapt to changing circumstances. Subordinates report that they can be more effective, because of time considerations, continuing with a second-best action plan than taking the necessary steps to get a better action plan approved. It is especially difficult to get changes made when one's boss has already publicly supported the second-best approach in initial discussions. A change now would require the supervisor to admit that he was wrong and that he wants the group to adjust to his subordinates' changed viewpoints. And, if the supervisor is a DA, the task becomes impossibly difficult. Most DAs, in fact, never risk it, unless they have something to gain personally, and only if they are sure the change will succeed. Leaders and Builders sometimes make up for the weaknesses in the MBO system by assuming the personal risk of unilaterally making changes on their own.

5. *"Will MBO hurt creativity and innovation? A supervisor may want his own objectives supported regardless of new thinking. Personal efforts will have to have a short-term pay-off, or no credit will be received at the next appraisal. Will there be less risk-taking as a result?"*

Theoretically, MBO should *help* creativity and innovation. They are included in the strategic plan and should therefore be reflected in long-term objectives. However, according to a frequent complaint from persons in technical and R&D groups (although production units are certainly not excluded), MBO, and especially the quarterly review process, discourages innovation. The fundamental problem is that upper managers are not able to judge the progress of a creative project when there are no obvious, periodic results. An Innovator may be doing an excellent job exploring alternatives, but his supervisor either can't judge the quality of his efforts, or can't convince his superiors of it. The DA, especially, will give the Innovator additional projects to make sure he is pulling his weight.

6. *"Some jobs just don't lend themselves to objective-setting. The most important parts of jobs are frequently not predictable or measurable. How do you specify in a mean-*

ingful way the objectives of a commercial artist, R&D scientist, or development engineer?"

In spite of claims to the contrary, many jobs *don't* lend themselves to quantitative objective-setting. The further a quantitative objective is removed from the direct control of the individual, and the less meaningful it is to his or her total impact, the easier it is for the supervisor to be arbitrary in judging performance. An MBO process loses its credibility when supervisors insist that such measures be fabricated in order to satisfy the system.

The six sets of employee concerns frequently reach fruition when an MBO program is installed. The mechanics of good management are in place, but the same old organizational problems remain: empire-building, avoidance of responsibility, orientation toward short-term results, lack of cooperation, and so on. But now the negative behaviors must somehow be documented as positives. Rather than improving, they are covered up—and in the process they become even more detrimental to organizational health.

Now we can add another hypothetical manager to our discussion (see Exhibit 6.1). Mr. Pseudo-Smith is a person who looks, talks, acts, and has the visible results of a Mr. Smith—but who isn't a Mr. Smith. Instead, Mr. Pseudo-Smith is the typical product of a poorly installed MBO program. The Pseudo-Smiths, the Smythes, and even a few Smiths in some bad circumstances, conclude that survival and promotion depend more on having the right documentation than on actual contributions to the organization.

There is no clear relationship between our five managerial types and the Smiths, Smythes, and Pseudo-Smiths. Depending on personal style and circumstances, Leaders, Builders, Innovators, and Mechanics could resemble either a Smith or a Smythe. In an especially bad environment, any of the four could partially become a Pseudo-Smith. DAs, however, are more likely to behave like Pseudo-Smiths, even when the organization does not require it. The DA learns how to model whatever behaviors senior management values. However, a formal MBO system hastens his or her development.

In an immature group (from an MBO standpoint, a group that did not evolve its own system), the very clarity and accountability of the system creates a short-term orientation, false impressions, and buck-passing. Suspicious of management's real intent, persons are cautious, guarded, and parochial—which creates more caution, guarding, and parochialism. As more persons decide to become Pseudo-Smiths, others are convinced that that is what the system requires.

If the results of MBO programs are so frequently negative, why are they so popular? Why are managements so enamored of them? One reason is the implied promise of making management easier, less judgmental, and more reliable. A *program* suggests a system that is easy to install and time-efficient. Objectives are quickly determined and recorded; the package is complete. However, an MBO program is a house of cards if there is no trust. In such a situation, objec-

Exhibit 6.1   Approaches to Management-by-Objectives

Factor	Mr. Smith	Mr. Smythe	Mr. Pseudo-Smith
Time-Span Orientation	Long-term objectives—this quarter, this year, next five years.	Short-term objectives—today, tomorrow, next week.	Long-term objectives—this quarter, this year, next five years.
	Long-term priorities—work is being done today on the five-year plan.	Short-term priorities—work depends on the most urgent perceived tasks.	Short-term priorities—work is designed to produce observable results by the next reporting period.
Commitment within the Management Team	Objectives and priorities are clearly understood and agreed to.	Objectives and priorities are intuitively understood and agreement is inferred.	Objectives and priorities are understood and superficially agreed to.
	Activities are designed to achieve organizational results.	Although competitive, activities are designed to achieve organizational results.	Activities focus on highly visible projects and are designed to achieve functional results.
Supervisory Style	Delegation—subordinates know their roles, their expected results, the big picture, and the sources of help. Through genuine participation they can significantly influence matters that affect their jobs.	Direction—subordinates' roles change with circumstances, must be able to turn on a dime and to react to clues from the supervisor. Participation may be genuine when it occurs but difficult to plan for and achieve.	Manipulation—subordinates understand that their first priorities are to accomplish those things that relate to their supervisor's most visible objectives, regardless of long-range organizational impact. Participation is a formality without substance.

Communication Environment	Open and significant—systematic effort is made to include all relevant persons in the communications chain.	Selective and significant—information is exchanged on a "needs to know" basis. Members compete to be the most informed.	Guarded and biased, yet extensive—communications document agreements, project status, and changes in expectations. Effort is made to minimize risk and to maintain image.
Evaluation of Performance	Judgment is based on both objective and subjective data. Predetermined objectives provide a systematic framework from which to view total performance, not only what was achieved, but how it was achieved under the circumstances.	Judgment is based primarily on subjective data. The supervisor arbitrarily selects the framework for evaluation: an intuitive collage of results, activities and impressions.	Under the guise of objectivity, judgment is minimized and evaluation is based on results. Supervisor arbitrarily ignores extenuating circumstances if upper management is disappointed with results.
Quality of Work Life	The classic motivational climate—pride in the group and its product, self-respect, mutual support both personally and professionally for a wide range of persons.	A fast-moving, competitive climate—exciting for those in the communications centers. Frustrating and demoralizing for those who are not.	Deadening—for most persons performance reaches the lowest acceptable level with the least risk. Functional self-interest leads to cynicism about the political climate. Image becomes more important than contribution.

tives between functions are subtly incompatible, unrealistic in terms of time and cost, don't represent the best use of resources, and are meant primarily to give upper management what upper management is perceived to want, regardless of what the organization actually needs. A false sense of clarity, objectivity, and control is created. Plans and decisions are based on documented reports and polished oral presentations (the "management-from-the-central-office" syndrome), rather than on a more direct involvement with the issues. Furthermore, the short-term results achieved by Destructive Achievers, although impressive, force the premature delivery of technical and production improvements. "Short-term" can be up to five years for an organization that is in good condition. It can take that long to "spend" the morale, personnel development, maintenance, and innovation that have been built up over the years. Documenting good management on paper is not the same as having good management in the workplace.

Remedial Action

Because everyone manages by objectives, MBO *is* management, and the required remedial actions to improve a formal system already in place can be as many and diverse as there are organizational problems. Distilled from these, however, are four considerations that relate specifically to MBO as a "unique" approach. Paradoxically, the first step is not to mention the term, since lower-level personnel have already seen it at its worst. Moreover, to mention it would draw attention to the wrong issue, that is, the *mechanics* of managing, rather than the *dynamics* of managing.

Second, managers must demonstrate by example that management is a process and that managerial integrity is more important than the management system used. The management process is dynamic, changing, adaptable, and flexible, and it requires the active participation of everyone involved. If the behaviors of all concerned are appropriate, the process can be educational and developmental for all and a major prerequisite for effective delegation.

Contrary to what many seem to believe, written objectives, action plans, standards of performance, and written reviews are the *records* of an MBO process, not the goals. The goals, attained through effective managing, are to achieve

- more objective, rather than arbitrary, planning of activities.
- clearer expected results.
- better coordination and cooperation with others.

- improved utilization of resources.

- more individual influence.

- greater control of events at all levels and functions.

It is far more important that the means picked to accomplish these goals make sense to each supervisor and subordinate pair than that they be documented in a standard written format. Each must see the other as conscientious, honest, and thorough in fulfilling his or her responsibility to the relationship. It is a continually evolving process leading toward a more sophisticated level of judgment in their technology, market, and personnel decisions.

Third, upper management must be able to sense that these goals have been achieved, rather than relying primarily on reports and brief impressions. To do this, there must be enough meaningful contact with persons at lower levels. Otherwise, the system simply won't be effective. It *will* get results, but it will become more political, and the organization will become more concerned with survival than excellence. Moreover, the excitement will be gone—a fertile environment for the Destructive Achiever.

And fourth, each subgroup must develop its own management process. Its format must be compatible with those of other levels or functions but need not be the same format. If the newly designed process comes from top management, its influence will be felt throughout the organization. However, real impact is achieved through the effective communication of issues of substance *between groups*. To try to make every function and level manage in the same way is unrealistic. If the results (including organizational health) are good, and if the process works at the top level, it doesn't matter what procedures are followed at lower levels. If problems at lower levels adversely affect results, then those problems need to be specifically addressed.

The key issue is this: Results, flexibility, and the involvement and innovation of subordinates are far more important than conformity to a formal MBO system. Rigid adherence to dictated procedures almost guarantees that important subtleties will be filtered out of the communication chain as they are passed up or across the organization.

Perspectives

Probably the most vulnerable advocates of MBO are those who have had good experiences with excellent managers. Their experiences were due to good management dynamics, not the mechanical procedure. But they have forgotten, or never realized, this. Hence, they expect to quickly reproduce the same good experiences with a new group by inculcating the same mechanical procedure. When criticized, doubted, and distrusted, they feel betrayed, bewildered by the defensiveness of their new group.

A case in point is the production manager[4] who was promoted to plant manager and transferred to a new location. Senior management informed him that he had just one year to demonstrate that he could turn the plant of 400 employees around, or they would shut it down. They produced two kinds of consumer electrical products and were experiencing severe foreign competition. They not only had to reduce costs, they also had to improve quality.

Although relatively inexperienced, the new plant manager was an advocate of MBO. When he first met his immediate staff, he clearly informed them of their goals for the coming year. He also met with employee groups throughout the facility and gave them the same message. In order to create a team effort, he made sure that appropriate subgroups throughout the facility cooperated in identifying common objectives and in developing action plans. To make a long story short, by the end of the year, the plant had demonstrated that it could succeed. Just about anyone in the facility would tell you that they had a very high regard for the plant manager and that the only way to manage was "by objectives."

Because of his success, the plant manager was promoted to a much larger facility of 1,200 persons. In this case, the product lines were different, and the plant was considered to be about average in performance. Again, the new plant manager met with his immediate staff. I'm sure he didn't use these exact words, but what he said went something like this:

> I know what professional management is all about. I know how to turn a plant around. There is no way that I can do the same kinds of things in a plant of 1,200 that I did in a plant of 400, so I'm going to teach you how to manage by objectives. In fact, I don't even intend to get involved with persons below your level.

After two years the plant manager was again transferred, but this time with a mediocre record at best. Throughout the facility, employees celebrated on the day he left. There was a lot of distrust between many different subgroups, and morale was low. The rumor was that it had become virtually impossible to get anything done, again, "unless it was on someone's list of objectives."

The experiences of this one individual help illustrate the nature of MBO systems and the persons who implement them. At the smaller plant, the manager's systematic style hastened his success. He gained power and control by actually improving the process of interpersonal and intergroup cooperation and commitment. It was a problem-solving, evolutionary process in which persons improved their understanding of the technology, each other as professionals, their challenges, and their opportunities. Improved collective judgment and more precise planning and execution was the *actual result,* not the plan on paper that so many MBO programs start out with.

At the larger plant, because he already understood MBO principles and had confidence in his own abilities, he attempted to use his powers of position and expertise to mandate a program into existence. What should have been a problem-solving process became power politics. He was overconfident of his abilities to personally control events, and so the results he achieved were inevitably mediocre.

MBO, as the term is commonly applied to a body of theory, is the best way to manage; in fact, it is synonymous with *good* management. But it should not be institutionalized in a formal sense—precise forms, set procedures, and so on. When a managerial group tries to use an MBO "program" to correct by force problems of planning, coordination, cooperation, and distrust, they usually manage to make the problems worse. The process becomes more political than problem-solving, and, inevitably, the system becomes as inflexible as a straight-jacket at some lower levels and in some functions. As the program begins to grow away from its initiators, its emphasis changes from substance to form. Managements need to work on the substance of problems directly. If in doing so their behaviors convey both effectiveness and integrity, a process will evolve that can be called MBO—but only after the fact.

Footnotes

1. Much of the material in Chapter 6 has been adapted from an article and a training videotape: Charles M. Kelly, "Remedial MBO," *Business Horizons,* September-October 1983, 62–67; and Charles M. Kelly, *Remedial MBO,* Association for Media-Based Continuing Education for Engineers, Atlanta, Ga., 1986.
2. Harry Levinson, *The Levinson Letter.* The Levinson Institute, Belmont, Mass., September 15, 1981, 4.
3. Peter Drucker, *The Practice of Management.* Harper & Brothers, New York, 1954, 121–136. MBO practices were not "invented." Effective executives were practicing "MBO" before there was a term to describe the activity.
4. This is a paraphrase of an actual situation.

CHAPTER 7

Performance Appraisal and the Ethics of Promotion

In appraising the performance of lower-level subordinates, it's easier for a senior manager to observe their upward projection of power and image than their downward and lateral projection of integrity. The former is something DAs do well and Builders do poorly. In the case of the latter, the situation is reversed. This distortion in perception worsens as an organization gets larger, the mechanics of its appraisal system becomes increasingly important, and the Destructive Achiever's competitive edge grows. Senior managers base their evaluations more on formal review procedures than on direct personal contact. In some companies, they may be spending more time in meetings talking *about* staff members than in actually interacting *with* them. As a result, the career of a lower-level subordinate often hinges on senior managers' reactions to his or her sometimes too-far-removed personal image and on the persuasiveness of the immediate supervisor in review sessions. Written and oral reports, "success profiles," consultant evaluations, and mechanically weighted measures of performance become more important than the individual's performance itself. This is especially true when values, behavioral habits, and *how* one gets the job done must be evaluated. So as not to be unduly impressed by those whose major accomplishment is the political mastery of the appraisal system, senior managers need to take a proactive stance and improve their ability to "sense" the value of an individual's performance. If the individual is several organizational levels down, they will need to

- understand the inevitable strengths and weaknesses of any appraisal system, including their own.

- communicate their promotion policies, including management's values, appraisal philosophy, and expectations, to all levels of personnel.

- commit themselves to promotional standards that are highly ethical. This means planning developmental experiences for all types of personnel in order to
  discover and remove or correct Destructive Achievers.
  discover, develop, and promote Leaders and Builders.

Their success will depend on a staff that is well informed and committed to making the system work in accordance with the standards.

Appraisal Systems: Strengths and Weaknesses

Judging from the constant stream of articles on the performance appraisal process, the search for a system that will have a positive impact on employees is apparently endless—and, to a great extent, fruitless.[1] For every article that describes a new procedure designed to motivate better performance, there is another that surveys the systems in use and finds out that no one likes any of them and few last more than three or four years without being revised. The most popular approach is to make appraisal a joint effort between appraiser and appraisee. Theoretically, both can remain objective and jointly come to a conclusion. If not, the appraiser at least will have demonstrated an openness in being willing to hear the appraisee's point of view.

Although this approach is fundamentally sound, as actually practiced it has several flaws and many potential benefits are lost. When senior management in the very beginning tries to persuade managers and employees alike to have unrealistically positive attitudes about appraisals, it is setting itself up for failure. As a result, its credibility suffers and the confidence of subordinates at all levels is reduced.

Most managements choose to assume that the following ideals are achievable and so communicate them to employees as if they actually existed in their own systems:

The appraisal is a positive part of the management process, which motivates more effective individual performances.

The appraisal is systematically objective and therefore ensures an accurate evaluation of each person's performance.

The appraisal (in the more progressive companies) results from a joint evaluation by both supervisor and subordinate.

When such unrealistic assumptions are made and communicated to employees, many of them leave their appraisal interviews in disgust. Either their own boss was incompetent or dishonest (usually felt to be the case), or they must have been at fault for not reacting better to the process. In practice, communicating unrealistic assumptions never results in positive attitudes. In fact, the effect can actually be harmful. The DA is the ultimate winner when the true nature of the process is camouflaged. So management must admit to and confront the real problems, then strive to make the system as effective as possible.

Appraisal system refers to the mechanical process of determining how an employee is to be rated by the company, of evaluating the employee's performance, and then of communicating this rating. It does not refer to the overall dynamic process of defining goals, clarifying expectations, coaching, and all of the other aspects of human resource development. These latter can have a positive as well as a negative or neutral effect on performance.

APPRAISALS ARE NEGATIVE

The first assumption, that the appraisal is a positive part of the total management process, is contradicted by every objective study I know. The most comprehensive cross-sectional survey of employee attitudes toward organizational promotional practices showed those attitudes to be strikingly negative. Fifty-eight percent of all middle managers, 69 percent of all supervisory-level managers, and 71 percent of all technical employees in managerial positions believed that advancement and promotion were often based on "a largely subjective and arbitrary decision on the part of the corporate or organizational superiors."[2] In addition, the results suggested that promotional practices rewarded those employees who had a strong upward bias. Eighty-two percent of all respondents believed that the critical factor in determining promotability is a satisfactory working relationship with superiors, not satisfactory relationships with subordinates or peers.[3]

As a result of their own surveys of companies, staff writers of *The Wall Street Journal* and *Business Week* concluded that "formal job appraisals grow more prevalent but get more criticism. . . ."[4] and that ". . . no one knows how to factor out human error."[5]

One should not conclude from these studies that *all* employees are dissatisfied with their company's appraisal system. Attitudes tend to become more positive higher up on the organizational ladder. This is to be expected, because high-level persons are the ones who have been successful in the system and who are responsible for designing it. Even so, one finds many ousted company presidents, out-of-favor vice-presidents, and forceably retired plant managers who have become bitter about the injustices of the process.

There are three basic reasons why the appraisal process is inherently negative: (1) egocentric bias prevents objectivity in discussing one's own accomplishments; (2) strong emotions surround the entire process, from monitoring achievements to actions resulting from the evaluation; and (3) the act of judging performance is detrimental to the coaching or counseling relationship.

Egocentric bias refers to our tendency to exaggerate our own influence on events. It is very difficult to be objective. In one study, 70 percent of the adult male respondents placed themselves in the top 25 percent in leadership skills. Egocentric bias also works in the other direction; people sometimes blame themselves excessively when things go wrong. As might be expected, egocentric bias works in favor of the Destructive Achiever and against the Builder, especially in participative appraisal systems. DAs are inclined to blame others for failures. A DA may be consciously dishonest or simply blinded by defensiveness. On the other hand, the Builder is too willing to accept blame without protest. Builders may go so far as to accept undeserved blame, either to maintain a harmonious working relationship with a peer group or to preserve their image as trustworthy and reliable associates.

Second, the whole process can be very emotional. The supervisor may feel guilty "playing God" with a person's life. The subordinate may want to deny—or resist—threats to his or her ego and self-image or to income and financial security. Although both recognize the importance of the appraisal, each may avoid full and open participation in it.

The most important reason—and the most subtle—however, is the third one, the negative effect that evaluation has on any coaching or counseling relationship. It is axiomatic that when one person acts as a judge of another, objective problem solving is significantly hindered. The person being judged has three problems to overcome: separating *performance* quality from *person* quality ("I am good or bad"); dealing with aspects of the performance problem itself; and dealing with the judge's perception of the problem. In other words, appraisees not only have to be concerned with the best way to solve a problem, they must also consider how their supervisor will evaluate their solution. In this sense even a positive (favorable) appraisal can have a negative impact, because it reinforces the belief that one needs to continue to "please the boss," rather than do the right things, as the survey on employee attitudes indicated.

When supervisors penalize subordinates for not solving problems the "right" way, they are not necessarily being unfair. There is simply no other option open for them. Judgment is a crucial part of performance, and every supervisor must evaluate how well a subordinate exercises it—in all areas of responsibility from technology to management of human resources. A subordinate can make the "wrong" technical decision or assign the "wrong" person to a project only so many times before suffering personally in his or her own performance rating. Usually there is no purely objective way to determine whether a subordinate's judgment was excellent, good, or poor. If option "A" failed, would option "B" have succeeded? If "Fred" successfully completed a project, would "Sam" have done even better? Who can say? It's the supervisor who must ultimately evaluate the quality of the subordinate's decisions.

The evidence we have from surveys and from what we already know about the nature of the coaching and counseling relationship indicates that most people would be better off if appraisals were unnecessary. It would be better if employees didn't have to be evaluated for higher salaries, promotions, job assignments, or career and life planning.

APPRAISALS ARE SUBJECTIVE

The publicly stated premise that is most damaging to management's credibility is that its system makes appraisals objective. Management-by-objectives is the best system on which to base appraisals, but, even so, the evaluation is still subjective. MBO and "appraisal by results" have unfortunately been peddled by some as management tools that remove judgment—considered unreliable—from the appraisal process. This attitude is devastating to the supervisor/subordinate relationship.

In a study of 10 organizations, Stein found two common problems in MBO systems: writing realistic quantitative objectives and later adjusting the objectives to reality.[6] These are issues of judgment, and they are at the heart of such systems. What it takes to achieve results is certainly open to debate, as are the effects of any intervening variables between the time of setting the objectives and observing the final results. Even Leaders may consider the appraisal-by-predetermined-results an attractive substitute for a more comprehensive, but possibly more subjective, evaluation of performance. And, in the worst case, Destructive Achievers actively use the process to punish their subordinates for not contributing to their own personal records. This is one of the biggest factors in uncooperative behavior between departments. The most negative effect, however, is that the pretension denies the maturity and common sense of the appraisee and leads to a relationship-destroying game. The supervisor is the infallible superior looking at objective data and passing judgment on the subordinate. The subordinate is far more willing to accept a "wrong" decision systematically based on objective data than a fiat based on the same data. The former will entail more dialogue and an admission of fallibility on the part of the supervisor. But the latter causes more resentment on the part of the appraisee and a loss of respect for a system that refuses to admit that the appraisee could be right in a disagreement. Of course, the worst circumstance is the unquestionable fiat based on unsystematic data such as personality traits and "leadership qualities."

THE SUPERVISOR JUDGES

Supervisors and upper management traditionally determine performance ratings. Regardless of the procedures followed, upper management decides who gets a choice assignment, a raise or a promotion or who is dismissed. This is still the case today. But, with the advent of open personnel records and appraisals and the vulnerability of companies to lawsuits, it has become imperative that appraisals be honest. To develop appraisal systems that subordinates will see as fair and positive, some managements have taken a participative approach in which at least some of the responsibility is placed on the subordinate, who is expected to be pleased to have more input into the system.

This approach is probably the best, but there are some real dangers. For example, it can be demeaning. The appraiser may try too hard to convey the impression that the subordinate is actually participating in the decision, or even significantly influencing it. The appraisee knows that the evaluation has already been made and won't be changed. He or she knows that the supervisor is manipulating the discussion by asking questions and pointing out standards in such a way as to cause the subordinate to arrive at the supervisor's conclusion. The subordinate also knows that an unreceptive attitude toward the supervisor's opinions can itself negatively influence the performance rating. In any strong disagreement, according to the supervisor, the subordinate is "unwilling to face the facts." Of course, this may be the case. However, the point is that the supervisor *always* wins. The perceptions of the supervisor and the subordinate are not given equal weight and any distortion of reality is the subordinate's.

How open should a subordinate be? Identifying a problem may justify a poor rating. This puts the supervisor in a bind. If he uses the subordinate's negative input, he penalizes openness. If he doesn't, what is the purpose of participation? Here again, the Builder is more likely to suffer, especially when the supervisor is being pressured by upper management to lower the ratings. The Builder's attempt to be objective may justify a lower rating if the supervisor is weak. DAs, on the other hand, see the open appraisal as just another power event.

Even though a participative appraisal can have negative effects, the subordinate may not see the supervisor as being insincere or dishonest, although sometimes that is the case. Rather, the evaluation of performance is seen as an act of judgment, that is, in cases of disagreement, the supervisor must prevail. Indeed, it is the supervisor's responsibility to use his or her best judgment, not the subordinate's.

Communication: Policies, Values, and Expectations

Appraisals do have value. They let employees know how they stand with their organization and may even be vital to an employee's career development and life planning. From a subordinate's viewpoint, these considerations alone make the process necessary.

However, the impact on the supervisor-subordinate relationship can be quite negative, especially in regard to future problem solving. Therefore, appraisers and appraisees alike must have a realistic understanding of the ego and the interpersonal dynamics involved. There are two steps that management can take to make its system credible. In doing so management will become more aware of the deficiencies and biases of its own appraisal system.

As a first step, managers must develop and communicate their views of the appraisal process: what they think is possible, what their goals are, and how they intend to go about achieving them. Although policy statements in booklets, instructional manuals, and on bulletin boards are helpful, management's *presence* in orientation and instructional sessions is imperative. Lower levels of the organization need to know first-hand what management thinks and wants before they can believe and appreciate their philosophy.

Second, everyone needs to understand the rationale, procedures, and skills involved in the company's system. Much of this can be dealt with in a traditional training format that includes the subject matter found in standard textbooks and packaged programs. *Issues of values, however, are most effectively addressed in an interactional format.* People need an opportunity to bring their own issues to the surface, to vent criticisms, frustrations, and anger, and to understand the key concepts described below. This must occur when upper management is present. If subordinates have to wait until they are safely outside the conference room to talk about their problems, the problems will never be solved.

THE APPRAISAL IN PERSPECTIVE

Even when communications are good, most employees are confused about what happens in the appraisal process at levels above their own. They are apt to be filled with all sorts of dark imaginings; therefore, it's important to show how the appraisal fits into the company's total human resource process and to explain the specific procedures.

To begin with, the appraisal itself (the communication of the rating to the employee) is but a small part of the whole process, which includes setting objectives, career counseling, and the kinds of activities that are normally associated with a traditional work-planning-and-review approach (see Exhibit 7.1).

Exhibit 7.1   HUMAN RESOURCE SYSTEM

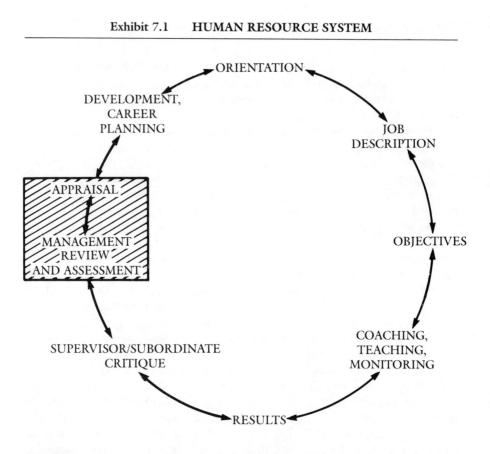

 These phases of the HR process are properly management's responsibility and should be acknowledged as nonparticipative. Although they are necessary activities, they have negative connotations for most employees. Their evaluational character is antithetical to the counseling/problem-solving relationship, whether performance is viewed as good or bad.

Other phases of the system should be genuinely participative. They can have a positive or negative impact on individual performance and development, depending upon supervisor/subordinate chemistry, interpersonal skills, and organizational competence, and the company's needs and conditions.

These are interactive steps and do not necessarily follow in sequence. They also occur throughout the year and so cannot be lumped into one or two sessions. To try to combine appraisal with any other part of the process, such as setting objectives, would automatically depreciate both activities.

Employees should be told at the start that the appraisal process is a negative part of the supervisor-subordinate relationship and has some significant disadvantages and weaknesses. But they should also be told that appraisal is necessary, and that no one has yet figured out how to make it a positive experience for most people. There are many published reports to document this; moreover, those in the discussion group who have worked for other companies will substantiate it. Of course, the next step is to point out to employees that every reasonable precaution has been taken to ensure that the company has one of the best systems currently in use. If this cannot be said sincerely, then management should re-evaluate its system.

THE APPRAISAL IS BRIEF

Some say that the appraisal should be long enough (hours) for employees to share points of view and develop plans for the future. These recommendations are ill founded. This sort of thing should occur in another session either prior to or following the appraisal itself. If someone is upset, or elated, to do more than communicate the rating accomplishes little. Therefore, the session should be brief (minutes), but certainly not coldly abrupt. Barring unusual circumstances, there should be no attempt to resolve any significant issues during appraisal.

Both the supervisor and the subordinate need to remember that the primary objective of the session is to make sure that the subordinate clearly understands how the company evaluates his or her performance and that the supervisor understands the employee's reaction to the evaluation. It is not a debate; and the longer the session, the greater the likelihood that it will become one. Builders are likely to be the worst debaters; Destructive Achievers, the best.

In either case, an emotional debate always degenerates quickly into a win-lose argument. Once this begins to happen, the subordinate focuses only on those facts that are favorable to his or her performance. And the supervisor emphasizes everything that the subordinate does wrong—hardly a situation conducive to a good relationship. Unless a subordinate has new data, or a new way to objectively evaluate his or her performance, which is unlikely, it does no good to argue about a rating if the supervisor and the supervisor's superior have already made their decision. Nor does it do the supervisor any good to dwell on the subordinate's deficiencies in an attempt to beat his or her self-concept to a pulp. A supervisor can easily fall into this in trying to justify a lower-than-expected rating.

It is to the advantage of both parties in a disagreement to postpone the discussion to a later time when cooler heads prevail. Then they can explore the causes and try to discover more objective ways to view performance. There may be no satisfactory remedy for the disagreement, and a lateral transfer may be in the best interests of both persons. An appraisal is appropriate only when it is part of an ongoing, comprehensive program. Otherwise it is impossible to appraise performance sensitively, regardless of the length of the session.

MANAGEMENT IS FALLIBLE

For management to insist that it doesn't make mistakes in rating employees injures its credibility with them, as well as its own commitment to objectivity. Of course, managers do not claim this directly, because they know that that would be patently wrong. Instead, they do it indirectly, by reacting defensively when employees question or criticize the appraisal system, or when an individual disagrees with his or her own performance rating. The prevailing opinion is that an admission of fallibility is a sign of weakness, an open invitation to challenges or lawsuits, and an admission that the management team is not a good judge of performance. Management seems not to realize that an admission of fallibility would indicate a willingness to constantly work at improving the system. It is, in fact, a sign of strength. Any debate that might come up would be more logical, less emotional. Instead of attacking a subordinate (to justify a low rating) at an impasse, a supervisor's honest response would be "I'm sorry we disagree, and you could be right. But, right or wrong, I have to go with my best judgment, and based on my observations and discussions with other managers, this was my (our) conclusion." Moreover, for management to admit that it could be wrong relieves the subordinate, whose ego has suffered. He or she may have to accept the fact that the performance rating was lower than expected, but is not forced to admit that his or her own judgment of self-worth was unrealistically high.

If top management would develop and model a philosophy that recognizes human error in personnel decision making, it would reduce the likelihood of legal actions against the company. Instead of stonewalling behind what has been advertised as a flawless system (a false security), all levels of management would constantly work to improve the system—not just to meet legal requirements, but because they recognize that the appraisal system is the key to the future of the company, and that it is never good enough.

PERFORMANCE IS RELATIVE

Fundamental to any appraisal system is the company's philosophy with regard to the way it categorizes and rewards performance. For example, ratings of (1) "outstanding," (2) "superior," (3) "good," and (4) "not rated" suggest relative values, but, specifically, what do they mean, and how are they applied? How good, for example, is "good"? Is one person's performance considered in and of itself, or is it compared to the performance of others? Are there pressures to keep individuals' ratings low? What makes ratings comparable from department to department or location to location? These are unpleasant questions to answer and lead to all kinds of "wordsmithing," in which management tries to remove the problems from the minds of employees without actually addressing them.

A task force once attempted to define in words (versus numbers) the meanings of five different performance categories, from "not acceptable" to "outstanding." It took five persons two full days before the words were finally composed and later published. Although they were probably necessary as a philosophical statement, as a practical guide they were useless. Essentially, the words were just an elaborate way of saying that a "3" met expectations, a "1" greatly exceeded them with more creativity, innovativeness, and/or effort, and a "5" didn't come close.

Exhibit 7.2 demonstrates the meaning of performance codes in a far more understandable way, and in a way that answers several other questions that are tough to deal with. These are the published expected performance distributions of two Fortune 500 companies. Whether a company publishes its expected distributions or not, everyone knows that they exist. All organizations must ultimately promote persons, or give them salary increases on some basis. In looking at the two distributions, one can immediately appreciate the difficulty of describing in words the meaning of performance categories. How would a verbal description of a "1" ("Outstanding") in company A differ from a "1" in company B, except that in B it is simply a smaller numerical proportion? By publishing its expected numerical distribution, and by explaining its rationale, a management team can answer questions and objections with a degree of realism that will at least be respected. Employees still may not like the system or the activity itself, but at least they know that they are being leveled with. Without going into detail, the following realities of organizational life now can make sense.

An objective, though certainly not error-free, way for management to maintain fairness and consistency between departments and locations is to monitor their adherence to the distribution guidelines. Some will observe that "no two groups are the same," and smaller subgroups may not fit the distribution at all. But a sufficiently large group, say 30 to 60 persons, should approximate the published distribution, unless it can show that its performance was unusually different. It follows from this that upper management *does* exert pressure on

Exhibit 7.2   PERFORMANCE CATEGORIES*

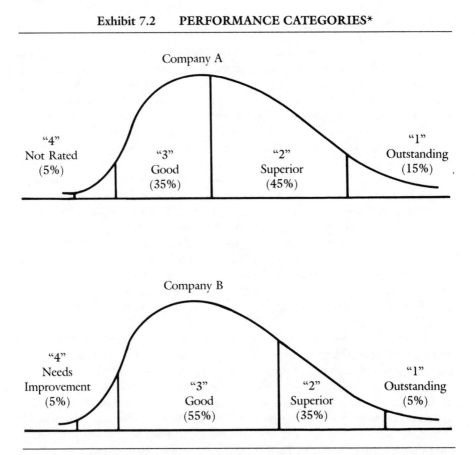

Company A

| "4" Not Rated (5%) | "3" Good (35%) | "2" Superior (45%) | "1" Outstanding (15%) |

Company B

| "4" Needs Improvement (5%) | "3" Good (55%) | "2" Superior (35%) | "1" Outstanding (5%) |

*Expected performance rating distributions of two Fortune 500 companies, as indicated by internally published guidelines.

many supervisors to lower their ratings of subordinates. There is a natural and constant upward bias on ratings as supervisors strive to improve the morale of their own staff, and as they see real improvement over time in subordinate performance. Yet all aspects of performance must improve throughout the company if it is to remain competitive and to advance with the state of the industry. Therefore, if the system is to maintain its integrity, it must be monitored, and over-raters need to become aware of their biases. So do under-raters, but there are far fewer of them.

It would be comforting to believe that individuals should be rated solely on their own merits without regard to what others are doing. However, if ratings are to have any meaning—to let persons know how well they are doing in their own company, to be used as a basis for monetary and status reward, and to serve as an aid for personnel decisions—they must be relative to the ratings of others in the same kind of position. Destructive Achievers instinctively know this; Builders must be reminded that they are in a competitive arena.

The relativeness of performance and the fallibility of management are important aspects of an appraisal *system*. However, they are almost insignificant when compared to management's demonstrated ethics in administering it. Employees will view the appraisal system as ethical only if managers take all reasonable precautions to be objective and fair.

The Ethics of Promotion

Interviews with appraisees indicate that they consider an ethical manager to be one who bases his or her evaluation on a genuine understanding of the subordinate's accomplishments and abilities. This means that, throughout the year, the supervisor must assess all relevant aspects of a subordinate's job and be thorough in discussions about them. Also, subordinates must have assignments that allow them to demonstrate their abilities. In the interview, the ethical supervisor is sincere in his or her evaluations but can be considered wrong by the subordinate, *and sometimes is*. The supervisor clearly lets the subordinate know where he or she stands, not only with the supervisor, but with senior management as well. The subordinate also is confident that the supervisor is communicating the same evaluation, including all the nuances, to senior management.

Appraisees consider a supervisor to be unethical when he or she has made little effort throughout the year to appreciate the true value of the subordinate's performance. Typically, this supervisor appears to be concerned only with his or her own problems, career, and objectives, and neglects managerial responsibilities to subordinates. The unethical supervisor is not involved with the subordinate in meaningful ways. Unethical supervisors may seem to be arbitrary, unorganized, or unsystematic in appraising performance. They don't seem to want to level with subordinates because they don't want, or aren't able, to defend their evaluations. Subordinates think these supervisors tell them one thing, but report something else to upper management.

Sometimes the interview is too casual. More than a few subordinates report that they weren't even sure that they had had an official appraisal. It is not unusual for supervisors to claim to have told a subordinate of a performance problem and the subordinate to deny ever having heard such a message. Of course, it is always unfortunate when a supervisor "lowers the boom" after an entire year has passed with no hint of the number or importance of a subordinate's deficiencies. In such a case, it is easy for the subordinate to suspect that the boss is reacting to upper management pressures, unwilling to risk fighting for the subordinate's interests.

Mistakes occur in all promotional systems. But their negative effects can be minimized if it is clear that management has high ethical standards for appraisal and promotion. This can be achieved only by making sure that all employees are conscientiously evaluated and have equal opportunities for success. To strengthen the appraisal system, management must:

- first detect, and then correct or remove, Destructive Achievers as soon as possible.

- give Builders developmental opportunities early in their careers.

These two goals deserve our utmost attention, because our culture's preference for a "can do" image over an ethical one comes at a high cost to genuine effectiveness in the future

DISCOVERING AND MANAGING THE DA

Contrary to popular belief, sophisticated Destructive Achievers are *not* usually found out. Frequent news reports of management misdeeds demonstrate that it is possible for a dishonest DA to reach the very top level of the executive suite, except when the organization is young and there are close and frequent hands-on, work-related contacts among several levels of the organization. Some DAs are even recognized as being among our best managers, legally successful in a short-term environment in which the ethical values of their organizations, or of society, are eroding. Potential DAs range from the incorrigible power seeker to the natural Leader. The former obviously should be detected and removed as soon as possible. But the Leader needs to have his or her values reinforced, or at least not contradicted, by the organizational leadership.

Unfortunately, DAs are difficult to spot when they are relatively new, even by conscientious managers. They are industrious and appear to be covering all the bases, including their relationships with peers and subordinates. At this point, to spot a potential DA, the first thing to do is consider the individual's reputation. A peer wonders why management didn't know,

> The word from [previous location] is that he antagonized a lot of people and was not seen as being promotable. He was a disruptive influence in the

managerial group. Now he is having a dysfunctional impact here also, and we're wondering why he was promoted to this job.

The following first impression is common among subordinates, especially when they are benefiting from the DA's actions:

> Before he came here, the book on (DA) was that he was a real political animal. He also now has an image problem with some of his peers here, but not with his subordinates. He's gotten our department some deserved recognition that it has needed for some time. We now have a stronger role than we had before, and he looks out for our interests. Some of his peers aren't used to having a strong person in that position.

Later the DA's subordinates came to understand why his reputation was deserved. Within one year, they wished they had never heard of him.

There are an unlimited number of ways in which subordinates and peers can observe the positive and negative consequences of the DA's behaviors. And, as the DA gains more power and confidence, his or her interactions with associates become even more blatant, sometimes to the point of recklessness. However, unless it wants to establish an undercover network, upper management must be sensitive to more subtle signs. Three of the most common ones relate directly to the DA's manipulation of the appraisal system. They are fairly easy to investigate: finger-pointing, strained relationships with certain peer groups, and arbitrary personnel appraisal.

Finger-pointing is due to the DA's unusually strong absolutist or legalistic orientation. According to the DA, if there is a crisis or major error, someone must have done something wrong, or violated an agreed-upon expectation. The DA may, rather than solve the problem itself, either avoid or assign blame, preferably to another department. If this is not possible, a subordinate will do. As one explains:

> One of (DA)'s problems is that he is right 90 percent of the time and he wants to be right 100 percent of the time, and that is unrealistic. He's a strong go-getter—wants a promotion very much. A couple of times he has made me look bad in public. He needs to give us more support when things go wrong. We agree to do something a particular way, and if it falls through, I get left holding the bag.

Evaluating a manager's relationships with peer groups can be a mixed bag. It is important to distinguish between the groups that *support* the DA's unit or *compete* with it and the ones that his or her unit *serves*. DAs whose unit provides a service will overcommit to whatever groups will influence their performance evaluation the most. Because their efforts are geared to the convenience of the groups they serve, their communications can be quite open and honest. The impression can be extremely favorable, as was that of this peer:

> (DA) has no reluctance at all to ask for others' opinions and discuss things openly, and has very free associations with others. He is quite good in meetings, very logical and persuasive, and also good technically. He delegates well to his subordinates and believes in a participative style of management.

The DA's own subordinates strongly disagreed, especially with the last two items. They saw him as a single-minded dictator who put on a show when dealing with client groups.

It's quite a different story when DAs are dealing with units that either compete with or support them. Then communications are designed to maximize career advantages. DAs use "urgency" as a manipulative device to get preferential treatment from a support group, regardless of the expense to other groups competing for the same resources. And when things go wrong, the finger-pointing begins. A subordinate relates:

> (DA) isn't sensitive to problems the other guy has and can be very tough with his peers. He has much more of a sense of urgency than others, and when he sees something going wrong, he wants it corrected right away. He goes to others and says, "It's your fault." If this doesn't accomplish what he wants, he'll go to higher levels of management to get it done. And maybe that's the way it should be done. But it upsets a lot of people, and when they leave his office, they're talking to themselves as they walk down the hall.

Most senior managements analyze a manager's outputs (physical results) very thoroughly. Unfortunately, they rarely analyze the effect on competing units, or how well the manager uses his or her support groups. Alarm bells should also ring whenever a manager, assuming a new assignment, announces that his or her predecessor was a poor manager who allowed incompetents to survive. This *may* be true. But when large numbers of employees are suddenly found to be incompetent, yet they originally came from a reasonably representative slice of the larger organization, a safe assumption is that they *are* being managed by a DA (the present incumbent) or they *have been* managed by a DA, or a Mechanic. In such a situation the chance is probably something like four to one that the person doing the rating is a DA, especially if the individuals being found deficient had good records with prior managers.

By lowering performance ratings, or threatening to, the DA gains at least two benefits. First, the fear of punishment via appraisal will force subordinates to prematurely deliver the bottom-line improvements that are already on-stream.

Second, the DA will be able to replace incumbents with persons who are more compatible with his or her personal goals.

Other signs of a possible DA are, although commonly known, often ignored or denied: remarkably good results (was something else sacrificed?); large turnover (departing employees are always found to have "good reasons"); poor opinion survey results (which are "too confusing to interpret"); large numbers of resumés on the market ("these are especially bad, or good, times"); and informal feedback. As the danger signals begin to surface, management must be willing to question its earlier performance assessment and, if necessary, admit its error. It should especially guard against its inclination to identify the DA's subordinates, peers, or situation as the problem. As soon as a suspected DA is identified, the supervisor must actively ensure that the problem is correctable. Many DAs are unable to understand or acknowledge the nature of their problems, so they must demonstrate that they are receptive to coaching and capable of doing more than just achieving readily observable, short-term results.

Supervisors who are respected, especially by upper management, are the only ones who can effectively coach the potential DA. Almost by definition, DAs are predisposed to be influenced only by those they perceive as having power and control. Therefore, those whose values are most in doubt should be assigned to strong managers who will model and insist on constructive behaviors. Since managers need to be sensitive to the potential DA's impact on others, they must be the type noted for being involved with subordinates and peers alike.

Simply assigning a potential DA to a totally different function—for example, a subservient support staff—is useless unless it is well planned. Both the DA and the supervisor need to clearly understand the DA's development needs and the purpose of his new position. During a coaching session a DA gave the following account:

> This is the third time I've heard it, and now I finally believe it. I think I got off on the wrong track when my first boss actually rewarded me for not cooperating with the technical department. He used to laugh about it when we would meet in his office. When I got transferred and worked for (Builder), he said that I needed to change. Instead of believing him, I thought he was just a weak manager in that regard—he wouldn't fight for his unit. Now my new boss is telling me the same thing and you've confirmed it.

Of course, when new managers do any of the things described here, it's no proof that they're DAs. The point is that, because management often investigates only after a great deal of damage has occurred, no other conclusion seems reasonable. As long as management temporarily benefits from the destructive behavior, it will rely on the normal evolution of events to resolve any doubts. And, as DAs are allowed to progress through management ranks, and as they become more entrenched in the organizational systems, it becomes progressively more difficult to correct the mistake.

DISCOVERING AND DEVELOPING BUILDERS

For whatever reasons, many Builders are self-effacing. They give the wrong impression, especially to managers higher than their immediate supervisor. They may feel that it is unseemly to advance one's own self-interests in interactions with others. They may have too many doubts and insecurities about their abilities, then later are amazed to find out that far less talented persons, even in the area of human relationships, have passed them by and become successful. They may be too fearful of making mistakes, not realizing that honest mistakes, even costly ones, are acceptable in the total context of performance. They are especially amazed when they later see that costly *dishonest* mistakes are accepted by the organizational leadership.

Senior managers traditionally have reservations about Builders, so to develop Builders will require unusual courage. They must be willing to take the same risks on Builders that they take on DAs. They must assign them to challenging, high-visibility projects. And, at the completion of the projects, they must pay as much attention to each individual's ethical standards and use or abuse of power as they do to his or her bottom-line results. If not, management's promotional choices will later be limited to recognized Leaders, undetected Destructive Achievers, and Builders who are underdeveloped and inadequately tested.

High-flying Leaders and DAs are promoted and moved from project to project so often that it is impossible to differentiate between the two on the basis of long-term impact. At the same time, many potential Builders do not get their opportunities early enough or often enough to give them the experiences they need to be competitive later in their careers. In addition, if those with low self-esteem are not given appropriate recognition early enough in their careers, many of their talents may be permanently lost to the organization. They get discouraged, because apparently they have officially peaked out, and they resign themselves to a lesser role. However, sometimes a job offer from another company is all it takes to revive their ambition.

On the other hand, a proactive management recognizes that the talents of some of the organization's most solid and reliable performers are almost deliberately hidden from view, especially early in their careers. They must be actively sought, discovered, and reinforced as early as possible, if they are to develop the political skills that are necessary at higher levels. As one direct supervisor noted:

"(B)'s impact is not recognized. You have to dig to see that what he does is very good." To do this, senior management needs to acknowledge its traditional biases and rely more on lower-level managers when they recommend Builders for promotion. The instincts of direct supervisors are usually good. They can sense the strengths and weaknesses of both DAs and Builders. But the instincts of superiors one level removed are not as good. They heavily favor the DA. As a result, senior managers are more likely to veto a subordinate's recommendation of a Builder for promotion than a recommendation of a DA. There have been cases in which lower-level managers actually had to conspire with each other in order to get senior management to promote a deserving Builder.

As one senior manager once said of a Builder, "I couldn't sleep well at night if I knew that he was in charge of that plant." Yet managers who had been the Builder's immediate superiors knew that he got more stability and reliability out of a manufacturing group than the typical DA. The Builder was quietly effective because he motivated others to get things done and under control. The DA's apparent effectiveness depends upon his own urgent, and disruptive, needs to stay on top of every situation likely to come to management's attention. Of course, direct supervisors need to verify their instincts about both DAs and Builders—by giving them projects that will test those qualities in doubt.

Perspectives

One should not be naive enough to expect that these kinds of improvements will ever be completely satisfactory to any level of management or to employees. The title of an article in *Psychology Today* is all too true: "Performance Review: The Job Nobody Likes."[7] The reasons for, and the ways groups go about, selecting persons to lead the appraisal and promotion process have always been deficient and an unpleasant aspect of organizational reality. When employees complain about "the system," management naively tends to blame them for their negative attitudes and itself for its own incompetence in evaluating performance and communicating the evaluation. Over the years, since appraisals have become popular, rather than refine its basic assumptions management typically relies on training in interpersonal skills to make the process a positive one. As a result, it has been dealing with the problem as it prefers it to be, rather than as it is.

Maturity requires that we live with a degree of uncertainty and with some aspects of life that can only be described as distasteful. If properly explained and confronted, a realistic approach to the appraisal process will help to create a more mature and hence positive and effective human resource system. It won't be perfect, but, if administered with integrity and a commitment to ethical standards, it will be an improvement over most systems in use today.

Footnotes

1. Much of the material in this and the following section was originally published in Charles M. Kelly, "Reasonable performance appraisals" in the *Training and Development Journal*, January 1984, 79–82.
2. Dale Tarnowieski, *The Changing Success Ethic*. An AMA Survey Report, New York, 1973, 33.
3. Ibid., 34.
4. Harry B. Anderson, "The rating game." *The Wall Street Journal*, May 23, 1978, 1.
5. "Appraising the performance appraisal." *Business Week*, May 19, 1980, 154.
6. Carroll I. Stein, "Objective management systems." *Personnel Journal*, October 1975, 526.
7. Berkeley Rice, "Performance review: The job nobody likes." *Psychology Today*, September 1985, 30–36.

CHAPTER 8

Communications and Culture

An organization's communication system is its culture in action, and it permeates all other systems. Pressures to conform to professional communications standards may improve the quality of information presentation—but at the same time reduce the quality of problem solving and decision making. Again, a paradox. As with other issues, there is an interaction between the communication culture and the Destructive Achiever. To the degree that the environment rewards *form* (presentation skill, persuasiveness, image), the DA will thrive and the Builder will suffer. And as DAs dominate the communication networks, they require greater system conformity from others.

Peters and Waterman (*In Search of Excellence*) described the situation well when they referred to the casual and informal communication climates of the excellent companies versus the more procedural and formal climates of the poorer companies.[1] The former are more concerned with the substance of communication, the latter with style and decorum. This is not a new issue. Throughout the ages there have always been two fundamental orientations to speaking: the orators versus the sophists in ancient times, the public speakers versus the elocutionists of more recent years, and the "conversational mode" versus the "performance" advocates of today.

In each case the issues are the same, and they depend upon how the communication event is valued. Is the activity itself intended to be a communication among people, or an individual performance? Is the objective to be the enhancement of society or the organization, or the enhancement of the individual? Of course, these considerations are not mutually exclusive, and a pure choice can never be made. However, they *do* relate to the kind of climate that any group chooses to pursue—the norms it values and advocates, the kinds of behaviors it rewards or punishes.

Management's values and the substance of the issue at hand should be the first considerations for each and every communication event. The skill of the communicator should be irrelevant, or at least secondary. Although this condition is obviously a desirable goal, there are powerful forces that prevent it: our sophistic standards for communication, as now advanced in superficial training programs; and the organization's demand that individuals rigidly conform to its communication norms, even when those norms are counterproductive.

TRAINING: A CAUSE OF POOR COMMUNICATION

An R&D scientist was sent to two different effective presentation workshops and was regularly coached by his supervisor prior to giving actual oral reports. After two years, "poor speaking skills and meeting participation" were still listed as his main development needs. Upper management then strongly suggested to his supervisor that his performance rating be downgraded from "superior" to "good" if he did not improve by the next appraisal period. What a message to send! Not only to the person who got a bad case of stage fright every time he was in front of senior management, but also to his associates, who recognized his contributions to the company as truly superior. At best, this event was a misguided attempt to "put teeth" into management's commitment to develop its human resources.

A performance problem should not be ignored if it significantly reduces results. However, people come in packages, and the strengths of some will be in areas other than communication. It is shortsighted to penalize or badger such persons just because their listeners have to work a little harder to make use of their significant contributions. It is also symptomatic of the problems some managements have in getting genuine employee involvement and participation: *you* must work to participate with *us* in our (unofficially) prescribed style; if we don't hear you, it's your fault.

In a training program, if the communication event is presented as a performance and the emphasis is on style, the message to participants is "How you say something is more important than what you say." Unfortunately, in many corporate cultures, this message is quite valid, as participants in public workshops will attest. Itself a victim of its own training concepts, management reinforces this lesson by making communication performance an overly important factor in rating job effectiveness. It holds the view that all personnel should exhibit the ideal qualities unrealistically advanced in either its own or outside "effective presentation" workshops. Management thus believes it has a right, even the obligation, to penalize those, or the message of those, who don't exhibit the desired qualities. (Of course, communication performance *is* important, even crucial, in some jobs, and to some degree in every job.)

On the other hand, if the communication event is seen primarily as one person talking with a group, the message is that a person can retain his or her uniqueness, within reasonable social conventions, and still get ideas across to the organization. Management reinforces this lesson when it gives its verbal and nonverbal support to those who have more-than-average problems in communicating. In other words, management must demonstrate that it is more interested in the substance of ideas and decisions than in the apparent sophistication of the presenter. (The problem that I am describing may well be an adjunct to another recently discovered deficiency of American management that was previously thought to be a strength: the rapid movement of personnel. Because managers often are judging subordinates in unfamiliar functional areas, it is much easier to be influenced by the superficial factors of personal demeanor than technical competence.) In fact, an overly "polished" communication climate increases stage fright among lower levels of the organization and further handicaps some speakers. Even "good" speakers can appear stilted, acting out the same prescribed movements, gestures, and mannerisms, and utilizing visual aids in exactly the same way.

In one case, a team of three consultants made a sales presentation to a management group in which they reported the results of preliminary exploratory interviews. Each one seemed to be a clone of the others. They all moved around the front of the room in exactly the same way, and they all turned the projector on and shut it off for *each* of the numerous slides (as recommended in the current literature). In short order, the "performance" called attention to itself, and members of the audience were seeing stars in front of their eyes. The consultants were not asked back (for other reasons, however). A more informal and conversational mode would have stimulated greater spontaneity, openness, and overall effectiveness—the kind of climate Peters and Waterman described.

My own experience suggests that there are at least three readily identifiable communication cultures: the Abusive, the "Professional," and the Confrontational/Supportive.

THE ABUSIVE CULTURE

In survey feedback sessions at one organization, many lower-level managers felt that those who were less talented in making oral reports were at a significant disadvantage. Some even stated that they would not hire the most technically qualified applicant for a job if he communicated poorly because "'they' would eat him alive." In responding to the feedback results, a senior-level manager did not see this as a problem: "They're right, and anyone who comes to work here had *better* be a good communicator or *learn* to be. We have our own climate, and it is *their* job to adapt to *us*. We can't adapt to hundreds of different persons."

Realistically, from the viewpoint of the employee's welfare, the senior manager was correct. It is in the best interest of every employee to adapt to the climate that management has created and to demonstrate the same values. If contact with upper levels is brief, if involvement is primarily through formal meetings and reports, and if many issues compete for attention or a decision, then one *must* be a good and energetic communicator. But, from the company's viewpoint, he was dead wrong. This philosophical position communicates to the organization that the company values form more than substance. *How* something is said becomes more important than *what* is said. To the degree that this conclusion permeates the organization, it becomes more political, and greater attention is paid to "image" than to genuine achievement. Granted, managment must do all it can to improve the communication skills of its personnel, even to the point of making them a part of the appraisal process. But if it is to remain competitive, it also has a corresponding obligation to utilize the talents of the wide range of persons who can best contribute. Therefore, communication skill should not be too heavily weighted in the performance of persons whose considerable strengths lie elsewhere.

The abusive culture goes well beyond the confrontational ideal in which there is free and open disagreement about ideas and recommendations. As one group of managers put it, "It's a way of testing the level of confidence that the speaker has in his ideas." In other words, if management can't base its decisions on the merits of a case, it will base them on how well the speaker is willing to withstand severe disapproval and, with style, rise above it.

This strategy may have some validity on occasion. But it contributes to the cynicism of those knowledgeable employees who see poor recommendations of associates, Destructive Achievers or otherwise, accepted in such an irrational climate. It also creates an automatic decision bias within the organization. Certain managerial types, especially Builders, Innovators, and Mechanics, do not have a viable forum for their ideas. The more reflective, thorough, and conservative persons open themselves to attack by virtue of being too ready to admit to the disadvantages of their proposals, and even to doubt the validity of their own conclusions.

This is especially true of long-term issues, which have more unpredictable variables. Those who can speak with confidence and self-assurance about the future have a tremendous advantage. Of course, the decision bias is further aggravated as conscientious persons withdraw or leave the company for more rational environments. A Builder in an abusive environment gave this view:

> I'm not sure I even want to be in management any more. They've deliberately put me in some high-visibility spots and I haven't enjoyed it. My supervisor has tried to convince me that I'm doing very well and that I have what it takes to succeed, but I just don't fit in with the kind of game they're playing here.

This is not just a problem of inexperienced, lower-level managers. It is not unusual to be asked to work on a one-to-one basis with high-level managers who "don't come across well at corporate headquarters." An individual can be quite fluent when talking in a supportive environment, yet not be able "to put two words together sensibly" when being abusively attacked in a cross-examination format. Although skill and performance in such a situation may be improved slightly with coaching and practice, it is much more a symptom of a poor management philosophy than one of lack of individual skill. Effective communications are tough enough to achieve without this kind of added burden being placed upon personnel.

The abusive culture can become so extreme that it makes objective problem solving virtually impossible. A subordinate describes his supervisor:

> In disagreements, (DA) simply doesn't discuss dissenting opinions and gets personal in his remarks, especially in subjective matters. If the issue is sensitive, he can become angry and attack the person who is presenting the information, or he will criticize the originator of the criticism. He will dismiss a counter viewpoint as a weakness of the person making it. People prepare for a discussion by getting ready to defend themselves versus studying the problem. They have to go around looking over their shoulders. His first reaction is "who is to blame."

THE "PROFESSIONAL" CULTURE

According to an Associated Press release, a human resources specialist in a Fortune 500 company "says he looks for individuals who are good communicators, regardless of specialized professional skills. 'Communication is a highly complex talent,' he said. 'Graduates who have mastered it have a head start in succeeding at their professions.'"[2]

The specialist's message has an important lesson for individuals with poor communications skills. It also has an ominous implication: some persons, especially managers, have their positions by virtue of having better communications skills than others, in spite of the fact that they may be less sophisticated in the technology, service, or business.

In some organizational cultures, professionalism is equated with the modern symbols of sophistication. In the case of oral communication, this might include a three-piece suit, leather shoes (with no metal), an erect bearing, self-confidence, an efficient, businesslike manner, and commercially made color slides. Woe to the person who wears white sox, is on the portly side, sweats a lot, sometimes loses his train of thought under questioning, or blurts out his comments. It doesn't matter that he may have a Ph.D. in physics and, by virtue of practical experience, knows more about the subject than anyone else in the room.

The cost of an overemphasis on professionalism in communication is much more than just a deterioration of the genuineness of the environment. It results in a loss of both time and money.

> In a case where an executive could have had his secretary make overhead transparencies at almost no expense, he had slides commercially made for $9,500—for an in-company presentation. This practice had become the standard for reports to top management, which not only was oblivious to the cost but deemed them appropriate to their status.

> A plant manager and his staff spent the better part of two weeks preparing and rehearsing presentations to be made at a meeting with just five members of top company management. They couldn't trust management's reaction to a more informal give-and-take discussion.

> A grammatical mistake was made on handouts for an in-company meeting. Instead of letting it go, the prevailing standards of professional image required that it be corrected. It was easier to throw out 25 sets and order new ones from the duplicating department than to correct the offending page, duplicate it, and recollate it.

These examples may seem relatively harmless, but the list is endless, and it can permeate all aspects of the communication climate. *The problem will continue to grow in an organization,* unless its leadership demonstrates that it values the substance of performance far more than the image of performance.

As might be expected, some Builders fail to appreciate the importance of communication professionalism. They and their subordinates are the ones who are most likely to suffer from an overprofessionalized environment. A subordinate describes his supervisor:

(B)'s meetings lack organization and preplanning. He doesn't demonstrate to individuals that he wants them to prepare for meetings. All we know is that we're going to a meeting, and just sort of ramble, in terms of letting others know what is going on. Once I was supposed to give a report at a meeting with [senior management] and (B) didn't even tell me until we were in the car to go the meeting.

The communication performance of this Builder was truly deficient, and he did need to improve. The example was chosen because the Builder was seen by associates as an ethical, *effective* manager whose net impact on the organization was positive, despite poor communications skills. Yet many senior management groups will take offense at his lack of sensitivity to their high status and not give him a conscientious hearing. In penalizing him, they penalize his group and ultimately the organization itself.

Obviously there are times when professionally made slides *should* be used rather than overheads, when extensive preparation *should* be made for important meetings, and when grammatical mistakes *should* be corrected. The point, however, is that the company culture, in many cases, has rewarded average or even poor performers for superficial reasons—their communications skills—and punished fundamentally sound performers for reasons equally superficial. This is not only expensive, it is degenerating to long-term organizational effectiveness. It may well be a major reason for the steady decline of many large corporations, and even of industries. Managers lose genuine and open contact with their own personnel and hence with their technology, their service, and their customers.

THE CONFRONTATIONAL, SUPPORTIVE CULTURE

Too many persons consider confrontation as synonymous with attacking (in an abusive sense) and as being detrimental to the supportive environment. Quite the contrary. In the supportive environment, assumptions, ideas, conclusions, and recommendations *must* be confronted if members of a work group are to respect each other. People can make effective contributions to a group only if they are aware of the impact their actions and ideas have on the group. However, the confrontational group is supportive in the sense that it always helps individuals present *their* side of the confrontation. In the communication sense, confrontation relates to ideational substance, not personalities.

I once attended a plant management staff meeting in which a technical group leader gave an oral report, recommending that one of two options be selected in handling production waste. A key element in his presentation was a series of numerical comparisons of several sets of data. In addition to being very nervous, the speaker began his talk by presenting unreadable data (too small and cluttered) from an overhead projector. The plant manager realized that the speaker had lost his audience, so he stopped the presentation in a very supportive way. He *respectfully* let the speaker know that his message was important, that it was not coming across, and why, and suggested that he just sit down and explain the situation. Through a series of questions, and discussion involving other members of the meeting as well, the individual composed himself and the subject was covered. The result was that the speaker knew that his "performance" was a flop but that he and his ideas were valued. The message from the plant manager to the rest of the members of the meeting was that substance of the communication was more important than the individual's presentation skill or personal sophistication.

I have seen similar situations in which individuals or groups were publicly humiliated, verbally as well as nonverbally, for communication inadequacies. This is especially disastrous to organizational values when, at the same meeting, less functionally or technically competent persons are rewarded (by outward management reaction), simply because they present themselves and their ideas well. In such events, individuals and groups realize that presentation skill and apparent personal sophistication are more important than substance or the communication message. They must take great care to ensure that the organization's self-image—of having high professional standards—is not damaged, no matter how artificial that image might be, or how irrelevant to the problem being discussed. In summary, in the supportive climate, the personal worth of each person is an accepted value, and the effective communication of substance is the objective. In the abusive or "professional" climate, personal value is conditional. It depends on how well the individual exhibits the sophistic standards of the leadership, and this determines what substance is allowed to be communicated.

Improving the Culture

As a starting point for improving the organizational culture for effectiveness, top management must choose which philosophical goal they wish to pursue: Is the effective presentation a performance or a means of communication? This choice is not as easy as it sounds. Although everyone espouses the latter, management's actions and in-house training programs more frequently support the former, and with reason. "How you say it is more important than what you say" has become a training axiom about human relationships in organizations. It is also likely to be a reason why organizations repeatedly fail to avoid bureaucratic stagnation. As form and style overcome substance and become entrenched in the corporate culture for taking action, they become an accepted aspect of reality that is impervious to change. As many are currently *teaching,* "That's the way it is." And now most persons seem to agree. A company can become excellent only to the extent that it can rise above the reality of what *is,* and become better. A lack of such a commitment *automatically* abdicates to the tyranny of the present.

As a second step, management should analyze its own actions and ensure that these support the philosophy they have chosen. In most cases, when the opportunity is provided, they are quite surprised and dismayed at the way lower-level personnel describe the communications practices that are rewarded or punished. And third, top management should participate in the design of its own in-house "effective presentation" workshop and become actively involved by *attending it.* The fact that this recommendation will shock most readers is indicative of the true importance presently given to the subject: It is a course in techniques and requires little management attention or interest.

The reason management must attend its own seminars is, they must be experienced if they are to be appreciated and influenced. One cannot judge the content or affect the development of a communication course merely by talking about its brochure or its training materials. These usually are properly compre-hensive and include just about every desirable aspect of the communication

Important questions to ask of any communications philosophy are:

How does it present the concept of "How you say it is more important than what you say"? As a guide to speakers? Or as an organizational defi-ciency and a challenge for the listener or decision maker to overcome?

Is the communication event presented as a performance to be judged in itself, or simply as an individual talking to a group?

Is the emphasis on enthusiasm and showmanship, or on accuracy and integrity?

Do persons leave the training session with a value commitment to fairness and objectivity in relationships with others (both as speaker and listener)? Or do they leave feeling that they are entering the competitive arena where they are pitted against each other in image projection, enthusiasm, and positive thinking?

An important qualification: This discussion is intended for the improvement of an organization's internal effectiveness through the full, fair, and objective use of its human resources. The speaker *must* be a performer (as much as is any actor) when

- competing with other speakers in situations where time is limited by the "buyer."

- the buyer is known to be influenced by the entertainment value of the presentation as much as by objective data.

- the buyer is known to be an impulsive decision maker.

- "image" is important.

One will immediately note that these same conditions are found *within* every organization to some degree. It is to that degree that the morale of the solid contributors will be low, debilitating political activity will be high, and the group will be unrealistic in meeting its challenges. In other words, Builders will be suffering, and Destructive Achievers will be thriving.

Perspectives

If a company's communication environment, including its training philosophy and values, is controlled too much or in the wrong way, form can take precedence over substance. Communication becomes an end in itself and too heavily weighted in the performance of persons whose strengths lie elsewhere. In the same way, the responsibility for the communication event becomes overloaded on the side of the speaker, employees become more preoccupied with image than with genuine organizational impact, and management reduces its effort to listen and to become actively involved. Management's concern for the communication skills of its employees is legitimate. The complexity of modern organizations and the competing demands for time make effective communication essential. Although training in effective communication is even more necessary now than in simpler times, care must be taken to ensure that it is of the right kind.

Even more importantly, management must analyze its own actions and make sure that it is appropriately rewarding the behaviors of those who make genuine contributions to the group's welfare. Obviously, these would include those who present themselves well and are good communicators. But management should make *doubly* sure that it does not reward the ones who are Destructive Achievers. Their development of communication skills, and the time and effort devoted to using them, are primarily designed to exploit organizational weaknesses in decision making and performance evaluation.

Footnotes

1. This chapter is an adaptation of an article originally published in *Sloan Management Review,* Spring 1985, 69–74.
2. "Job seeker's asset." *Associated Press,* Bartlesville, Okla., November 27, 1985.

PART THREE

Designs for the Future

Moral courage is a rarer commodity than bravery in battle or great intelligence. Yet it is the one essential, vital quality for those who seek to change a world that yields most painfully to change. . . .

Our future may lie beyond our vision, but it is not completely beyond our control. It is the shaping impulse of America that neither fate nor nature nor the irresistible tides of history, but the work of our own hands, matched to reason and principle, will determine our destiny.

—Robert Kennedy

CHAPTER 9

Power and Prejudice

Thus far we have considered the Destructive Achiever's behaviors, education, and development in the corporate context. Such an isolated view depreciates the significance and pervasiveness of the problem and the difficulty in correcting it. Contrary to what many would like to believe, the conflicts between status levels within the organization are not just a matter of poor communication, misunderstanding, or unclear organizational purpose. At least one major cause is far more basic: attitudes of superiority that naturally lead to the righteous abuse of power.

The reason why otherwise conscientious and well-intentioned persons can promote or elect a Destructive Achiever to power is partially explained by the very nature of power and prejudice. The DA apparently represents the group's values and furthers its interests, and his or her excesses are rationalized or considered acceptable. This also helps to explain how those who should become Leaders and Builders develop into Destructive Achievers instead. Those who do not have the personality of the DA, but find themselves pressured to conform to anti-ethical or unethical standards, need to justify their behaviors. In one way or another, their rationalization requires an attitude of superiority. Even though the attitude is an assumed one, it still suggests that they really feel they have the right to take actions, and to realize personal benefits, beyond the normal bounds of ethical restrictions. The superiority they feel might simply be that of the "insider," or it could be a complex, subconscious prejudice against those they presume to be inferior in some way—conceivably even the general public. Deliberately destructive behaviors in organizations depend on this attitude of inherent superiority, coupled with the power to act.

In *Re-inventing the Corporation,* John Naisbitt and Patricia Aburdene wrote that "the corporation is the analogue for the rest of society."[2] In many ways, human behaviors in political and social structures also serve as an analogue for the corporation. Because of their universality, human behaviors can be used to clarify some aspects of human values better than isolated corporate examples.

According to Manuel Velasquez, Kant's "categorical imperative" incorporates two criteria for determining moral right and wrong in a person's reasons for acting:

> Universalizability: The person's reasons for acting must be reasons that everyone *could* act on at least in principle.
>
> Reversibility: The person's reasons for acting must be reasons that he or she would be *willing* to have all others use, even as a basis of how they treat him or her.[1]

Obviously, these criteria require that the individual respect the fundamental equality of all persons and be able to feel empathetic toward others—place oneself in another's shoes, so to speak. This is very difficult to do because of the apparently logical nature of prejudice.

The situation in Northern Ireland is an excellent example of how Kant's categorical imperative extends beyond corporate walls. It has been so well publicized and is so far removed from us that we can discuss it fairly objectively. It is analogous to our own prejudicial problems—not only racial and sexual, but organizational as well. It also demonstrates the pervasive nature of bias and why bias develops so easily and is so difficult to change once it is entrenched in the system. (If we fail to accept the fact that prejudice—or bias—is natural, we will never be able to come to grips with its potency and ultimate malignancy.) Several years ago I watched a television documentary on the causes of the violence there. In one scene, a Protestant cab driver gave the reporter a tour through a Catholic neighborhood. "Just look at how these people live. Look at the trash in the streets—the way they let their yards go. Look at those garbage cans turned over. This is the way they want to live. They don't want any better."[3] The scene had a strong effect on me because it reminded me of when I was 15 years old, riding along with a relative of mine through a black section of Kansas City. My relative was "educating" me about the fundamental nature of blacks, how *they* want to live. I find it incredible how anyone could attribute poor living standards to race or religious preference. The two just don't seem to go together. But, 40 years ago, it seemed to make sense that Blacks somehow wanted substandard living conditions. Such is the nature of prejudice.

The Power–Superiority Link

The documentary on Northern Ireland continued with an interview with a social scientist who made two observations: (1) The problems in Northern Ireland have nothing to do with religion; and (2) the root cause of all prejudice is economic. But it just so happens that the most significant, readily available source of economic discrimination in *that* particular country is *religious* difference. I would qualify this in only one way: The root cause of prejudice is power, not only economic, but social and psychological as well. We discriminate economically in order to advance ourselves materially, socially to advance in personal status, and psychologically to maintain interpersonal dominance (for example, "nature designed a wife's mentality for household drudgery").

COGNITIVE DISSONANCE

Prejudices are reinforced. A psychological concept called "cognitive dissonance" explains how this happens. "Cognitive dissonance" means that we can't have two conflicting values at the same time without incurring some form of distress. In other words, I am distressed because I can't take advantage of you and think well of you at the same time. I therefore have internal pressures to reconcile these incompatible values, and so achieve harmony within myself. For these reasons, there are dangers in either the pursuit of power or its application. Attitudes of superiority can lead to the destructively selfish use of power, and the selfish use of power can lead to attitudes of superiority.

So, the con artist is convinced that the sucker deserves to be taken, the wife beater is convinced that his wife is cheating on him, the executive fails to promote the underpaid, non-degree-holding worker because, if he had the required gumption, he would have gone to college. These are vicious cycles. We take advantage of someone. We must then think ill of that someone who has allowed us to take advantage of them—which justifies even worse treatment . . . and so on. And, in the worst cases, people end up killing each other.

After a few years, the real cause of the prejudice (desire for power and superiority) fades into the background, and new generations are born into a world in which differences, although artificially induced, appear real. People now fail to understand that their present perceptions are the result of actions by past leaders. That is, previous leaders created conditions that ensured that the prejudice would continue.

For example, we traditionally have not trained hourly workers beyond the most basic operator skills. We certainly haven't trained them in interpersonal skills. Now we observe that many hourly workers are difficult to deal with, show little initiative, and seem not to be committed to the success of the organization. Then we conclude that training in advanced technology or interpersonal skills would be wasted on them.

THE ORGANIZATIONAL SETTING

Progressive management and behaviorial science theories often fail when managers don't come to grips with their own biases. For example, an engineer may begin his managing career with a commitment to treat others as equals and to develop a participative management style. But when he tries out his theories in a power-oriented environment, others take advantage of him and perceive him to be a weak and unrealistic leader. Failing to realize that he has stepped into a unique situation created by someone else, he quickly assumes a more traditional, control-oriented style, even when he later enters a more progressive environment. He had made an intellectual commitment, but his actual experiences prevented him from adhering to it. In other words, we all know reality in two ways: the way it appears to be, and the way we know it really is through testing. Unfortunately, testing our biases can be painful and personally costly, especially if it means confronting the prevailing organizational values. Therefore, most of us are content to retain our expedient biases and enjoy their short-term benefits—even if others suffer in the process.

The same is true on a larger scale. A management group may agree intellectually to experiment with progressive leadership practices based on a premise of employee equality. However, they do not have a genuine conviction based on experience. They implement the forms, procedures, and techniques of participation and involvement, but they still retain the old values of power and superiority. Others then inevitably learn that managers cannot be trusted to maintain their commitments to objective problem solving. The managers haven't done so in the past, and their present behaviors seem not to have changed, despite their apparent sincerity.

THE NATURE OF BIAS

The emphasis on improving conditions for protected classes distracts us from realizing that bias is a fundamental defect in the human thought and feeling process that is devastating to objective problem solving. Bias is a category of mental error that is detrimental to both the individual and the group, with significant psychological, financial and productivity costs. To quite an extent, bias has not much to do with the specific group differences themselves, and so can be as much a factor in a superior-subordinate relationship as in any other. American organizations tend to address issues of bias as unique problems of protected classes and so have designed programs specifically to benefit those classes. The emphasis has properly been on the social and moral reasons for taking corrective action in restoring a climate of fairness in the workplace. Unfortunately, the spotlight on protected classes can accentuate differences. Educational, sensitivity, and awareness sessions typically are focused on specific minorities. When differences surface, they are seen as unique to *specific* groups, for example, blacks and whites. To some extent, this makes the differences seem even more irreconcilable than before. That is, the problems are perceived and treated as being the result of the specific (e.g., racial) differences themselves.

We now need to improve the way we view bias itself. Bias is the tendency of an individual to seek reasons to believe in his or her own superiority, be it of one's own talents, accomplishments, race, sex—or managerial level or function. Recognizing this general tendency may be the best way to eventually reduce bias toward both individuals and groups.

Experiments in Reducing Bias

This theory was tested in a series of experimental workshops that originated in a Fiber Industries manufacturing plant in Greenville, S.C. It was reported in detail in the American Management Association's *Management Review* in 1983.[4] The experimental process was designed to deal with bias as a general problem— a problem for whites, males, managers, and nonmanagers—as well as members of protected classes. To demonstrate the universal and debilitating effects of bias, participants in each workshop engaged in three different types of confrontation: management/nonmanagement, male/female, and black/white.

"Confrontation" is used in a unique way and needs to be defined. It is the process of addressing significant, but potentially ego-threatening, issues for the purpose of objective problem solving. Persons identify, define, and attempt to reduce specific problems between groups. The emphasis is on issues that directly affect job satisfaction or work effectiveness—not on increasing sensitivity and understanding. Although our definition implies more conflict than simple discussion, it does not suggest "attacking," and is in fact more of a supportive process than a conflict process. (The confrontational process is discussed in greater detail in Appendix A.)

In later sessions, the management confrontation was called an "exempt/nonexempt" confrontation, because salaried professionals were included in the management group. There were no significant differences in results; the same issues surfaced between different organizational levels, regardless of where the dividing line was placed. For simplicity, all references hereafter will be made to the management and nonmanagement subgroups.

The management/nonmanagement confrontation added a new dimension to the way we normally think of bias. If bias has the same characteristics and results in this domain as it does in others, it adds a great deal of credibility to the view that it is an acquired and changeable attitude. People aren't born into these categories. They acquire biases simply by having different organizational positions and experiences. And how a specific person reacts to these influences is an *individual* matter—it is not inherent in the position, title, or experience.

Because of the initial group's success, over 100 groups of employees eventually participated in the process, in two different manufacturing plants and two R&D facilities, in the states of North Carolina, South Carolina, and New Jersey. As expected, in all three types of confrontation (male/female, black/white, management/nonmanagement), the issues turned out to be the same, regardless of location or type of facility: respect, fairness, involvement, trust, opportunity, being listened to, being treated as an individual, and the like (see Exhibit 9.1).

Exhibit 9.1   COMMONLY SHARED ISSUES BETWEEN GROUPS

All confronting subgroups—male/ female, black / white, management/ nonmanagement—tend to give the same answers to the following questions: "In what ways do (e.g., managers and nonmanagers) share the same opinions of themselves?" "What problems do we share?" "What do we both feel good about?"

Opinions we share about ourselves:
Want appreciation to be shown
Want respect and fair treatment
Want to be informed
Like to be part of a team
Want to be treated as an individual
Want to have a feeling of accomplishment

Problems we share:	*Good feelings we share:*
Lack of recognition	Friendly atmosphere
Lack of cooperation	Pay raise and promotion
Lack of trust	Being successful
Lack of communication	Teamwork
Pressure and frustration	Positive strokes
Inflation	Being shown respect
Job security	Being listened to
Job changes	Knowing you did a good job

DIFFERENCES THAT *MAKE* A DIFFERENCE

Although stated in different words in each session, the conclusion of participants was that the differences that make a difference, whatever the particular
confrontation, are *individual* attitudes and actions. These are what create an
atmosphere of either equality and teamwork or inequality and factionalism. Depending on these qualities, the end results are: fairness or favoritism, open communication or restricted communication, involvement or isolation, opportunity
or stagnation, and team identification or subgroup identification.

The fact that these are based on individual attitudes or actions is repeatedly
demonstrated in two ways. First, when criticisms are made about attitudes or
actions of members of "the other group," they are almost always qualified.
Either "this is only true of some of you," or "we can see progress being made
in this area by some of you." Second, when females, blacks, or nonmanagers
complain about many of the specific behaviors of their opposite group, they
frequently find that members of that group also experience the same problems
with their own members.

The first point is self-explanatory; examples of the second are both numerous and enlightening. Nonmanagers felt that managers left them out of the
communications chain and said that they would like more involvement in matters that affected their jobs. They gave examples of suggestions that were not
responded to, and of confusing, unwise, or arbitrary management actions that
were never explained. Managers accepted the criticisms, but also expressed their
own frustrations at having the same problems *with each other* and with levels of
the organization above them. They described the time restraints they felt, and
the problems of coordinating complex objectives between levels and functions.

A female group reported that women typically were not given challenging
assignments because the predominantly male management lacked confidence in
their abilities; this, in turn, hurt their chances for promotion. As an example, an
engineer described her own assignments over a period of four years. This was
quickly responded to by the male group, who gave comparable accounts of
perceived inequities. In fact, the perception of the males was that the females
received better assignments and had the better chance for promotion. The discussion then became oriented to the general subject of project assignments and
how they could be more objective.

In another session, black operators felt that they were discriminated against
by white foremen in the ways they handled requests for time off. One operator
explained that he had been denied permission on a few occasions and that white
co-workers had received permission for less valid reasons. With apparently the
same level of frustration, white operators said that they felt exactly the same
way, but the cause was *favoritism,* not *racism.* Foremen entered the discussion
by trying to describe how difficult it was to handle a specific request under the

guidelines of a general personnel policy. The black operators learned from this exchange that perceived favoritism is not necessarily racism. The foremen got a much greater appreciation of the strong operator resentment of inconsistently applied personnel policies. Operators learned of the confusion foremen experienced in trying to fairly apply policy guidelines to specific cases.

It would be wrong to conclude that the above situations are *not* the result of bias. Sometimes they are, and sometimes they are not. Whatever the case, the lesson to be derived is that perceived bias is a product of *individual* attitudes and actions. It is not an *inherent* result of male/female, black/white, or management/nonmanagement differences.

Although the terms "Destructive Achiever" and "Builder" were never used in the sessions, these categories of persons were well described in all three confrontations. White men who are Destructive Achievers make an unethical use of their membership in the more powerful group, whether that group is white, male, or management. They exhibit an attitude of superiority and are unfair and arbitrary. Moreover, the net long-term effect on the relationship is negative. Their behaviors create a high level of resentment, which can then be directed even to the Leaders and Builders in the group who are trying to behave ethically.

DIFFERENCES THAT *DON'T* MAKE A DIFFERENCE

The predominant, and often repeated, conclusion of the confrontation sessions was that race, sex, and organizational level differences are not the *causes* of unfair, unsatisfactory, or ineffective relationships. They are, rather, the *targets* of individual attitudes of superiority. Of course, the specific confrontation issues are stated in terms of race, sex, and organizational level—that is, although the cause of a problem may be an individual's air of superiority, it will be described in terms of that person's race, sex, or level. Women will describe a man's attitude of superiority as "male chauvinism," blacks will refer to a white person's "racism," and nonmanagers will refer to a manager's "elitism." Although the references are to specific persons the issue is seen as being a general one: a perceived attitude of superiority and the negative results that accompany it—lack of involvement, trust, confidence, communication, identification with the total group, and opportunity for the discriminated person to contribute.

To say that a difference in race, sex, or level "doesn't make a difference" in a relationship does not suggest that there won't be some important and unique issues to be perceived between people. To be sure, there are all kinds of background issues that affect one's orientation toward another. Many of these issues, commonly found in society, include such things as the nature and tradition of black/white relationships in a community, the effects of power (organizational level) on human perceptions and actions, and even males' learned attitudes toward the female members of their families (see Exhibits 9.2, 9.3, and 9.4).

For example, females have discovered that, if they are talking together in the hallway, males tend to assume that they are gossiping. If a group of blacks are in the hallway, whites may suspect that they are talking about or plotting against them or the system. If nonmanagers are together in a group, they are "goofing off." Again, the references are specific, but the problem is general: the observer who fails to understand the distortions in perceptions caused by an attitude, conscious or subconscious, of inherent superiority. Yet, after discussing many such potentially disruptive perceptions, participants conclude that psychologically well-adjusted persons will be able to reach a satisfactory, even rewarding, relationship with others who are different—in ways "that don't make a difference." Leaders and Builders, who are obviously committed to ethical behaviors toward members of their opposite group, demonstrate that this is possible. Of course, many remain essentially unchanged by the confrontational

---

Exhibit 9.2   MALE/FEMALE PERCEPTIONS

The following perceptions tend to be repeated in confrontations between male and female groups.

How females tend to see males	*How males tend to see females*
ATTITUDE OF SUPERIORITY	Exact and neat in tasks
Dependable	Stick to rules and guidelines
Good sense of humor	More safety conscious
Kind	Show more kindness
Willing to cooperate	Stand up under tedious work
Less sensitive to people	More sensitive to people
Helpful	More conscientious on the job
Honest	Better at (job) housekeeping
Courteous	Courteous
Boastful	Less competitive
Impatient	More jealous
Physically strong	Weaker physically
Stubborn	More health problems
Think they should be first to get the good jobs	Get into jobs they can't handle physically
Less understanding	Quick to get emotional
Resent women and their pay	Quicker mood changes
Don't gossip	Talk more
Over-dominating	Tend to give up too easily

Exhibit 9.3   BLACK/WHITE PERCEPTIONS

The following perceptions tend to be repeated in confrontations between black and white groups.

How blacks tend to see whites	*How whites tend to see blacks*
ATTITUDE OF SUPERIORITY	Feel we owe them something
Ambitious (career advancement)	Ambitious (desire to improve position in society)
Able to overcome prejudice	More open with feelings
Prompt and dependable	Live one day at a time
Ability to organize	Willing to give assistance
Feel they should have own way	Feel we are all prejudiced
Less tolerant toward any minority	Want more job security than whites
Generally helpful	Family oriented
Quick to apologize	Easy-going
Reserved, quiet	Have more fun
Different recreational tastes (water sports, golf, etc.)	Different tastes in music, dancing
React negatively when they see blacks in a group	Stick together
More afraid to deal with racial problems	Want to be accepted in community and by peers
Different cultural background	Different cultural background
Have fewer obstacles, a more positive attitude	More coordinated physically

procedure. But they don't advertise it in the meeting room or in the subgroup discussions; it is too closely identified with personal deficiency.

This is an important point, because it means that the group-generated norms are positive ones, and unreasonable comments are immediately attacked by peers, as well as by the co-confronters. For example, when an operator suggested that supervisors avoid favoritism by following rigidly defined procedures for allowing time off, the other operators defended the importance of allowing the supervisors to exercise judgment. In another situation, a white female observed that dogs and cats don't associate with each other; therefore blacks and whites, being different species, probably don't naturally associate with each other either. The looks of disbelief and the audible groans of other whites were all that was needed. The discussion continued on its original course as if nothing had been said.

After a female group cited examples of sexual harassment, a male described an incident in which he was "chewed out" by a female. He said he felt that all

Exhibit 9.4   MANAGEMENT/NONMANAGEMENT PERCEPTIONS

The following perceptions tend to be repeated in confrontations between manager and nonmanager groups.

How nonmanagers tend to see managers	*How managers tend to see nonmanagers*
ATTITUDE OF SUPERIORITY	Don't have the "big picture"
Under more pressure, stress	Jobs can be dull and boring
"Kill the messenger" syndrome	Some want special treatment
Plan well	Resourceful
Stick to company policy	Pride in work
Too company-oriented, hang-ups on production	Only see own problems and not the department's or facility's
More quantity- than quality-conscious	Don't understand reasons for change
Feel they are always right and you are always wrong	Won't move from home to pursue job or career
Some feel you work for them and not the company	Some feel rules were made to bend or break
Publicly make negative comments about nonmanagers	Publicly make negative comments about supervisor
Assign work in a demeaning manner	Career is not as dependent on project success
Clearly let you know who is in charge	Dedicated to jobs
	Self-directing

---

the young women knew that he was harmless and just trying to be friendly by calling them Honey. He quickly learned that he was partially right, but mostly wrong. Some females said they were highly offended by being treated that way, but a few said they enjoyed harmless flirting. The group concluded that it was the wrong thing to do unless the two persons knew each other well.

In discussing subtle forms of bias, a white engineer explained that he usually went to one of the white operators for information, not because he felt they were superior to black operators, but simply because he felt more comfortable doing so. The black group explained that this really hurt. The black operator may have five years of experience, and the white be newly hired—yet the black operator sees the unknowing engineer automatically going to a white operator for ideas and advice. The total group conclusion: Learn to speak comfortably with everyone, or risk unintentional offense.

When managers reported that nonmanagers did not have a personal interest in the success of technical projects, they were almost shouted out of the room. In the highly charged discussion that followed, the nonmanagers severely criticized many specific words, actions, and standard practices of managers. Some examples were referring to "my technician" instead of the name of the person; introducing a visitor to other project professionals but not to the technician; not including technicians in project reviews with upper management.

In this and many similar exchanges managers learned that nonmanagers *do* take pride in a project's success, that they deeply resent being thought of and treated otherwise, and that many of the (then) current norms of behavior were significantly demotivating. So, even though there are prejudiced points of view expressed in the workshop, they are quickly corrected in open discussion, and receive no support. In effect, the total group develops its own norms—the best basis for lasting cultural change. The norms described in Exhibit 9.5 are examples of generalized norms that are common to all groups. Not listed are the numerous norms that had special relevance to individual groups: for example, not allowing personnel records to be viewed by unauthorized personnel; informing lower-level personnel of significant changes in business conditions; criteria for listing names on patent applications; and so on.

Exhibit 9.5   GROUP–GENERATED BEHAVIORAL NORMS

In each confrontation session the final question was "What would most improve (e.g., management/nonmanagement) relationships in this facility?" All groups answered along the same lines, with the following behavioral norms being the most frequently mentioned.

Accept people as individuals, on their own merits
Show real teamwork
Treat everyone equally
Accept the fact that there are good and bad in both groups
I'm OK — You're OK
Show an adult-to-adult attitude
Recognize we are all here for the same purpose
Perform our jobs in a professional manner
Have more mingling
Identify and eliminate barriers
Have a problem-solving attitude
Have two-way communication
Recognize our similarities
Show empathy; put self in others' shoes
Show no favoritism
Set an example of individual effort
Demonstrate sincerity with follow-through

Negative Impact of Bias

Most persons were unaware of how much bias affects interpersonal effectiveness until they participated in these discussions. Whatever the bias is based on, the effect on its target is essentially the same: withdrawal, deliberate withholding of observations, data, and ideas, malicious compliance, and sabotage. *The effects are especially destructive when the source is located at a higher organizational level, since the feeling is extended to the company itself—its equipment, products, and other personnel.*

The most frequently mentioned reaction to bias is *withdrawal*. In the more extreme cases, a person may request a transfer, or even leave the company. Probably most persons withdraw merely by mentally "checking out." For example, females may become less active in communications because they feel intimidated. One female group reported that, in one-to-one discussions with males, the male usually tries to dominate simply because he assumes that that is the male role. A personnel representative explained that whenever she had a joint assignment with one of her male colleagues, the male almost always tried to take over control of the project. Sometimes she felt she only had two choices: either sacrifice her better ideas to the male ego or harm the relationship by creating a conflict. (The male group responded by agreeing that men try to dominate, but usually *not because the other person is a female*—and added that the same thing occurs when two males interact in a discussion. One or both may try to dominate until a satisfactory relationship is established. Here again, superior ideas are sometimes lost to the more aggressive person.)

The *withholding* of observations, data, and ideas is the same as withdrawal, except that it relates to situations where involvement is optional. Persons who communicate superior attitudes simply don't get all the help that normally is available for the asking. Deliberate withholding is a completely safe way of getting even, probably the second most frequent reaction to bias. *Malicious compliance* is closely associated with withholding and is most frequently mentioned in the management/nonmanagement confrontation. Persons follow orders or instructions, knowing full well that the issuer made a mistake or didn't consider an important factor. The project fails and again the score is evened a bit more. By following literally directions that frequently don't describe all the proper conditions for a trial or production run, nonmanagers can create false data, generate waste, or ruin a time schedule. If an engineer doesn't specify that a new roll of graph paper is to be put into a measuring instrument, it isn't done, and four hours of machine data are lost. If he or she makes an obvious mistake in specifying a control setting, it isn't corrected. Not only that, if the engineer isn't particularly thorough, he or she may never discover that the error was made and may later assume that the results of the trial were valid.

Sabotage is much less frequently mentioned, but its extent is still a surprise to participants. Because of the sensitive nature of the subject, descriptions are usually general and relate to the broad kinds of things individuals have seen other persons do. Discussions about sabotage focus more on *why* people do it. When participants report sabotage, they seem to be struggling with some internal conflict. They are embarrassed for their own group's actions yet they have a strong desire to get the other group's attention. The message they project is: We're willing to take the risk of telling you we do it—because we don't *want* to do it, we would like the conditions changed so we don't feel we *have* to do it (to maintain a sense of equity), and hey, these problems are important to you, too!

In one session, after a group reported that sabotage was a way of getting back at the Destructive Achiever (although they didn't use the term), an incredulous engineer asked, "But *why* do you do it?" The nonmanagers then explained that they had been treated without consideration and respect, and that there was a limit to their tolerance. Example: an engineer may suddenly schedule a production trial to be started late on a Friday afternoon, which could have been scheduled much earlier with better planning. If the engineer has a bad reputation, and the operators have come to perceive the engineer as arbitrary—and have other plans for the weekend—"every mechanic and most of the operators know how to make the equipment break down due to normal causes."

They went on to explain that they willingly suffer inconvenience and "go the extra mile" to make a project successful when the person running it "treats us like human beings." It is interesting to note that the significant and immediate cost to the company is hardly a consideration. The deciding issue is the perceived attitude and behaviors of the engineer. Examples of sabotage include damaging manufacturing equipment, testing equipment, tools, company products, and personal possessions. The targets of sabotage are individuals, both supervisors and co-workers, who are directly responsible for the equipment or its products. Clearly the greatest costs to the organization are the least observable: withdrawal, withholding, and malicious compliance. The costs of sabotage, once discovered, get everyone's attention, but probably don't come close to that of dysfunctional or deliberately counterproductive personnel.

Unexpected Lessons

Many of the above lessons were expected, or at least hoped for, in the confrontational sessions. However, there were some surprises, and they have implications for managers that go well beyond issues relating to protected classes.

First, virtually everyone was surprised that participants found the management/nonmanagement confrontation to be the most stressful of the three. In the beginning, participants as well as facilitators had been most concerned about what was going to happen in the black/white session. Although there may be a combination of reasons for the especially threatening nature of the management/nonmanagement session, a significant one is that the power differentials between these two groups are always *official* and *constant*—day-in, day-out, without let-up and usually without any realistic recourse. At least there is some respite from the problems felt in male/female and black/white relationships.

Second, managers were surprised at the extent to which they were seen by nonmanagers as having superior attitudes and as behaving arbitrarily. They had expected many disagreements about specific issues, but were unaware of the many kinds of things they did that lower-level persons considered demeaning to them.

Third, upper management's "shoot-the-messenger" attitude seemed to be pervasive. Those who challenge the holders of power, or bring them bad news, sometimes are taking a significant risk, even when they are sincerely trying to improve conditions. When this issue was raised in the male/female and black/white sessions, most males and whites seemed to be able to understand the problem. Some even indicated a high level of respect for the blacks and females who were willing to risk confronting them. On the other hand, when the issue was raised in the management/nonmanagement sessions, higher-level persons were almost always shocked by the degree to which lower-levels felt they couldn't communicate bad news upward without suffering for it. Managers had to hear examples of the kinds of things that were going on before they could appreciate the extent of the problem. As one technician said, "When the director (three or four levels up) describes my project on his monthly report, I don't even recognize it."

And fourth, those who were considered "obstructionists" or "troublemakers" before the workshop were frequently considered conscientious persons with genuine concerns afterward. Some individuals became well known throughout a facility for the alternative they had to withdrawal: They challenged every perceived injustice to themselves, their colleagues, or their organization. It got to the point where their counterparts—whites, males, or managers—avoided them if possible. "Susan," a black female machine operator is a case in point.

Susan was one of the most outspoken participants during one entire workshop, expressing serious concerns with a combination of humor and anger. In the male/female session, she told how she had seen men help each other with a difficult task who would not help a woman; in a humorous but half-serious vein she observed that, any time a man did help a woman on the job, "he just wants something from you" (laughter). In the black/white session, she described her resentment of whites who seemed to cover their real feelings and tried to act nice to her simply because it was the "in" thing to do. It was in the management/

nonmanagement confrontation that she made her greatest impact: "I resent it when a foreman keeps giving assignments to his favorites and I keep getting dirty jobs. And I'm a very sensitive person" (explosive laughter). She pounded the table with her fist and exclaimed, "I *am* sensitive!" The laughter stopped immediately. In a broken voice she went on: "I just don't show it like you all do. But it deeply *hurts* when you have a legitimate complaint and nothing happens." The warmth she engendered came from all directions. Later on some of the foremen were talking. One expressed their shared surprise: "I couldn't believe it, I saw Susan in a whole new light. I used to think that she would argue with a fence post. Now I see her as a very concerned person who has some legitimate gripes. She just wants to be heard and to see some changes made."

In supportive confrontation, so-called troublemakers are seen in a much more favorable light; moreover, their on-the-job behaviors sometimes improve significantly. This is probably due to three reasons. First, when someone like Susan expresses a legitimate complaint in a free and open forum, the others, who normally remain silent or withdrawn, will support her. Susan's co-confronters realized that the problem was real and that Susan had the courage to risk confrontation. Second, when her complaint didn't make sense, not only was it not supported by her peers, it was disagreed with, and Susan actually moderated her own opinions. (Several women derided Susan's pessimistic view of men. One observed that, in her own area, she had helped a man with a difficult task and "never had a problem since.") And third, Susan recognized the changes in the ways people related to her and enjoyed the new level of attention and respect she got.

The example of Susan has some broad implications concerning the causes of prejudice. Consider the case of a white foreman who may *or may not* be perceived to be biased by a black operator. If the black operator complains about the way job assignments are handled, the foreman then considers the complaint to be a veiled charge of bias, because he knows the effort he made was fair. (Some black groups report that many of them have deeper voices than whites and tend to speak more gruffly. As a result, even when expressing a normal complaint, they may come across to whites as threatening or aggressive. "Black boys have this problem with white teachers all the time.") The foreman is now defensive against the operator and sends out more nonverbal clues of a negative attitude; the operator becomes more dissatisfied, and so on. At this point, the negative feelings between the individuals are real and may be expressed in racial terms.

Much the same thing can happen in a manager/nonmanager interaction that has no other significant differences. The manager may have had courses in leadership, interviewing, interpersonal skills, sensitivity, and so on, but may not be used to dealing with persons who have not and so don't appreciate the importance of the skills associated with them. As a result, a sincere objector, who has never learned *how* to properly object (with the proper image), may come across as malicious or as a malcontent, especially if he or she has the courage to keep at it.

If the above is, in fact, what sometimes happens, it could explain such quick and favorable changes in perceptions as sometimes occur after only a three-day workshop (a day and a half is spent getting the group ready for the confrontations). The workshop provides an environment that is rarely found on the job: a forum where confrontation is legitimate and protected, where free and open discussion moderate extreme opinions, and where people have the time to get a better sense of the sincerity and good intentions of others.

It seems logical to conclude that we need to become better aware of the dynamics involved in power differentials and the effects of superior attitudes. Otherwise, stressful problems in an organization can easily create prejudice where none existed before. In a negative version of the Pygmalion effect, we perceive and therefore create the negative behaviors we expect to see in others.

Perspectives

Bias is an individual attitude of inherent superiority which destroys the sense of obligation to be fair and leads to arbitrary behaviors. The resulting discrimination hurts the victim, the discriminator, and the group. The victim loses opportunities for communication, involvement, contribution, and advancement, as well as for self-esteem and the other necessities for well-being.

Discriminators unwittingly lose the benefits of the best efforts of others, even to the point of being actively sabotaged. They also suffer the loss of the psychological benefits that characterize an atmosphere in which there is free and open communication and a genuine spirit of teamwork. More important, their descendants, or the successors in power, will eventually have to deal with the progressively degenerating reality that they have created.

Organizations suffer an unquantifiable but significant expense, because abilities and performances are not objectively evaluated and utilized, and because of interpersonal conflicts that result in some form of cost: lost time, ideas, equipment, or product. These are only symptoms, however. The ultimate cost to the organization is that both managers and employees identify with their own special interests rather than with the total organization.

At some point in this process Destructive Achievers *must* predominate. If the organization insists on not reversing the degenerating ethical trend, only DAs will be willing to maintain its security. It will require expedient actions based on power to overcome the Leaders' and Builders' (sometimes referred to as troublemakers, whistleblowers, or subversives) resistance to ethical violations of the system.

Harry Levinson noted that senior managements will increasingly ". . . need assistance in: (1) Managing people who are less inclined to identify with their organizations; (2) Managing people who are more inclined to identify with personal special interests, and bring the conflicts that go with that identification into the workplace; and (3) Managing people who are more openly aggressive. . . ."[5] The results of these confrontation sessions support his view, especially if we include senior management itself in the three categories Levinson mentioned. Virtually all employees, including managers, identify with various special interest groups. Destructive Achievers' personal interests *can* be so strong that they mentally isolate themselves completely from the larger organization. On the other hand, it is obvious that some employees, of every category, identify with their organization as well as with their special interests. They see themselves as equals, and exhibit pride and a strong identification with efforts to enhance the organization's effectiveness and success.

The challenge to management is not to try to get subgroups of employees to stop identifying with special interests, which would be impossible, and likely counterproductive, but rather to give them the opportunity to discover that their interests aren't very different from each other in the first place. Part of this discovery is management's realization that a major threat to a group's self-interest is a problem common to all groups: prejudice based on a superior attitude, and the resulting abuse of power.

Footnotes

1. Manuel G. Velasquez, *Business Ethics*. Prentice-Hall, Englewood Cliffs, N.J., 1982, 91.
2. John Naisbitt and Patricia Aburdene, *Re-inventing the Corporation*. Warner Books, New York, 1985, 1.
3. This is a paraphrase of the cab driver's remarks. The program was not taped.
4. Charles M. Kelly, "How to reduce bias on the job—and increase productivity." *Management Review,* February 1983, 14–18.
5. Harry Levinson, *The Levinson Letter*. The Levinson Institute, Belmont, Mass., July 15, 1982, 1.

CHAPTER 10

Confrontational Problem Solving

Chapter 9 dealt with the prejudice that results when an unwarranted superior *attitude* leads to an abuse of power, and vice versa. Now we need to consider why the natural Leader, whose superior *status* is legitimate, is vulnerable to becoming a Destructive Achiever. It is likely to happen if the individual does not understand the nature of legitimate superiority, along with its limitations and dangers, and fails to take adequate measures to monitor his or her own actions. Both of these acts usually require objective critiques by others, and in ways that ensure that the holder of power listens conscientiously.

In considering the natural Leader, we need to acknowledge that our need for leaders who legitimately exercise power is both material and psychological. Leaders make concerted group action possible. At the same time, they provide the support and control necessary for a sense of community. The positional leader, therefore, rightly has superior power in controlling and enforcing the management practices of the formal organization. In the same way, the expert has superior power with regard to his or her technology; and the charismatic leader has superior power in the social group.

If leaders rightly have superior power, what, then, is *equality*? "Equality" is much more than equal access to the parking lot, uniforms for all, and the chance to talk in meetings. If equality is to have any meaning at all, it must include three conditions: (1) the *right* to confront those who have superior power; (2) realistic *opportunities* to do so; and, most importantly, (3) a commitment by those entrusted with power to *ethically respond* to confrontation. If this last condition is missing, the first two have no significance.

Even Leaders and Builders are more inclined to repress confrontations than to encourage them. It is an area of human behavior in which there are few guidelines for success, and conventional wisdom seems to suggest that avoidance is the better choice. However, when managers fail to acknowledge and ethically respond to confrontations, they suffer three serious negative consequences. First, their failure to respond legitimizes unethical behaviors for others throughout the organization. Self-interested decisions and actions become the norm instead of the exception. Individuals' identifications narrow down to their smaller groups and away from their larger groups (department versus facility; black, white, male, female versus "employee"; family wealth versus community welfare). Second, it leads to later confrontations that *can't* be avoided, that are harder to control, and that are destructive in nature. Unreasonable and adamant compensatory demands, sabotage, debilitating strikes, and even personal violence may occur.

Third, the failure to identify and correct unethical conditions makes the problems themselves harder to correct at a later time. Once an unethical decision has been made, the conditions will never again be the same, and the specific opportunity is permanently lost. For example, some organizations now are faced with two ethically unsatisfactory alternatives: dismiss recently hired blacks and continue the injustices of the past, or dismiss tenured whites and violate their legal rights of seniority.

A definition is in order at this point. In his book *Business Ethics,* Manuel Velasquez wrote that ethics involves a study of the moral judgments involved in moral decisions. These concerns are often centered around three general moral standards: *rights, justice,* and *utility.*[1] "Rights" are concerned with policies, institutions, and behaviors in terms of the protection they provide for the interests and freedoms of individuals. They include the right to be honestly informed of performance rating, to work in a reasonably safe environment, to buy products that have been honestly represented, and so on.

"Justice" is of three types. *Distributive* justice relates to fairness. Do the benefits allocated to the individual match his or her contribution to the group? Are the burdens of supporting the group appropriate to the position? *Retributive* justice refers to blame and punishment. Are disciplinary actions consistent and proportionate to the wrongs? *Compensatory* justice concerns the fairness of restoring to a person what the person lost when he or she was wronged by someone else. What is the appropriate compensation to the victim of unsafe working conditions? "Utility" requires that one determine what alternative actions are available, estimate the benefits and costs that each action would have for *each* and *every* person affected by the action (including the decision maker), and choose the alternative that produces the greatest *sum total* of utility (not just the utility for the person making the decision).[2]

Of course, the principles of rights, justice, and utility are interrelated. They never exist in isolation. They simply give different perspectives from which to view moral considerations in decision making. Of the three perspectives, however, utility is the first principle the natural Leader or Builder violates in the process of becoming a Destructive Achiever. The genuine values of natural Leaders and Builders (and those of society) force them to appreciate the necessity for maintaining ethical conditions in the areas of rights and justice. Identifiable individuals are involved, and obvious injustices are subject to moral blame. On the other hand, utility involves "only money," and to manipulate money is an acceptable part of the game of becoming successful. This is currently an anti-ethical arena; that is, the manipulation of money is not a moral issue. The determination of future workloads, scope of personal control, resource allocations, amount of technical and maintenance support—all these are legitimate prizes for the manager who seeks or gains power. The fact that inefficiencies are not allowed to be confronted, and that inefficiencies eventually lead to more obvious violations of rights or justice, are not, according to today's standards, grounds for moral censure.

Actually, utility is the most constant (not to be confused with important) *ethical* standard relating to objectivity in problem solving. It affects the long-term welfare of our "excellent" organizations more than any other. Deliberately inefficient utility decisions are ultimately more harmful to more individuals than are obvious and direct violations of rights and justice. Such decisions are extremely easy to rationalize; in fact, current books and periodicals actually recommend them. They are central to the whole concept of gaining personal power in organizations. As a result, managers will repress any confrontation that might challenge their utilization of resources in terms of the optimum benefit for the total group (their unit, department, facility, corporation, society—or humanity). In too many environments, and to their advantage, it is easier for DAs to become successful if their efforts are focused on the utility benefits for the *smallest* group specifically identified with them. This may become at best an anti-ethical orientation in the Leader or Builder; at worst, a blatantly unethical one in the Destructive Achiever.

A hypothetical example illustrates how an anti-ethical orientation causes a situation to naturally degenerate. It goes something like this. The president of a waste disposal company tells a department manager that he must: (1) stay within budget restrictions, and (2) obey all laws relating to toxic waste disposal. The department manager objects that the budget does not allow enough money to meet legal requirements. The president then responds that additional funds are unavailable because of expansion plans—and if the department manager can't handle the job, he'll get someone who can. Of course, sometimes, the manager quits. Too often, however, he stays within budget by violating the law, endangers public health, and covers up his actions. Under current standards, the president can only be accused of poor administration—he either allocated resources unwisely, or he selected the wrong department manager. The manager, on the other hand, is unethical to the point of criminal culpability—he endangered human lives.

The same thing happens day in and day out on a smaller scale. Usually it's nothing more than a second-best option being selected over a best option, because of the consideration of which persons would benefit most. Since the differences are not great, deliberate malfeasance cannot be proved. However, it inevitably leads to dishonest analyses of problems, distorted reporting of unfavorable results, including outright falsification, and other kinds of behaviors that sometimes reach the headlines. In the latter stages of inefficient operations, the wrong persons are being punished or promoted in order to maintain control of shaky operations, and managers' survival becomes more important than rights or justice for persons further down the line.

This seems to be the usual course of events in organizations that get larger and more complex. The seductive degeneration begins when the utility standards of mundane operational decisions are not allowed to be challenged by knowledgeable third parties. These third parties are usually lower-level or ancillary personnel, whose confrontations have been filtered out of the system by intervening managers. This is extremely unfortunate, because they are often *the* most important source of objective critique. They don't have the self-interest of those who are the usual participants in the decision-making process—the managers of directly competing organizational subgroups. Control without confrontation leads to the perpetuation and magnification of error. Systems without an active commitment to utility standards always degenerate. (A commitment to rights and justice is a given.)

A Case Study in Confrontational Problem Solving

The experiment at the Fiber Industries plant, described in Chapter 9, was designed to establish a better interpersonal climate of equality at lower levels of the production organization (shift supervisors, foremen, and operators). The emphasis was on the ethical principles of individual rights and justice. The results of the first two workshops were so positive, resulting in observable and significant improvements in behavior, that plant management decided to consider the same process as the main building block for its new productivity effort—a "bottom-line" utility issue. (Details of this case were first reported in 1983 in the *National Productivity Review*.[3]) The manmade fibers industry was experiencing a severe cost-price squeeze, and plant management viewed the necessity for quality and productivity improvement with a heightened sense of urgency. The confrontational process appeared to have significant value in quickly creating an open communication environment and in building commitment. Senior management therefore decided to use an expanded version of it as a vehicle for task force interaction. It was felt that the confrontational process could quickly surface the most important issues, expose and reduce intergroup conflicts, especially between managers and nonmanagers, and increase the facility's "sense of community."

Five important characteristics of this modified approach were a *direct* emphasis on equality of team membership; *multiple confrontations,* to generate data for discussion and to demonstrate the nature of subgroup biases; a *full diagonal slice* of participation, in order to ensure an objective diversity of views; *group-generated norms* for commitment; and, as a result of the previous four, a *significant emotional event* for genuine attitude change. The major differences between this approach and most other change efforts were: (1) the emphasis on equality; and (2) a confrontational approach to problem solving.

FULL DIAGONAL SLICE

The first group to participate in the modified process included a diagonal slice of *all* levels of the largest plant production department and its support staff. This meant that there were direct personal contacts among the production manager, engineers, superintendents, operators, foremen, mechanics, technicians, technical group leaders, and so on. They began by discussing issues of equality and the interpersonal climate. They then progressed to specific production and quality improvement systems.

The extent of this slice could not be maintained with later groups, as upper-level managers were "used-up." This appeared not to have a negative effect. It was well known throughout the plant that all members of upper management had attended a workshop, and that eventually *every* member of the plant would attend. This effectively communicated the message that the process was for everyone, and actual practice demonstrated the truth of the commitment. The eventual complete absence of management was compensated for by showing videotapes of previous confrontations and by having senior management meet with each workshop group on its last day to deal with participants' questions or objections.

EMPHASIS ON EQUALITY

Confrontation itself is a commitment to equality in the workplace. Effective interpersonal relationships are all based on a mutual respect between individuals and an assumption of equal *personal* status. Organizational status differences are an accepted factor, but they need not have a negative effect on relationships.

Listening, two-way communications, objectivity in problem solving (versus ego defense), group identification (versus identification with special interests), involvement ("I *want* to help you with your problems"), a sense of community ("I *belong* with this group"), and pride of accomplishment depend far more on the climate of equality than on any interpersonal skills that can be taught. Skills are certainly helpful, but they are irrelevant when a condescending attitude is communicated to another.

MULTIPLE CONFRONTATIONS

Relevant negative issues must be taken into account if objective problem solving is to occur. Therefore, confrontations between groups ensured that important problems were raised and discussed. This is far superior to situations in which top management, a consultant, or a facilitator determines subject matter. Even if top management does an excellent job of selecting relevant topics, it can't match the group's sense of accuracy and ownership when it develops its own issues. The sense of urgency and the heightened emotions also increase motivation to follow through when conclusions are drawn or decisions are made. There is a greater sense of commitment once a group has successfully faced and progressed through ego-threatening issues that are potentially damaging to interpersonal relationships. Although some problems are not solved, there is at least a greater mutual appreciation of different points of view.

An important point to make here is that there is a synergistic effect in *multiple* confrontations. People learn that the causes and negative effects of superior attitudes are fundamentally the same, whatever the specific relationship. Just as the management/nonmanagement confrontation helped persons to better understand the nature of male/female and black/white issues—the reverse is also true. Male/female and black/white confrontations help persons to better understand and address management/nonmanagement issues. A paradox is that the *worse* conditions are in a group, the *more* important it is to have multiple confrontations. Most managers instinctively seek to avoid "confounding" issues. However, multiple confrontations help to defuse the sensitivity to any one category of differences. If there are serious morale problems in management/nonmanagement relationships, it becomes more important to spend some prior time on the general problems of bias, superiority, and unethical use of power.

If a group has already dealt with important issues in other relationships, it is in a much better position to address its most ego-threatening relationship: management/nonmanagement. (Multiple confrontations need not include male/female or black/white; these just seemed to be appropriate in this specific case.

In other cases, a maintenance/production or operations/support staff confrontation was used in combination with management/nonmanagement.) Multiple sessions simply increase the overall sophistication and comfort levels of participants. They are more accepting of the confrontation process itself, and are willing to take greater risks, by speaking openly and forcefully about problems in the work group.

Senior managers are almost always shocked when lower-level personnel passionately describe the destructive behaviors of "certain managers" (specific names are not allowed, although some ears undoubtedly redden). They learn first-hand that some card-carrying members of the team are deliberately being counterproductive in order to achieve personal objectives. Of course, any group is likely to learn from multiple confrontations. However, if a group already has high morale and good interpersonal relationships, it can usually concentrate on its most sensitive issues without the necessity of multiple confrontations.

GROUP-GENERATED NORMS

Although management statements of philosophy, values, and norms are necessary and influence member behavior, they pale in significance when compared to group-generated norms. Periodically throughout the process, each training group developed its own norms of behavior, based on previous discussion. The areas dealt with were male/female, black/white, and management/nonmanagement relationships, the plant "psychological contract," and productivity and quality performance standards. This meant that eventually almost every member of the plant actively participated in generating documents on accepted standards of behavior and performance. In addition, many counterproductive behaviors, which had been rationalized by some as "acceptable under the conditions," were clearly labeled *wrong* by the group, with specific reasons why. Sound management axioms in textbooks suddenly took on meaning when employees described such things as the importance of maintaining quality work standards (or deliberately violating them), not allowing favorites or "obnoxious" employees to get away with poor work habits, fairly promoting the persons who deserve it, and so on.

It's reassuring that every group reaches essentially the same conclusions, although they may be stated in different words each time: treat everyone equally, regardless of race, sex, or organizational position; don't stereotype people, treat each one as an individual; listen; produce quality products; be a team member; and so on. One could take the data from almost any session and compile a fairly decent booklet on effective interpersonal relationships for productivity.

SIGNIFICANT EMOTIONAL EVENT

Dr. Morris Massey, a recognized authority on value and attitude change, concluded that it takes a significant emotional event to bring about fundamental and lasting attitude change.[4] The diagonal slice, the confrontations, and the generation of group norms assure a significant emotional event for most persons. This is of particular importance to management in their efforts to implement a productivity improvement system, because it helps to translate *form* into *substance*. Managers internalize the system, they *live* it, instead of just designing, administering, and monitoring it. Instead of merely *understanding* the use of mission, philosophy, and value statements in a productivity improvement effort, production managers *value* them. They have sat next to operators who were concerned over foreign competition, frustrated at being required to produce poor quality products to meet production goals, and confused about the reasons for certain personnel policies. Instead of just *learning* that mutual respect is important in intergroup relationships, upper levels *experience* respect for the conscientiousness and ability of those at lower levels. In turn, lower levels experience respect for the good intentions and skill of upper levels in dealing with very complex problems and conflicting demands.

In sum, all types of employees experience the confrontations over sensitive and potentially explosive issues. They discover to their pleasant surprise that relationships are significantly better as a result. With a greater sense of group identity, they are in a much better position to mobilize themselves for whatever productive effort they have designed.

The Productivity (Utility) Dimension

In order to incorporate a productivity dimension into the modified confrontational process, management at the Fiber Industries plant commissioned the first group as a "Quality-of-Work Life/Productivity Task Force." The group's added objective was to recommend the type of program that would be most effective for the plant, given the nature of its problems and resources. At the beginning of the last day of the workshop, the members of the task force were asked to review the advantages and disadvantages of the productivity and quality-of-work-life efforts that had been tried in the plant during the last five years. They were then to come up with their own hybrid version of a program that would represent the optimum improvement in both the quality of work life (QWL) and quality of work (productivity, or PTY). Recommendations for the program had to meet these broad standards, as set by the plant manager and immediate staff:

1. It must be time-effective: "We need to *focus* our efforts so that we spend a minimum of time spinning our wheels."

2. It must have staying power, making improvements that will be permanent: "We want to develop a system that enables us to effectively improve conditions that *stay* improved."

3. It must include all the appropriate functions (personnel) of the team: "Participation and involvement help bring about job satisfaction and commitment."

4. It must stipulate measures of success: "This is essential if we are to know that we *are* improving, and that we are maintaining our improvements. A lack of feedback (measures) makes it difficult to monitor and ensure continued success."

Unexpectedly, in just one day, the group was able to progress far enough to not only make general recommendations but recommend a specific next step: appointment of a full-time Quality-of-Work-Life/Productivity coordinator with support staff. An interesting note: In many ways, the recommended structure (see Exhibit 10.1) for this QWL/PTY Committee is ideally suited to a theoretically sound sociotechnical effort—and it was first suggested by a mechanic.

The task force recommended that the coordinator have direct access to all levels and functions in the plant. It recognized the debilitating nature of subgroup political infighting, and wanted to structurally encourage its reduction. They also recommended that all members of the committee must have participated in the basic confrontational workshop, and that membership rotate on a staggered basis throughout the department and its support staffs. This should stimulate a more confrontational, yet more cooperative and less political, approach to problem solving.

The QWL/PTY Committee was eventually selected, and it evolved its own procedures and working relationships on the plant floor. Its members periodically received help from consultants in understanding sociotechnical and behavioral modification methodologies. (Richard Walton observed that "the dynamic nature of the interaction between human and technical factors indicates that system design for particular applications should generally be pursued within an evolutionary model of design. This contrasts with a more typical assumption that the design can and should be completely conceived before implementation."[5])

The coordinator's responsibility was to bring together the appropriate members of the problem-solving group, depending upon the specific issue. For example, one problem may be best handled by a single foreman or engineer, another by 10 persons drawn from across three shifts, yet another by a unique

Exhibit 10.1   THE QWL/PTY COMMITTEE

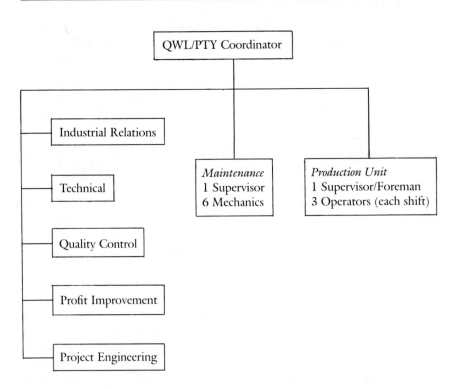

task force. This *ad hoc* approach seems to have some advantages over typical quality circles. Instead of a fixed group attempting to find problems they can solve, the nature of each problem determines the ideal composition of the group to solve it, and the group exists only as long as the problem does. (Of course, in many cases, the problem-solving group would be of the same composition as a typical quality circle.) This strategy automatically reduces the deterioration of utility standards that is the result of subgroup self-interests. The entire problem-solving process—from analysis to implementation—is conducted within the single *ad hoc* group. Since supportive confrontation was an established norm, any person at any level had official permission to fully participate.

The QWL/PTY Committee dealt with three fundamental types of issues:

1. Specific problems affecting either job satisfaction or productivity.

2. Work group redesign.

3. Managerial and technical skills training.

Anyone could present problems and proposals in these areas, and, in fact, the Committee heard from people at all levels: operators and mechanics, first-line supervision, and senior management. All input received consideration, and, in each case, the Committee made a genuine effort to achieve the best solution for the long-term viability of the plant.

The success of the program was immediate, and documented savings were more than $2 million by the end of the first year. This did not include a large number of small projects that were never evaluated, nor did it take into account improvements due simply to greater worker conscientiousness. In some cases, waste was reduced, and better conversion efficiency occurred, but with no identifiable causes.

The types of productivity improvements were typical of those reported in other studies, but a few examples may clarify the range of circumstances in which changes occurred. In one case, an operator suggested that one person could attend to four machines (instead of the standard three) if certain equipment problems could be corrected and if operational procedures were improved. The plant professional staff, with operator input, studied the situation and agreed. As a result, the labor requirements were reduced, and the operators felt pride in the accomplishment.

In another case, unit management concluded that manning for a certain process made the plant noncompetitive and therefore had to be reduced. The QWL/PTY Committee was approached, told that the manning was noncompetitively high, and asked if it would participate in finding the best ways to accomplish a reduction. It was explained that Committee input was necessary in order to make the workload as equitable as possible and to avoid jeopardizing quality or productivity. The Committee expressed reservations about the reductions, but nevertheless went about the task of making several pertinent recommendations, which were successfully adopted. As a result, management achieved a necessary objective with minimum negative impact on morale. The operators looked upon the changes as an unpleasant but necessary fact of life, and appreciated the opportunity to participate in what was a difficult solution.

These examples demonstrate that productivity improvements originate at all levels. Although many improvements were operator- or team-generated, others were initiated by management. In all cases, however, an attempt was made to give all viewpoints a full hearing, and to give those closest to a problem the opportunity to solve it.

The QWL/PTY Committee brought about several changes in training. Operator and mechanic workshops were expanded, to improve understanding of equipment characteristics and procedures and how they affect quality and productivity. These workshops delved much further into the theories of operations than had previous training. A very important aspect was that there were more planned interactions between different organizational levels and functions because of the range of levels present. These workshops were also sources of ideas for improvements.

Because of better interpersonal relationships, coaching and information sharing improved. For example, a source of production downtime and high maintenance costs was winder malfunctions. When an operator had difficulty with a production position, he or she usually took the winder out of service and summoned a mechanic. The mechanic often found no evidence of malfunction, and bickering over the real source of the problem developed. When members of the Committee met to consider the problem, a mechanic suggested that he be allowed to work one-on-one with each operator and explain winder operation, the most common cause of failure. This was done, and downtime was reduced by 50 percent. But the cost reduction was only half of the benefit. There was, in addition, an improvement in overall cooperation, and fewer egocentric behaviors. The mechanics felt that the operators were treating the equipment with more respect, and the operators now felt that the mechanics were providing good service.

Although similar productivity programs and results are not unusual these days, the addition of the confrontational process, and the emphasis on equality and basic human values, did at least three things for the plant. First, the "clearing of the air" had a remarkably quick and favorable impact on every group that went through it. This is especially important for any task force that must establish open communications and get genuine nonpolitical commitment to a project's success. All common approaches to productivity improvement—sociotechnical systems, quality circles, quality-of-work-life—are based on recognized behavioral science theory. Their typical structures and procedures are both valid and useful. However, they are only empty forms. They only become effective when they are infused with the substance of skillful individuals who are committed to make them work for the benefit of the *total* group.

Second, the confrontational process caused the discussion of management's philosophy, goals, and values to be very persuasive, not only for lower-level employees, but *for the managers themselves*. As research has indicated, effective top management groups are often characterized by a clear sense of mission, knowledge of their own goals and values, and an ability and willingness to communicate these to the rest of the organization. A sociotechnical approach may therefore begin with top management developing statements of mission, philosophy, values, and goals.

With the help of a qualified consultant, the top managers may spend many agonizing days analyzing who they are and what they want the organization to be. But while such an effort will produce the *form* of an effective sociotechnical effort, too often it will not create the *substance*. Some managers may develop a clear statement of mission, goals, and values in an open confrontational meeting with their personnel. Others may develop what appear to be identical documents, but in the isolated security of a small conference room. The motivation of the former is to ethically refine their own and the organization's planning, implementation, and quality-of-work-life. This is the substance that makes any theoretically sound system work. The latter is likely to be an anti-ethical attempt to improve bottom-line results by taking whatever steps are necessary to gain a greater control of production. This is the mechanical, even manipulative, approach, which turns any system into merely a fad. The full diagonal-slice confrontational experience can help make the procedure one of substance. It results in a more gut-level (and more lasting) commitment by management to the values that it previously espoused only on an intellectual level.

And third, the group norms generated by the process were the result of in-depth discussions of important issues—even ego-threatening ones—that are not often surfaced by most groups. In *every* group, management's destructive attitudes of superiority—as demonstrated by specific, observable behaviors—were listed as important issues to be addressed. When behavioral norms are developed from such discussions, there is a greater awareness that specific destructive behaviors must be changed if an organizational development effort is to succeed. In addition, eventually *everyone* in the plant participated in the discussions, not just a representative task force charged with making an official "statement of philosophy." Although a representative task force may develop a philosophy statement (or code of ethics) for the whole organization, how well others implement it will depend upon the *demonstrated* sum total of attitudes and values of the entire membership.

The reason it is so difficult to distinguish substance from mere form during a change effort is that managements, and even experienced consultants, fail to appreciate the manipulative nature of some of their own actions. Fundamentally, this manipulative orientation results from the belief that systems and procedures (including philosophy and values) can be designed at various levels or by designated groups and passed down (or over) to others for implementation.

In effect, this mechanical kind of arms-length management sends the same old message to the organization: "You should identify with us, our mission, and our system, but we won't make the effort to identify with you." Without management's own involvement in *all* parts of the organization, establishing a task force is seen only as a token gesture. Essentially, the organization is *still* not a community of equals, and lower levels and various functions realize that status and the traditional hierarchy are alive and well.

Perspectives

As yet, no one knows how to reliably or permanently improve the performance of human groups, not only corporations and societies, but even small organizational units. This case study is provided as an example of a confrontational approach that worked well for one organizational unit. It is not presented as the only answer to the problems of inequality and power politics, but rather as an illustration of considerations that may assist others in addressing the same issues.

Even though it was later announced that the plant would be shut down because of unavoidable product-line restructuring, management honored its commitment to put all employees through the process. They felt that the benefits to employees and the plant itself justified its continuance. Besides facilitating a more orderly, gradual, shut-down, at the time of this writing, savings are estimated to be over eight million dollars. As with all case studies (such as the Topeka Project described in Chapter 5), each observer will interpret these results in light of his or her own ideology. Many will see it as proof of the values of equality and confrontational problem solving. Others may feel that the same results could have been achieved more directly with less idealistic procedures.

There are undoubtedly limits to the applicability of such a process. This plant management was supportive from its inception; indeed, it was at management's request that the process was initiated as a *productivity* effort. It probably would be of little value to organizations that are already dominated by Destructive Achievers, or that are hopelessly mired in power politics—which leads us to Chapter 11.

Footnotes

1. Manuel G. Velasquez, *Business Ethics*. Prentice-Hall, Englewood Cliffs, N.J., 1982, chapters 1 and 2.
2. In this section on Velasquez's three moral standards, I try to define certain ethical considerations: Some phrases and sentences are exact quotations; others have been paraphrased. My purpose is to present some idea of the *scope* of ethical considerations, not a complete explanation.
3. Charles M. Kelly and James M. Norman, "The fusion process for productivity improvement." *National Productivity Review,* Spring 1983, 164–172.
4. A vivid description of this concept is given by Dr. Morris Massey in his videotape, *What You Are Is Where You Were When*. Magnetic Video Corporation, Farmington, Mich., 1976.
5. Richard E. Walton, "Planned changes to improve organizational effectiveness." *Technology in Society,* 2, 4, 1980, 409.

CHAPTER 11

Operational Ethics and the Entropy of Power

Contradictory ideologies "work" insofar as the success of each is defined by each competing advocate, and all tend to degenerate along the lines predicted by their critics. Hence, whether a group's systems are based on capitalism or communism, Adam Smith economics or Keynesian economics, power or problem solving, it is important that group members appreciate the unique quality of life that their own ideologies encourage, and their vulnerabilities to destructive influences. We fear imagined threats from without, but the most destructive force is usually within the organization: that is, the undetected Destructive Achiever. Because the DA uses the system for his own advantage, he "protects" it from discovering and correcting its weaknesses. It goes without saying that DAs also protect the system from discovering their own contributions to those weaknesses.

In addition, there sometimes appears to be an unholy alliance between the Leaders of a system (or ideology) and its protective DAs. Although any responsible management is concerned with long-term organizational welfare, its first line of defense against internally generated threats is instinctive, short-term, and usually wrong. Because the Destructive Achiever apparently contributes to the short-term achievement of the Leader's "vision," threats to the DA are considered to be threats to the Leader. Therefore, there is a strong bias to label the internal objector who is critical of the DA or the system as malicious, immature, or unrealistic.

The experience of Morton Thiokol's Allan McDonald is very instructive (see p. 23). At a closed-door hearing, he testified that his willingness to be openly critical of upper management had cost him his job as chief of Thiokol's booster program.[1] Newspapers reported that senior Thiokol management had a different view: "Although Morton Thiokol officials insist McDonald was not demoted, they have conceded he was reassigned to reduce his contact—and 'potential friction'—with officials at NASA."[2] In either case, whether the demotion was real or imagined, malicious or well-intentioned, the fact that the action was even taken demonstrates bureaucracy's blind defensiveness. It has a constant sense of urgency to reduce the "friction" of those who confront its managers in any significant way. Bear in mind that McDonald's reduced status in the company occurred in the glare of world-wide attention, and created an outrage in Congress.

A classic description of management reaction to internal confrontation was recently reported as a result of a Greensboro, N.C., lawsuit. (Of course, when organizational personnel confront management in meaningful ways, they are often placed into the category of "outsiders.") Court papers alleged that Duke University president Keith Brodie knew for years that nurses and doctors feared that a University Medical Center psychiatrist, Bernard Bressler, was having sex with female patients and was treating patients with dangerous combinations of drugs. A former psychiatric nursing supervisor said under oath that she had told Brodie and other medical officials about sexual allegations involving Bressler and about widespread concerns over his drug-prescribing practices, and that she got no results.

> That's been something that haunts me. . . . Was there something else I could have done or should have done? Was there someone else I should have spoken to? I felt at the time that I was doing everything that I could with such frequency that I was an annoyance to those people. I mean, they didn't want to hear from me anymore, the things that I was saying. They really were wishing very strongly that I wouldn't say them anymore.[3]

She went on to say that Brodie told her that Bressler was the main income producer for the Duke-based psychiatric clinic, and that that made it difficult to discipline him.[4]

In both of these cases the confronters appear to have made strenuous and conscientious efforts, without fanfare or publicity, to get their managers to correct problems that eventually led to disaster for each organization. Instead of the bureaucracies giving them credit for their rare courage and loyalty when problems were developing, McDonald's responsibilities were reduced—and the nursing supervisor reported: "[I] was in tears so much. I hurt so badly over the Bressler situation and all of his patients there the whole time that I was there that it was really a very, very unhappy two years for me."[5]

Of course, these two incidents are not unusual; many similar cases have been well publicized. What has *not* been so well publicized is that the "anti-whistleblower phenomenon" is only the tip of the iceberg. It is merely the culminating step of the more general problem. Management has a natural inclination, not only to suppress bad news, but also to resist acting on it when it does come into the open. Note that McDonald and the nursing supervisor were not whistleblowers until *after* the damage had been done. Until that time they were merely conscientious upward communicators who were either ignored or repressed. As is usually the case, *the management created their whistleblowers* by their own actions or inactions. They seemed almost *obligated* to rationalize the potential dangers and ethical discrepancies that were brought to their attention.

Our Unexamined Biases

Our chronic biases against those who would challenge the prevailing system and its actions, and our natural defensiveness against ethical constraints to exercising personal power, have been very well publicized. There appears to be something disloyal in those who criticize the ruling elite. And the holders of power feel that the worthiness of their mission—or their inherent superiority—justifies the expediencies of their actions. It is generally acknowledged that these biases are among our most obvious problems in operating organizations today. However, very sophisticated versions of them can also be found in graduate schools of business and behavioral science—and their significance has been largely ignored or unappreciated.

THE ORGANIZATIONAL OBJECTOR

The editorial staff of *The Levinson Letter,* a psychologically oriented business publication, gave an excellent description of the plight of the confronter when it answered the question, "Why do some individuals stick their necks out?"

> Managers who decide to publicize their company's misdeeds are rarely chronic "troublemakers." Most have excellent records. But they have extremely powerful consciences and cannot sit by while something immoral, illegal, or unfair is occurring. The same is true of many Christians who helped Jews during World War II. They had never been heros or taken a stand against Nazism, but when they saw what was happening in their own towns they had to act.

> It is a fact of life that whistleblowers are held in contempt in most organizations. When the beans are spilled, the organization will attack in return. The person's character may be inpugned, with no guarantee of vindication.

A long period of frustration, family problems, and depression may follow. It's an experience that most people would avoid. But the whistleblower is in a unique and uncomfortable position, suffering guilt for remaining silent, or suffering isolation and reprisals for speaking out.[6]

Contrast this view of the whistleblower as hero with what would be considered a progressive view from the business establishment. David W. Ewing, managing editor of the *Harvard Business Review* at the time of his article, "How to Negotiate with Employee Objectors," made the following observations:

Variously referred to as whistleblowers, dissenters, or dissidents, more often than not they are able and well-intentioned people. Not surprisingly, they stir the concern of many executives and fellow employees. To separate them from the traditional recalcitrants, I refer to them as *employee objectors*.

The increasing number of employee objectors is one reason for concern. While no quantitative studies of the trend have been made, increases in the frequency of lawsuits, complaints handled by employee assistance departments, executives' personal experiences, and other information suggest that employee objectors are at least ten times more numerous today than they were, say, 10 or 15 years ago.

What is more, objectors may be costly to an organization. One well-known Eastern company has spent millions of dollars defending itself against the legal attacks of a single determined objector, and its struggles may be far from over.[7]

One can hardly argue with the denotative sense of these observations, as well as of the material in the rest of the article. It gives a tip of the hat to the possibility of positive contributions of employee objectors, and recommends that management be aware that their concerns may have validity. However, the connotative sense of the article is condescending to the general run of recalcitrants, traditional or otherwise. Ewing's major persuasive appeal is directed at management's desire to control the costs of destructive confrontation: lawsuits, labor unrest, bad publicity, and loss of managerial credibility. Instead of addressing fundamental causes, which are more costly than the resulting legal problems, he emphasizes the symptoms. He focuses on controlling the objector and the confrontation, rather than on the effectiveness of upward communications or the counterproductive behaviors of middle-level Destructive Achievers.

This in itself is not a criticism. One cannot deal with all dimensions of a subject in a single article. The implied ideology is the issue—this is what managers will incorporate into their later behaviors. Even Ewing's recommended procedures, which are fundamentally sound, demean the conscientious objector:

1. Draw out the objector's personal concerns.

While you may see the facts, arguments, and positions advanced by an employee objector as inaccurate or misconceived, you would be wise to treat the person's fear and worries as important facts. . . . Get the objector's personal reactions into the open. Consider them respectfully, even if you find the person's arguments foolish and irritating. Especially if the objector is wrought up, it is wise in the beginning to keep the focus *off* substantive ideas and positions taken. . . .

In addition, objectors are not likely to be as impressed as you are with your understanding of the situation, nor are they likely to appreciate the value of your sources of information—your discussions with other managers, perhaps, or your knowledge of confidential company information, or possibly your familiarity with a new company plan that alters that situation.[8]

The first step is certainly a good one in principle, but, as described, it is based on power and maintaining control, and is hardly a mental set conducive to objective problem solving. It also reinforces a manager's inclination to presume a superior understanding of lower-level problems. In some ways it *is* likely to be superior, but in many other ways it is likely to be distorted. Ewing continues:

2. Find out what motivates the objector.

Even after drawing out the objector's fears and concerns, continue to avoid a head-to-head hassle over the merits of his or her argument. Don't get led into a debate. Instead, try to understand the person's purposes and interests in challenging management. Ask yourself what this person is trying to achieve and what gain he or she seeks. This is much more likely to help you progress than focusing only on the position taken.[9]

This again is a good step if it is a genuine effort to understand the full ramifications of a problem. However, the absence of appropriate qualifications implies that the motivations of the senior manager, and intervening managers, are superior to those of the objector. Note objector McDonald's description of the way a senior NASA manager began a discussion: "Mr. Mulloy came into my office and slammed the door and . . . was very intimidating. He was obviously very disturbed and wanted to know what my motivation was."[10] Of course, the motivations of an objector *may* be deceptive, but to enter a discussion with that kind of expectation hardly contributes to a climate of objectivity.

Ewing's last two steps are sound: (3) Think up mutually beneficial options; and (4) propose objective tests to determine outcomes.[11] However, the excellence of the fourth step as a problem-solving effort is significantly diminished with another reliance on superior power *vis-à-vis* the lower-level confronters:

Remember that, as management's representative, you generally have an important advantage in these discussions. For instance, as a top executive of BART, you're known to the business and government communities, whereas the engineers have no public identities. The visibility and authority of your office enable you to communicate to many more people—and faster—than the engineers can. As a manager of the chemical company, you have instant access to support and resources that are beyond the reach of the chemist. As an executive of the airline company, you have similar advantages.[12]

In other words, senior managers always have a fall-back position in the back of their minds. If the confrontation goes against their wishes, they can overwhelm objectors by manipulating communications with the news media or by calling on friendly resources in the power establishment.

Ewing concludes the article by writing in the last paragraph that any agreements should be wise ones and that they should improve or at least not damage the relationship between the parties. However, the general tone of the article is best reflected by the first three paragraphs of his four-paragraph conclusion:

In most organizations in most states, managers are not legally compelled to negotiate with employee objectors who are not union members. Managers can thumb the objectors out of the company, if they want to. As a practical matter, however, it is becoming less desirable to do so. Especially with employees who have served the company for a while and whose capabilities are proved, it pays to seek a mutually advantageous solution.

Still, even the most adept managers do not always find employee objectors responsive to negotiation. What should be done then? If the objector seems to be more interested in being a gadfly or rabble-rouser than in being cooperative and helpful, then, if you have done your best to negotiate and assuming you comply with any company policies on the subject, it may be time to fire the person.

If anyone questions you, offer your notes on the discussions. If one or two others joined you at some stage, they can attest to your efforts. If company policy or government regulations require you to submit to a hearing procedure, your duly noted discussions should serve as documentation for your decision.[13]

Again, the connotative sense of Ewing's conclusion betrays an almost grudging acknowledgement that, after all, a manager should respect loyal employees by listening to them and trying to reach a satisfactory compromise. But does this recommendation prepare managers to receive the valid criticism of employees who don't fit management's biased image of employees who have "served the company well"? Will an unsophisticated Builder get the same hearing as a systems-oriented Destructive Achiever?

I selected Mr. Ewing's article for analysis because he is a recognized expert on employee rights and the author of *Freedom Inside the Organization, Bringing Civil Liberties to the Workplace,* and *Do It My Way or You're Fired.* Also, the steps he recommends in the article itself are procedurally sound and reflect what many would consider to be a progressive approach. Most managements would improve their handling of objectors if they would do half of what he recommends. His approach, however, is designed for organizations as they exist today, and, in my view, it does little to make them better. In fact, in the words of Peters and Waterman, it may only enable organizations to be "losing less fast" as employees begin to recognize the condescending nature of such a process. It is an attempt to develop a problem-solving situation, which requires equality and confrontation, with a legalistic process based on superiority and power.

THE DEFENSIVENESS OF POWER

Whenever a subgroup (or an individual) separates itself from its community in a superior way, its value orientation becomes either absolutist (we have the right answers and therefore the right of coercion) or subjectivist (what's in it for us).[14] It doesn't matter what the superiority is based on: sex, race, age, organizational level, or organizational function. Once a group has done this, its first line of defense against ethical intrusions on its decision making is the suppression of confrontation. Its members either prevent it from happening at all, or they make sure that it is not effective when it occurs. In this way, protests about the inconsistencies between publicly stated values and actual behaviors of the powerful are muted. This allows them to retain their ignorance, at least outwardly, of their own negative behaviors, and their cognitive dissonance is comfortably minimized.

In "Is Organization Development Possible in Power Cultures?", Peter Reason discounted the effectiveness of various forms of traditional organizational development methods based on objective problem solving. He concluded his article with a series of disconcerting questions:

> We must recognize that there is a major arena in which OD has nothing to offer. . . . If we in OD are going to have a theory of the use of power it must be a radical theory that shows us how to use power in the service of change, rather than how to simply survive in a political environment. But does OD, *should* OD encompass more revolutionary, even subversive inter-

ventions? Do we have the skills, the courage, and above all the commitment? Should we work where collaboration and consensus are readily available, or should we move in to participate in a more revolutionary arena? Should we, and do we want to?[15]

Along similar lines, in "Top Management Politics and Organizational Change," Larry Greiner reviewed four types of "political resolutions of power in top-management groups." He referred to one group as the "covert resistance" type, in which the assertive CEO is resisted by subordinates as a tyrant, and which "contains the seeds of a consulting disaster." Greiner rejects confrontation as a realistic option for this kind of group:

> Another naive intervention, in my opinion, would be to initiate a "confrontation meeting" for the top group. I doubt that the proposal would be accepted, but should it be, the personal conflicts seem so deeply rooted that only a more severe crisis would be precipitated.[16]

Greiner goes on to suggest that one of the more realistic solutions would be for the change agent to recommend the removal of obstructive subordinates and their replacement with more constructive ones.

This new direction of organizational development may be correct for ethical consultants in some organizations today. However, what appears correct for a short-term solution may only add to the increase of future unethical behaviors. Any experienced executive knows of consultants who have developed shady reputations as "king makers" in the organizations they were supposed to serve. Their recommendations or actions seemed designed to assure a continued consulting arrangement rather than to address problems. They were more concerned with discovering and allying themselves with the powerful than with challenging them to improve their behaviors.

In other words, when consultants or internal executives use power politics—withholding information, attacking and removing nonsupporters, and the like—they are continuing the same kind of environment they are committed to change. The difference is that the cast of characters is different. Our only hope is that these newly minted consultants will have ethical standards that now seem to be so rare among those seeking greater recognition, money, or whatever personal goals they have. Will their personal blueprints for the ideal organization be an improvement over those of other experts of historical note: the advisors who helped bring about the conglomerates of the sixties; the economists of the seventies that encouraged farmers to expand their operations; and the present gurus who force organizations to become more competitive via reduced operating costs—while skimming off huge amounts of "excess cash" for themselves as a reward? Without an open, problem-solving environment, how can we evaluate what the power-wielding experts, of any kind, are doing to our long-term quality of life?

Despite these concerns, Greiner's assessment of the success record of previous large-scale OD efforts based on more traditional problem solving appears to be accurate:

> For many years I have puzzled over why even the best-planned change efforts by management consultants and experts in organization development seem to go awry. Despite numerous reports of success in the research literature and in the anecdotes of bravado consultants, I would be willing to wager that the private failures far outnumber the public successes—say in a ratio of 20 to 1.[17]

The new generation of consultants are correct about the failures of problem-solving approaches to make large-scale improvements at top levels of organizations. They are becoming even *more* correct as Destructive Achievers filter upward into the apparently impenetrable positions of power. Organizations are increasingly getting into situations where power, personal interests, and political alliances are more important for getting things done than is objective problem solving. Not only that, as the political realities of top management teams become more publicized, power politics becomes the normative behavior at lower levels of organizations.

Our "new" solutions to power politics—the counter use of power—are as old as humankind itself. The fundamental assumption is that power will be neutralized if everyone is taught how to use it. To some extent this may be true. Unfortunately, the more likely result is that the use of power will only reach new heights, with the players in the same relative positions, and objective problem solving the loser. Upward or confrontational communications are difficult to achieve even in a healthy, cooperative organization. In a power-oriented environment, communications only get worse, in spite of ethically intended efforts to control the process.

Our Philosophical Failures

We already know what makes organizations effective—and what maintains them. We just haven't learned how to act on the knowledge. People with power must make disciplined decisions that they instinctively know are ethically correct, even when those decisions may require short-term sacrifices for themselves or for their special-interest subgroups. Although Leaders in small organizations often develop this discipline very well, it is difficult to achieve on a large-scale basis. As churches, businesses, or entire societies grow and become more complex, Leaders tend to drift away from their initial commitments to ethical behaviors. They become more concerned with solving and controlling immediate day-to-day problems, and are tempted to deny the significance of moral wrongs. The study of ethics is disparaged as being unrealistic, impractical, or ineffective.

DENIAL OF MORAL WRONG

Karl Menninger, who equated mental health with moral health, asked the question, *Whatever Became of Sin?* He noted that ". . . as a nation, we officially ceased 'sinning' some twenty years ago."[18] (Since 1953—Menninger's book was published in 1973—no U.S. president has mentioned sin as a national failure, instead referring to problems of "pride," "self-righteousness," and "shortcomings.") After referring to Arnold Toynbee's observation that "all the great historic philosophies and religions have been concerned, first and foremost, with the overcoming of egocentricity," he went on to conclude:

> Egocentricity is one name for it. Selfishness, narcissism, pride, and other terms have been used. But neither the clergy nor the behavioral scientists, including psychiatrists, have made it an issue. The popular leaning is away from notions of guilt and morality. Some politicians, groping for a word, have chanced on the silly misnomer, permissiveness. Their thinking is muddy but their meaning is clear. Disease and treatment have been the watchwords of the day and little is said about selfishness or guilt or the "morality gap." And certainly no one talks about sin![19]

If the clergy and behavioral scientists are leaning away from notions of morality, what can one expect of those whose professions are more concerned with the materialistic aspects of existence? Menninger chastises those who deny personal responsibility for immorality, and who use the excuse that wrongdoing is caused by external circumstances.

In contrast to Menninger, in "Politics of a Process Consultant," Andrew Kakabadse saw counterproductive executive conflict as an understandable clash of wills, and chastised those who would advocate organizational change based on rational problem solving:

> The element of irresponsibility in the philosophies of prominent behavioral scientists is that the painful processes of change can turn into a guilt-

ridden experience, since the "experts" set standards that never were attainable.[20]

What standards? That persons with incomes in the high six figures conscientiously make decisions on the basis of the balanced interests of all their constituencies? That they exhibit the same concern for objective problem solving as do their own subordinates? That they reward subordinates and associates for organizational performance, rather than for personal loyalty? That they actually adhere to the standards of their own religions or publicly expressed ethical standards?

Our historic failures to create a perfect, ethically based society, or a specific organization, should encourage us to seek better approaches and answers, not to abandon the search. The excuses that ethical perfection is unattainable, or that feelings of guilt have no place in management, corrupt the free market concept. Its major premise is that forcing persons to behave ethically is always inefficient. Regulations based on legalistic power may perversely work to the eventual advantage of the Destructive Achiever as rules become more bureaucratic and cumbersome. Society is caught in a double bind. Unethical managers can ruin the free market system, and regulations to force ethical behaviors tend to be self-defeating. If it is now unrealistic to expect top management groups to behave ethically, then we need to develop better ways—other than through the use of power—to convince them that they should.

DISPARAGEMENT OF ETHICS

As a beginning, we need to increase our respect for the study of ethics. Peter Vaill, editor of the American Management Association's *Organizational Dynamics,* gave three possible reasons why the subject of ethics has received at best very occasional and uneven attention in the business school curriculum ("It is just a field that has never gotten going, so to speak"). First, business schools would have to question the ethical practices of some of their own benefactors—the managers who "hire our graduates, read our books, let us speak at their conferences, pay us consulting fees." Second, "very, very few academics in management schools are qualified to offer a really serious course on ethics anyway." And third, "the subject has a pressing reality for only a very few people—people who grasp the significance of the subject for civilization whether there are current 'hot' issues around or not."[21]

Vaill's third reason is the essential one, and probably contributes to the previous two. We believe in ethical behaviors in the abstract future, but fail to appreciate the almost irreversible impact of unethical behaviors in our present reality. Closely associated with this is the fact that a first priority of most educational programs is to teach future leaders how to maintain power and control in order to protect the prevailing system. The assumption is that ethical behavior will evolve naturally as system-educated persons assume their responsibilities. The study of ethics, which mandates close scrutiny of system failures and weaknesses, is seen as less necessary and possibly even dangerous to those currently in power.

John Long, then acting dean of the School of Business, Indiana University, called attention to a common self-conscious reservation of so many toward the study of ethics: "I regard myself as a poor role model in exemplifying the ethical values that I like to profess. As a result, I often back off from doing any professing at all. I suspect that numerous other faculty members feel the same tugs and hesitancies . . . in wondering whether or not to address ethical aspects inherent in business and professional decisions."[22]

If college professors are hesitant to address ethical issues in the abstract (and I agree with Long, I think they are), their concerns don't compare with those who find themselves face-to-face with ethical dilemmas within their own operational group. Besides being branded as a pretentious moralist, there are other penalties for the organizational objector. These would include ostracism, loss of the material benefits that come with full group membership, and a vague sense of shame at possibly having threatened the subgroup's status within the larger community.

It is surely more difficult for group members to confront real and present ethical issues if they don't understand the terminology and rationale for ethical behavior, or if they have had no experience in the disciplined discussion of such issues. As Long also observed, "Sometimes . . . we are reminded that other matters—such as wisdom, honor, and earnestness—over the long run are much more important than many of the subjects to which we give explicit curricular attention (such as marketing, finance, production)."[23] It is ironic that a subject that has traditionally received our highest espoused priority has been neglected in our formal education.

Clients of business schools undoubtedly have little tolerance for ambiguity and a high need for certainty in their exercise of control. However, an obligation of educators is to disabuse executives, and possibly themselves, of the naive assumption that there are discoverable absolutes of human behavior that will allow executives to manage and make decisions based on methods, systems, and controls—independent of judgment, wisdom, and ethics.

In addition to rediscovering the wisdom of the ages, we may have to seek unheard-of solutions. What are the breakthroughs in the development of leadership ethics that compare to the material achievements of the medical and physical sciences—even of the behavioral sciences? Or do we already know, but are unwilling to risk using them?

Developing Operational Ethical Standards

Although vitally important, academic instruction in ethics is only the beginning. It can certainly help persons understand the necessity for ethical behaviors and prepare them to discuss specific issues more effectively. However, an *operational* ethical use of power cannot be taught; *it must be developed within each and every system.* Ethical behaviors do not necessarily transfer from organization to organization. For example, a lawyer may comply with a legal code of ethics in dealing with his or her firm's clients, yet unethically capitalize on inside information as a member of a corporate board of directors.

Besides being a philosophical ideal, the development of operational ethics is the best defense of the positive qualities of an established system, and of the managers who are truly Leaders and Builders. It is axiomatic that neither power nor problem solving can be effectively utilized for the good of the total group if (1) the group doesn't have an identifiable, agreed-upon ideology; or (2) the behaviors of individuals are fundamentally inconsistent with the group's ideology. As with most issues, the way a group attempts to achieve these conditions is all-important. Unfortunately, our traditional approaches to operational ethics have overemphasized the power approach.

OPERATIONAL ETHICS: THE POWER APPROACH

A CBS News/*New York Times* poll found that 55 percent of respondents believed that most executives are not honest. Asked for his reaction to the poll, Patrick McGuire of the Conference Board said that management has to convince employees that unethical practices will not be tolerated, and that policy statements "don't mean a fiddler's damn if the corporate culture means that people who transgress this code do so with impunity."[24] McGuire's suggested solution merely reflects what gave rise to the problem. We have become overly dependent on the traditional formula that says senior management must develop organizational policy, persuasively communicate its policy to personnel, and police and punish those who violate its published standards. The problem with this formula is that efforts to *force* ethical behavior are extremely costly and usually ineffective. In addition, they have little or no influence on those who *set* the ethical standards for the organization: senior management itself. Unaware of its own biases, and regardless of its communicated intent, it bases its actions on what it perceives to be the values within its own ranks and society at large.

Too many successful managers have resigned themselves to the ethics of the day, and have decided that realities of organizational life dictate that outdated standards of integrity and ethics are properly left in the churches and classrooms.

(How else could spokespersons for the American Tobacco Institute present their united attacks on common sense?) In decisions involving employees, stockholders, and the public, little or no time is spent on what *should* be done; instead, the focus is on the apparent practicalities of immediate personal or organizational needs. Instead of solving the ethical causes of their problems, they prefer to rely on self-serving communications and sophisticated management systems and controls. *Statements of values, philosophy, and mission, recognized "tools" of organizational change, become perfunctory steps in the process of implementing the Compleat Power Formula.* As a result, values *must* suffer.

As stated earlier, when top leadership levels isolate themselves from their own organization or society, their communications and decisions inevitably model the absolutist and subjectivist value orientations. As other levels adopt the same values, the prophecy of the imperfectability of humankind is fulfilled again, and counterproductive behaviors are legitimized throughout the organization.

OPERATIONAL ETHICS: THE PROBLEM-SOLVING APPROACH

Our challenge, therefore, is to make ethical ideals a genuine part of the total system, *especially* for senior managers, as well as for lower-level personnel. Instead of limiting ourselves to the way we are, we must strive to become better. It would seem, then, that if we are to incorporate ethical standards into an operational ideology, we need to develop experiences that will allow managers to vitalize, or revitalize, their ethical instincts. I suspect that this can best be achieved by developing better ways of capitalizing on what we already know.

First, persons at all levels must contribute to the development of the organization's ethical values. If we rely solely on positional, expert, or charismatic leaders to develop and maintain our values, we're in trouble. Their egocentric interests, trained incapacities, and the pressures of cognitive dissonance can insulate them from the insights of others. The lower levels of an organization and its minorities best understand the deficiencies or inequities of the dominant system or ideology. Leaders certainly have important and powerful roles to play in the process, but they represent only a portion of the cross-sectional grass roots of a human group.

Second, participation and involvement increase employee commitment to organizational goals. The resulting group-generated norms are the most effective determinants of actual behavior. To the extent that subgroups (or individuals) integrate themselves with other subgroups (or individuals) as equal members of a community, their value orientations are "objectivist." That is, all subgroups and their members are part of the objective reality and contribute to the community's understanding of its problems and challenges. Rather than coercion or manipulative persuasion between individuals and groups, the *modus operandi* is information/idea sharing and mutual problem solving.

If these two assumptions are correct, senior managers need to re-establish their presence in organizations in such a way that they can learn (or relearn, if they have forgotten) from their own personnel that

- Employees almost universally desire to work in an organization with high ethical, as well as quality and production, standards.

- The desired ethical standards of all levels and functions are essentially identical and are the same values we universally endorse. (In fact, top management's proclamations of policy and philosophy, frequently developed after long and agonizing discussions, usually are only polished versions of these same values.)

- Unethical behaviors by managers have tremendous long-term costs to the organization.

Senior managers must interact with a wider range of employees and demonstrate a desire to solve problems, not just amass and wield power. This will require that they confront and openly discuss the deficiencies of their own values, behaviors, and systems for operating the business. When managers experience first-hand the obvious pain of employees who want to do a quality job in a company they can be proud of—and who feel that their managers are preventing it—the impact can be very positive. Also, the learning can be two-way. Lower-level employees also get a better appreciation of management's problems when they are sincerely trying to make the right decisions.

Three fears of senior managers about value-based problem solving discussions are unwarranted. First, one need not feel like a hypocrite when participating in discussions of ethics. Fortunately, the public discussion of and commitment to ethical standards do not require that managers be saints. Certainly there are overlaps in ethical behaviors from one area to another, and there are individual performance variations within a single area. In other words, people come in unique combinations of moral strengths and weaknesses, and, within any work group, there may be a spouse abuser, a tax cheat, or a moderately well-developed Destructive Achiever. However, for their mutual benefit, they all can engage in discussions of behavioral standards, and make public commitments to adhere to them in the future.

In fact, it might be a good idea to acknowledge that we all are inclined to behave like Destructive Achievers to some degree. That is, we do some things for our own benefit that may possibly be at the expense of the total group. At a top-level meeting of a corporate management group, a respected executive presented the opening address. At one point he said, "I was once a plant manager myself, and I know that there were times when I made decisions that made the plant and myself look good, but that were not good for the corporation. We all know that we do this occasionally, but, dammit, we've got to recognize that it's not in our best interests. If we're going to turn this situation around, we're going to have to do things better than we have in the past." Listeners were obviously surprised at the speaker's candor. Later discussion indicated that they agreed with his views and were glad the issue was raised. As one observed, "It was like a breath of fresh air."

There is a paradox here. Destructive Achievers spend much of their energy defending themselves against confrontations (or against their own consciences) about their motives or behaviors. Hence, to the extent that managers are willing to monitor and acknowledge their tendencies to behave like DAs, they likely will be Leaders and Builders, *not* DAs. What is more important, lower-level employees will be able to sense these distinctions in their managers.

Second, there is a concern that relatively uneducated employees, or lower-level managers, may not be able to handle the discussions well and will embarrass themselves. Actually, the factors that will determine success will be the demonstrated integrity and good intentions of the participants—not organizational level, formal education, or pretensions of top management saintliness (even though the last factor is more characteristic of the power approach to operational ethics than it is of the problem-solving approach). Of course, typical operational groups can't definitively answer difficult philosophical questions (neither can a panel of philosophers), but *seeking* the answers can help them create normative standards for right and wrong behaviors, and can increase member commitment to uphold agreements. Moreover, the most common ethical problems in business and industrial organizations are not all that difficult to discuss. Most groups are not concerned with esoteric speculations as to what constitutes "happiness" or "the good life" as defined by Aristotle or Epicurus, although these are certainly relevant issues.

Most ethical problems are not complex issues of *what* is right or wrong. Rather, the issue is to what extent we as individuals are going to agree to *do* what is right, and are willing to allow appropriate confrontation of our behaviors by other relevant persons. My experience suggests to me that any grass-roots sample of 20 or more persons from a typical organization will reach common agreements very quickly, and they will be quite consistent with the values we universally endorse (as described in Chapters 9 and 10).

A third concern of senior management is that the results of such discussions will not last. This is a valid concern, because commitments to ethical behaviors never last if left unattended, and they usually *are* left unattended. It is essential that Leaders and Builders not be lulled into complacency by virtue of their organization's good health—that is exactly the kind of environment that gives the competitive edge to the Destructive Achiever. A sound ethical base must be jealously guarded against *internal* attack, or it will be lost.

Perspectives

Our recognition of currently publicized organizational realities may lead us to the unwarranted conclusion that certain counterproductive behaviors are normal and even acceptable for the person who seeks power. This in effect becomes an egocentric *anti*-ethic: an abandonment of idealistic goals, a lack of commitment to the long-term welfare of the organization or society, destructive special-interest conflicts, and a decreased "sense of community" in institutions as they grow larger and more complex. Leaders cannot allow this trend to remain undisputed by virtue of their absence or inaction. The most important aspect of leadership *presence* is its active participation with others in the mutual development of behavioral norms, both by example and by the explicit discussion of values and mutual expectations. If this is missing, the creation of documents of ethical intention are little more than exercises in creative writing.

Leadership's most important *action* is the vigilant stewardship of its human resources, with the resulting development and promotion of those who support its values, and the correction or removal of those who don't. These behaviors simply will not occur in systems already dominated by Destructive Achievers. Individuals in such groups are addicted to power politics regardless of the costs and will react only to superior power. They are a new breed of materialistic Darwinists as epitomized by the greenmailers. Their ideology is that anything not illegal is moral, and the material enrichment of those who take advantage of weaknesses in the system somehow benefit it. Of course, their "benefits"— exposing system weaknesses by exploiting them—come at the expense of others or the environment and have nothing to do with an exchange of value or with providing a product or service. Destructive Achievers never willingly allow themselves or their organizations to be confronted by those who are victimized by their actions: employees, shareholders, consumers, or the public at large. Their behaviors will be changed only by the application of a superior or a different kind of power, of a nature suggested by writers such as Kakabadse, Reason, Greiner, and others.

Objective problem solving—as a means of developing and maintaining an effective, agreed-upon ideology—requires equality of membership, appropriate confrontation, and ethical commitment. Therefore, it can be effective only at the level at which there are Leaders or Builders who can control these factors in their areas of responsibility. Of course, as with all behavioral classifications, there are degrees of Destructive Achievers, and many should be capable of improving their own behaviors and their organizations to some degree. Whether or not this can occur will depend upon greater public insistence on ethical behaviors among its members, especially its leaders. Unfortunately, this will be difficult to bring about. Never before have the appeals to egocentrism and materialistic values been so persuasive.

We publicly worry about the negative impact of popular TV on our children's immature values. Should we not also worry about the impact of "Lifestyles of the Rich and Famous" on the upwardly mobile middle executive? What is the level of maturity of the power-consumed executive whose study of ethics is limited to an occasional textbook reference—citing the importance of personal integrity in getting control over others?

As Thomas Merton writes, "Every moment, and every event of every man's life on earth, plants something in his soul." We must give serious thought to the ways we are programming our society's values—for our 65-year-old CEOs, as well as for our 12-year-old emerging adults. Above all, we must not impose limits on our capacity to improve our behaviors by assuming that the currently known realities of human nature are unchangeable. Possibly the variability of human behavior *is* unchangeable, but surely the proportion of moral to immoral behaviors can be favorably influenced if we are willing to make the effort.

We've never been able to figure out how to reliably prevent the cyclical decline of great civilizations—or corporations. As a matter of fact, the ethics of some of our present leaders may be no worse than they always have been. Robber barons, oppressive bureaucrats, and unprincipled role models have always been with us. However, their values, methods, and misdeeds have never before been so thoroughly publicized and accepted as normal. And maybe anti-ethics *is* the natural course of societies or organizations as they achieve success. Average citizens become more "sophisticated" about the apparent nature of man and organizations, and materialism becomes the criterion for the good life, instead of virtue or wisdom.

There are, however, also signs of hope. The Iran-Contra affair has produced a public reaffirmation of our country's traditional values, and a reminder that the ends never have and never will justify the means. Although some Americans were briefly enamored by the power and heroic image of Oliver North, public opinion eventually condemned the powerful elite who, with him, functioned outside of the legal process. Behavior that had, in recent memory, been considered a necessity of foreign policy and international affairs was revealed as destructive both of our internal political process and our relationships with other countries. The stock market crash of October 1987, similarly, revealed a financial community deeply disturbed by the behaviors of some of its celebrated idols of the recent past. Unrestrained greed, which Wall Street had declared a virtue, was again revealed as the destructive and unethical quality it is.

Unfortunately, it still seems to take crisis and tragedy to force our leaders to take a serious interest in operational ethics. Although all members of society collectively determine our values, only our leaders have the authority to realize those values in the institutions, organizations and systems that constitute society as a whole. An *intrinsic* commitment to ethics on the part of our leaders and managers is the essence of, and the first requirement for, organizational effectiveness.

Whether or not we can ever achieve this ideal condition in each specific organization is not the issue and never has been. As has often been observed, humankind is divided into two parts: those who are part of the problem and those who are part of the solution. We may not be able to believe that we—and our organizations—are perfectible. But we can at least strive for more perfect organizations, and work to improve our own areas of responsibility according to some set of ethical standards. To do otherwise is to abdicate to the natural entropy of power.

Footnotes

1. Reported by Sandy Grady, "President can help man who blew whistle on NASA." *Knight-Ridder* news release, appearing in the *Charlotte Observer,* May 14, 1986.
2. Reported by William C. Rempel and Rudy Abramson, "Firm's engineer offered redesign role." *Los Angeles Times* news release, June 1, 1986.
3. Gary Wright, John Wildman, and David Perlmutt, "Suit: Duke chief knew of charges." *Charlotte Observer,* Sept. 6, 1986, 1B.
4. John Wildman and David Perlmutt, "Virginia board probes psychiatrist." *Charlotte Observer,* Sept. 9, 1986, 3C.
5. Wright, Wildman, and Perlmutt, op. cit. 6A.
6. Harry Levinson, "A special calling." *The Levinson Letter.* The Levinson Institute, Belmont, Mass., Sept. 16, 1985, 1–2.
7. David W. Ewing, "How to negotiate with employee objectors." *Harvard Business Review,* January–February 1983, 103.
8. Ibid., 104–105.
9. Ibid., 106.
10. Reported by Sandy Grady, "President can help man who blew whistle on NASA." *Knight-Ridder* news release, May 14, 1986.
11. Ewing, 108–109.
12. Ibid., 109–110.
13. Ibid., 110.
14. To understand the relationship between value orientations and the exercise of power, see Harold D. Lasswell and Abraham Kaplan, *Power and Society: A Framework for Political Inquiry,* Yale University Press, New Haven, Conn., 1950; and Abraham Kaplan, *American Ethics and Public Policy,* Oxford University Press, New York, 1963.

15. Peter Reason, "Is organization development possible in power cultures?" In Andrew Kakabadse and Christopher Parker (Eds.), *Power, Politics, and Organizations: A Behavioral Science View,* Wiley, 1984, 201.
16. Larry E. Greiner, "Top management politics and organizational change." In Suresh Srivastua & Associates (Eds.), *Executive Power: How Executives Influence People in Organizations.* Jossey-Bass, San Francisco, 1986, 174.
17. Ibid., 155.
18. Karl Menninger, *Whatever Became of Sin?* Hawthorne Books, New York, 1973, 15.
19. Ibid., 227–228.
20. Andrew Kakabadse, "Politics of a process consultant." In Andrew Kakabadse and Christopher Parker (Eds.), *Power, Politics, and Organizations: A Behavioral Science View,* Wiley, 1984, 182.
21. Peter B. Vaill, "Why we don't teach ethics in B school." Presented at the Conference on Management Training and Development, the Conference Board, New York, February 1986.
22. John D. Long, "The responsibility of schools of business to teach ethics." *Business Horizons,* March-April 1984, 4.
23. Ibid., 2.
24. "Low marks for executive honesty." *New York Times,* June 9, 1985, 3, 1f–6f.

APPENDIX A

Confrontation

"Confrontation" needs to be defined when it refers to a group process. Dictionary definitions range from "bringing face-to-face" to "conflict." In addition to being vague, the term has negative connotations for most persons, suggesting an unpleasant, undesired, or unplanned encounter, especially for the party being confronted. Because of this, many suggest that another term be used, such as "problem-solving discussion," in order to make the process more acceptable to more people.

Although it is tempting to yield to this persuasive view, it would dilute the process from the beginning. Confrontation is more than mere discussion. Problem-solving *discussions* are the recreational aspects of organizational interaction—these are what persons *enjoy,* and are hardly at issue. Instead, confrontation is the process of addressing necessary but potentially ego-threatening issues for the purpose of objective problem solving. Those in charge of the process must acknowledge at the outset that they are committing themselves to the most meaningful form of organizational equality, with commensurate benefits, challenges, and risks. In fact, arriving at that acknowledgment is an important step in the process. Without the courage of this kind of commitment, successful problem solving of sensitive issues is unlikely.

On the other hand, confrontation does not suggest violent or destructive kinds of interaction in which the parties seek to destroy or demean their co-confronters. A more extensive definition would include five qualifications (see Exhibit A.1):

Exhibit A.1   CONFRONTATION

Is	*Is Not*
An involvement process	A sensitivity process
Constructive criticism	Personal attack
Accentuating a sense of community	Accentuating differences
A discussion	A debate
Mutual problem solving	Persuasion

Confrontation is an involvement process, not a sensitivity process. There is no attempt to get inside anyone's skin or to expose personal weaknesses or deficiencies. The objective is neither personality therapy nor emotional venting. Rather, the intent is to wholly involve all members of a group by means of full participation in discussing significant issues. This *can* be a very emotional experience, but it comes from being recognized as a mature adult, and results from addressing the *substance* of sensitive issues.

The process is constructive criticism, not personal attack. The intent is not to put someone else down, to gain a personal competitive advantage, or to get revenge for previous slights. Yet there must be a willingness to criticize important assumptions, decisions, or actions of the group, regardless of the level or function of their initiator(s). The most important and potentially disastrous problems are the ones that are denied or diverted because of the ego-defensiveness of key individuals. Unless criticisms are given and taken in a constructive atmosphere, there is little chance that they will positively affect the group process.

Confrontation should emphasize a "sense of community," not a heightened sense of differences. Unfortunately, when groups meet to discuss problems, their various subgroup differences are often highlighted. Everything from seating arrangement to meeting agenda may emphasize factors that are unique to the subgroup differences themselves (for example, management/nonmanagement status). Although the implications of such differences may eventually need to be addressed, the primary emphasis should be on the welfare of the total group, and on the ethical solution of its problems. To the degree that the group doesn't have a sense of community, this in itself becomes an issue that must be addressed *before* effective confrontation of *de facto* differences can occur. Otherwise, solutions will be the result of power politics, not problem solving.

The required orientation is one of discussion, not debate. One enters a discussion with the intent to discover the realities of a situation and to openly contribute and receive information. One enters a debate with a preconceived position to defend or attack. Communications are selective in a debate, both in sending and receiving—which is hardly conducive to objective analysis.

Extending the previous qualification even further, *confrontation is objective problem solving, not persuasion.* The intent is to develop the most *effective* solution by means of the evolution of objective analysis. To the extent that group members are persuaded—versus educated or informed—their collective views will reflect the emotional biases about reality, not reality itself.

Two qualifications are in order. First, these requirements for effective confrontation do not deny the valid use of persuasion, debate, sensitivity sessions, or even attack and the accentuation of differences. These behaviors all have their place in human interaction. But *don't* confuse them with objective problem solving; when they occur, something else is going on. Second, in an effective confrontation process, these nonconfrontational behaviors are likely to occur, and sometimes *should* occur. Although a group may agree to adhere to the previous guidelines, members should be well aware that ego-sensitive issues easily lead to defensiveness and conflict. In such cases, the defensiveness and conflict may have to be acknowledged or worked out before returning to the original issue.

An operator once stood up in a meeting, pointed her finger at a foreman, and shouted, "You're exactly the kind of foreman that we just don't need around here!" (The foreman previously had said that if the operator didn't like the way management was treating her, she could get a job elsewhere.) This could have been a very negative event under normal circumstances, but the group, with facilitator assistance, turned the situation into a very positive experience. The operator explained how she felt about the foreman's comment, the foreman explained his reaction to the operator's outburst, and the total group discussed the effect of the incident on the problem-solving process. In other words, by confronting the operator's and foreman's nonconfrontational behaviors, the group turned the personal attacks into another problem to be objectively discussed.

In fact, there seemed to be a consensus that the operator's emotional attack actually helped the confrontational process. It may seem a contradiction that a nonconfrontational personal attack can help the confrontational process, but it isn't. Although the operator and the foreman avoided confronting the original issue, they caused the identification of another issue that was more important at the time: the lack of respect between foremen and operators. This was brought into the open and discussed, and the quality of later confrontation noticeably improved.

The analysis of this situation is more than an exercise in semantics. It is crucial that we be able to recognize when we are confronting an issue under discussion and when we are *seeming* to confront it, but are actually avoiding it. When we recognize that we are avoiding an issue, for example, by personally attacking someone, then it becomes important to confront that digression before we attempt to return to the initial issue. In other words, nonconfrontational behaviors need to be confronted and negated so that the group can return to confronting its more job-related technical issues. By having a heightened awareness of ethical concepts, a healthy sense of community—and an understanding of what true confrontation is all about—a group will be able to come much closer to realizing this ideal. If any of these elements are missing, nonconfrontational behaviors are likely to be improperly handled and, hence, have a destructive impact on the group.

Confrontation Participation

Successful confrontation depends upon the same kinds of effective behaviors as described in popular textbooks on management, discussion participation, and conference leadership. There are certain behaviors, however, that are especially relevant when dealing with sensitive, ego-threatening issues.

DESTRUCTIVE BEHAVIORS

Although the following negative behaviors may be observed in successful confrontational groups, especially when they are under unusual stress, the groups will be able to recognize and overcome them. On the other hand, negative behaviors seem to be prevalent in groups that consciously or subconsciously discourage confrontation. They have become part of the standard operating procedures, and there is a tacit agreement not to acknowledge or to try to overcome them.

1. *Communicating personal disapproval* is probably the single behavior that is most destructive to confrontation. It is especially destructive when someone with superior power communicates that "you have a negative attitude," or "you are incompetent." Although these attitudes are sometimes expressed in words, more often than not they are communicated by tone of voice or facial expression.

Lower-level personnel know that such labels can mean the end of a promising career. As soon as dissent becomes associated with incompetence or a negative attitude, management can forget the effective upward communication of its most important issues—it just won't occur. The exception to this is when some personnel decide to take on the organization, either because they feel they have already lost in the political arena and have nothing more to lose, or because they feel that an ethical issue is more important than personal career risk. Confrontation under such conditions usually becomes internally disruptive, and is positively acted upon only as a result of influences outside the immediate group.

Of course, expression of personal disapproval works both ways. Lower-level personnel express disapproval when they claim that conditions or decisions were much better under previous managers, when they charge that present management is uncaring, or that the managers of another organization are superior.

2. *Talking down* is usually evident in one of two ways: "You wouldn't say something like that if you knew the problems we have to deal with in the executive suite." Or, from the other direction, "You wouldn't make that kind of decision if you knew what was going on out in the marketplace." These sorts of comments suggest a deficiency in the other person: he or she isn't in the communications chain, or not broadly experienced in the total environment, or has lost touch with an area of responsibility.

In both cases, persons are criticizing other individuals' opinions by suggesting that they lack the communicator's superior understanding of the problem. It is much better to assume that it is quite normal for competent persons to have different perspectives, and to educate others when they lack understanding. For example, executives should describe the problems they *are* facing. Or lower-level personnel should explain what *is* occurring in the marketplace.

3. *Probing for names of troublemakers* should obviously be off-limits in a problem-solving confrontation, but it is easy to slip into. When a person is shocked at some especially bad news, an automatic reaction is to blurt out, "Who is doing that sort of thing?" or "Who is saying that . . . ?" Or, from the opposite direction, "Who was responsible for making that stupid decision?" These sorts of reactions are the first step in a misguided effort to find out who is to blame, rather than to solve the problem.

Of course, one function of management is to detect and correct those who violate established policies and procedures. But a confrontation session is not the place for it. As soon as personnel discover that management used an open, problem-solving process to assess blame, even inadvertently, the meeting becomes ritual, and little substance is discussed. If an environment is healthy, the roles of various persons in a problem situation usually become known, but by their own acknowledgment and the usual course of events, not through witch-hunting.

4. *Abruptness* suggests nonsupportive attack and tends to make others defensive, especially if they are not very fluent themselves. Quick-thinking executives who spend much of their time in verbal communications among themselves may carry their abrupt style over into discussions with persons less comfortable with rapid-fire verbal exchanges. Whether intentional or not, their quickness in asking questions, or in responding, gives the impression of a trial lawyer trying to rattle the witness. As one lower-level person put it, "There's no way we can debate with them [upper-level managers]. That's what they do all day long."

(The person was correct. The particular event was a *debate* and not a discussion. The managers "won" the debate, considered on a point-for-point basis. But in the minds of other personnel, they lost the discussion; they never were able to really appreciate the realities of the situation.)

5. *Gloating or displaying arrogance* kills rapport. Yet it is easy to do when you win an argument. Arrogance can be expressed in many subtle, but unmistakable, ways—a smug look or a condescending smile, leaning back in the chair and folding your arms, finger-pointing in an authoritative way, or heavily emphasizing certain words. Such gestures proclaim, "I won the point" and emphasize the contentious aspects of the discussion. Although it may temporarily be ego-building for the winner, the debate-persuasion-power qualities of the session become more dominant and the problem-solving potential is reduced.

6. *Talking only when the discussion is favorable* to an individual or his or her group ruins credibility. Some people participate freely when compliments are being exchanged but, when severe criticisms are being made, rely on associates to respond. As a result, members of the confronting group will discount the positive things the individual said. Because such persons neither accept nor reject criticism, their real positions on important issues remain unknown, and they cannot be trusted.

7. *Forcing an outward expression of agreement when it isn't there* is symptomatic of bureaucrats who cannot handle confrontation, and further drives dissent under the surface. They mistakenly believe that *esprit de corps* depends on the outward expression of consensus—or that total agreement is possible when reasonable persons get together and understand the wisdom of management. As a result, participants soon learn that those in control aren't going to allow the discussion to end until they get agreement, or will end it in such a way that the dissenters lose status.

It is better for the leadership to acknowledge and accept disagreements and accurately include them in any summaries or rationale for decisions. This would permit a genuine consensus, in which persons agree to support decisions that they disagree with but that were made with their reservations duly noted. Of course, this kind of consensus also depends upon the ability of the leadership to convey ethical behaviors in making its decisions, and upon the ability of dissidents to perceive them.

8. *A "pep talk" at the conclusion of a confrontation backfires* more often than it helps, at least the way it is usually done. The prevailing view of the manager as motivator—not problem solver—easily leads to counterproductive behaviors. Too often the motivational pep talk takes the following form:

> Well, folks, this certainly has been an enlightening experience. I've learned a lot that I didn't know before, and I'm sure we all have. I want to thank

you all for being so open and taking the risks of expressing your genuine concerns about the problems we have. We obviously have a lot of problems to address and have our work cut out for us.

However, I know you're well aware that this is a less-than-perfect world, and there will always be problems and conditions that we don't like. And it is tough to get 24 different people to agree on any one issue. If we all would just come back to work on Monday and put a lot of these things behind us, and do a better job doing what we know we should be doing, a lot of these problems will just naturally work themselves out.

Such an attempt at motivation dilutes the importance of the preceding problem-solving discussion, and is sometimes bitterly resented. It betrays the manager's desire to continue the *status quo* without interruption. He or she sees the confrontation process primarily as group therapy, not problem solving. The message is that the deficiency is lower-level motivation, not management's values, systems, or procedures.

CONSTRUCTIVE BEHAVIORS

Each of the above destructive behaviors has a positive corollary. The following are especially helpful in developing a supportive climate for confrontation.

1. *A nondefensive attitude* is fundamental to supportive confrontation. Of course, criticism always hurts, and it is impossible not to *feel* defensive. When disagreements or disapproval are expressed, the natural tendency is to retaliate with defensive behaviors: questioning the other person's competence or motivation, cross-examining the critic, one-upmanship, depreciating the importance of the issue, and so on.

However, with experience, one can learn that genuine openness and demonstrated receptivity to confrontation accomplishes two important goals: (1) They serve as models for others to be open and receptive, and (2) they take the personal sting out of the criticisms. They make criticisms more objective and less personal. The emphasis is on behaviors, not the goodness or badness of an individual. When the criticized person demonstrates reason, the critics become more understanding—even helpful. They may go so far as to acknowledge the excesses of their criticisms, or their own roles in the problem.

As an aid to a nondefensive attitude, it helps to remember that, in a really open session, inexperienced confronters will create an overly negative impression. They are not used to confrontation and are prone to get nervous and emotional—to be "rough" and "loud." They don't know how to repair the damage after they have made a tactless mistake, and may even tend to exacerbate the situation when their error is pointed out. Experienced confronters need to recognize that these behaviors are normal and anticipate them. It is easy to confuse anger with personal malice when inexperienced persons criticize specific conditions.

With experience, however, persons learn that emotional outbursts are frequently followed by good feelings, even welcome laughter, when the confronted person reacts well. People appreciate honest answers and can recognize conscientiousness—or a snow job. When the confrontation is over, they will give credit for good and honest effort, even though most issues were not "solved" to their satisfaction.

2. *Participants should prepare in advance to handle tough issues,* even though the confrontation is a discussion and not a debate. Each individual is obligated to represent reality in the most accurate way possible. For example, if a discussion is going to be about quality problems, the production manager must refine his or her thinking about the proper blend of output and quality improvement projects. What are the realities of the state of the art, production demands, and consumer requirements? One should not wait until the meeting itself to develop a coherent view of one's own area of responsibility.

3. *Supporting valid points of the "other" group* builds a sense of community. Instead of "we-them," the discussion is among "us." One should be positive about one's own position when it is justified, but should also admit to valid problems. This is especially difficult to do when one manager may wish to disagree with prevailing management opinion or with senior management. It appears to violate the rule of supporting senior management, or of presenting a united management consensus.

However, if the confrontation is to be a true group problem-solving effort, disagreements among all members—including individual managers—must be acceptable. If there are any "off-limits" issues that require management unity, they should be announced at the very beginning. Of course, a sign of group maturity is when issues are discussed freely without fear that dissension will be equated with disloyalty. If lower-level employees feel that even managers cannot disagree with the prevailing opinion—what can they believe about their own chances of influencing the *status quo*?

4. *Humor can be very useful, but it should not distract the group.* Humor is much appreciated as a welcome break in a stressful discussion, but it must be judiciously used. If a person uses it inappropriately, it may be seen as an unfair attempt to lessen the impact of a valid viewpoint. That is, when serious progress is made in clarifying a sensitive issue, a group will justly resent humor that depreciates its importance. Of course, in some cases, a little self-directed humor can actually *add* to the significance of an issue, and can be especially effective in improving rapport.

5. *Questions or objections can occasionally be turned around.* When there seems to be no solution to a problem, a criticism can sometimes be turned around: "We've come up with the best answer we can, how would *you* handle it?" Discussion may then demonstrate that all suggestions have undesirable qualities. The critics themselves cannot agree on a single solution. However, this technique can easily be overused. A responsible person must still give his or her best assessment of a situation and explain the rationale behind it. If too many issues are turned around, a person may develop a deserved image of defensiveness or unwillingness to express a controversial viewpoint.

6. *The discussion should be cut off before it gets tedious.* Some attempts to achieve an idealized standard of openness, or a consensus, are misguided. Although most inexperienced confronters tend to end discussions of sensitive issues too early—or when they get emotional—some experienced persons ruin the confrontational process by running it into the ground. For most, there are reasonable limits for productive discussion. When these are exceeded, in order to pacify a few unreasonable persons, the group's respect for the confrontational process will be lost. Therefore, when the group realizes that *productive* discussion has ended, it is important that the leader or facilitator officially close the meeting. This can be done in a participative way, by checking out the opinions of group members. Obviously, it must *not* be a devious excuse to silence minority opinion.

7. *A public commitment to improve behaviors* is a good way to end a confrontational session. This form of "pep talk" can be quite productive. But it is directed at one's self or one's *own* subgroup. For example, a senior manager may positively make a commitment to address the problems that he or she and the management group are apparently causing. This demonstrates that the communications not only have been heard and appreciated, but also will be acted upon. This is the ultimate form of motivational communication. It is directed at the managers, but all personnel hear and appreciate it.

Perspectives

Upward communication is publicly valued by managers but avoided in practice. The most important upward communications are those that are unexpected. This is true of *all* communication; an expected communication is only a ritual, not a meaningful message.

Because communications of unexpected or unfavorable events are often seen as threatening to those in power, genuine confrontation of ego-sensitive issues is actively discouraged. In some cases, this discouragement is deliberate and conscious; in other cases, unintentional and subconscious. Either way, obvious negative behaviors discourage confrontation, just as positive behaviors can bring them about. The confrontational process has never received the attention it deserves. Developing it in an organization is a worthy goal of any management group that is truly committed to objective problem solving, not power politics.

APPENDIX B

Work, Job Satisfaction, and a Sense of Community

Work is not recreation for most people in large organizations, nor is it ever likely to be. Some enjoy it very much; some hardly at all. A good way to sabotage the long-range credibility of an effort to improve the organizational environment is to make exaggerated claims. Quality circles, job enrichment, quality-of-work-life, and other programs are too frequently initiated with the actual or implied promise that

- jobs will become more interesting.

- work will become more enjoyable.

- impediments to job satisfaction or effectiveness will be permanently removed.

- persons at all levels will have much more influence about their areas of work or responsibilities.

When an organizational development effort is effective, these benefits do result to some extent. But, in most cases, after the more obvious issues are discussed and the easier problems solved, many employees find that their jobs still are tediously repetitive, boring, dirty, disruptive to home life, or dangerous. Job satisfaction is not "fun" in the recreational sense, although there are certainly opportunities for enjoyment in just about every job. Instead, job satisfaction relates to a person being in a job that "makes sense," and to his or her need to

213

be an accepted, deserving, and respected member of the work group. Therefore, no actual or implied promises should be made that jobs will be easier, cleaner, or more enjoyable as a result of a change effort. To the extent that these benefits are realized, so much the better—that is a bonus. Instead, we should aim to improve the organizational climate, so that each employee gains the personal satisfaction that results from feeling that

- I am respected by my associates (supervisors, subordinates, and co-workers) for what I do.
- my job "makes sense," as it is designed, and in terms of its results.
- I deserve the compensation I am receiving.
- I am making a worthwhile contribution to my associates, my organization, and my society.
- if I have an idea or other contribution to make, I am given the appropriate opportunity.
- I am personally just as good as anyone else, in terms of basic human rights.
- I am proud of what I, my associates, and my company produce.
- my associates make every reasonable effort to ensure that my job is as safe, satisfying, and effective as possible.

In other words, we are trying to create and maintain a sense of community, in which in every person the basic human need to be recognized as a worthwhile individual is met. Membership in the grooup is felt by each individual to be deserved. All jobs are necessary and important, and everyone has an equal opportunity to appropriately contribute his or her best efforts. A happy circumstance is that personal and organizational effectiveness are both directly related to the extent that this kind of job satisfaction is realized.

In such a climate, employees' solid identification with their co-workers is extended to their equipment and their organization, and has immeasurable benefits. Thomas J. Peters, co-author of *In Search of Excellence,* observed that the best companies reflect "a bone-deep belief in the value and potential of every individual in the organization" and that when people feel this belief, "the improvement that comes from having a turned-on team is 400%, 500%, 800%—not 2% here or there."[1] Even managers who think they really believe that "people are our most important product" don't fully appreciate the significance of the slogan.

Consider the tremendous benefits when

- persons throughout the organization have a "do-it-right-the-first-time" philosophy—avoiding the ruinous expenses of rework, handling returns, production of waste, missed schedules, delays, *ad infinitum* (versus a "do-it-any-way-to-get-through-the-day" philosophy).

- persons have a cooperative attitude, rather than merely complying with direct orders. Example: salespersons who help out in busier departments rather than standing around while furious customers kept waiting go elsewhere.

- persons at all levels know how to do their jobs. They are constantly learning, because they are constantly interested (versus having little interest beyond the initial training period, "just enough to get by—this may be *your* company, but it isn't *my* career").

- employees at all levels control their own costs. Paper clips, photocopies, nuts and bolts, gasoline, lunch and break periods, forms, tags—when all added up—can be very expensive. (Management attempts to control such expenses are usually ineffective, even counterproductive.)

- management models a conscientious attitude—respect for people, time, supplies, equipment, and the company (called "managerial leverage").

- work associates model the same kinds of attitudes (called group norming, peer pressure).

- after a problem-solving discussion is concluded at the end of the day, people linger around, still giving their viewpoints.

- employees meet after work at the local tavern and discuss better ways of solving problems (versus playing "ain't it awful," commiserating with each other about how lousy the company is).

- an employee comes to work on a Monday with the answer to a tough scheduling problem (she had been thinking about it over the weekend, versus doing her best to blot the work environment out of her mind).

- a drive belt rips during an important production run, and an employee takes it home and fixes it himself (the local repair shop "couldn't get to it today," it would have cost $200 to fix, and the employee owns a heavy-duty textile sewing machine).

The best part about all this is that everyone benefits, because everyone is doing it for his or her own satisfaction. To the degree that people are "living" on the job, they have reclaimed another third of their active adult lives as truly their own. And the difference in results is what Peters was referring to. It is there for the asking, if done with integrity and a sense of equality for all members of the organization.

Footnote

1. Special Report: Excellence in Business," *Behavioral Sciences Newsletter,* Roy Walters & Associates, Mahwah, N.J., October 10, 1983, Book 12, Vol. 19, p. 4.

Index

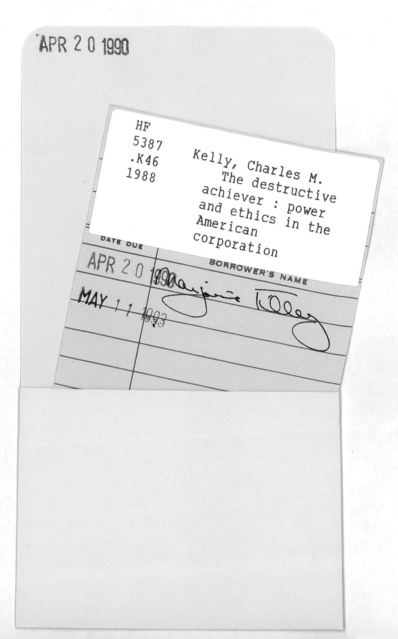